Biocommunication of Phages

Guenther Witzany
Editor

Biocommunication of Phages

Springer

Editor
Guenther Witzany
Telos-Philosophische Praxis
Buermoos, Salzburg, Austria

ISBN 978-3-030-45884-3 ISBN 978-3-030-45885-0 (eBook)
https://doi.org/10.1007/978-3-030-45885-0

This Springer imprint is published by the registered company Springer Nature Switzerland AG
The registered company address is: Gewerbestrasse 11, 6330 Cham, Switzerland

Contents

Chapter 1
What Does Communication of Phages Mean?

Guenther Witzany

Abstract Phages have serious effects on global energy and nutrient cycles. Phages actively compete for host. They can distinguish between 'self' and 'non-self' (complement same, preclude others). They process and evaluate available information and then modify their behaviour accordingly. These diverse competences show us that this capacity to evaluate information is possible owing to communication processes within phages (intra-organismic), between the same, related and different phage species (interorganismic), and between phages and non-phage organisms (transorganismic). This is crucial in coordinating infection strategies (lytic vs. lysogenic) and recombination in phage genomes. Therefore it is essential to investigate what communication of phages means and to identify the difference of the biocommunication approach to investigations that are restricted to the molecular biological perspective.

1.1 Introduction

Bacteria are evolutionarily one of the most successful living organisms, originating nearly since the beginning of life. Besides archaea, bacteria were the dominant cellular organisms in the first 2 billion years of biological evolution. In at least the last 3.5 billion years, they colonized nearly every ecological niche on earth. They are essential symbionts of all eukaryotic organisms and are required for their survival. On the other side, they cause diseases of even epidemic scales and not only modern medicine is in a permanent struggle with the consequences of bacterial infections worldwide. More recently, multidrug-resistant bacteria have necessitated the search for other ways to fight bacterial infections than mainstream research on antibiotics. Future-oriented researchers are currently projecting the post-antibiotic era.

Bacteria are important ecosystem determinants in the soil and oceans globally, with 1 ml seawater containing ca. 1 million bacteria (Williamson 2011). Bacteria

G. Witzany (✉)
Telos-Philosophische Praxis, Buermoos, Salzburg, Austria
e-mail: witzany@sbg.at

G. Witzany (ed.), *Biocommunication of Phages*,
https://doi.org/10.1007/978-3-030-45885-0_1

are undoubtedly the best-adapted organisms on earth. An abundance of rapid genetic variations, together with genetic adaptations, occurs conferring resistance to environmental circumstances, including intense heat and radiation. Bacteria communicate in various ways (Witzany 2009). The most popular means of bacteria communication is the quorum sensing/quorum quenching research and cheating phenomena by which bacteria coordinate and organize their essential life strategies (Kaiser and Losick 1993; Schauder and Bassler 2001; Bassler and Losick 2006; Ben-Jacob 2014). The acquisition of complex genetic datasets by virally-derived infection events seems to be the main source of evolutionary adaptational processes, and this was not in the focus of bacterial research in the last decades. Thus, the main genetic resources and gene word order in bacteria genomes which determine their behavioural patterns primarily did not evolve from cellular predecessors or the genetic lineage of the bacterial population horizontally but are results of vertical natural genome-editing activities of viruses (Witzany 2011; Argov et al. 2017; Meaden et al. 2019).

The whole range of bacteria lifestyles is in constant interaction with its predators and co-evolutionary partners, the phages. Phages are the most abundant living agents and outnumber bacteria 10 times. Phages are also the most diverse inhabitants on earth. To date, they are completely underestimated in their number, skills and competences and remain the dark matter of biology (Youle et al. 2012; Hatfull 2015). They have serious effects on global energy and nutrient cycles. They determine bacterial virulence, eukaryotic fitness and the global carbon cycle (Díaz-Muñoz and Koskella 2014). Bacteriophages are found in nearly all ecospheres including sea and freshwater, the soil, polar regions, deserts and within other organisms (Abedon 2011; Armon 2011; Batinovic et al. 2019; Kavagutti et al. 2019; Warwick-Dugdale et al. 2019). Phages actively compete for hosts, hunt prey, sense their environments and make choices (Rohwer et al. 2014). They can distinguish between ‘self’ and ‘non-self’, which means they complement some and preclude others (Villarreal 2009). Phages process and evaluate available information and then modify their behaviour accordingly. Additionally, they are the evolutionary ancestors of eukaryotic dsDNA viruses (Koonin et al. 2015).

These diverse competences indicate that the capacity to evaluate information is possibly due to communication processes within phages (intraorganismic), between the same, related and different phage species (interorganismic), and between phages and non-phage organisms (transorganismic). Additionally the show typical reaction patterns to abiotic influences of the environment. This is crucial in coordinating infection strategies (e.g., lytic vs. lysogenic) and recombination in phage genomes.

We begin by asking what communication of phages means. Is it beneficial to study communication processes of phages instead of pure physical interactions? How can one define “communication” of phages?

1.2 Communication Means Interactions Mediated by Signs

In contrast to former definitions of communication (mathematical theories of communication, information theory, systems theoretical approaches or other mechanistic attempts to encompass the phenomenon of communication), the most recent empirically-based definition of communication is: Interaction of at least 2 living agents mediated by sign(al)s. Therefore, communication is basically a social event.

The crucial difference between biocommunicative interactions and interactions in abiotic environments is that biocommunication depends on signs, i.e., sign mediated. This means that interactions occur by the recognition and reaction (generation, submission and uptake, decision making) of signalling substrates, which include chemical molecules (soluble, airborne,) electric, tactile, or as in animals, vocal and visible signals (Witzany 1993, 2000, 2010b).

In contrast to abiotic interactions, where no signs are present, the use of signs in sign-mediated interactions follows three levels of rules (not laws): combinatorial rules - how to correctly combine single signs to sign sequences (syntax), content coherence rules - how to correctly combine signs with meaning (semantics), and contextual rules - how to correctly combine signs with the real-life context by the sign-using agent (pragmatics). If one level of rules is missing, no natural communication process occurs (Witzany 2019). In abiotic interactions – if e.g., water freezes to ice – no such rules are present.

Biocommunication of Phages is the first book that will compile contributions in the following sections (Fig. 1.1):

- Trans-organismic communication: interactions between phages and non-phage organisms (infection/defence strategies, host genome editing, symbiogenetic cooperation etc.)

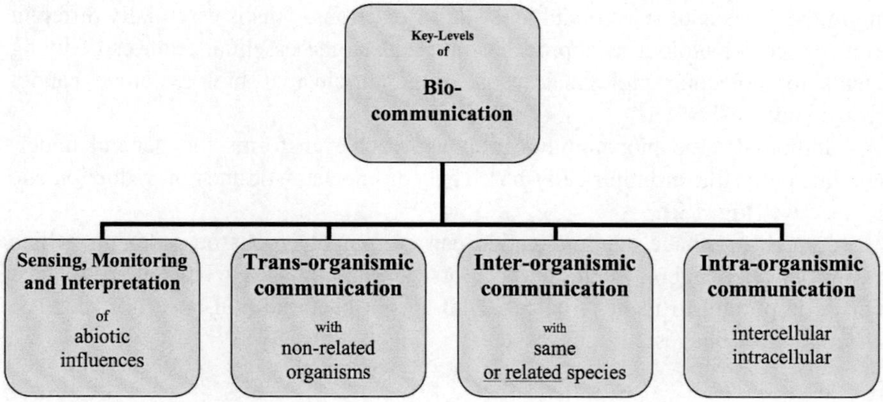

Fig. 1.1 The biocommunication approach identifies 4 levels involving living beings

- Inter-organismic communication: phage-phage interactions (competition, cooperation, etc.)
- Intra-organismic communication (reproduction, recombination, memory, learning, etc.).
- Vocabulary of phages (enzymes, signalling).
- Phagetherapy

At this point, it is important to also differentiate the signs into 3 different kinds: index, icon and symbol.

(a) There may be signs with a solely indicating function (*index*). This means the sensed sign represents something relevant to the sensing agent, a structure or chemical specificity, e.g. in plants, gravitational or dry/wet circumstances. In the case of phages, the density of chemical compounds was applied. It indicated that the phage could not attach to bacteria but remained in a persistent (temperate) style, followed by some internal phage regulatory processes.
(b) Another kind of sign is the function of an *icon* that represents a detectable similarity, e.g., certain colours of flower structures which look like female insects for male insects, the contour of a wolf which is a warning sign to dogs.
(c) The third important sign is *symbols* that do not mean what they represent but are signs that biotic agent groups use and share through conventions that the group members agree to by social learning experiences. This is seen in the dialects in the language of the bees or the main representative of symbol use, the human languages.

The biocommunication approach investigates all kinds of sign-mediated interactions in all domains of life (Witzany 1993, 2000, 2010a, b, 2011, 2012a, b, 2014, 2016a, b; 2018; Witzany and Baluška 2012; Witzany and Nowacki 2016), according to the biocommunication method: to identify interactions mediated by signs, according to the 3 levels of rules on all levels of interactions. This is essentially different from molecular biological approaches and adds some essential features of living agents to molecular biological results which molecular biology alone cannot investigate.

Additionally, the biocommunication approach transforms the general understanding of life into an empirically-based but non-mechanistic and non-reductionistic perspective (Fig. 1.2).

Research on phage communication can exclusively focus on indexing, which means that chemical molecules or receptors can cause phages to react appropriately. This is important to identify bacterial host from bacterial non-host, host defence system, helper phages, virophages, etc.

Different Paradigms: Explainung and Understanding „Life"

Concept of	**Molecular Biology**	**Biocommunication**
„dead"	pre-biotic chemical reactions	no sign-mediated interactions
„living"	replication/biological selection (molecular reactions)	sign-mediated interactions (social events)
determinants	natural laws (thermodynamics)	semiotic rules
RNA-ensembles	molecular assembly	agent-groups integrate or preclude non-self agents
viruses	escaped selfish parasites	essential agents of life
genetic variation	error replication	RNA interaction based innovation generation
genetic novelty	random mutations	viruses and subviral RNA-networks edit code
biological selection	fittest type	fittest consortium
genetic code	genetic material	semiotic text (according syntax, pragmatics, semantics)
communication	information transfer via coding/decoding mechanisms	agent-based social interactions mediated by signs according semiotic rules

Fig. 1.2 Different paradigms investigating living agents and defining life: the molecular biological paradigm explains all life processes primarily by the physical-chemical properties; the biocommunication approach explains life as a social event of cellular and sub-cellular (viral and RNA-network) agents that communicate, i.e. based on interactions that are sign-mediated according to 3 levels of semiotic rules, syntax, semantics, pragmatics

1.3 Phage Behaviour Motifs

Phages are not pure molecules like molecules in an abiotic planet. They share biotic features such as biotic behaviour. Forest Rohwer assembled some of the most obvious examples of behavioural strategies in a "lexicon of phage behaviors "and identified the concerned protein families (Rohwer et al. 2004). The following list is taken from this lexicon:

(A) Prowling for a Partner

Before finding a host, a phage responds to environmental assaults by resistance, genome defence, hide or succumb, or even dormancy.

- phages find some hosts by electrostatic intersections and dipole moments
- in prowling for some hosts, phages interact with the capsid and surrounding milieu
- virions search for a host cell surface in a bid to recognize and sense an appropriate host and identify its appropriateness for the touch-down movement. This indicates a relatively stable interaction between the capsid and host surface.

(B) Court Host

If phages bind irreversibly to a host-cell surface court host, a virocell is created. This includes several interaction motifs such as

- penetrate the host, infiltrate, extend appendage, plug membrane hole, digest, burrow or merge
- fertilize the host and deliver capsid content by secretion, remodel capsid to extend the tube, translocate the genome
- adsorption, which means acceptance of the host by attachment, latch onto pilus, hold on, grasp, ride flagellum

(C) Ensure Virocell Viability

After the takeover of the host cell, phages ensure virocell viability by

- dispatching host defences by camouflaging, hiding, avoiding, mimicking or disguising
- defending the virocell by recognizing kin, allowing superinfection or, in another context, preventing a super-infection
- maintaining metabolism by photosynthesis or preventing suicide, nursing the virocell
- garnering resources by hijacking, commandeering, seizing, redirecting, devouring
- coercing the host by outwitting or outshining, rendering helpless, killing or threatening, or injuring

(D) Hedge Life History

In various motifs, the main source or main motif is ensuring the life history of the species. This means to decide whether to replicate the genome now or later, which needs some fine-tuned and coordinated steps.

- temporally coordination by mutating the host, killing competitors, integrating into the host, excising, jumping around, defending the host and inactivating.

Also, increasing virocell fitness by enabling colonization of a new niche or production of a toxin

- linearize the genome or circularize the genome and/or expressing proteins
- maintain pace in the red queen race by evolution, going extinct, mutation and diversification (by capsid coat, modularity, modifying nucleotides, tropism switching)
- lytic replication by fixing errors, mass production of genome copies, open helix, genome replication initiation, innovating sequence structure, protecting single strands
- recombination by outcross (illegitimate recombination) or inbreed (homologous recombination)
- obtain materials by cannibalizing, reusing, thieving, stealing, recycling

(E) Morphogenesis of Progeny

Offsprings of phages are not passive molecules but are active agents coordinating and organizing survival. An ensemble of virions

- care for offspring development by baseplate assembly, capsid accessorization, scaffolding, tail assembly, care for symmetry, tail length measurement, DNA processing, shell building and head and tail joining.
- package DNA by pumping, stuff prohead by terminase, coat with protein
- obtain components by coercing, stealing, synthesizing, scavenging, cheating and quality control
- building the virion factory by managing mass production, building the nest, coordination and organization, preparing the tools

(F) Wean Progeny

The last step in the replication cycle of phages we consider is the process of freeing virions from the virocell. The virocell concept – originally developed by Patrick Forterre – states that the bacterial host cell does not represent its original identity (Forterre 2013). The identity is captured and manipulated according to the phage life strategy. If the mass production of virions is finished, the bacterial cell membrane is lysed. This includes several steps of active behaviour such as

- annihilate outer membrane
- build pyramids
- degrade cell wall
- light fuse
- set timer
- sabotage

Although manipulated into another identity, the virocell is still involved in the rich social life and metabolic activities of the microbial host population until host cell bursts.

1.4 Trans-Organismic Communication: Communication of Phages with Non-phage Organisms

If a bacterial strain is persistently infected by a phage and caused to compete with a bacterial strain which is not infected by the same virus, the uninfected strain will undergo lysis. This means that infection and colonization of bacteria are connected with the acquisition of an immunity function which does not allow the destruction of the infected one by the uninfected one. Infected bacteria share a common immunity which is absent in uninfected bacteria. Phage colonization in a nonlytic but persistent lifestyle has a symbiotic function which protects host cells and host strains (Villarreal 2016).

It is well known that half of the bacteria in the oceans are killed daily by phages. The remaining half is the result of the incredible reproduction rate, which is the main reason for their survival. The adaptive immune system, known as CRISPR/Cas, which copies and pastes relevant sequences out of phage genomes and integrates them into the genome of infected bacteria, also plays a critical role, serving as indicators for identifying similar infecting phages that may trigger a restriction protein to kill the invading agent (Koonin et al. 2019; Koonin and Makarova 2019). CRISPR–Cas systems integrate phage DNA sequences into CRISPR loci on the host genome. This leads to heritable immunity against invading agents. This is a complex reaction motif: the integration of the sequence tool of the phage genome into the bacterial genome is not random. It must fit the remaining genome structure, should not damage the previous functional structure of the bacterial genome and should fit into the order of previously integrated sequences. The syntax rules to achieve this are currently unknown.

The defence system of CRISPR Cas is a rather complex immune system, which includes a process to prevent self-targeting and destruction processes, that defends bacteria from phages and plasmids by recognizing invading DNA (Harrington et al. 2018). Besides that, CRISPR Cas serves as signalling within the bacterial host (see below) and can guide sequence-specific transposition (Strecker et al. 2019; Dimitriu et al. 2019). Therefore CRISPR Cas contributes relevantly to the evolutionary variants of bacteria in certain adaptation processes through the generation of new genetic identities (Westra et al. 2014).

1.4.1 Interaction Motifs with Far-Reaching Consequences

The interaction motifs of phages and their prey bacteria are very complex because, through their intensive impact on bacteria, phages are relevant for bacterial distributions, populations and communities in all known ecospheres on earth (Brüssow 2018). This includes the way bacterial communities are successful in competition with other bacterial communities, how they establish equilibrium with other bacterial communities in symbiotic ecospheres, such as the human oral cavity with 700 different bacterial communities. The well-balanced equilibrium of bacterial communities is the main source of oral cavity health (Kohlenbrander et al. 2002, 2005). Asides infections, diet or daily hygiene procedures may disturb this equilibrium.

A more powerful relationship exists in the human gut (Manrique et al. 2017; Guerin et al. 2018; Shkoporov et al. 2019; Sausset et al. 2020). These phage bacterial interaction profiles are important for humans and all eukaryotic organisms with essential symbiotic relationships with bacteria (Bondy-Denomy and Davidson 2014; Carroll-Portillo and Lin 2019). About 80% of the fecal waste of animals is bacteria. Faecal waste thus represents an excellent habitat for phages. In all cases, phages determine how this symbiotic interaction can function or may even be disturbed and unbalanced. Any impact of bacterial populations on the whole range of eukaryotic organisms thus strongly depends on how these bacterial communities are affected, infected by phages and how phages remain in a persistent or lytic lifestyle with relevant impacts on the competing bacterial communities (Villarreal 2005; Feiner et al. 2015). Clearly, this persistent lifestyle of phages cannot be described as predatory against bacteria but as cooperation in most cases with co-evolutionary relevance (Borges et al. 2018; Fillol-Salom et al. 2019; Argov et al. 2019).

Phages also relate with bacteria in the carrier state in which phages cause chronic infection in bacteria. Here, the phages do not integrate into the host genome or have lytic consequences. The phage remains in this infection state, and its progeny is passed to daughter cells asymmetrically after division (Cenens et al. 2013). This relationship between phage and bacteria is a co-existence lifestyle (Roux et al. 2019). It affects the host of bacteria in that it alters some relevant ecological consequences (Siringan et al. 2014).

More recently, another kind of persistent infection has a defence which is termed "Hibernation". The persistent state of phages in the host is reversible and is regulated by the availability of appropriate nutrients (Bryan et al. 2016), such as host DNA after host DNA breakdown and glucose.

The way phages respond to bacterial defence strategies is rather interesting (Ofir and Sorek 2018). If bacteria are infected by phages, they mobilize defence activities such as e.g., CRISPR Cas. Phages may counteract these defence activities by changing attachment sites by e.g., modifications in phage protein, or as shown more recently find ways to cooperate to overcome phage resistence (Landsberger et al. 2018; Stanley and Maxwell 2018). Another reaction motif is importing a degradation system into the host that destroys bacterial nucleic acid sequences (Seed et al. 2013). More recently, research has shown that phage infection may abolish the swarming motility of host bacteria and induce the release of signalling molecules that warn uninfected subpopulations to move towards uninfected areas and promote the survival of the overall population (Bru et al. 2019).

It is noteworthy that a key role in this transorganismic communication is played by the various toxins such as holins, endolysins, bacteriocins, pyocins and colicin which are involved in generating pores - clear features of bacteria that derived from phage infections (Riley 1998; Young 2002; Nakayama et al. 2000; Bull and Regoes 2006; Villarreal 2009). This means that without phage infections bacteria would not have these features.

1.5 Inter-Organismic Communication: Communication of Phages with Other Phages and Viruses

Nearly all behavioural motifs of phages are commonly shared within phage communities (Turner and Chao 1999; Lima-Mendez et al. 2011; Stedman 2015). The prophages, i.e. phage sequences within the bacterial genome interact with other prophages. Prophages also interact with lytic phages. In such cases, the prophages serve as signs (indicators) for lytic phages to remain in a persistent lifestyle or become lytic if other indicators of the bacterial environment transport that information (Gallego Del Sol et al. 2019; Argov et al. 2019). In contrast to prophage, i.e., integrated state of phages within bacteria genome, we also know non-integrated persistence such as episomes that resemble plasmids and replicate independently from the host genome (Villarreal 2005).

Also, helper viruses such as satellite phages are common in infection cycles if phages deficient for capsid and/or tail production and lytic virulence need helper prophages that help the phage with carrying out specific functions (Liu et al. 1997). Such phage-phage interactions have relevant implications for bacterial communities in their symbiotic macroorganisms, such as in the human gut (Moelling 2016). It is important here to note that social interactions of phages are usual und determine viral fitness (Abedon 2009; Bernheim and Sorek 2018). Coinfection increases the complexity of interaction patterns for both phage communities, bacteria communities and their affected macroorganisms in shaping their community compositions.

Additionally, we should mention the dual lifestyle of genome-integrating virophages. Virophages act as a parasite of giant viruses (Paez-Espino et al. 2019). For example, they coinfect with Mimivirus and reduce burst size. This means virophage coinfection increases the survival of infected populations (La Scola et al. 2008; Berjón-Otero et al. 2019). In other behavioral motifs virophages play key roles as target of host defence in the interaction network of host cell, giant virus and virophage (Koonin and Krupovic 2017; Mougari et al. 2019).

1.5.1 Addiction Modules: Complementarity of Transorganismic and Interorganismic Communication

Addiction modules represent at least two competing genetic parasite clouds, which try to invade host genomes. This represents a complementary interaction event (transorganismic and interorganismic). Addiction modules can be defined as features that consist of a stable, toxic component, which is counterbalanced by an unstable component inhibiting and suppressing the toxic component. This behavioral motif of phages originally was described by Lehnherr and Yarmolinsky and later generalized by Villarreal (Lehnherr et al. 1993; Lehnherr and Yarmolinsky 1995; Villarreal 2012a, b; Villarreal 2015; Villarreal 2016).

This is an important impact of phages on bacterial hosts: the integration of toxins of various kinds, even more than one, so that bacteria may integrate toxins that do not harm the host because the toxin (T) is counterbalanced by an antitoxin (A) derived from a competing phage infection (Harms et al. 2018). If such T/A module-based bacteria contact other bacterial populations that don't possess the whole module, it may kill these populations. The same happens with restriction/modification enzyme modules that are relevant for several interaction motifs (Kobayashi 2001; Yahara et al. 2007; Mruk and Kobayashi 2014).

It is particularly interesting that mixtures of cryptic prophages, i.e., defective and silent, may take up a large portion of the genome, as documented in E.Coli, where K12 may represent 20% of the total genome. This may be up to 35 sets of Toxin/Antitoxin sets out of several cryptic prophages (Wang et al. 2010). If we remove these viral TA sets, prokaryotic cells may become more sensitive to various stressors, such as antibiotics, osmotic, oxidative and acid stress. Deletions of such TA modules in K 12 phage lead to the loss of the ability to form biofilm in small media and therefore resembles a kind of group effect phenotype (Villarreal and Witzany 2015). The persistent (temperate) lifestyle of phages in host genomes may lead to high-density regions which do not serve as genetic fossils but a kind of spread community of active agents. Importantly, this demonstrates how host genetic identity depends on how phages or defective parts of phages determine host gene structures (Hambly and Suttle 2005; Villarreal 2009).

The emergence of genetic identities, first in prokaryotic organisms, then in the evolution of multicellular communities such as whole tissues or organs may start with such addiction modules. Such identity networks are constituted by the cooperation of self-harm and self-protection, without amalgamating them into one feature. They remain two capacities, that are temporally counterbalancing, but may provide other cooperation networks that compete and/or destroy the other one in certain circumstances (context) and are, therefore, open to generate new identities, even evolutionary ones. It should be noted that all these interaction motifs may vary according to methylated properties, i.e. epigenetic imprintings that change expression patterns and variable protein meanings without changing the primary DNA sequence.

Importantly, from a "virus first "perspective, this interaction motif of addiction modules is a dominant motif of genetic counter-regulation in all domains of life not restricted to T/A or R/M motifs and a key element in the generation of diversity and host genetic identities (Villarreal 2012b). Addiction module generation integrates competing (counter regulating) genetic information of parasites (present in organisms of all domains of life) into host genetic identities, which non-infected hosts from the same or related species do not possess (van Sluijs et al. 2019). This may alter host phenotype by complex genetic information in a single event, not depending on error replication (chance mutations) and their selection over long periods. Such complementing – in most cases defective – viruses or virus-derived parts may cause disease or even kill the host if counter-regulation becomes unbalanced or recombined with other remaining defective minorities of former viral infections (Villarreal 2005, 2011). This indicates that defective minorities do not remain as waste or "junk "but as important re-usable module like tools (Villarreal and Witzany 2019). Additionally, such addiction modules serve as immune functions against the same or related infective genetic parasites. This means it will protect infected hosts and kill non-infected ones (Fig. 1.3).

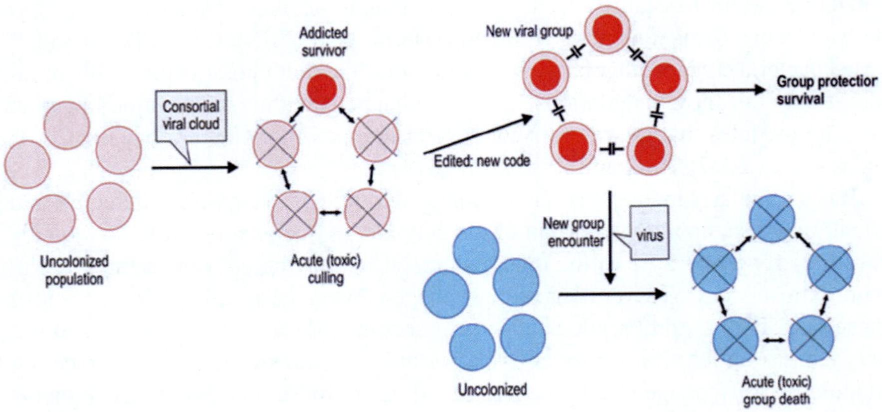

Fig. 1.3 Generalized interactional motif of infection-derived group identities: the addiction module as a result of counterbalanced (addicted survivor), infection-derived, and persistent genetic parasites that initiate evolutionary inventions (new viral group) by natural genetic engineering of host genetic identities, some we can find as toxin/antitoxin, restriction/modifcation, or insertion/deletion modules. (From Villarreal, L.P. Viruses and host evolution: Virus-mediated self identity. In López-Larrea, C. (ed.), Self and Nonself, Springer Science+Business Media, Austin, LandesBioscience, New York, pp. 185–217, 2012b. With permission)

1.5.2 Addiction Modules and the Evolution of Programmed Cell Death (PCD)

P1, formerly seen as a plasmid not virus is such an example: Eventually, it became recognized as a real virus that persists as a stable episome that does not integrate into the host genome. If E. coli is colonized by P1, it reaches at least one T/A module by which expresses an immunity function. But if P1 colonized E. coli and is infected by other phages such as e.g., T4, the antitoxin gene is destabilized and the remaining stable toxin will kill this infected cell, thereby preventing T4 transmission to uninfected bacterial neighbours. This means, that a P1-colonized bacteria will start suicide to prevent T4 replication and transmission of a harbouring colony (Lehnherr and Yarmolinsky 1995). Therefore, P1 plays an essential role in group identity formation of bacteria and represents a general model in that persistence generates group identity by the integration of addiction modules. The addiction module has been first seen at P1 and the post-segregation killing of the host if the host lost the virus (Lehnherr et al. 1993; Engelberg-Kulka and Glaser 1999; Hazan et al. 2001).

Single cells die to protect the remaining group identity. If reminded that development of cells, tissues, organs and whole organisms depends on certain developmental stages in that parts of the whole organisms follow expression stages determined by epigenetic imprintings that differentiate the growing tissue cells and allow some cells die and others to grow, we can imagine that certain gene regulatory pathways from early times in biology evolution such as bacteria and phages may be co-opted and exported to later-evolved organisms.

Programmed cell death is a consequence of an unbalanced addiction module (T/A or R/M). If the balance is stable, it determines the genetic identity of a host and temporally imprints the host population. This imprinting is absent in host populations that are not infected by such balanced modules. Such imprinting may become unbalanced by several defectives of former infections such as introns, transposons or hyperparasite invasions (intene, retroposon) (Villarreal 2009). The new identity of a bacteria population (with its certain group behaviour and preferred host organisms) by such R/M modules on the genetic level (R/M addiction genes) and epigenetic level (methylated DNA). The interesting aspect here is that this focuses on infection techniques via addiction module building and its consequence, the imprinting of host genetic identities and integrates complementary features, the colonization, new immunity and new identity.

Now we can consider the crucial feature for the process of acquiring all the genetic content of bacteria and the processes that caused their diversity (group identities). This has serial, cumulative and episodic consequences which are important aspects in understanding the cumulative evolution of complexity also.

Interestingly, most evolutionary biologists engage fitness determinants only from the role of cellular genes. The roles of persistent genetic parasites and their interactions within the host genome on survival have been overlooked for decades. Additionally, the roles of defective genetic parasites as key players in communication (sign-mediated interaction) with non-defective parasites are also essentially absent from evolutionary biology. Indeed, the crucial role of communication per se is poorly developed in evolutionary biology.

The fundamental interplay of addiction modules through their temporal stability and immunity functions and their harmful consequences to excluded individuals via altered addiction module status results in stabilized group survival as a key feature of most organisms, including eukaryotes.

Phage-bacteria and phage-phage interactions are the most pragmatic concepts to coherently explain the regular control of bacterial populations and their harmfull and beneficial roles to eukaryotic host organisms after understanding the evolution and emergence of bacterial genetic identities and diversity of interactions of bacteria, their persistent viruses and their eukaryotic host organisms (Witzany 2011; Guglielmini et al. 2019).

1.6 Intra-Organismic Communication: Communication Within Phages

All relevant interactions of phages on the transorganismic and interorganismic level need a variety of internal organisation and coordination processes such as that used for their genetic expression. The genome of the phage must be coherently expressed in most cases by hijacking the bacterial transcription process to reach these goals, although some giant phages do not depend on a bacterial host for replication (Ceyssens et al. 2014). The activity of the bacterial transcription process by RNA polymerase (RNAP), is regulated by a variety of small phage-encoded proteins (Tabib-Salazar et al. 2019). In considering the infection of bacteria in the animal gut, we should remember that the genetic expression pattern of the phage may also react to some satellite phage that is currently present as well as the signalling molecules of the other phages that induce a persistent lifestyle. Such signaling may change if the environmental conditions (context) of the host changes rapidly, or/and the bacterial adaptive immunity in certain strains is increasingly strong, because bacteria-host interactions are relevant in a certain stage of gut microbiome situation (context). This means the phage (and host) genetic expression is rather context-dependent and needs some context coherent reaction modes.

The reaction patterns of phage expression is also dependent on the status for methylation of certain sequence in host strains which is enhanced following repeated similar situations (infections), a feature that seems to be a relevant memory/learning behaviour of phages to increase successful interaction. The interaction within phages is thus very complex and has to sense (interpret) complex incoming information from other parasites in a rather fast and appropriate way. Otherwise, the phage response would not be successful for survival (Casadesús and D'Ari 2002).

1.7 Phage Communication Vocabulary: Examples

Persistent phages in bacterial host genomes may switch between lytic and lysogenic life cycles. Recently it was shown that *Bacillus subtilis* SPbeta phage produce a peptide (AimP) which serves as a signal within a communicative interaction during phage infection (Erez et al. 2017; Abedon 2017). AimP reduces the expression of the negative regulator of lysogeny (AimX) by binding to the transcription factor (AimR) promoting lysogeny. Thus, persistent phages have to decide every time they infect a bacterial cell on starting the lytic cycle or lysogenic cycle (Weitz et al. 2008). Several phages and infection-derived mobile genetic elements encode peptides (e.g. arbitrium) that serve as signs in communicative interactions. This is counter-regulated by a non-coding RNA that serves as peptide response and controls regulation of the lysogenic state (Stokar-Avihail et al. 2019).

In the type III CRISPR Cas system, the recognition of foreign DNA leads to the production of a small molecule (cyclic oligodenylate) which activates a CRISPR-associated RNAse which cleaves cellular RNA nonspecifically (Amitai and Sorek 2017). Phages may produce a specific response: an anti-CRISPR protein (acr) that inhibits CRISPR Cas immune function (Landsberger et al. 2018). This is the result of phage-phage cooperation to overcome CRISPR resistance (Borges et al. 2018). First, phages block the CRISPR Cas immune system of bacteria: this allows a second infective phage to replicate within the host. The success of this cooperation, however, depends on the density of phage populations.

Additionally, bacteria defence systems also rely on communicative interactions within the bacteria as documented in the GMp-AMP synthase –STING pathway, an immune response of animals that acts as a sensor of cytosolic viral DNA and produces a cyclic FMP-AMP signalling molecule that activates the immune response by binding to the STING protein. This can be found in bacteria as an antiphage defence also (Cohen et al. 2019). Additionally, bacteria may produce various other molecules that can block a successful viral infection (Clokie 2018).

1.8 Phagetherapy

Bacterial multidrug-resistance is a serious medical problem with rapid progression worldwide. The common use of broad-range antibiotics leads to antibiotic crisis and resistance of many human bacterial pathogens. Bacteriophages infect and kill bacteria, even multidrug-resistant ones and so-called superbugs (Rohde et al. 2018a, b; Pirnay et al. 2018; Sakr et al. 2018). Their noninfectious nature to humans should make them safe for human biomedicine (Rehman et al. 2019).

1.8.1 Historical Notes

Phages were applied as anti-bacterials first by Felix D'Herelle in Paris in the early 1920s. D'Herelle collaborated with Georgi Eliava at the Pasteur Institute and, later on, after Eliava returned to Tbilisi, Georgia, both founded the Institute of Bacteriophage Microbiology and Virology in 1933. Phages had to be isolated from the environment, cultivated on bacterial hosts, and purified with technology that was available at that time. In the following decades, this institute developed to the world's biggest phage research and production centre. During and after world war II, this research increased, especially in Russia. Many patients visited and still visit this centre for individualized phagetherapy, mainly using complex phage cocktails (Ajuebor et al. 2018; Rohde et al. 2018a, b). The best-documented applications stemmed from former Russia (in Georgia) in the 1960s, of successful prophylaxis against Shigella dysentery and E. Coli diarrhoea.

Another "hot spot" of phagetherapy was established in Poland with high efficiency in appropriate phagetherapy applications, especially in investigations of immune response under phagetherapy. In contrast to the phagetherapy practice in Georgia, the Polish used monophage preparations exclusively (Górski et al. 2018; Górski et al. 2019).

In the following decades, in Belgium (Astrid Military Hospital, Brussels), France (Pasteur Institute), Switzerland (ETH Zurich), and Germany (Charité University Hospital, Berlin), promising clinical applications of phagetherapy were established and officially supported (Sakr et al. 2018; Wienhold et al. 2018). A more recent trial in Bangladesh failed to show successful applications, but a second trial was more promising (Sarker et al. 2017; Bolocan et al. 2019).

Soon after the discovery of antibiotics in the 1940s, interest in using phages for therapeutic purposes was nearly lost, except in the Tibilisi Institute (Georgia). The public health crisis from multidrug-resistance of bacterial strains led to increasing interest in alternatives to antibacterial agents, integrating a variety of therapeutics and prevention strategies. In 2018, 100,000 persons died from antibiotic-resistant bacterial infections definitively, of the 700,000 persons estimated to die from antibiotic-resistant bacterial infections annually.

1.8.2 Reasons for Multidrug-Resistance

The increase in multi-resistant bacteria is attributed to the excessive use of antibiotic substances in human medicine and argriculture. This abuse of antibiotics has been seen in human and veterinary practices as well as industrial and agricultural practices that have increased the prevalence of drug resistance among many bacterial strains.

Animal farming also contributes to bacterial resistance, creating a vast reservoir of antimicrobial drug resistance in combination with zoonoses. Unfortunately, major zoonotic bacteria belong to the ESKAPE group. The ESKAPE group includes *Enterococcus faecium, Staphylococcus aureus, Klebsiella pneumoniae, Acinetobacter baumannii, Pseudomonas aeruginosa, Enterobacter spp*. They possess increased resistance to antibiotics such as penicillin, vancomycin, methicillin, carbapenems and others. (Santajit and Indrawattana 2016; Mulani et al. 2019).

The human microbiome inhabits very different sites. The skin is an important site, with staphylococci as one of the most prevalent residents. If the skin is damaged, the microbiome may colonize the skin injury and cause infection. Many problems with bacterial biofilms occur on implantation material, skin transplantation, several injuries in the oral cavity or lung, chronic wounds such as diabetic foot ulcers and unbalancing of the intestinal gut microbiome. More different indications for both chronic and acute conditions, including bone-and-joint, urogenital, respiratory, wound, cardiac, and systemic infections are reported meanwhile. Here phagetherapy looks very promising because phages may degrade biofilms that cause diseases (Rohde et al. 2018a, b).

1.8.3 Current Plans to Fight Multi-Drug Resistance by Phagetherapy

Phages penetrate biofilms much better than antibiotics substances because they express exopolymer-degrading enzymes, such as polysacharide depolymerases, which are more efficient in multilayer biofilms than antibiotics. The efficiency of antibiotics decreases through the various layers of biofilms. Furthermore, phages may cause beneficial immune responses.

Therapeutic use of phages as a future-oriented alternative to conventional antibiotics is a relevant focus, according to WHO assembly resolution (68.7.3.) from 2015 which called for national action plans by May 2017 to combat the antimicrobial drug resistance crisis ("...inappropriate use of antimicrobial medicines in all relevant sectors continues to be an urgent and widespread problem in all countries..."). The G7 Health Ministeries initiated a global "One health" approach in 2015 and 2017. The EU initiated such a strategy by the Horizon 2020 program. If nothing changes, ten million persons have been predicted to die from AMR in 2050.

European One Health Action Plan against Antimicrobial Resistance (AMR) (2017/2254(INI)) is a further step followed by Antibiotic Resistance Threats in the United States, 2019 (Atlanta, GA: U.S. Department of Health and Human Services, CDC; 2019) which states that more than 2.8 million antibiotic-resistant infections occur in the U.S. each year, and more than 35,000 people die. There are a variety of judicial cases in several states regarding research on the subject. Mostly, they derive because national laws follow suggestions of definitions that are congruent with antibiotic economics.

The realization that phages and bacteria have a variety of interaction patterns by various signalling ways, i.e. communicative interactions, invites research on how humans can apply this competence of phages to kill bacteria with dangerous impacts on human health (Wienhold et al. 2018; Rohde et al. 2018a, b). This includes bacteria immune systems and reactions to phage infection and colonization and the various ways phages can counteract these immune functions. At this time, phagetherapy seems like a curious therapy method besides the mainstream antibiotic era (Domingo-Calap and Delgado-Martínez 2018). In the next 20 years, this will surely turn around to protect millions of humans that are confronted with the consequences of multidrug resistance.

1.9 Conclusion

The rich variety of communicative interaction, i.e. interactions between phages and bacteria, phages and other phages and phages on other living organisms that depend on sign(alling) undoubtedly demonstrates that phages are essential agents within the animate nature. The rich interaction motifs of phages also indicate that viruses are essential agents within the roots and stems of the tree of life. Bacterial genome construction is under the crucial influence of phages, and the disease-causing toxins of bacteria, in most cases, are results of former viral infections that transferred toxins and antitoxins to host bacteria. Interestingly, the communicative competences of phages are often found in a complementary way. This means transorganismic communication is intertwined with interorganismic and intraorganismic communication as demonstrated by the most important result, the addiction modules. It is thus essential that we better understand phage interactions if we are to succeed in developing phage therapy to bacterial multi-drug resistance.

As a further result, the old question of whether viruses are alive or not remains a historical curiosity, because it is impossible to examine bacteria without the co-evolutionary role of phages and the constant interactions of phages on bacterial populations. Defining viruses not to be alive is based on a wrong definition of life in general because it ignores communication and the inability of communicated viruses to self replicate is not a deficiency. Meaningful and consequential communication has occurred. And it results in effective adaptation to a planetary cellular world that ensures virus replication without dependence on its own replication apparatus which would contradict low energy cost determinants.

Acknowledgement I want to thank Luis P. Villarreal for helpfull comments.

References

Abedon ST (2009) Bacteriophage intraspecific cooperation and defection. In: Adams HT (ed) Contemporary trends in bacteriophage research. Nova Science Publishers, New York, pp 191–215

Abedon ST (2011) Communication among Phages, Bacteria and soil environments. In: Witzany G (ed) Biocommunication of soil microorganisms. Springer, Dortrecht, pp 37–65

Abedon ST (2017) Commentary: communication between viruses guides Lysis-Lysogeny decisions. Front Microbiol 8:983

Ajuebor J, Buttimer C, Arroyo-Moreno S et al (2018) Comparison of Staphylococcus phage K with close phage relatives commonly employed in phage therapeutics. Antibiotics (Basel) 7(2):E37. https://doi.org/10.3390/antibiotics7020037

Amitai G, Sorek R (2017) Intracellular signaling in CRISPR-Cas defense. Science 357(6351):550–551

Argov T, Azulay G, Pasechnek A et al (2017) Temperate bacteriophages as regulators of host behavior. Curr Opin Microbiol 38:81–87

Argov T, Sapir SR, Pasechnek A et al (2019) Coordination of cohabiting phage elements supports bacteria-phage cooperation. Nat Commun 10(1):5288

Armon R (2011) Soil Bacteria and bacteriophages. In: Witzany G (ed) Biocommunication in soil microorganisms. Springer, Dortrecht, pp 67–112

Bassler BL, Losick R (2006) Bacterially speaking. Cell 125(2):237–246

Batinovic S, Wassef F, Knowler SA et al (2019) Bacteriophages in natural and artificial environments. Pathogens 8(3):E100

Ben-Jacob E (2014) My encounters with bacteria--learning about communication, cooperation and choice. Phys Biol 11(5):053009

Berjón-Otero M, Koslová A, Fischer MG (2019) The dual lifestyle of genome-integrating virophages in protists. Ann N Y Acad Sci 1447(1):97–109

Bernheim A, Sorek R (2018) Viruses cooperate to defeat bacteria. Nature 559(7715):482–484

Bolocan AS, Upadrasta A, Bettio PHA et al (2019) Evaluation of phage therapy in the context of enterococcus faecalis and its associated diseases. Viruses 11(4):E366. https://doi.org/10.3390/v11040366

Bondy-Denomy J, Davidson AR (2014) When a virus is not a parasite: the beneficial effects of prophages on bacterial fitness. J Microbiol 52(3):235–242

Borges AL, Zhang JY, Rollins MF et al (2018) Bacteriophage cooperation suppresses CRISPR-Cas3 and Cas9 immunity. Cell 174(4):917–925

Bru J-L, Rawson B, Trinh C et al (2019) PQS produced by the Pseudomonas aeruginosa stress response repels swarms away from bacteriophage and antibiotics. J Bacteriol 201:e00383–e00319

Brüssow H (2018) Population genomics of bacteriophages. In: Polz M, Rajora O (eds) Population genomics: microorganisms. Population genomics. Springer, Cham, pp 297–234

Bryan D, El-Shibiny A, Hobbs Z et al (2016) Bacteriophage T4 infection of stationary Phase E. coli: life after log from a phage perspective. Front Microbiol 7:1391

Bull JJ, Regoes RR (2006) Pharmacodynamics of non-replicating viruses, bacteriocins and lysins. Proc Biol Sci 273(1602):2703–2712

Carroll-Portillo A, Lin HC (2019) Bacteriophage and the innate immune system: access and Signaling. Microorganisms 7(12):E625

Casadesús J, D'Ari R (2002) Memory in bacteria and phage. BioEssays 24(6):512–518

Cenens W, Makumi A, Mebrhatu MT et al (2013) Phage-host interactions during pseudolysogeny: lessons from the Pid/dgo interaction. Bacteriophage 3(1):e25029

Ceyssens PJ, Minakhin L, Van den Bossche A et al (2014) Development of giant bacteriophage ϕKZ is independent of the host transcription apparatus. J Virol 88(18):10501–10510

Clokie MRJ (2018) Bacterial defence molecules target viral DNA. Nature 564(7735):199–200

Cohen D, Melamed S, Millman A et al (2019) Cyclic GMP-AMP signalling protects bacteria against viral infection. Nature 574(7780):691–695

Díaz-Muñoz SL, Koskella B (2014) Bacteria-phage interactions in natural environments. Adv Appl Microbiol 89:135–183

Dimitriu T, Ashby B, Westra ER (2019) Transposition: a CRISPR way to get around. Curr Biol 29(18):R886–R889

Domingo-Calap P, Delgado-Martínez J (2018) Bacteriophages: protagonists of a post-antibiotic era. Antibiotics (Basel) 7(3)

Engelberg-Kulka H, Glaser G (1999) Addiction modules and programmed cell death and antideath in bacterial cultures. Annu Rev Microbiol 53:43–70

Erez Z, Steinberger-Levy I, Shamir M et al (2017) Communication between viruses guides lysis-lysogeny decisions. Nature 541(7638):488–493

Feiner R, Argov T, Rabinovich L et al (2015) A new perspective on lysogeny: prophages as active regulatory switches of bacteria. Nat Rev Microbiol 13(10):641–650

Fillol-Salom A, Alsaadi A, Sousa JAM et al (2019) Bacteriophages benefit from generalized transduction. PLoS Pathog 15(7):e1007888

Forterre P (2013) The virocell concept and environmental microbiology. ISME J 7(2):233–236

Gallego Del Sol F, Penadés JR, Marina A (2019) Deciphering the molecular mechanism underpinning phage arbitrium communication systems. Mol Cell 74(1):59–72.e3

Górski A, Jończyk-Matysiak E, Łusiak-Szelachowska M et al (2018) Phage therapy in prostatitis: recent prospects. Front Microbiol 9:1434. https://doi.org/10.3389/fmicb.2018.01434

Górski A, Międzybrodzki R, Węgrzyn G et al (2019) Phage therapy: current status and perspectives. Med Res Rev 40:459–463. https://doi.org/10.1002/med.21593

Guerin E, Shkoporov A, Stockdale SR et al (2018) Biology and taxonomy of crAss-like bacteriophages, the most abundant virus in the human gut. Cell Host Microbe 24(5):653–664.e6

Guglielmini J, Woo AC, Krupovic M et al (2019) Diversification of giant and large eukaryotic dsDNA viruses predated the origin of modern eukaryotes. Proc Natl Acad Sci U S A 116(39):19585–19592

Hambly E, Suttle CA (2005) The viriosphere, diversity, and genetic exchange within phage communities. Curr Opin Microbiol 8(4):444–450

Harms A, Brodersen DE, Mitarai N et al (2018) Toxins, targets, and triggers: an overview of toxin-antitoxin biology. Mol Cell 70(5):768–784

Harrington LB, Burstein D, Chen JS et al (2018) Programmed DNA destruction by miniature CRISPR-Cas14 enzymes. Science 362:839–842

Hatfull GF (2015) Dark matter of the biosphere: the amazing world of bacteriophage diversity. J Virol 89:8107–8110

Hazan R, Sat B et al (2001) Postsegregational killing mediated by the P1 phage “addiction module”phd-doc requires the Escherichia coli programmed cell death system mazEF. J Bacteriol 183(6):2046–2050

Kaiser D, Losick R (1993) How and why bacteria talk to each other. Cell 73(5):873–885

Kavagutti VS, Andrei AŞ, Mehrshad M et al (2019) Phage-centric ecological interactions in aquatic ecosystems revealed through ultra-deep metagenomics. Microbiome 7(1):135

Kobayashi I (2001) Behavior of restriction-modification systems as selfish mobile elements and their impact on genome evolution. Nucleic Acids Res 29(18):3742–3756

Kohlenbrander PE, Andersen RN, Blehert DS et al (2002) Communication among Oral Bacteria. Microbiol Mol Biol Rev 66:486–505

Kohlenbrander PE, Egland PG, Diaz PI et al (2005) Genome-genome interactions: bacterial communities in initial dental plaque. Trends Microbiol 13:11–15

Koonin EV, Krupovic M (2017) Polintons, virophages and transpovirons: a tangled web linking viruses, transposons and immunity. Curr Opin Virol 25:7–15

Koonin EV, Makarova KS (2019) Origins and evolution of CRISPR-Cas systems. Philos Trans R Soc B 374:20180087

Koonin EV, Krupovic M, Yutin N (2015) Evolution of double-stranded DNA viruses of eukaryotes: from bacteriophages to transposons to giant viruses. Ann N Y Acad Sci 1341:10–24

Koonin EV, Makarova KS, Wolf YI et al (2019) Evolutionary entanglement of mobile genetic elements and host defence systems: guns for hire. Nat Rev Genet 21:119–131. https://doi.org/10.1038/s41576-019-0172-9

La Scola B, Desnues C, Pagnier I et al (2008) The virophage as a unique parasite of the giant mimivirus. Nature 455(7209):100–104

Landsberger M, Gandon S, Meaden S et al (2018) Anti-CRISPR phages cooperate to overcome CRISPR-Cas immunity. Cell 174(4):908–916

Lehnherr H, Yarmolinsky MB (1995) Addiction protein Phd of plasmid prophage P1 is a substrate of the ClpXP serine protease of Escherichia coli. Proc Natl Acad Sci U S A 92(8):3274–3277

Lehnherr H, Maguin E, Jafri S et al (1993) Plasmid addiction genes of bacteriophage P1: doc, which causes cell death on curing of prophage, and phd, which prevents host death when prophage is retained. J Mol Biol 233(3):414–428

Lima-Mendez G, Toussaint A, Leplae R (2011) A modular view of the bacteriophage genomic space: identification of host and lifestyle marker modules. Res Microbiol 162(8):737–746

Liu T, Renberg SK, Haggård-Ljungquist E (1997) Derepression of prophage P2 by satellite phage P4: cloning of the P4 epsilon gene and identification of its product. J Virol 71(6):4502–4508

Manrique P, Dills M, Young MJ (2017) The human gut phage community and its implications for health and disease. Viruses 9(6):E141

Meaden S, Capria L, Alseth E et al (2019) Transient CRISPR immunity leads to coexistence with phages. bioRxiv. https://doi.org/10.1101/2019.12.19.882027

Moelling K (2016) Nutrition and the microbiome. Ann N Y Acad Sci 1371:53–64

Mougari S, Sahmi-Bounsiar D, Levasseur A et al (2019) Virophages of Giant viruses: an update at eleven. Viruses 11(8):E733

Mruk I, Kobayashi I (2014) To be or not to be: regulation of restriction-modification systems and other toxin-antitoxin systems. Nucleic Acids Res 42(1):70–86

Mulani MS, Kamble EE, Kumkar SN et al (2019) Emerging strategies to combat ESKAPE pathogens in the era of antimicrobial resistance: a review. Front Microbiol 10:539

Nakayama K, Takashima K, Ishihara H, et al (2000) The R-type pyocin of Pseudomonas aeruginosa is related to P2 phage, and the F-type is related to lambda phage. Mol Microbiol 38:213–231

Ofir G, Sorek R (2018) Contemporary phage biology: from classic models to new insights. Cell 172(6):1260–1270

Paez-Espino D, Zhou J, Roux S et al (2019) Diversity, evolution, and classification of virophages uncovered through global metagenomics. Microbiome 7(1):157

Pirnay JP, Cooper I, Caplin J et al (2018) Silk route to the acceptance and re-implementation of bacteriophage therapy-part II. Antibiotics (basel) 7(2):35
Rehman S, Ali Z, Khan M et al (2019) The dawn of phage therapy. Rev Med Virol 29(4):e2041
Riley MA (1998) Molecular mechanisms of bacteriocin evolution. Annu Rev Genet 32:255–278
Rohde C, Wittmann J, Kutter E (2018a) Bacteriophages: a therapy concept against multi-drug-resistant Bacteria. Surg Infect 19(8):737–744. https://doi.org/10.1089/sur.2018.184
Rohde C, Resch G, Pirnay JP et al (2018b) Expert opinion on three phage therapy related topics: bacterial phage resistance, phage training and Prophages in bacterial production strains. Viruses 10(4):E178
Rohwer F, Youle M, Maughan H, et al (2014) Life in our phage world. A centennial field guide to the earth's most diverse inhabitants. Wholon, San Diego
Roux S, Krupovic M, Daly RA et al (2019) Cryptic inoviruses revealed as pervasive in bacteria and archaea across Earth's biomes. Nat Microbiol 4(11):1895–1906
Sakr Y, Jaschinski U, Wittebole X et al (2018) Sepsis in intensive care unit patients: worldwide data from the intensive care over nations audit. Open forum. Infect Dis Ther 5(12):ofy313. https://doi.org/10.1093/ofid/ofy313
Santajit S, Indrawattana N (2016) Mechanisms of antimicrobial resistance in ESKAPE pathogens. Biomed Res Int:2475067
Sarker SA, Berger B, Deng Y et al (2017) Oral application of Escherichia coli bacteriophage: safety tests in healthy and diarrheal children from Bangladesh. Environ Microbiol 19(1):237–250
Sausset R, Petit MA, Gaboriau-Rothiau V et al (2020) New insights into intestinal phages. Mucosal Immunol 13:205–215. https://doi.org/10.1038/s41385-019-0250-5
Schauder S, Bassler BL (2001) The languages of bacteria. Gen Develop 15:1468–1480
Seed KD, Lazinski DW, Calderwood SB et al (2013) A bacteriophage encodes its own CRISPR/Cas adaptive response to evade host innate immunity. Nature 494:489–491
Shkoporov AN, Clooney AG, Sutton TDS et al (2019) The human gut Virome is highly diverse, stable, and individual specific. Cell Host Microbe 26(4):527–541
Siringan P, Connerton PL, Cummings NJ et al (2014) Alternative bacteriophage life cycles: the carrier state of campylobacter jejuni. Open Biol 4:130200
Stanley SY, Maxwell KL (2018) Phage-encoded anti-CRISPR defenses. Annu Rev Genet 52:445–464
Stedman KM (2015) Deep recombination: RNA and ssDNA virus genes in DANN virus and host genomes. Annu Rev Virol 2(1):203–217
Stokar-Avihail A, Tal N, Erez Z et al (2019) Widespread utilization of peptide communication in phages infecting soil and pathogenic Bacteria. Cell Host Microbe 25(5):746–755
Strecker J, Ladha A, Gardner Z et al (2019) RNA-guided DNA insertion with CRISPR-associated transposases. Science 365:48–53
Tabib-Salazar A, Mulvenna N, Severinov K et al (2019) Xenogeneic regulation of the bacterial transcription machinery. J Mol Biol 431(20):4078–4092
Turner PE, Chao L (1999) Prisoner's dilemma in an RNA virus. Nature 398(6726):441–443
van Sluijs L, van Houte S, van der Oost J, Brouns SJ et al (2019) Addiction systems antagonize bacterial adaptive immunity. FEMS Microbiol Lett 366(5):fnz047
Villarreal LP (2005) Viruses and the evolution of life. ASM Press, Washington
Villarreal LP (2009) The source of self: genetic parasites and the origin of adaptive immunity. Ann N Y Acad Sci 1178:194–232
Villarreal LP (2011) Viral ancestors of antiviral systems. Viruses 3(10):1933–1958
Villarreal L (2012a) Viruses and host evolution: virus-mediated self identity. Adv Exp Med Biol 738:185–217
Villarreal LP (2012b) The addiction module as a social force. In: Witzany G (ed) Viruses: essential agents of life. Springer, Dortrecht, pp 107–145
Villarreal LP (2015) Force for ancient and recent life: viral and stem-loop RNA consortia promote life. Ann N Y Acad Sci 1341:25–34

Villarreal LP (2016) Persistent virus and addiction modules: an engine of symbiosis. Curr Opin Microbiol 31:70–79
Villarreal LP, Witzany G (2015) When competing viruses unify: evolution, conservation, and plasticity of genetic identities. J Mol Evol 80(5–6):305–318
Villarreal LP, Witzany G (2019) That is life: communicating RNA networks from viruses and cells in continuous interaction. Ann N Y Acad Sci 1447:5–20
Wang X, Kim Y, Ma Q et al (2010) Cryptic prophages help bacteria cope with adverse environments. Nat Commun 1:147
Warwick-Dugdale J, Buchholz HH, Allen MJ et al (2019) Host-hijacking and planktonic piracy: how phages command the microbial high seas. Virol J 16(1):15
Weitz JS, Mileyko Y, Joh RI et al (2008) Collective decision making in bacterial viruses. Biophys J 95(6):2673–2680
Westra ER, Buckling A, Fineran PC (2014) CRISPR-Cas systems: beyond adaptive immunity. Nat Rev Microbiol 12(5):317–326
Wienhold SM, Lienau J, Witzenrath M (2018) Towards inhaled phage therapy in Western Europe. Viruses 11(3):E295. https://doi.org/10.3390/v11030295
Williamson KE (2011) Soil phage ecology: abundance, distribution, and interactions with bacterial host. In: Witzany G (ed) Biocommunication in soil microorganisms. Springer, Dortrecht, pp 113–136
Witzany G (1993) Naur der Sprache – Sprache der Natur. In: Sprachpragmatische Philosophie der Biologie. Koenigshausen & Neumann, Würzburg
Witzany G (2000) Life: the communicative structure. BoD, Norderstadt
Witzany G (2009) Bacteria and viruses: communal interacting agents. In: Chauhan A, Varma A (eds) A textbook of molecular biotechnology. I.K. International Publishing, New Dehli, pp 905–914
Witzany G (2010a) Uniform categorization of biocommunication in bacteria, fungi and plants. World J Biol Chem 1(5):160–180
Witzany G (2010b) Biocommunication and natural genome editing. Springer, Dordrecht
Witzany G (ed) (2011) Biocommunication in soil microorganisms. Springer, Heidelberg
Witzany G (ed) (2012a) Biocommunication of Fungi. Springer, Dordrecht
Witzany G (ed) (2012b) Biocommunication of animals. Springer, Dordrecht
Witzany G (2014) Biological Self-Organization. IJSSS 3(2):1–11
Witzany G (2016a) The biocommunication method: on the road to an integrative biology. Comm Integr Biol 9(2):e1164374
Witzany G (2016b) Key levels of biocommunication. In: Gordon R, Seckbach J (eds) Biocommunication: sign-mediated interactions between cells and organisms. World Scientific, Singapore, pp 37–61
Witzany G (ed) (2018) Biocommunication of Archaea. Springer, Cham
Witzany G (2019) Communication is the main characteristic of life. In: Kolb V (ed) Handbook of astrobiology. CRC Press, Boca Raton, pp 91–105
Witzany G, Baluška F (eds) (2012) Biocommunication of plants. Springer, Berlin/Heidelberg
Witzany G, Nowacki M (eds) (2016) Biocommunication of ciliates. Springer, Dordrecht
Yahara K, Horie R, Kobayashi I et al (2007) Evolution of DNA double-strand break repair by gene conversion: coevolution between a phage and a restriction-modification system. Genetics 176(1):513–526
Youle M, Haynes M, Rohwer F (2012) Scratching the surface of Biology's dark matter. In: Witzany G (ed) Viruses: essential agents of life. Springer, Dortrecht, pp 61–81
Young R (2002) Bacteriophage holins: deadly diversity. J Mol Microbiol Biotechnol 1:21–36

Chapter 2
Phage-Phage, Phage-Bacteria, and Phage-Environment Communication

Stephen T. Abedon

Abstract Besides obtaining and utilizing resources, organisms have three basic ecological tasks: to survive, to reproduce, and to move. Survival is necessary for reproduction, reproduction increases numbers, and movement at a minimum assures that all of an organism's 'eggs' are *not* found in the same spatial 'basket'. For bacteriophages (phages), these facets can be differentiated into mechanisms that operate within the context of bacterial hosts (intracellularly) versus less so, i.e., instead extracellularly. Survival of phages, or their inactivation, thus can occur in the course of infection of bacteria, or instead as free virions, that is, as phages which have not yet adsorbed and infected a bacterium. Reproduction by phages can range from that which is more closely linked with normal bacterial metabolism and which ends with the phage intracellular (as in the course of lysogenic cycles) to that which involves substantial modification of normal bacterial metabolism and which ends with the phage now extracelullar (as in the course of lytic cycles). Movement of phages can occur most familiarly as diffusing virions, but also while phages are infecting bacteria. In the course of undergoing these and other processes, materials and information can flow from one entity to another, such as resulting in, for example, detection by infecting phages of multiple virion adsorptions to the same cell, transfer of genetic material from phages to bacteria, or extracellular factors influencing phage infection abilities. Broadly, such flow of materials and information from one entity to another can be described as different processes of communication, including between phages, between phages and bacteria, or between phages and the environment more generally. In this chapter I explore such phage-associated communication including, particularly, in terms of its impact on phage ecology.

S. T. Abedon (✉)
Department of Microbiology, The Ohio State University, Mansfield, OH, USA
e-mail: abedon.1@osu.edu

G. Witzany (ed.), *Biocommunication of Phages*,
https://doi.org/10.1007/978-3-030-45885-0_2

2.1 Introduction

Numerous viruses exist whose hosts are either single-celled organisms or whose hosts otherwise can be described as microorganisms (Hyman and Abedon 2012, 2018). Among these "Viruses of microorganisms" are the bacteriophages, or phages, which are viruses that infect members of the cellular domain, *Bacteria*. These contrast with the *viruses* of domain *Archaea*, which generally are not described as phages but instead as archaeal viruses (Abedon and Murray 2013), or contrast instead with the viruses of domain *Eukarya*. All of these viruses, as obligate intracellular parasites, interact especially with their hosts, which they infect to produce new virions. Viruses, however, also interact with the broader, non-host, extracellular environment, as well as with other viruses. Considered here are many of these various interactions with hosts, with non-hosts, and with fellow viruses.

Ecology is the study of interactions of organisms with their environments. Those interactions, broadly speaking, may be depicted as forms of communication – the movement of something (i.e., 'signals') between or among entities. Reception of signals can result, ecologically, in organism modification, including in terms of modification of behavior or physiology, though such modifications may or may not be strictly beneficial to either the sending or receiving entity. Signals and their reception also can be viewed from the perspective of hereditary material, resulting in evolution. Communication, broadly defined, thus represents movements of something from one entity, biotic or abiotic, to another entity, also biotic or abiotic, as resulting especially in some sort of change in the receiving entity. For example, the presence of lactose provides a signal that can result in alteration of gene expression by *Escherichia coli*, as in Monod's classic operon model. This change can be viewed as an environment-induced reconfiguration of the bacterium's 'behavior', and certainly of the bacterium's physiology, as results in lactose being used by *E. coli* as a carbon and energy source.

Here I consider many ways in which phages can communicate with biotic as well as abiotic components of their environments, and *vice versa* (see Table 2.1 for overview). The results are changes in phage phenotypes, as well as changes in environments, which then differ from what would exist given an absence of phage action. I concentrate mostly on behavioral, ecological, or evolutionary ecological (the latter, i.e., adaptive) aspects of communication as well as phage-associated DNA movement, and this is rather than genomic, metagenomic, or phylogenetic considerations. For an overview of communication as it can occur within the viral but not just phage world, see Witzany (2019). Note that this review represents an explicit modification of a previously published review (Abedon 2011b, c) that predominantly considered phage-involved communication as occurring especially within soils.

2.2 General Concepts

In this section I review basic principles of communication involving microorganisms (Sect. 2.2.1). I then discuss variations on what can constitute the hosts of viruses (Sect. 2.2.2). These are provided to help inform the narrower emphasis of this chapter, which is on forms of communication involving bacteriophages, i.e., as summarized in Table 2.1.

2.2.1 Communication and Microorganisms

Communication is an important concept with many facets, as can vary depending upon both one's discipline of study and personal proclivities (Witzany 2019). A lay definition, anchoring one end of a spectrum, might entail two-way exchanges involving verbal or written language. That is, for example, where one individual speaks, a

Table 2.1 Categories of communication involving phages

Section	Communication	Context	Form[a]
2.3.1	Phage-to-Phage	Intracellular[b]	Various[c]
2.3.2	Bacteria-to-Phage	Intracellular	Trans-organismic
2.3.3	Phage-to-Bacteria	Intracellular	Trans-organismic
2.3.4	Phage-Infection-to-Environment	Extracellular[d]	Trans-entity[e]
2.3.5	Phage-Virion-to-Environment	Extracellular	Trans-entity
2.3.6	Environment-to-Phage-Infection	Intracellular[f]	Abiotic influence[g]
2.3.7	Environment-to-Phage-Virion	Extracellular	Abiotic influence

[a]As derived from Witzany (2017, 2019) though with broader considerations
[b]Phage-to-Phage communication generally involves an intracellular component, though as emphasized here can involve an extracellular component as well, such as free phages or quorum-sensing autoinducer molecules. Note that all intracellular phenomena also may be described as Intra-organismic, as defined by Witzany (2017, 2019)
[c]Includes Intra-organismic, Inter-organismic, and Trans-organismic, depending on the phenomenon. These respectively are communication as occurring within cells (Intra-organismic, as covered by most of Sect. 2.3.1) or among similar species or strains (Inter-organismic, also as covered by Sect. 2.3.1), but also as can occur between phage types that may not be considered to represent similar species, the latter including such as between-species recombination or interference between different phage types upon coinfection (though with such interference being not just Inter-organismic but also Intra-organismic)
[d]Extracellular here is referring to the location of impact on the environment, though the phage infection origin of this communication implies an intracellular aspect as well
[e]Not found in Witzany (2017, 2019), "Trans-entity" is meant to expand upon "Trans-organismic" to include communication not just with unrelated environmental biotic entities but environmental abiotic factors as well
[f]Intracellular here is referring to the location of the environmental impact on phage infections, though the origin of this environment-sourced communication in this case is extracellular
[g]In addition to abiotic influences, local environment influences may be the result of biotic factors, e.g., such as the action of extracellular enzymes

second (or more) listens, then ponders, responds, and so on, with perhaps verbal and non-verbal interactions occurring simultaneously. At its basis, though, communication involves simply signals emanating from one or more entity that are received by other entities, where the latter might respond in some manner by modifying their own physical, chemical, physiological, behavioral, or even morphological states.

In addition to visual as well as vocal interactions between animals, communication can involve touch, plus extensive forms of communication exist that occur via chemical signals, whether airborne (e.g., pheromones, the sweet smell of a flower, etc.), fluid-based, or as secondary metabolites such as associated with soil bacteria (Karlovsky 2008). Chemically mediated cell-to-cell signaling also allows coordination of the growth and behavior of multicellular organisms, i.e., as intra-organismic communication (Witzany 2019). Indeed, extensive communication continuously takes place within our own bodies – as part of a process collectively described as homeostasis – that maintains a perturbation of our bodies away from an otherwise inevitable decay.

Though interactions within ecosystems are not necessarily coordinated or predominantly cooperative, they nevertheless are communicative. Whether through soluble chemicals or entities physically touching, signals of various sorts can emanate from one individual and be received by others. The result can be either a more passive response by the signal's recipients, as can be the case when damaging agents are conveyed from one individual to another, or a more active response by the recipient, the latter, e.g., as seen with quorum sensing (Waters and Bassler 2005).

One also can view the *generation* of signals as passive versus active, such as the passive release of chemicals upon organism decay versus the active exporting of signal molecules to the extracellular environment (the latter, again, e.g., as seen with quorum sensing). Situations also can occur where both signals and responses are intentional and occur expressly for the sake of conveyance of information, thereby being phenomena that might be more easily appreciated as forms of communication. At its most basic, however, a signal is simply a change that in some manner is potentially perceivable, e.g., as might modify an organism's or an environment's functioning. Thus, from Oppenheim and Willsky (1997) as quoted by Chakravorty (2018), "…signals, which are functions of one or more independent variables, contain information about the behavior or nature of some phenomenon." The sun setting thus, for example, is a signal.

This perspective of just what a signal can consist of goes beyond the definition provided, for example, by Diggle et al. (2007), in which a signal essentially is both intentional and beneficial. That is, as used here, a signal simultaneously may be neither intentional nor beneficial, but instead simply is something ('A') that potentially impacts something else ('B'), and particularly something ('A') that can vary in some manner either within an environment or across the spatial range that an organism or population of organisms can inhabit. Nevertheless, Chakravorty (2018) notes that for a signal to "convey an observable change" it must vary over time rather than just space. In biology, however, there is more than one way to consider time. For example, there is ecological time, with signals as environmental changes potentially experienced by an organism during its lifespan, but also there is evolutionary time, with signals also as environmental changes but as experienced by a population potentially over multiple lifespans. In one instance the signal might be communicated to a single organism whereas in the other instance the signal may be communicated to the shared gene pool of a group of related organisms.

Thus, for example, excessive environmental presence of the element arsenic ('A'), as a signal if not necessarily an immediately useful signal, can modify body functioning ('B') toward reduced health (thereby impacting the organism ecologically). It also, however, can reduce the evolutionary fitness of individuals that happen to exist within that local environment versus other individuals making up the same gene pool but which happen to live elsewhere. Arsenic as a signal, therefore, is communicating its presence and doing so without intention to one or more organisms. In this case, though, those organisms happen to be responding in a manner that is not beneficial to them, either ecologically nor inevitably even evolutionary, unless they can achieve some form of arsenic resistance. Communication at its most basic, that is, can involve signals which are neither necessarily intentional nor necessarily beneficial to any of the involved parties.

Irrespective of such details, natural selection favors those organisms whose survival and reproduction tend, at a minimum, to decline least in response to environmentally common signals, or even improves. Thus, we can envisage microbiomes, for example, as consisting not just of myriad microorganisms, but also of myriad communicative pathways, pathways which to recipients can be beneficial, benign, or detrimental. Here I consider especially those communicative pathways that involve phages.

2.2.2 *Ecological Variations on the Concept of Virus Life Cycle*

The concept of virus *infection* can be differentiated into infection of individual cells versus infections that span across multiple, clonally related cells. The latter includes cells making up individual, multicellular organisms such as plants or animals. In both cases, individual cells versus multicellular organisms, the virus life cycle consists of host acquisition, host infection, and transmission from hosts. For infection of free-living cells, virus adsorption to a cell corresponds to host acquisition while transmission corresponds to a diffusion-mediated extracellular search for new cells to adsorb and then infect. Those viruses that infect organisms that consist of isolated cells, such as individual bacteria, can be considered to be examples of parasites of unicellular organisms, or UOPs (Unicellular Organism Parasites). Those viruses that instead infect multicellular organisms, such as ourselves, may be described as MOPs (Multicellular Organism Parasites; Table 2.2) (Abedon 2008).

In addition to viruses, UOPs can include, for example, bdellovibrios such as *Bdellovibrio bacteriovorus*, which are Gram-negative bacteria that infect other Gram-negative bacteria. Among viral UOPs are those that infect, as noted, members of domain *Bacteria* (Lehman 2018), called bacterial viruses, bacteriophages, or phages, but also those that infect members of domain *Archaea*, which as noted most commonly are described as archaeal viruses (Abedon and Murray 2013; Abedon 2018c). In addition are those viruses that infect protozoa (Reteno et al. 2018),

Table 2.2 Parasite lifecycles as differentiated ecologically in terms of host properties

Abbreviation	Stands for	Description
UOP	Unicellular Organism Parasite	Hosts cells for these parasites are relatively isolated from other such hosts and are unicellular; these hosts are a typical emphasis of the study of phage biology especially as inhabiting environments generally lacking spatial structure, e.g., *E. coli* growing in well-mixed broth culture; parasite movement between cells in the case of UOPs is equivalent to parasite movement (transmission) to new hosts
MOP	Multicellular Organism Parasite	Hosts for these parasites consist of multiple, usually highly differentiated cells, e.g., as those making up animals or plants but also making up multicellular algae and fungi; generally these hosts are a typical emphasis of virology as well as of medical microbiology and individually represent 'environments' that possess substantial spatial structure; movement of intracellular parasites such as viruses between cells in this case often does not represent movement between individual hosts, i.e., for MOPs intra-host parasite dissemination can be distinguished from inter-host dissemination, with the latter but not the former representing parasite transmission to new hosts
COP	Colonial Organism Parasite	Hosts for these parasites consist of multiple, minimally differentiated cells, e.g., bacterial biofilms or cellular arrangements but also, for example, minimally differentiated colonial algae; generally these hosts collectively represent 'environments' that possess often substantial spatial structure, i.e., as colonies or microcolonies, but nevertheless which are not a typical emphasis of virology or of medical microbiology; COP movement between cells may or may not represent movement (transmission) between individual colonies

unicellular algae (Short et al. 2018), or yeasts (Vainio and Hantula 2018) (collectively, viruses of microbial eukaryotes). The viruses of what traditionally have been considered to be unicellular organisms constitute the bulk of what can be described as viruses of microorganisms (Hyman and Abedon 2012, 2018). Notwithstanding this diversity of microbial hosts, nevertheless among UOPs the best studied by far are the bacteriophages, with over 100 monographs published over approximately the past century that have emphasized various aspects of phage biology. Though focusing on bacteriophages, much of this review nonetheless may also be applicable to communication involving these other categories of UOPs.

Cellular organisms also may be distinguished into those that are neither unicellular nor multicellular but instead colonial. An additional category of parasites based on that host configuration thus might be introduced: "COPs", standing for Colonial Organism Parasites. Included would be viruses of various eukaryotic organisms that band together into groups of cells that traditionally have been described as colonial rather than multicellular, such as perhaps the viruses of the various minimally differentiated volvocine green alga (Herron et al. 2009). Prokaryotes, however, also can exist as colonial groups of cells. Prominently for prokaryotes, these include colony-like cellular arrangements (Abedon 2012b), such as streptococci, or instead prokaryotes existing as multi-celled microcolonies

(Abedon 2011a, 2012b, 2017a, d), 'microcolony' being a commonplace descriptor for such entities especially in the mixed-species biofilm literature (Abedon 2015c). Bacterial colonial forms are what is seen particularly within spatially structured environments, and it is within spatially structured environments, e.g., such as within biofilms, that communication between individual microorganisms generally should be most easily achieved.

2.3 Pathways of Bacteriophage-Associated Communication

Various forms of communication within environments are summarized in Fig. 2.1. In this section I distinguish among phage-associated communication processes in terms of directions of communication (Table 2.1), i.e., from specific categories of originators of signals (such as phages) to specific categories of recipients (such as bacteria). I begin, though, with consideration of phage-to-phage intercellular communication. Note that where appropriate I substitute the narrower case of 'lytic'

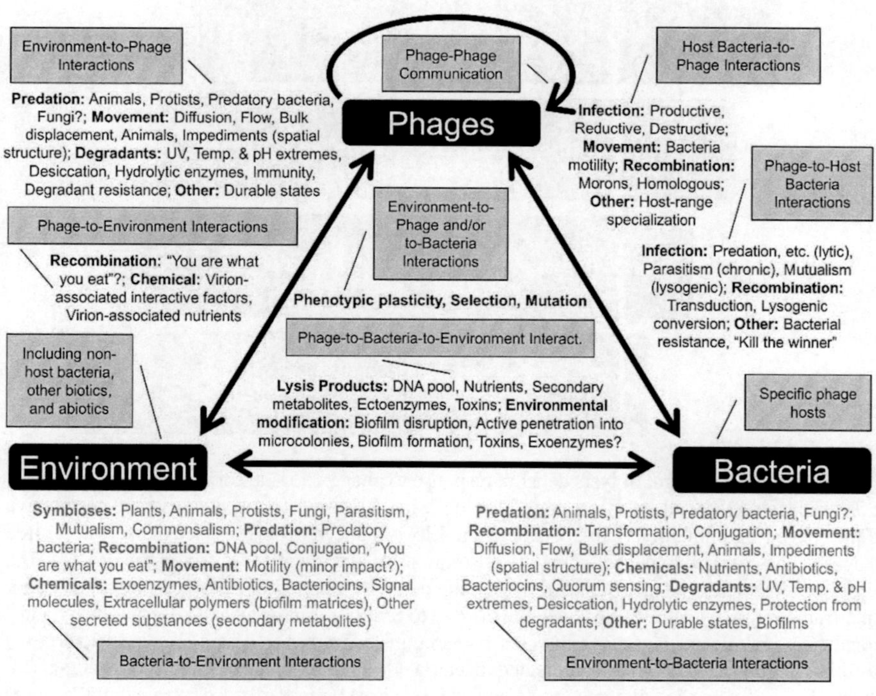

Fig. 2.1 Various pathways of communication within environments including as involving phages. Text in gray (bottom) does not involve phages and is not otherwise discussed in the main text. (This figure has been updated from that presented in Abedon 2011b, c)

infection or cycle with the broader case of 'productive' infection or cycle, the latter, i.e., so as to simultaneously describe both lytically and chronically releasing phage infections (Hobbs and Abedon 2016).

2.3.1 *Phage-to-Phage Communication*

Mechanisms of phage-to-phage communication, as considered in this section, by and large involve signals being released from one or more phage-infected bacterium that are then received by one or more additional phage-infected bacterium. Reception of these signals prominently gives rise especially to lysis delays (Abedon 2017c, 2019a; Igler and Abedon 2019). See Fig. 2.2 for summary of the underlying

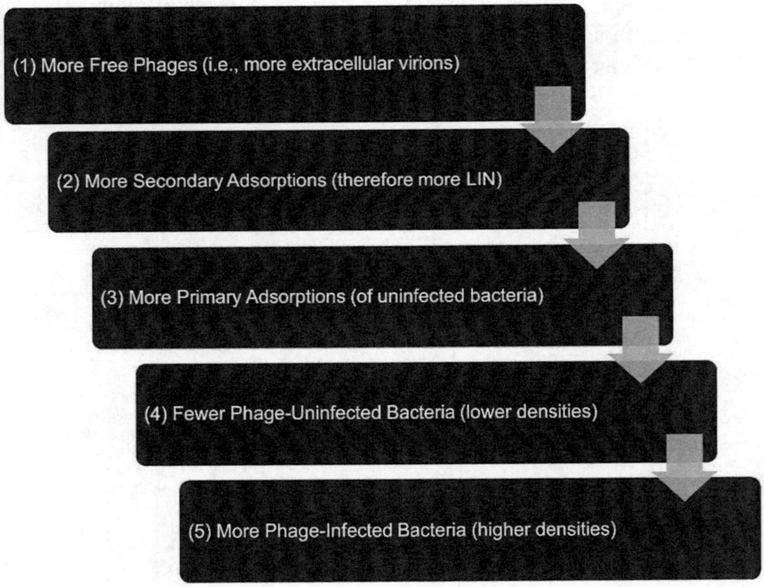

Fig. 2.2 Underlying logic of lysis delay in response to phage virion adsorption of already phage-infected bacteria (secondary adsorption). This logic, in the figure, is presented from the perspective of the phage lysis-inhibition (LIN) phenomenon. LIN is induced in the presence of excesses of free phages (1) as give rise to phage secondary adsorptions of already phage-infected bacteria (2). The resulting LIN phenotype involves phages 'holding on to' the phage-infected bacterium over longer time frames (lysis is delayed). Meanwhile, as due to other, 'primary' phage adsorptions by additional free phages, not yet phage-infected bacteria are converted to phage-infected bacteria (3) resulting in both fewer phage-uninfected bacteria (4) and more phage-infected bacteria (5). Delayed lysis can protect intracellular phage virions, at least temporarily, from 'suicidally' adsorbing to already phage-infected bacteria. LIN also increases phage burst sizes, the latter as especially useful given diminishing numbers of bacteria available for subsequent phage infection. The communication in the case of lysis inhibition is from lysing phage-infected bacteria (releasing free phages) to not-yet-lysed phage infected bacteria (secondary phage adsorptions)

ecological logic behind this response. In order of discovery, these phenomena are lysis inhibition (LIN; Sect. 2.3.1.1), what I call high multiplicity lysogeny decisions (HMLDs; Sect. 2.3.1.2), and, most recently, arbitrium systems (ASs; Sect. 2.3.1.3). Two additional mechanisms are discussed here as well, both of which result instead in lysis acceleration, one of which is considered under the heading of LIN (Sect. 2.3.1.1) while the other is discussed in association with bacteria-to-phage communication (Sect. 2.3.2.1.3). Phages can also communicate with other phages entirely within phage-infected bacteria, e.g., such as crosstalk between prophages infecting the same bacterium (Espeland et al. 2004; Lemire et al. 2011), but that is not considered here. Also not considered in much detail is phage-to-phage communication such as via genetic recombination between phages (Hendrix 2002, 2008; Hendrix et al. 2002, 2003; Pedulla et al. 2003; Brüssow and Desiere 2006; Mavrich and Hatfull 2017) or as involving various sorts of intracellular antagonistic interactions (Abedon 1994; Turner and Duffy 2008; Abedon et al. 2009).

2.3.1.1 Lysis Inhibition (LIN)

LIN is a mechanism of latent period extension that presumably serves as an ecological response to a combination of high environmental densities of phage-infected bacteria and correspondingly low environmental densities of phage-uninfected bacteria, where in both cases these explicitly are bacteria that are susceptible to the LIN-displaying phage (Abedon 2019a). What distinguishes LIN from HMLDs and AS as phage-associated communication phenomena, however, is that LIN involves latent period extensions of phage lytic cycles rather than a display of lysogenic cycles instead of lytic cycles. In addition, so far as is understood, and contrasting pseudolysogeny (Sect. 2.3.2.1.2), all through the LINed latent period these individual phage-infected bacteria continue to produce new phage virions. Thus, LIN is associated not just with longer phage latent periods and associated delayed lysis but also with larger phage burst sizes.

What is the signal that results in infecting phages displaying these longer, LINed latent periods? The answer is the adsorption of phage virions to already phage-infected bacteria, a mechanism first described by Doermann (1948). Thus, other phage-infected bacteria lyse, releasing phage virions. These virions can then adsorb to still phage-infected, that is to still unlysed infected bacteria, thereby inducing the LIN phenotype. These 'secondary' phage adsorptions occur with greater probability the higher phage infection environmental concentrations and thereby, as a result, also the higher free phage environmental concentrations. With LIN, more phage virions are produced (larger burst size) just as phage-susceptible, phage-uninfected bacteria are becoming scarce within environments, the latter as due to new phage adsorptions.

A LIN-related phenomenon that is also at least communication-like is so-called synchronized LIN collapse. This is an accelerated lysis of LINed populations of phage-infected bacteria. That faster lysis may, to a degree, help to spare extracellular virions from adsorbing to still phage-infected bacteria, where such adsorptions result in inactivation of these subsequent (secondarily) adsorbing phages.

Synchronized LIN collapse, like LIN, thus arguably represents a phage virion-mediated form of phage-to-phage intercellular communication (Abedon 1992, 1999, 2019a). Both LIN and synchronized LIN collapse are phenomena that are associated especially with the T-even phages of *Escherichia coli*, i.e., phages T2, T4, and T6 (Abedon 1994, 2000).

2.3.1.2 High Multiplicity Lysogeny Decisions (HMLDs)

A few years after it was found that multiple *adsorptions* were able to induce LIN in certain phage infections (secondary adsorption), it was found also that multiple phage *infections* (coinfections) can bias temperate phage infections toward lysogenic cycles rather than productive infections (Boyd 1951; Lieb 1953; Levine 1957; Fry 1959; Six 1961; Brooks 1965; Hoffman Jr and Rubenstein 1968; Kourilsky 1973; Avlund et al. 2009; Joh and Weitz 2011). The ecological utility of this HMLD phenomenon presumably is similar to that of LIN. That is, multiple phage adsorptions of a single bacterium would imply that other bacteria in the environment are being similarly affected, resulting especially in reductions in the number of bacteria that are phage uninfected. A difference from LIN, however, is that with HMLDs it is more than phage adsorptions that are generating lysis delays but instead (as noted) multiple phage *infections* (with LIN, secondary phages by and large do not also infect, that is, they do not deliver their DNA to the bacterial cytoplasm, which is a consequence of primary phage-expressed superinfection exclusion; Abedon 1994). In addition, with HMLDs the lysis delay is in terms of increased tendencies toward phages displaying lysogenic cycles rather than toward extended lytic or productive cycles, the latter as seen with LIN.

2.3.1.3 Arbitrium Systems (ASs)

The latest form of phage-phage intercellular communication to be discovered, which generated much enthusiasm for the existence of such communication, is that of arbitrium systems (Erez et al. 2017). These were first described in *Bacillus subtilis*, though see also (Hargreaves et al. 2014; Dou et al. 2018; Gallego Del et al. 2019). Like HMLDs, ASs result in biases toward lysogenic cycles rather than lytic cycles, as seen at the start of infections by these temperate phages. The underlying natural selection logic also is equivalent to that for HMLDs, with greater levels of signal emanating from greater concentrations of phage-infected bacteria resulting in extended phage latent periods (lysogenic cycles) just as presumably more bacteria are becoming phage infected (Abedon 2009a; Abedon et al. 2009). Potentially, then, more phage bursts take place sometime in the future following subsequent prophage induction, i.e., as when perhaps phage-infected bacteria have become less prevalent. Unlike HMLDs, however, the signals for ASs are quorum sensing-like autoinducer molecules produced by phage-infected bacteria.

2.3.2 *Bacteria-to-Phage Communication*

Mechanisms of bacteria-to-phage communication, as considered in this section, by and large are associated with phage infections of bacteria. These may be differentiated into a number of distinct aspects including (i) bacteria-mediated modifications of phage phenotype (Sect. 2.3.2.1), (ii) the transfer of DNA sequences from bacteria to phages (Sect. 2.3.2.2), and (iii) the impact of bacteria on phage movement through environments (Sect. 2.3.2.3).

2.3.2.1 Bacteria Impact on Phage Phenotype

Phage infections can occur in a number of distinguishable modes (Hobbs and Abedon 2016). Drawing in part from usage by others, I have described these infections collectively as phage *productive*, phage *reductive*, or phage *destructive* (Abedon et al. 2009) (Table 2.3). 'Productive' refers to the generation of new phage virions during bacterial infections; 'Reductive' (Lwoff 1953) refers to phage entrance into more quiescent, temporarily non-virion productive states such as lysogeny (Howard-Varona et al. 2017), but also pseudolysogeny (Miller and Day 2008) (see Sect. 2.3.2.1.2). 'Destructive' (Sect. 2.3.2.1.1) refers to permanent loss of a bacterium-infecting phage's ability to produce new virion progeny, that is, rather than necessarily also 'destruction' of the phage-infected *bacterium*. Often, though not always, phage-destructive infections will be a consequence of bacteria displaying phage-resistance mechanisms of various sorts.

Which infection mode is displayed by a given phage, infecting a given bacterium, is determined in the course of communication between molecules produced by phages and molecules produced by bacterium, though this can involve also communication between phages (Sect. 2.3.1). All non-defective phages nevertheless should be able to display productive infections. Reductive infections for some phages (temperate phages) or under some circumstances instead destructive infections (presumably for all phage types) may however occur instead of productive infections. Note, though, that a phage's ability to productively infect can also be impacted by environmentally mediated modifications of phage virions as occur prior to virion adsorption of bacteria, something that would fall under a heading of environment-to-phage-virion communication (Sect. 2.3.7), or instead may occur due to more direct environment-to-phage-infection communication (Sect. 2.3.6).

2.3.2.1.1 Destructive Infections: Antagonism, Deception, and Primitive Immunity

There are a number of obvious situations in which bacteria negatively impact phage phenotype. These include when bacteria display phage restrictive or abortive infections, which are phage-destructive infections that may be distinguished in terms of whether bacteria survive or instead don't survive, respectively (Abedon 2008;

Table 2.3 Modes of infection: phage productive, phage reductive, and phage destructive

Infection mode	Description	Examples[a]
Productive	Infection production of new phage virions as initiated either immediately following phage adsorption ('purely productive') or immediately following prophage induction ('induced productive'); productive infections can result in either lytic or chronic phage virion release depending on the phage type	Purely lytic infection; Induced lytic infection; Purely chronic infection; Induced chronic infection
Reductive	Infection which is delayed in its production of new phage virion particles, but in principle is not indefinitely delayed nor resulting in loss of virion production ability; certain reductive infections can resolve into either reductive or productive infections	Lysogenic infection[b] Pseudolysogenic infection[c]; Infection of sporulating bacteria; Hibernation mode[d]
Destructive	Infection is indefinitely not able to produce new phage virions (or, less strictly, infection is less able to produce new phage virions), i.e., an infection that lacks a virion production ability; destructive here does not refer to the phage impact on the bacterial host but instead, for example, the bacterial host impact on the infecting phage	Restrictive infection[e]; Abortive infection[f] Phage growth limitation; (Reduced infection vigor[g])

[a]For discussion of a number of variations on phage infection types, see Hobbs and Abedon (2016)
[b]See Lwoff (1953) for use of the term, "Reductive", as applied to lysogeny. Lysogeny is displayed only by temperate phages and may be resolved, following induction, as either lytic or chronic productive infections, depending on phage type
[c]Pseudolysogeny, the term, is used here in the Miller and Day (2008) sense. See also Abedon (2009c) and Los and Wegrzyn (2012). Pseudolysogenic states may be entered into either by temperate or non-temperate phages and may be resolved into either lysogenic or productive infections
[d]See Bryan et al. (2016) for introduction of the concept, hibernation mode
[e]Restriction defined here is as resulting in permanent loss of phage ability to produce new virions but without loss of viability of the phage-infected bacterium
[f]Abortive infection defined here is as resulting in permanent loss of phage ability to produce new virions in combination with permanent loss of the viability of the phage-infected bacterium
[g]Reduced infection vigor refers, for example, to declines in phage burst size as relative to under more ideal phage infection circumstances (Hyman and Abedon 2010). Note that phages displaying reduced infection vigor, unless reduced to zero productivity, are by definition also displaying productive infections, hence the parentheses

Abedon et al. 2009; Hyman and Abedon 2010; Labrie et al. 2010; Dy et al. 2014; Doron et al. 2018; Azam and Tanji 2019) (see also Sect. 2.3.3.1). These are clear examples of bacteria-to-phage communication, though particularly ones with negative consequences for the infecting phage.

In addition to destroying phages, restriction-modification systems can also modify phage genomes in a manner that alters their susceptibility to cutting by the same restriction-modification system. Notably, this phage modification is a rare occurrence in terms of its ability to forestall phage restriction, though it is important in that it imparts on so-modified phages an ability to infect previously resistant bacteria bactericidally (Korona and Levin 1993). These inadvertent phage modifications can be viewed as imparting on phages the ability to communicate a *lie*, that is, to mislead subsequently infected bacteria into treating a phage genome as self.

There also exists a type of destructive infection that may be viewed as a form of phage-mediated bacterium-to-bacterium communication (Sumby and Smith 2002). In this instance, phages are modified during productive infections such that subsequent infection of a clonally related bacterium, here the soil bacterium, *Streptomyces coelicolor*, results in phage restriction. That is, the previously infected bacterium is communicating, through the released phages, that these phage genomes represent foreign DNA which should be eliminated. This "Phage growth limitation" system may be viewed as analogous to animal immune system functions in which the molecules of parasites are tagged as foreign such as by antibody binding to be subsequently eradicated, though this occurs for animals within an infected organism (thus acting against MOPs) rather than between infected organisms (acting instead against UOPs). Also analogous to animals, it is the passage of parasites among clumped, clonally related tissues, in this case within bacterial 'tissue', which fosters the utility of this mechanism, i.e., with phages in this case acting more as COPs than as UOPs (Table 2.2). For more on analogies between mechanisms of bacterial resistance to phages and immune systems as found in multicellular organisms, see Abedon (2012a).

It also is possible for bacteria to reduce the productivity of phage infections but without blocking these infections altogether. Such mechanisms I have described elsewhere as reducing a phage infection's 'vigor' (Hyman and Abedon 2010). Note that reductions in phage infection vigor might serve as a means of delaying the progress of phage exploitation of bacterial microcolonies found making up, for example, bacterial biofilms (Abedon 2017d). Reduced infection productivity, however, can be a consequence not only of bacteria-mediated anti-phage mechanisms, or instead phage-bacterium genetic incompatibilities, but also can be a consequence of environment-to-phage-infection communication (Sect. 2.3.6).

2.3.2.1.2 Reductive Infections

Phage lysogenic cycles were deemed as "Reductive" by Lwoff (1953), as in "the genetic material of the infecting phage… is '*reduced*' into prophage" (p. 272, emphasis his). Phage lysogenic cycles serve as intervals during which phages are not producing new virions, though their genomes are being replicated in association with bacteria replicating their own genomes. To the extent that bacterial properties may modify a phage's likelihood of displaying a lysogenic cycle, such as by modifying phage lytic/lysogenic decisions at the start of infections, then such modifications may be described as a form of bacterium-to-phage communication. The propensities for bacteria to give rise to prophage-inducing signals presumably also can vary as a function of bacterial genetics, thereby potentially representing another means by which bacteria communicate to phages, in this case providing signals that might contribute to the stability of lysogenic cycles (prophage induction represents conversion of a lysogenic phage cycle to a productive phage cycle). Often such prophage induction occurs especially as a consequence of environmental signals that are transduced through the bacterium to the phage (i.e., Sect. 2.3.6), particularly as resulting in bacterial DNA damage that induces the bacterial SOS response (Little 2005; Campbell 2006).

Pseudolysogeny, as defined here (Miller and Day 2008; Abedon 2009c; Los and Wegrzyn 2012), is a mechanism of delay in phage infection activity that is a consequence especially of poor bacterial growth conditions, i.e., particularly nutrient limitations. See also the concept of phage "Hibernation modes" (Bryan et al. 2016). Phages can also be packaged during sporulation into dormant bacterial endospores, thereby presumably increasing phage survival abilities (Pantastico-Caldas et al. 1992; Sonenshein 2006). Since bacterial genes are responsible for the sporulation phenotype (Errington 2003), their expression too may be viewed as modifying phage infection outcomes. Though mediated through bacteria, nevertheless these phage-reduction phenomena can also be considered to represent environment-to-phage-infection forms of communication, especially communicating environmental nutrient limitations (Sect. 2.3.6).

So long as phage-infected bacteria are more durable than free virions, then any of these phage reductive states may be advantageous to phages, that is, under conditions where virion survival is less likely while bacterium survival is more likely. With this emphasis on phage survival, those cases described in the previous paragraph may represent modified phage infection behaviors that are useful especially given environmental "Hard times", i.e., conditions disfavoring phage virion-mediated population growth (Stewart and Levin 1984); see also (Williams et al. 1987; Miller and Day 2008). By contrast, consider the concept of phage banking (Sect. 2.3.7.2.1).

2.3.2.1.3 Phages Listening in on Bacterial Quorum Sensing

As discussed in Sect. 2.3.1.3, phages can generate quorum-sensing autoinducer molecules and also respond to them. With the phage arbitrium system (Sect. 2.3.1.3), the consequence of autoinducer detection by an infecting phage is an increased tendency to display lysogenic rather than lytic cycles. Alternatively, Silpe and Bassler (2019) describe a system in *Vibrio cholerae* in which phages respond not to phage infection-expressed autoinducer but instead to *host*-generated autoinducer. In this latter system, autoinducer reception results in prophage induction, that is, a loss of the reductive state toward instead a productive infection, and this is rather than increased tendencies toward lysogenic cycles; see also (Ghosh et al. 2009; Hargreaves et al. 2014; Harms and Diard 2019; Maxwell 2019). Note that the ecological utility to phages of listening in on bacterial quorum sensing is not certain (Abedon 2019a; Igler and Abedon 2019). Nevertheless, this system certainly seems to represent an example of bacteria-to-phage communication that results in phage phenotypic modification, particularly modification in terms of phage-infection characteristics.

2.3.2.2 Bacteria Impact on Phage Genotype (Evolution)

Any bacterial mutation that affects phage infection characteristics, or the existence of any differences in phage phenotype when infecting different bacterial strains – such as increased propensity to display lysogenic rather than lytic cycles – can be indicative of a bacterial gene's impact on phage phenotype. Such impacts therefore are illustrative of a form of bacteria-to-phage communication. It seems logical also that any interactions between phage- and bacteria-encoded molecules might be products of phage evolution, particularly natural selection, unless those interactions are either antagonistic to phages (in which case the interactions would be selected against among phages) or instead (or also) are products of bacterial evolution such as during lysogenic cycles (in which case the evolution is occurring within the same genome as the phage, just not within the phage itself). The potential for variations in bacterial genes to modify phage phenotypes thus underscores the obvious potential for bacteria-to-phage communication to impact phage evolution.

2.3.2.2.1 Antagonistic Coevolution and Evolutionary Arms Races

A large fraction of bacterial impact on phages can be described as resulting from bacterial mechanisms of resistance to phages, e.g., restriction-modification systems as well as evolved abortive infection mechanisms (see "Destructive infections…", Sect. 2.3.2.1.1, and Table 2.3). Phages, however, can respond by displaying mutations or other mechanisms by which bacterial resistance is overcome (Samson et al. 2013; Stanley and Maxwell 2018, Burrowes et al. 2019). A result can be phages and their bacterial hosts evolving, but not doing this evolving independently of each other, e.g., (Buckling and Rainey 2002; Levin and Bull 2004; Brockhurst et al. 2007; Scanlan et al. 2015; De Sordi et al. 2019). The latter includes especially what can be described as evolutionary arms races between phages and bacteria, a phenomenon also known as antagonistic coevolution.

2.3.2.2.2 Antagonistic Pleiotropy and Phage Host Range

Phages also may specialize on specific bacterial strains. This specialization can be viewed as a form of bacteria-to-phage-communicated phage evolutionary adaptation. Evolutionary adaptations, however, often involve tradeoffs (Sect. 2.3.3.4.1). Tradeoffs are where improvements in one character give rise to reduced function in another. The tradeoffs resulting from specialization can also be described as antagonistic pleiotropies. Antagonistic pleiotropies especially are tradeoffs that do not occur at the same time or under the same circumstances as the selective payoff, the latter, e.g., such as increased phage burst sizes while infecting a preferred host (Williams 1957; Holder and Bull 2001; Elena and Sanjuán 2003; Duffy et al. 2006; Heineman et al. 2008; Presloid et al. 2008; Rodriguez-Verdugo et al. 2014). Specifically, that is, phage specialization is something that is occurring toward benefit while a phage is infecting one host but toward detriment while the same phage type is infecting a different host, the latter, e.g., such as reduced burst sizes when *not* infecting the preferred host.

An extreme of such antagonistic pleiotropy is seen with phage host-range switching, where phage interaction with one bacterial molecule, e.g., the phage receptor molecule found on a bacterium's surface, is lost upon phage acquisition of affinity for a different bacterial molecule (the new phage receptor molecule). Alternatively, some phages have been found that display fairly wide host ranges, spanning multiple bacterial genera (Hyman and Abedon 2010; Ross et al. 2016; Hyman 2019; de Jonge et al. 2019). These broader host ranges presumably would facilitate phage access to greater host numbers, but this could be at the expense of more effective infection of any one host type, i.e., as might be manifest as a generally reduced infection vigor (Chan and Abedon 2012).

2.3.2.2.3 Moron Accretion and Phage-Prophage Recombination

Phage evolution also occurs as a consequence of acquisition of bacterial genetic material. The "Moron-accretion hypothesis" in fact posits that all phage genetic material had its ultimate origin from bacterial DNA (Hendrix et al. 2000). Bacterial DNA that phages have acquired, which serves to increase phage evolutionary fitness, or at least that does not decrease phage fitness too greatly, may be retained within phage genomes with some reasonable likelihood. This DNA is then subject to mutational modification, which is then subject to additional natural selection (Hendrix 2008).

A more specific form of recombination between phages and bacterial chromosomes is the homologous recombination that can occur between infecting phages and resident prophages when both are found together in the same bacterium; see Abedon (2009a) for discussion and references. Here phage-evolved genes can be transferred intact to a second phage, though it is a matter of semantics whether this represents bacteria-to-phage communication versus phage-to-phage communication of DNA sequence, that is, depending on whether or not a prophage DNA-donor is considered to be a component of the carrying bacterium versus instead representing a more bacterium-independent entity.

2.3.2.3 Bacteria Impact on Phage Movement

Bacteria can impact phage movement, either enhancing or obstructing that movement. Phages can also move, especially non-diffusively, in association with various non-bacterial aspects of environments, e.g., such as wind. Phage movement also can occur while infecting bacteria that similarly are moved by non-bacterial environmental aspects, though these latter more environmental aspects of phage movement are covered in a different section (Sect. 2.3.6).

2.3.2.3.1 Hitchhiking on Bacterial Motility

Enhanced phage movement presumably can occur during infection of bacteria if bacteria themselves are motile (Abedon 2009d). Such self-propelling bacteria should be able to move these infecting phages along faster than phage virions can move by diffusion alone. Generally, one therefore should always be cautious regarding the degree of relative phage immobility, and thereby degree of environment spatial structure (lack of mixing), when within experimental microcosms there is a potential for phage infection of motile bacteria, i.e., bacteria that are capable of moving substantial distances on their own accord.

Phages likely would be more adept at moving in association with motile bacteria as reductive rather than lytic infections, given the longer retention by phages of bacteria during non-lytic infections. Such movement presumably can occur especially during lysogenic cycles (Igler and Abedon 2019) given that lysogeny – contrasting pseudolysogeny, or bacterial sporulation or hibernation modes (Table 2.3) – can occur and persist in association with fully active rather than more quiescent bacteria. Such bacterial motility should also readily occur during chronic phage infections, and indeed could very well represent a key advantage of chronic phage release over lytic phage release: that is, a combining of both active infection motility and ongoing virion release in order to efficiently spread phage progeny throughout environments. Lytic infections of motile bacteria, at least while those infections persist pre-phage-induced lysis, nevertheless should be motile as well, just not motile for as long.

Motility by phage-uninfected bacteria, by contrast, may allow these bacteria to escape localized regions of phage exploitation of bacteria, i.e., regions that may therefore contain higher concentrations of phage virions (Gill and Abedon 2003; Abedon 2017d). Such regions may be described as foci of phage infection (Abedon 2011a), and infection foci presumably, for example, can occur within soils or within bacterial biofilms. Alternatively, motile bacteria may be more likely to encounter randomly dispersed, diffusing phage virions than will a more stationary bacterium (Stent and Wollman 1952; Koch 1960). This latter impact of bacterial motility, however, may be less substantial than the impact on virus encounter rates of motility that is associated with larger cell types such as protozoa (Murray and Jackson 1992).

2.3.2.3.2 Bacterial Immobility

If bacteria are immobile, such as bacteria found in association with biofilms, then infection by a phage can have the consequence of temporarily halting virion diffusive movement. This temporary phage immobility presumably is equivalent to that which is thought to occur upon bacterial infection during phage plaque formation (Yin and McCaskill 1992; Abedon and Culler 2007; Abedon and Yin 2008, 2009; Gallet et al. 2009; Abedon and Thomas-Abedon 2010). That is, while phage rates of movement within agar may be slower than phage diffusion rates within more fluid media, or especially as within well-mixed broth, those rates of virion movement nevertheless will tend to be even slower over the course of phage infection of immobile bacteria. Such alternating immobility with virion production, on the one hand, and subsequent virion diffusion but without virion production on the other, is analogous to what are known as 'reaction-diffusion' processes (Krone and Abedon 2008), a concept that more typically is invoked to describe the dynamics of spatially constrained chemical reactions.

Bacteria also might obstruct phage movement by being tightly packed together (Yin and McCaskill 1992). See though Abedon and Culler (2007) for suggestions that non-host bacterial cells alone are unlikely to have a substantial impact on phage virion diffusion. Alternatively, there are bacteria-excreted extracellular polymers. Especially if it is potential bacterial hosts of phages that are doing the polymer excreting, then these polymers may inhibit phage access to phage-adsorptive bacterial cell surfaces (Wilkinson 1958; Lacroix-Gueu et al. 2005; Briandet et al. 2008; Hu et al. 2010; Abedon 2011a). On the other hand, if bacteria are immobile but filamentous, then phage adsorption to one end of a bacterial filament presumably could result in virion release also at the other end (Abedon 2011b), thereby allowing phages in effect to 'move' from one location to another in the course of bacterial infection even if the infected bacterium remains stationary (Eriksen et al. 2018). Such phage infection-associated movement could be highly relevant if virion diffusion is otherwise highly constrained such as by extracellular polymers, assuming that phages can first reach a bacterium to adsorb despite the presence of these polymers. See, however, Sect. 2.3.3.5 for discussion of phage production of extracellular polymer-degrading enzymes.

2.3.2.3.3 Sorptive Scavenging

Bacteria also may limit phage movement by serving as virion-adsorption sinks, a concept that has been termed "Sorptive scavenging" by Hewson and Fuhrman (2003). Earlier this idea was suggested by Yin and McCaskill (1992) in terms of a potential for a slowing of phage plaque growth due to phage adsorption to still-intact bacteria, that is, rather than virions somewhat more rapidly diffusing past those bacteria. Sorptive scavenging also is to some extent equivalent to the concept of 'self shading' (Boots and Mealor 2007), that is, as analogous to a leaf preventing sunlight from reaching another leaf found on the same plant. Here, though, it would be clonally related bacteria making up the same microcolony that are interfering with the ability of phages to reach their underlying sister bacteria.

If bacteria are already infected by the same phage type as the phages that they are adsorbing, the latter therefore secondarily adsorbing, then those secondarily adsorbing phages not only will be delayed in their diffusion but in fact will be killed due to display by the initially infecting phage of superinfection exclusion (Abedon 1994). Alternatively, in the absence of phage-expressed superinfection exclusion, secondarily adsorbing phages may still be greatly reduced in their potential to contribute genetically to the resulting phage burst, as due a variety of mechanisms: the first phage monopolizing the cell's replication machinery, the first phage altering that machinery such that it is less appropriate for the secondarily infecting phage, or instead the first phage simply having had a head start in terms of genome replication over the later arriving phage genome. In addition, generally a bacterium can support only one phage burst irrespective of display of superinfection exclusion, thereby reducing a phage population's productivity given multiple phage adsorptions to individual bacteria (Abedon 2015a). While lysis inhibition may allow certain phages to counter this reduced productivity by increasing the burst sizes of individual phage-infected bacteria (Sect. 2.3.1.1), this burst-size enhancement occurs at the expense of increasing latent periods, which in turn should increase the potential of individual phage-infected bacteria to adsorb and thereby inactivate free virions (Abedon 1990, 2009b, 2019a).

One consequence of sorptive scavenging, independent of the survival of individual phages, can be a delay in the penetration of phages into bacterial microcolonies and biofilms (Abedon 2017d). Such delays, regardless of the cause, possibly could contribute to a persistence of bacterial microcolonies or biofilms despite their also supporting ongoing lytic phage infections (Simmons et al. 2017; Eriksen et al. 2018). It is possible also that bacteria entrance into stationary phase or maturation of a biofilm's matrix, i.e., as resulting in more 'mature' phage-infection targets, could also contribute to such phage penetration delay and/or to microcolony or biofilm survival despite phage attack (Abedon 2016). The latter has been observed by Darch et al. (2017), who find that larger, older, or matrix-producing microcolonies (which they call "Aggregates") are more tolerant of phage treatment. So too potentially might "Debris" generated upon bacterial lysis (Rabinovitch et al. 2003; Aviram and Rabinovitch 2008; Bull et al. 2018) or membrane vesicles released from bacteria outer membranes (Reyes-Robles et al. 2018, and references cited) contribute to forms of sorptive scavenging.

2.3.3 Phage-to-Bacteria Communication

Phage-to-bacteria communication varies in terms of the costs as well as the benefits that are experienced by recipient bacteria. These impacts range from reductions of bacterial fitness to zero, i.e., bacteria are killed by phage action, or even worse (Sects. 2.3.3.1 and 2.3.3.4), to instead where bacterial fitness is enhanced, the latter for example as a consequence of lysogenic conversion (Sect. 2.3.3.2). In addition, just as phages can receive DNA from bacteria (Sect. 2.3.2.2.3), so too can bacteria receive DNA from phages, i.e., as via phage-mediated DNA transduction (Sect. 2.3.3.3). Though not covered here, Hansen et al. (2019) review how phages can increase propensities of bacteria to display biofilm growth, but also phages can actively degrade biofilms, even without necessarily infecting bacteria (Sect. 2.3.3.5). Not covered as well is the acquisition of spacer sequences by CRISPR-Cas systems from the genomes of infecting phages, e.g., see Abedon (2011), nor the use of these sequences to restrict subsequently infecting phages of the same type, though both clearly are forms communication, first from phages to bacteria in the form of new DNA and then from bacteria back to phages in the form of causing a phage-destructive infection. See also Fernández et al. (2018) for review of numerous specific, ecologically relevant phage impacts on and interactions with bacteria.

2.3.3.1 The Many Costs to Bacteria of Phage Infections

The worst thing that could happen to a bacterium, as effected by a phage, is a non-induced lytic infection, that is, a productive infection terminating in lysis that does not follow lysogeny. This is because not only does a bacterium die as a consequence of the lytic phage infection, but so too, upon subsequent lytic phage infections may the bacterium's fellow clone-mates die as well. With prophage induction, by contrast, bacterial clone-mates generally do not also die despite the release of new phage virions, and this is a consequence of those fellow lysogens expressing superinfection immunity (Sect. 2.3.3.2.1). An abortive infection, by contrast, is one that is lethal to the phage-infected bacterium, but which, at least as strictly defined, also does not produce new phages. With abortive infections, at least to a first approximation, therefore only the initially lytically phage-infected bacterium is negatively impacted rather than also related, closely physically associated bacteria being impacted as well (Abedon 2012b; Fukuyo et al. 2012; Berngruber et al. 2013; Iranzo et al. 2015). With restrictive phage infections, by contrast, the phage dies but the bacterium survives, thereby suggesting minimal negative phage impact on the host bacterium (see Sect. 2.3.2.1.1 for additional discussion of phage-destructive infections).

Chronically releasing phage infections, because they do not result in bacterial lysis despite progeny virion release, presumably are preferable to bacteria versus lytic phage infections. Though not necessarily lethal to bacteria, however, chronic phage infections still have a potential to slow bacterial growth and therefore reduce bacterial fitness (Abedon 2006). In addition, chronic phage infections may increase a bacterium's susceptibility to environmental toxins such as antibiotics (Hagens et al. 2006). In terms of impact simply on bacterial metabolism, lysogenic cycles despite not producing phage virions also should be costly to bacterial lysogens. Often this metabolic cost is not assumed to be substantial, however, and indeed may be made up for by prophage expression of useful gene products, i.e., as via lysogenic conversion (Sect. 2.3.3.2.1). See Fig. 2.3 for a recapitulative summary of these and other productive, reductive, and destructive phage infection types.

2.3.3.2 Phage Infection as Symbiosis

The fact that phages develop within an organism that is at least physiologically still living, i.e., the bacterial cell, makes phage-bacterial interactions inherently symbiotic. Keep in mind, though, that not all symbioses are beneficial to both parties. Nevertheless, with reductive infections neither the infecting phage nor the phage-infected bacterium are killed, at least not while the reductive infection is ongoing. In addition, phages while still reductively infecting do not produce virions. The more familiar of these reductive infections are lysogenic cycles (Table 2.3), where infecting phages exist over long periods as intracellular prophages (Little 2005; Campbell 2006; Miller and Day 2008; Abedon 2009c; Howard-Varona et al. 2017; Los et al. 2019). With lysogeny, and so too with phage chronic infections, phages

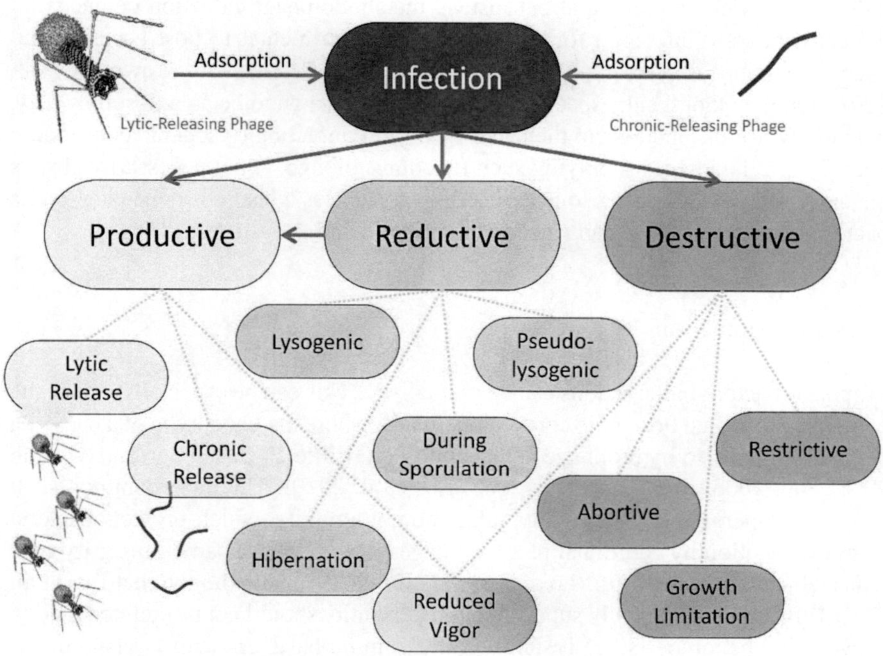

Fig. 2.3 Phage-productive versus phage-reductive versus phage-destructive infections. Gray, dotted lines indicate specific sub-categories. Not shown is that pseudolysogenic infections along with perhaps other reductive infections (Table 2.3) can also in principle resolve to lysogenic infections. In addition, the arrow going from reductive to productive infections is indicative of any one of these reductive infection types (lysogeny, pseudolysogeny, hibernation, and during sporulation) potentially resolving to a productive infection. Note that growth limitation systems ("Growth Limitation") are destructive during a second round of phage infection of a so-displaying host rather than during the first round, thus making the first round in fact productive rather than destructive, though that earlier virion productivity is not indicated in the figure. The curved lines in the upper-right and near the lower-left corners of the figure are representations of filamentous phages, which are members of phage family *Inoviridae* (Hay and Lithgow 2019). See the main text for additional discussion of the various indicated productive, reductive, and phage-destructive phage-infection phenomena

and bacteria nevertheless exist as at least potentially independent organisms. These infection types thus are possibly beneficial to both parties in the symbiosis, phage and bacterium, thus representing mutualisms.

Ecologically, then, to what extent are these extended symbiotic interactions between phages and bacteria either mutualistic (host and phage benefit) versus commensalistic (host neither benefits nor is harmed while the phage benefits), or even parasitic (host is harmed, while again the phage benefits)? More generally, myriad channels of intraspecific communication can occur within a single long-term symbiotic relationship, with some of these interactions beneficial while others can be antagonistic to the relationship. The latter is the case, at least conceptually, even if the symbiosis continues to be ongoing (Fellous and Salvaudon 2009). Prophage induction as well as non-lethal but instead metabolic phage-infection demands on the cells they are infecting, for example, can be detrimental to host bacteria, and bacteria possibly as a result may possess anti-prophage mechanisms (Lawrence et al. 2001). On the other hand, especially prophages are known to be able to supply new, potentially useful functions to the bacteria they are infecting (lysogenic conversion; Sect. 2.3.3.2.1). Phages supplying such functions, though, is not necessarily always the case, and perhaps doing so is less necessary the less a phage is dependent on its bacterial host for survival and reproduction (Levin and Bergstrom 2000).

2.3.3.2.1 Mutualism

Prophage-bacterial interactions can be mutualistic. This can be especially within the context of so-called lysogenic conversion, that is, where the phenotype of a bacterial lysogen is modified by prophage gene expression (Little 2005; Hyman and Abedon 2008; Paul 2008; Miller and Day 2008; Los et al. 2010). The most common such benefit is superinfection immunity, a.k.a. homoimmunity, which protects bacteria from exploitation by additional phages, phages that are of the same immunity type (Blasdel and Abedon 2017; Mavrich and Hatfull 2019). Note though that Los et al. (2010) argue that technically superinfection immunity should not be counted among conversion phenotypes since it stems from the more basic effort of a phage maintaining its prophage state. Still, superinfection immunity supplies to the host a useful function, protection from especially lytic infections by certain phage types.

In another example of lysogen phenotype being affected by the infecting phage, though not necessarily resulting in a mutualistic interaction, prophages can downregulate the metabolic activities of bacteria, which possibly provides lysogens with increased survival potential (Chen et al. 2005). As perhaps an extreme example of such tendencies, infection by various phages of soil bacteria have been shown to have positive impacts on host sporulation rates (Silver-Mysliwiec and Bramucci 1990). Indeed, see Argov et al. (2017) for a number of examples where prophages in various ways modify the phenotype of their hosts, in many or most cases presumably to the lysogen's benefit.

Prophage induction itself can lead to the destruction of neighboring, potentially competing bacteria, thereby providing a benefit to unlysed lysogen clone-mates. A phage of *Bacillus aneurinolyticus*, ϕBA1, in fact displays bacteriocin-like activity on some hosts (an apparently phage-DNA-independent bacterial death that results as well in a lack of phage survival) but normal phage infection on other hosts (Ito et al. 1986). This process of temperate-phage induction and subsequent killing of neighboring, non-lysogenic bacterial competitors has been be dubbed as a form of allelopathy by Stewart and Levin (1984) and subsequently described as "Kill the relatives" by Paul (2008); see also (Abedon and LeJeune 2005; Brown et al. 2006; Gama et al. 2013; Hargreaves et al. 2014).

2.3.3.2.2 Parasitism

Contrasting prophages, chronically infecting phages (Hay and Lithgow 2019) may be described as classically parasitic organisms since they display ongoing, potentially bacterial host-detrimental, but typically sub-lethal infections. Lytic phages typically are viewed as parasites as well, as in 'obligately intracellular parasites', but due to their propensity to kill their host bacteria, they often may be described instead as predators. In ecology, however, predators typically represent a higher trophic level, i.e., a feeding position that is one level above that of prey. Phages, however, fail to achieve levels of molecular assimilation of bacteria or indeed ingestion of bacteria that "feeding" would imply, since most of the consumed bacterium is discarded as waste without first becoming a part of the phage either physically, chemically, or even spatially (Thingstad et al. 2008) (Sect. 2.3.4.1). Phagotrophic protists, by contrast, clearly can be viewed as predators that feed on bacteria.

It has been argued, alternatively, that lytic phages may be described as parasitoids (Forde et al. 2004), which are parasites that consume still-alive hosts, often from the inside, ultimately resulting in host death. The analogy is that lytic phages similarly consume their bacterial hosts from the inside, also ultimately killing the infected bacterium. The same low levels of molecular assimilation that can be used to criticize the labeling of phages as predators similarly could be applied to the labeling of phages as parasitoids, however. It is important, though, not to lose sight of the fact that lytic phage infections are equivalently detrimental to their bacterial hosts regardless of how we choose to label those interactions ecologically.

2.3.3.3 Phage-Mediated Horizontal Gene Transfer (Transduction)

The above considerations of symbioses are ecological in terms of the consequences of phage-to-bacteria communication. Alternatively, one can view phage-to-bacteria communications from an evolutionary perspective, that is, in terms of mutation, sampling error (i.e., genetic drift), and migration (movement of genetic material between populations), as well as, of course, natural selection (Duffy and Turner 2008; Abedon 2009a). In this section I emphasize phage impact especially on the migration of genetic material between populations of bacteria, e.g., between different bacterial strains.

Migration can be viewed as the physical movement of organisms or, more pertinently regarding evolutionary biology, the movement of individuals into or out of populations. While bacteria can physically migrate into and out of populations, it is important to realize that bacterial genes, independent of bacteria themselves, also are capable of such movement. In microbiology one typically describes this migration of genes between populations as horizontal gene transfer (HGT) or lateral gene flow, though another term which seems applicable is introgression (Cohan et al. 1991; Campbell 1994; Lawrence and Ochman 1997; Brown et al. 2001; Colegrave 2002; Johnson et al. 2004; Abedon 2009a). Introgression, that is, is low-level gene flow between otherwise minimally genetically interacting populations. Such gene flow, as movement of genetic sequences from one organism into another, is readily mediated by phages in a process termed transduction, which is the packaging of bacterial DNA into phage capsids and subsequent delivery of that DNA to the cytoplasm of a second bacterium (Schneider 2017). The recipient bacterium both survives and integrates the received DNA into its genome via various forms of recombination. A number of studies have addressed the question of to what extent phage-mediated transduction likely takes place within natural environments (Jiang and Paul 1998; Ogunseitan 2008; Colavecchio et al. 2017; Popa et al. 2017; Ghosh et al. 2008).

Bacteria additionally may be recipients of what are better described as phage rather than bacterial genes, i.e., genes which normally are found within a phage's genome rather than ones that are accidentally packaged into phage capsids. Lysogenic conversion is the most familiar context within which such genes are observed (Sect. 2.3.3.2.1). More generally, phages might serve as "the proving ground of choice for evolutionary innovation" for potential bacterial genes, "the critical motive force for the evolution of the entire biosphere" (Krisch 2003). Phages also can inactivate bacterial genes, most prominently via prophage insertion into bacterial chromosomes, e.g., as mediated by phage Mu (Paolozzi and Ghelardini 2006). Adsorbed virions, if subject to digestion rather than eventually lysing their bacterial hosts, in addition presumably can contribute nutrients to the adsorbed bacterium (Fuhrman 1999).

2.3.3.3.1 Some Transduction Asides per Popa et al. (2017)

Considerations of Transductive Host Ranges Popa et al. (2017), with regard to phage-mediated transduction, considered what might be described as a phage "Bactericidal" or "Productive" host range, particularly as being broader than a phage's "Lysogenic" host range. It is from the latter they based their phylogenomic networks. We might also consider, however, a phage's DNA "Penetrative" host range, i.e., the range of bacterial hosts over which a phage may be able to deliver DNA, which in turn is potentially equivalent to a phage's "Transductive" host range. It is especially this transductive host range that may be broader than a phage's bactericidal or productive range since for transduction to be possible then phage-carried DNA only has to enter a bacterium's cytoplasm, rather than then also rapidly and substantially express associated phage genes (Hyman and Abedon 2010).

That is, in terms of DNA transduction, a phage's host range and thereby a phage's potential to mediate transductive communication between bacteria (transductive host range) might be wider than would be anticipated by most standard phage host-range determining assays such as spot or plaquing assays, and also wider than determinations of potential for lysogen formation might suggest (i.e., "Spotting" host range, "Plaquing" host range, and "Lysogenic" host range, respectively; Hyman and Abedon 2010). As Popa et al. nevertheless point out, there likely are strong limits on the breadth of phage host ranges within which phage-mediated transductive communication may occur. The latter parallels suggestions of greater likelihoods of recombination between phages, i.e., as a form of phage-phage communication that can result in phage mosaic genome evolution. That is, recombination between phage genomes more likely occurs between phages whose host ranges are more similar (more overlapping) versusdonor and recipient are phylogenetically, while Lawrence and Hendrickson less similar (less overlapping) (Hendrix et al. 1999). Phage host range thus should place limitations on the movement of phage-carried genetic material not only into bacterial genomes but into the genomes of other phages as well.

Considerations of Transductive Gene Duplication Also regarding Popa et al. (2017), I point the reader to the work of Lawrence and Hendrickson (2003) and their concept of a "Zone of Paralogy", which postulates that to a degree *retention* of new genes acquired during horizontal gene transfer (HGT) events can be more likely the greater the phylogenetic distance between donor and recipient. This is a notion that appears to be partially challenged by Popa et al's observation that "…genetic similarity, rather than ecological opportunity, is a driver of successful transduction during microbial evolution." That is, Popa et al. rightly suggest that HGT events should be more likely to occur the closer DNA donor and recipient are phylogenetically, while Lawrence and Hendrickson suggest that the more similar the donor and recipient then the more difficult for newly acquired DNA sequence to be retained by natural selection. In particular, HGT events between closely related organisms resulting in allelic replacement should fall outside of Lawrence and Hendrickson's Zone of Paralogy.

Popa et al. also point out that transduction-mediated bacterial gene acquisition instead via illegitimate recombination can result in high degrees of genetic homology, thereby resembling products of gene duplication events. These pseudo within-genome duplications of genes would not, however, necessarily always "confer separate functions" as per Lawrence and Hendrickson's (2003) Zone of Paralogy. In other words, per Popa et al. (2017), as genetic similarity increases between bacterial DNA donors and DNA recipients, then the potential for retention of illegitimate recombination events, at least as mediated by transduction, apparently does not drop off to zero, and this is despite the presumed functional redundancy of the now duplicated genes. Nevertheless, there can be little basis for selective retention explicitly of these new genes the more similar they are to genes that are already present within a bacterium's genome, as truly redundant genes should fall outside of Lawrence and Hendrickson's Zone of Paralogy. See Popa and Dagan (2011) for further discussion of "Functional barriers" to DNA acquisition by organisms.

Xenology, Autology, and Auto-transduction Acquisition of regions of homology between bacteria as mediated by transduction can be described as an example of a "Xenology" if due to DNA transfer between species (Gogarten 1994; Fitch 2000). Popa et al., by contrast, use the term, "Autology" to describe DNA transfer as occurring within a single lineage, i.e., as meaning "Phage-mediated gene duplication" by transduction. Related to "Autology" is "Auto-transduction" as described by Haaber et al. (2016), which is phage-mediated DNA transfer by temperate phages that comes back to the originally phage-spawning bacterial lysogen. That is, phages generated by lineage 'A' pick up DNA from neighboring, related bacteria (lineage

'B'), only to then transduce that newly acquired DNA from lineage B back to lineage A. Thus, 'phage-as-lysogen-to-bacteria' bactericidal communication – communication from lineage A to lineage B as equivalent to kill the relative (Sect. 2.3.3.2.1) – may then be followed by 'phage-to-phage-as-lysogen' transductive communication, i.e., DNA movement from lineage B back to lineage A.

The three phenomena, xenology, auto-transduction, and autology, thus might be distinguished, approximately, as referring to between-species, within-species, and within-strain (as a clonal lineage) gene transfer, respectively. Popa et al. also refer to autology as resulting from "Selfing", which presumably is as analogous to the tendency, particularly among some plants, for fertilization to occur within the same flower rather than between different plants (selfing, that is, is having reproductive sex with self as contrasts with outcrossing which is having sex with something other than self). Thus, phage-mediated transductive communication explicitly can be between otherwise genetically identical bacteria (autology) as well as between relatively genetically similar bacteria, though with a 'twist' (auto-transduction). Simply due to limitations in phage host range, both auto-transduction and autology, however, are likely to be more common than transduction between genetically dissimilar bacteria (xenology).

2.3.3.4 On Tradeoffs and Killing the Winner

In this section I consider two negative impacts that phages can have on bacteria, both as viewed from a bacterial evolution perspective. Both are derived from phage predation of bacteria. These are tradeoffs and frequency-dependent selection.

2.3.3.4.1 Tradeoffs

A tradeoff represents a barrier to the breadth of circumstances an organism or associated adaptation is able to excel in. That is, improvements in functioning of one aspect of an organism often by necessity are associated with reductions in the functioning of a different aspect of an organism. Tradeoffs thus can simultaneously impact functioning under the same environmental circumstances or instead can affect non-concurrent functioning, with especially the latter described as antagonistic pleiotropies (Sect. 2.3.2.2.2). Antagonistic pleiotropies, that is, generally are defined other than in terms of simultaneous functioning, e.g., as occurring in different environments, at different ages for an organism, or while infecting different hosts. For giraffes, for example, long necks make it easier to outreach other browsers for access to canopy leaves, or to better spot predators, but this at a slightly different time makes it difficult for giraffes to access water for drinking. More extreme, whales are great swimmers but terrible runners. Prominent for bacteria, in terms of phages, are tradeoffs between phage resistance and some other aspect of bacterial fitness, such as bacterial growth rates as may be measured in otherwise identical environments but in the absence of phages (Levin and Bull 2004; Kerr et al. 2008; Thingstad et al. 2008; Hyman and Abedon 2010; Azam and Tanji 2019).

The consequence of tradeoffs on bacterial evolution of phage resistance is that in a world lacking in phages, the dominating bacteria will tend to be those that display the least phage resistance. Thus, there can be costs to bacteria for evolving phage resistance, and these costs can lower the potential for a bacterium to dominate or perhaps even just successfully compete within a given environment, unless the phages to which they are resistant are also found at relatively high titers in those same environments. Furthermore, one basis of phage therapy as an anti-infection strategy (Sect. 2.3.6.1) could be the lower potential for evolved phage-resisting bacterial mutants to persist as infection-causing pathogens relative to phage-sensitive bacteria. That is, phage resistance is not necessarily always compatible with a bacterium's ability to persist as an invasive pathogen. Phage resistance thus can often be viewed as an impairment imposed on bacteria by attacking phages. This is rather than mechanisms of bacterial protection from phagesal ways lacking in bacterial-virulence or bacterial growth-rate tradeoffs.

2.3.3.4.2 Frequency-Dependent Selection, a.k.a., Killing the Winner

Phage-mediated, intraspecific, frequency-dependent selection, i.e., as impacting a bacterial species' relative prevalence within an environment (Levin 1988; Abedon 2009a), typically has been described in the more recent phage literature as "Killing the winner" or "Kill the winner" (Thingstad et al. 2008; Rodriguez-Brito et al. 2010; Winter et al. 2010; Diaz-Munoz and Koskella 2014). That is, bacterial winners – those bacterial phage-susceptibility types displaying the greatest environmental densities (highest bacterial concentrations) – will be more susceptible to devastating phage attack simply due to attaining those higher densities within environments. The result is a potential for greater evolutionary fitness for lower-density bacterial phage-susceptibility types along with a resulting selection for greater overall bacterial diversity (that is, *low* diversities of phage-susceptibility types are less likely to dominate environments if those bacteria as a consequence are *more* likely to be decimated by phages). A distinction between killing the winner and frequency-dependent selection, however and as noted, is that the former is also an ecological phenomenon, that is, occurring across multiple species (many bacterial populations), whereas frequency-dependent selection is an evolutionary phenomenon, that is, occurring within only a single species (one bacterial population). As a complication, note also that the original, steady-state formulation of killing the winner described "Winners" as those bacteria which replicated fastest rather than necessarily those which were most prevalent (Jacquet et al. 2018). In any case, these processes involve phages that in a sense are lytically communicating to bacteria that the latter are too successful within a given environment.

One should always expect higher bacterial densities to support production of greater phage densities, and therefore a greater phage potential to impact bacterial populations, particularly so long as other factors are held constant. A shortcoming of the idea of kill the winner in terms of experimental validation, however, is that dramatic and therefore easily observed bacteria killing will only occur if relatively high phage densities can be achieved, such as in the range of 10^7 or more phages of a single type per ml (Appunu and Dhar 2008; Abedon and Thomas-Abedon 2010; Abedon 2014, 2018b). Such a high titer being associated with specific phage types, averaged over an entire environment, is simply not likely, especially given that virus total numbers in specific environments in many cases do not collectively reach these levels (Wommack and Colwell 2000). That is, if there are 'only' 10^7 phages/ml within a given environment, then it is unlikely that any one phage type will have a population size also of 10^7/ml, unless the environment in question is extremely simple in terms of what microorganisms are present.

Alternatively, we can view kill the winner as a more localized phenomenon, such as from the perspective of biofilm-associated bacterial microcolonies, i.e., as phage targets (Abedon 2011a, 2012b, 2017a, d). For two bacterial types that are otherwise identical, a microcolony that is larger overall – in a sense being a winner among microcolonies – will, given its larger surface area, be more likely to encounter a phage and therefore more likely to be reduced in size or eliminated by phage infection (Abedon 2012b) (again, when holding all other factors constant). Of course, the higher the number of microcolonies of a given phage susceptibility type that are present within an environment, then the more prevalent ultimately can be those would-be encountering phages. On the other hand, the more spatially structured an environment, then the less potential there is for diffusing phages to encounter a given microcolony, e.g., such as in soils versus within well-mixed broth (Tan and Reanney 1976; Pantastico-Caldas et al. 1992). Such spatial structure presumably would reduce the efficiency of kill the winner on more environment-wide scales, resulting in a lower potential for experimental validation, but with kill-the-winner effects still prominent locally, e.g., such as on millimeter or smaller spatial scales (Abedon 2017a).

2.3.3.5 Phages Impacting Biofilms and Bacteria Extracellular Polymers

Phages, by infecting bacteria, in principle can disrupt bacterial biofilms and microcolonies, or destroy cellular arrangements. Though such phage-mediated disruption is readily demonstrated in the laboratory (Azeredo and Sutherland 2008; Abedon 2011a, 2015c, 2018a; Sillankorva and Azeredo 2014), it is an open question how significant a role phages play in such disruption in natural environments; for speculation, however, see (Abedon 2012b, 2016, 2017d).

Bacteria also can secrete extracellular polymers as make up bacterial capsules or, more generally, can secrete extracellular polymeric substances (EPSs) as can surround bacterial cells. These polymers bind together bacterial biofilms, thereby constituting the biofilm matrix, and also may inhibit or at least slow the movement of phage particles that have diffused into this polymer (Sect. 2.3.2.3.2). Phages, however, can encode and produce EPS-hydrolyzing enzymes known as EPS depolymerases (Sutherland et al. 2004; Abedon 2011a). These enzymes, either with or without their associated phages, can be applied as anti-biofilm agents (Hughes et al. 1998; Sillankorva and Azeredo 2014; Chan and Abedon 2015; Lin et al. 2018).

Ecologically, phage EPS depolymerases could allow virions to more freely move through these polymers either toward bacteria to infect or away from bacteria once virions have been released from bacteria into the extracellular environment. A substantial fraction of these EPS depolymerases appear to be virion associated (Pires et al. 2016). Particularly, movement toward bacteria to infect may be enhanced by that virion association as it physically links diffusing phages to the enzyme (Abedon 2011a; Chan and Abedon 2015). Enzyme that is not virion associated, by contrast, will not necessarily be present when a phage virion is first encountering a capsule- or EPS-associated bacterium.

2.3.4 Phage-Infection-to-Environment Communication

In addition to phage-to-bacteria communication, phages also can impact aspects of environments that are separate from their hosts, but nonetheless affecting these non-host environmental aspects in a manner that is mediated through phage infection of bacteria. This phage-to-bacteria-to-extracellular-environment communication – or phage-infection-to-environment communication – is primarily accomplished via phage-induced bacterial lysis (Sect. 2.3.4.1), but also can be effected through prophage gene expression (Sect. 2.3.4.2). Among lysis-related effects is the likely phage impact on the quantity and quality of free DNA found within environments, i.e., as can contribute to horizontal gene transfer via the processes of transformation (also Sect. 2.3.4.1).

2.3.4.1 Lysis-Mediated Phage-to-Environment Communication

Other than in terms of their own existence as virions as found in the extracellular environment (Sect. 2.3.5), phages can influence environments specifically through bacterial lysis and do so in at least three ways: by releasing nutrients from bacteria, by releasing enzymes from bacteria, and by releasing potentially still-functional DNA from bacteria. In addition, phage-induced bacterial lysis could play a role in the release of secondary metabolites from bacteria, adding to those released through bacterial secretion (Karlovsky 2008).

2.3.4.1.1 Nutrient Liberation

By effecting bacterial lysis, phages convert bacteria-associated nutrients into soluble or at least less-protected forms. These nutrients are then available, particularly in terms of dissolved organic carbon, to heterotrophic bacteria (Suttle 2007; Weinbauer 2004) as well as to eukaryotic heterotrophic absorbers such as fungi. In addition, lysis-released inorganic nutrients will be available to all organisms that obtain their nutrients via absorption from the environment, versus via ingestion (Jacquet et al. 2018). This process of breaking down and to some degree mineralizing bacterial cells basically represents a form of bacterial cell decay, though one initiated as a process while the so-affected bacteria are still metabolically active. In broader terms, virus-induced release of nutrients into environments gives rise to what is known in aquatic environments as the viral shunt (Wilhelm and Suttle 1999) whereby photosynthetically fixed carbon becomes less available to higher trophic levels. That is, nutrients become less available to those organisms that consume intact bacteria, though at the same time nutrients become more available to chemoheterotrophic osmotrophs such as heterotrophic bacteria, i.e., organisms that are able to consume the components of no longer intact bacteria (Jacquet et al. 2018).

2.3.4.1.2 Enzyme Release

Another consequence of phages lysing bacteria is the release of enzymes into the extracellular environment, enzymes that otherwise would remain physically associated with bacteria. These, that is, are enzymes which normally are found in the cytoplasms and periplasms of bacteria, or even found as extracellularly facing enzymes present as membrane proteins. As a consequence of lysis-induced release, enzymes may be able to reach and hydrolyze otherwise indigestible nutritious substrate found in the environment. While I find it compelling that these enzymes probably play relevant roles particularly within relatively simple ecosystems (Chróst 1991; Morita 1997), their extracellular presence due to phage-induced bacterial lysis perhaps is less important the more complex the environment, perhaps particularly environments that already contain large diversities of nutrient-solubilizing microorganisms. Nonetheless, there remains at least a possibility that lysis-released bacterial enzymes could play important roles in all ecosystems.

2.3.4.1.3 DNA Release

A third mechanism of phage-infection-to-environment communication, as mediated through bacterial lysis, involves the release of DNA. While this DNA can be viewed as just another solubilized nutrient, lysis-released DNA also, if not too fragmented, can serve as a source of genetic material for bacterial transformation (Pietramellara et al. 2009). That is, certain bacteria are able to pick up environmental (naked) DNA into their cytoplasms and subsequently incorporate these 'snippets' into their genomes (Lorenz and Wackernagel 1994; Day 2004; Pietramellara et al. 2009; Blokesch 2016). It thus is conceivable that the DNA pool available for this transformation is larger than it would be absent phage-mediated bacterial lysis (Abedon 2009a). Given how closely associated bacteria can be within biofilms, phage infection of biofilm bacteria and subsequent DNA release might further increase, there, the potential for bacterial transformation (Li et al. 2001; Wuertz et al. 2001; Wang et al. 2002; Hendrickx et al. 2003; Molin and Tolker-Nielsen 2003; Hannan et al. 2010; Kouzel et al. 2015).

To the extent that kill the winner operates (Sect. 2.3.3.4.2), then the DNA available upon phage-mediated bacterial lysis may either be more diverse owing to a greater assortment of bacteria that are available to be lysed over time, or less diverse owing to a bias toward lysis of particular (winner) bacterial populations. Furthermore, not only is bacterial DNA potentially released upon phage-mediated bacterial lysis, but so too is unencapsidated phage DNA released upon this lysis, which also may be available to bacteria for transformation. In addition is the potential for DNA to contribute to a biofilm's extracellular matrix (Molin and Tolker-Nielsen 2003).

2.3.4.2 Prophage-Mediated Environmental Modification

Phages might also modify environments toward their own ends. This consideration is based on speculative analogies to the known phage encoding of bacterial virulence factors. These virulence factors, and encoding phages, are associated with many bacterial pathogens of animals (Abedon and LeJeune 2005; Hyman and Abedon 2008; Christie et al. 2012). Some phage-associated virulence factors, particularly exotoxins such as Shiga, diphtheria, or cholera toxins, can be viewed as environment-modifying enzymes. That is, the disease symptoms that these toxins effect are a consequence of their disruptive modifications of the body environment. In addition, the resulting environmental modifications might serve to enhance a phage's replication or dissemination (Abedon and LeJeune 2005).

It is conceivable that similar factors are encoded by phages infecting nonpathogenic, environmental bacteria, and then expressed perhaps chiefly during lysogenic cycles. These factors, that is, might modify environments and do so in some manner that benefits the producing phage, rather than solely modifying the hosting bacterium via lysogenic conversion (Sect. 2.3.3.2.1). For example, phages might encode enzymes whose release from bacteria can lead to the digestion of local substrate either for the sake of providing nutrients to host bacteria or to enhance the potential for virion diffusion, such as away from the lysed parental bacterium and toward not yet phage infected bacteria. Indeed, though lytic- rather than lysogenic-cycle associated, phage-encoded EPS depolymerases perhaps could be viewed as such enzymes (Sect. 2.3.3.5).

2.3.5 Phage-Virion-to-Environment Communication

I posit four mechanisms by which free phages, that is extracellular phage virions, may communicate their presence to the non-host environment. First, phages can be consumed by certain eukaryotes (Bettarel et al. 2005; González and Suttle 1993), and perhaps under some rare circumstances contribute to the genetic material of those organisms, i.e., the "you are what you eat" hypothesis (Doolittle 1998;

Andersson 2005; Keeling and Palmer 2008). Second, and as noted, there exist certain virion-associated hydrolytic enzymes that can disrupt bacteria-secreted extracellular polysaccharides (Sect. 2.3.3.5). Third, phages can serve as soluble nutrients following virion decay. Four, as part of that decay, virions can release DNA which then in principle could transform naturally competent bacteria.

2.3.6 Environment-to-Phage-Infection Communication

For phage biologists with an interest in phenotype, a fascinating aspect of communication as it occurs between phages, bacteria, and environments, is how modifying environmental parameters can result in modification of phage phenotypes. This concept we can loosely describe as a phage phenotypic plasticity, but also as a result of environment-to-phage-infection communication.

2.3.6.1 Phage-Infection Responses to Abiotic Environmental Factors

Perhaps the most familiar phage response to environmental stimuli is the induction of prophages following lysogen exposure to DNA damaging agents such as UV light or mitomycin C (Jiang and Paul 1996; Cochran et al. 1998; Weinbauer and Suttle 1999; Campbell 2006). This response happens to be one that is both stimulated and mediated while a phage is still associated with its bacterial host. Another and similar phenomenon is the resolution of the otherwise reductive pseudolysogenic-type state (Miller and Day 2008) in which increasing host metabolism – as due to increased environmental nutrient densities – results in increased phage-infection metabolic activity, producing a productive, lysogenic, or perhaps even destructive infection (Fig. 2.3). With both induction of prophages and activation of pseudolysogens, the phage thus is potentially displaying an adaptive response to changes in environmental conditions, in both cases seen as increasing infection activity.

The response to a worsening of environmental conditions can be a reduction in the metabolic activity of a phage infection, or even an avoidance of adsorption altogether (Kutter et al. 1994). Starvation conditions are also known to increase the likelihood of a temperate phage displaying lysogenic cycles upon infection (Herskowitz and Banuett 1984; Miller and Day 2008). At an extreme, the phage could display the mentioned pseudolysogeny, which is a non-lysogenic, non-productive, non-destructive, *and* non-reproductive delay in initiation of a more metabolically active infection (Abedon 2009c). Less extreme responses to reduced environmental densities of nutrients can involve decreases in phage burst sizes, lengthening in phage infection periods (i.e., latent periods), or both. This has been observed in laboratory cultures following nutrient limitation (Webb et al. 1982; Hadas et al. 1997) and represents an environmentally effected reduced phage-infection vigor. These responses may be relevant especially to the extent that environments are oligotrophic or, instead, if environments alternate between eutrophic and oligotrophic (feast or famine) (Williams et al. 1987). Williams et al. (1987) reviews similar results associated with changes in temperature.

2.3.6.2 Antibiotic Impact on Phage Infection Activity

Phage therapy is the use of bacterial viruses to treat bacterial infections (Abedon et al. 2011; Kutter et al. 2010; Abedon 2017b). More generally is what can be described as a phage-mediated biocontrol of bacteria, including of environmental bacteria (Abedon 2009e; Harper 2013). While there are many advantages to using phages as antibacterial agents (Kutter 2005; Häusler 2006; Loc-Carrillo and Abedon 2011; Curtright and Abedon 2011; Abedon 2015b; Nobrega et al. 2015; Wienhold et al. 2019), nevertheless in most countries, including in the U.S. and in Europe, the approach has not yet become popular among physicians (Kutter et al. 2015). Such low-level implementation of phage therapy is despite numerous examples of phage treatment success (Burrowes and Harper 2012; Loc-Carrillo et al. 2012; Abedon 2015d, 2018a, 2019c; Chang et al. 2018) including a number of successful, recent case studies (Fish et al. 2016; Schooley et al. 2017; Chan et al. 2018; Ferry et al. 2018; Aslam et al. 2019a, b; Dedrick et al. 2019; Gupta et al. 2019; Hoyle et al. 2018; Law et al. 2019) as well as one successful clinical trial (Wright et al. 2009). One possible explanation for this lag in more widespread introduction of phage therapy, particularly into Western medicine and especially given the critical need for new antibacterial agents (Brunel and Guery 2017; Martens and Demain 2017; Weledji et al. 2017; Bragg et al. 2018), is that insufficient effort has been made toward integrating phage therapy into standard antibacterial treatments, that is, in combination with antibiotics.

Combining phages with antibiotics as antibacterial agents is not necessarily as straightforward as superficially it might appear. As antibacterial agents, phages explicitly are bactericidal, whereas it is well known that bacteriostatic antibiotics often are antagonistic to bactericidal agents (Jawetz et al. 1951; Ocampo et al. 2014). Indeed, there exist a relatively large number of studies that have documented the negative impacts of various antibiotics on various phage types in terms of, for example, substantially reducing phage burst sizes, including to the point where no virions appear to be produced at all (Abedon 2019b). Thus, antibiotics can be viewed as an environmental factor that can impact phages, particularly as that impact is mediated through phage-infected bacteria (environment-to-phage-infection communication). There are exceptions to such negative impact, however. One prime example is when cell wall synthesis-inhibiting antibiotics are used rather than bacteriostatic agents. In this case, there typically can be a substantial delay between antibiotic addition and subsequent loss of bacterial viability. We can predict therefore that simultaneous phage and cell wall synthesis-inhibiting antibiotic addition might still allow for substantial phage functioning as an antibacterial agent (David et al. 1980).

In addition, and relevant to the idea of environment-to-phage-infection communication, is that the antibiotic "signal" may not always be efficiently conveyed to phages, particularly given bacterial infection tolerance to treatment antibiotics (Ceri et al. 1999; Jolivet-Gougeon and Bonnaure-Mallet 2014; Macia et al. 2014; Olsen 2015; Brauner et al. 2016, 2017; Fisher et al. 2017). In this case, it is possible that antibiotics may have less of an impact on the infection activity of some phages, thereby allowing for phage-antibiotic cotreatments in which these two agents target two distinct bacterial populations within bodies, even given use of bacteriostatic antibiotics. Specifically, and to a degree tautological, those bacteria that are more antibiotic impacted are more affected by the antibiotic treatment while those bacteria that are less antibiotic impacted may be substantially impacted instead by the phage co-treatment. For review of studies of the impact of antibiotics on phage infection activity, see Abedon (2019b) and see also (Knezevic et al. 2019; Morrisette et al. 2020; Segall et al. 2019; Tagliaferri et al. 2019).

2.3.7 Environment-to-Phage-Virion Communication

Environmental factors, ones that are other than a phage's host bacterium, can communicate their presence to phages by impacting virion movement (Sect. 2.3.7.1) as well as virion survival (Sect. 2.3.7.2). This too can be viewed as an environment-to-phage communication, but specifically with the phage present as a free virion rather than as a phage-infected bacterium.

2.3.7.1 Non-host-Environment Impact on Phage Virion Movement

Many things can contribute to phage movement. These include diffusion and fluid flow. Given the size of virions, however, especially diffusion can be relatively *limited* in comparison with diffusion of dissolved materials or even diffusion of soluble enzymes (McKay et al. 2002). By contrast, flow of surface waters (Ferguson et al. 2007) or of groundwaters (Bales et al. 1995; McKay et al. 2002) can allow virion movement over meters, at least. Both diffusion and flow, however, may be inhibited by various obstructions (Wong et al. 2008; Van Cuyk and Siegrist 2007; Davis et al. 2006). These obstructions include especially colloidal substances into which phages can be absorbed, such as clays (Bixby and O'Brien 1979; Duboise et al. 1979; Williams et al. 1987; Chattopadhyay and Puls 2000; Hassen et al. 2003; Day and Miller 2008).

Animals also can play a role in the movement of phages, in part if phages can survive digestion, but also should virions or instead phage-infected bacteria adhere to animal surfaces (Sisler 1940; Dennehy et al. 2006). Phages also have been found to reversibly attach to the mucus secreted by animals, thereby slowing virion diffusion (Barr et al. 2013) but presumably also potentially allowing virions to reversibly adhere to moving animals. Bulk movement of phages also should occur such as due to mechanical action, e.g., trees becoming uprooted. Phages also seem to spread from location to location through the air (Weinbauer 2004; Reche et al. 2018) and a number of papers consider such movement especially from the perspective of sewage aerosolization, e.g., Brooks et al. (2004). These latter authors, however, point out that (p. 8), "It is known that a bioaerosol is subject to intense physical pressures from the environment (specifically low humidity, ultraviolet and temperature extremes) which tend to inactivate microbes during transport of bioaerosols over long distances." For the latter, see also (Clark 2005).

2.3.7.2 Non-host-Environment Impact on Phage Virion Survival

Many environmental factors can impact phage virion survival. Perhaps most obviously, for those environments that are exposed to sunshine, are the negative impacts of sunlight and UV radiation (Murray and Jackson 1993; Wommack et al. 1996; Balogh et al. 2010; Fisher et al. 2011; Flannery et al. 2013). In addition is desiccation and also pH extremes (Williams and Lanning 1984; Song et al. 2005; Davies et al. 2006). The latter can occur especially in the course of phage ingestion by animals, as possibly can lead to virion digestion (Sect. 2.3.7.2.2). Environments also can supply mutagens, i.e., factors that can result in phage mutation (Drake and Ripley 1994). By contrast, however, is phage banking (Sect. 2.3.7.2.1).

2.3.7.2.1 Phage Banking

Virions can become trapped such as within biofilms (Briandet et al. 2008; Abedon 2011a) or within otherwise tight spaces such as within soils (McKay et al. 2002). But, is this entrapment permanent or instead only temporary? Regarding only temporary, note the concept of phage banking (Breitbart and Rohwer 2005), which is the idea that especially phage virions may remain dormant over long periods within environments in anticipation of future contact with a susceptible bacterial host. This terminology is analogous to the concept of seed banks, in which plant seeds remain dormant in soils in anticipation of future conditions that are more favorable to seed germination and subsequent plant growth. Consistent with but predating use of the phrase "Seed bank" as applied to phages, note this from Williams et al. (1987), p. 162: "There is evidence that most soil bacteria have only spasmodic periods of activity in micro-sites dispersed within the soil mass… Therefore it is likely that virulent phage in the absence of an active susceptible host must exist for considerable periods as free virions in the soil and be subjected to environmental factors and fluctuations." Williams et al. then go on to provide a review of the evidence from which this position is derived. Phages also may be 'banked' within reductively infected bacteria (Sect. 2.3.2.1.2).

2.3.7.2.2 Phage Digestion

Materials ingested by animals likely will contain phages, given phage ubiquity within nearly all environments. The ingested phages may be digested along with other organic matter, except to the extent that virions are resistant to digestive processes. One can envisage, for example, phage loss because of earthworm action as well as from protist engulfment. It seems unlikely, though, that phages would contribute greatly to either's nutrient needs. Predatory bacteria, such as *Myxococcus*, similarly can secrete hydrolytic enzymes as groups (bacterial 'wolfpacks') to obtain nutrients from soil-associated organisms, materials, and, potentially, even phages (Berleman et al. 2006; Evans et al. 2007). Of course, as *Myxococcus* spp. are bacteria, there also exist phages that are capable of infecting them, e.g., (Azuaga et al. 1990), and see also (Zeph and Casida 1986).

2.4 Conclusion

The intention of this chapter is not so much to be convincing that the concept of 'communication' may or may not be appropriate to describe the many phage-associated ecological interactions that are discussed. The point instead is to indicate that the variously described flows of substances and information, and their impacts, both exist and are relevant to a more complete understanding of phage ecology and its underlying principles. Phages thus interact with many things, resulting in modifications of phage properties, and those interactions result as well in modifications of host bacteria and extracellular environments.

References

Abedon ST (1990) Selection for lysis inhibition in bacteriophage. J Theor Biol 146:501–511
Abedon ST (1992) Lysis of lysis inhibited bacteriophage T4-infected cells. J Bacteriol 174:8073–8080
Abedon ST (1994) Lysis and the interaction between free phages and infected cells. In: Karam JD, Kutter E, Carlson K, Guttman B (eds) The molecular biology of bacteriophage T4. ASM Press, Washington, DC, pp 397–405
Abedon ST (1999) Bacteriophage T4 resistance to lysis-inhibition collapse. Genet Res 74:1–11
Abedon ST (2000) The murky origin of Snow White and her T-even dwarfs. Genetics 155:481–486
Abedon ST (2011) Facilitation of CRISPR adaptation. Bacteriophage 1(3):179–181
Abedon ST (2006) Phage ecology. In: Calendar R, Abedon ST (eds) The bacteriophages. Oxford University Press, Oxford, pp 37–46
Abedon ST (2008) Phages, ecology, evolution. In: Abedon ST (ed) Bacteriophage ecology. Cambridge University Press, Cambridge, pp 1–28
Abedon ST (2009a) Bacteriophage evolution and ecology. Adv Appl Microbiol 67:1–45
Abedon ST (2009b) Bacteriophage intraspecific cooperation and defection. In: Adams HT (ed) Contemporary trends in bacteriophage research. Nova Science Publishers, Hauppauge/New York, pp 191–215
Abedon ST (2009c) Disambiguating bacteriophage pseudolysogeny: an historical analysis of lysogeny, pseudolysogeny, and the phage carrier state. In: Adams HT (ed) Contemporary trends in bacteriophage research. Nova Science Publishers, Hauppauge/New York, pp 285–307
Abedon ST (2009d) Impact of phage properties on bacterial survival. In: Adams HT (ed) Contemporary trends in bacteriophage research. Nova Science Publishers, Hauppauge/New York, pp 217–235

Abedon ST (2009e) Kinetics of phage-mediated biocontrol of bacteria. Foodborne Pathog Dis 6:807–815

Abedon ST (2011a) Bacteriophages and biofilms: ecology, phage therapy, plaques. Nova Science Publishers, Hauppauge/New York

Abedon ST (2011b) Communication among phages, bacteria, and soil environments. Soil Biol 23:37–65

Abedon ST (2011c) Communication among phages, bacteria, and soil environments. In: Witzany G (ed) Biocommunication of soil microorganisms. Springer, New York, pp 37–65

Abedon ST (2012a) Bacterial 'immunity' against bacteriophages. Bacteriophage 2:50–54

Abedon ST (2012b) Spatial vulnerability: bacterial arrangements, microcolonies, and biofilms as responses to low rather than high phage densities. Viruses 4:663–687

Abedon ST (2014) Bacteriophages as drugs: the pharmacology of phage therapy. In: Borysowski J, Miedzybrodzki R, Górski A (eds) Phage therapy: current research and applications. Caister Academic Press, Norfolk, pp 69–100

Abedon ST (2015a) Bacteriophage secondary infection. Virol Sin 30:3–10

Abedon ST (2015b) Ecology of anti-biofilm agents I. antibiotics versus bacteriophages. Pharmaceuticals 8:525–558

Abedon ST (2015c) Ecology of anti-biofilm agents II. Bacteriophage exploitation and biocontrol of biofilm bacteria. Pharmaceuticals (Basel) 8(3):559–589. https://doi.org/10.3390/ph8030559

Abedon ST (2015d) Phage therapy of pulmonary infections. Bacteriophage 5:e1020260

Abedon ST (2016) Bacteriophage exploitation of bacterial biofilms: phage preference for less mature targets? FEMS Microbiol Lett 363:fnv246

Abedon ST (2017a) Active bacteriophage biocontrol and therapy on sub-millimeter scales towards removal of unwanted bacteria from foods and microbiomes. AIMS Microbiol 3:649–688

Abedon ST (2017b) Bacteriophage clinical use as antibactertial "drugs": utility, precedent. Microbiol Spectr 5:BAD-0003-2016

Abedon ST (2017c) Commentary: communication between viruses guides lysis-lysogeny decisions. Front Microbiol 8:983

Abedon ST (2017d) Phage "delay" towards enhancing bacterial escape from biofilms: a more comprehensive way of viewing resistance to bacteriophages. AIMS Microbiol 3:186–226

Abedon ST (2018a) Bacteriophage-mediated biocontrol of wound infections, and ecological exploitation of biofilms by phages. In: Shiffman M (ed) Recent clinical techniques, results, and research in wounds. Springer, New York

Abedon ST (2018b) Phage therapy: various perspectives on how to improve the art. Methods Mol Biol 1734:113–127

Abedon ST (2018c) Viruses of domain archaea. In: Hyman P, Abedon ST (eds) Viruses of microorganisms. Caister Academic Press, Norwich, pp 167–192

Abedon ST (2019a) Look who's talking: T-even phage lysis inhibition, the granddaddy of virus-virus intercellular communication research. Viruses 11:E951

Abedon ST (2019b) Phage-antibiotic combination treatments: antagonistic impacts of antibiotics on the pharmacodynamics of phage therapy? Antibiotics 8:182

Abedon ST (2019c) Use of phage therapy to treat long-standing, persistent, or chronic bacterial infections. Adv Drug Deliv Rev 145:18–39

Abedon ST, Culler RR (2007) Bacteriophage evolution given spatial constraint. J Theor Biol 248:111–119

Abedon ST, Duffy S, Turner PE (2009) Bacteriophage ecology. In: Schaecter M (ed) Encyclopedia of microbiology. Elsevier, Oxford, pp 42–57

Abedon ST, Kuhl SJ, Blasdel BG, Kutter EM (2011) Phage treatment of human infections. Bacteriophage 1:66–85

Abedon ST, LeJeune JT (2005) Why bacteriophage encode exotoxins and other virulence factors. Evol Bioinformatics Online 1:97–110

Abedon ST, Murray KL (2013) Archaeal viruses, not archaeal phages: an archaeological dig. Archaea 2013:251245

Abedon ST, Thomas-Abedon C (2010) Phage therapy pharmacology. Curr Pharm Biotechnol 11:28–47

Abedon ST, Yin J (2008) Impact of spatial structure on phage population growth. In: Abedon ST (ed) Bacteriophage ecology. Cambridge University Press, Cambridge, pp 94–113
Abedon ST, Yin J (2009) Bacteriophage plaques: theory and analysis. Methods Mol Biol 501:161–174
Anca MS, Dwayne RR, Steffanie AS (2019) Stronger together? Perspectives on phage-antibiotic synergy in clinical applications of phage therapy. Current Opinion in Microbiology 51:46–50
Andersson JO (2005) Lateral gene transfer in eukaryotes. Cell Mol Life Sci 62:1182–1197
Appunu C, Dhar B (2008) Isolation and symbiotic characteristics of two Tn5-derived phage-resistant *Bradyrhizobium japonicum* strains that nodulate soybean. Curr Microbiol 57:212–217
Argov T, Azulay G, Pasechnek A, Stadnyuk O, Ran-Sapir S, Borovok I, Sigal N, Herskovits AA (2017) Temperate bacteriophages as regulators of host behavior. Curr Opin Microbiol 38:81–87
Aslam S, Courtwright AM, Koval C, Lehman SM, Morales S, Langlais Furr CL, Rosas F, Brownstein MJ, Fackler JR, Sisson BM, Biswas B, Henry M, Luu T, Bivens BN, Hamilton T, Duplessis C, Logan C, Law N, Yung G, Turowski J, Anesi J, Strathdee SA, Schooley RT (2019a) Early clinical experience of bacteriophage therapy in three lung transplant recipients. Am J Transplant 19(9):2631–2639
Aslam S, Pretorius V, Lehman SM, Morales S, Schooley RT (2019b) Novel bacteriophage therapy for treatment of left ventricular assist device infection. J Heart Lung Transplant 38:475–476
Aviram I, Rabinovitch A (2008) Dynamical types of bacteria and bacteriophages interaction: shielding by debris. J Theor Biol 251:121–136
Avlund M, Dodd IB, Semsey S, Sneppen K, Krishna S (2009) Why do phage play dice? J Virol 83:11416–11420
Azam AH, Tanji Y (2019) Bacteriophage-host arm race: an update on the mechanism of phage resistance in bacteria and revenge of the phage with the perspective for phage therapy. Appl Microbiol Biotechnol 103:2121–2131
Azeredo J, Sutherland IW (2008) The use of phages for the removal of infectious biofilms. Curr Pharm Biotechnol 9:261–266
Azuaga MJ, Munoz J, Gonzalez F, Arias JM (1990) Isolation and characterization of bacteriophages from *Myxococcus virescens*. Microbios 61:83–88
Bales RC, Li S, Maguire KM, Yahya MT, Gerba CP, Harvey RW (1995) Virus and bacteria transport in a sandy aquifer, Cape Cod, MA. Ground Water 33:653–661
Balogh B, Jones JB, Iriarte FB, Momol MT (2010) Phage therapy for plant disease control. Curr Pharm Biotechnol 11:48–57
Barr JJ, Auro R, Furlan M, Whiteson KL, Erb ML, Pogliano J, Stotland A, Wolkowicz R, Cutting AS, Doran KS, Salamon P, Youle M, Rohwer F (2013) Bacteriophage adhering to mucus provide a non-host-derived immunity. Proc Natl Acad Sci U S A 110:10771–10776
Berleman JE, Chumley T, Cheung P, Kirby JR (2006) Rippling is a predatory behavior in *Myxococcus xanthus*. J Bacteriol 188:5888–5895
Berngruber TW, Lion S, Gandon S (2013) Evolution of suicide as a defence strategy against pathogens in a spatially structured environment. Ecol Lett 16:446–453
Bettarel Y, Sime-Ngando T, Amblard C, Bouvy M (2005) Low consumption of virus-sized particles by heterotrophic nanoflagellates in two lakes of the French Massif central. Aquat Microb Ecol 39:205–209
Bixby RL, O'Brien DJ (1979) Influence of fulvic acid on bacteriophage adsorption and complexation in soil. Appl Environ Microbiol 38:840–845
Blasdel BG, Abedon ST (2017) Superinfection immunity. In: Reference module in life sciences. Elsevier. https://doi.org/10.1016/B978-0-12-809633-8.90021-9
Blokesch M (2016) Natural competence for transformation. Curr Biol 26:3255
Boots M, Mealor M (2007) Local interactions select for lower pathogen infectivity. Science (New York, NY) 315:1284–1286
Boyd JS (1951) Observations on the relationship of symbiotic and lytic bacteriophage. J Pathol Bacteriol 63:445–457
Bragg RR, Meyburgh CM, Lee JY, Coetzee M (2018) Potential treatment options in a post-antibiotic era. In: Infectious diseases and nanomedicine III. Springer, Singapore, pp 51–61

Brauner A, Fridman O, Gefen O, Balaban NQ (2016) Distinguishing between resistance, tolerance and persistence to antibiotic treatment. Nat Rev Microbiol 14:320–330

Brauner A, Shoresh N, Fridman O, Balaban NQ (2017) An experimental framework for quantifying bacterial tolerance. Biophys J 112:2664–2671

Breitbart M, Rohwer F (2005) Here a virus, there a virus, everywhere the same virus? Trends Microbiol 13:278–284

Briandet R, Lacroix-Gueu P, Renault M, Lecart S, Meylheuc T, Bidnenko E, Steenkeste K, Bellon-Fontaine MN, Fontaine-Aupart MP (2008) Fluorescence correlation spectroscopy to study diffusion and reaction of bacteriophages inside biofilms. Appl Environ Microbiol 74:2135–2143

Brockhurst MA, Morgan AD, Fenton A, Buckling A (2007) Experimental coevolution with bacteria and phage. The *Pseudomonas fluorescens* – Φ2 model system. Infect Genet Evol 7:547–552

Brooks JP, Tanner BD, Josephson KL, Gerba CP, Pepper IL (2004) Bioaerosols from the land application of biosolids in the desert southwest USA. Water Sci Technol 50:7–12

Brooks K (1965) Studies in the physiological genetics of some supporessor-sensitive mutants of bacteriophages lambda. Virology 26:489–499

Brown EW, LeClerc JE, Kotewicz ML, Cebula TA (2001) Three R's of bacterial evolution: how replication, repair, and recombination frame the origin of species. Environ Mol Mutagen 38:248–260

Brown SP, Le Chat L, De Paepe M, Taddei F (2006) Ecology of microbial invasions: amplification allows virus carriers to invade more rapidly when rare. Curr Biol 16:2048–2052

Brunel AS, Guery B (2017) Multidrug resistant (or antimicrobial-resistant) pathogens – alternatives to new antibiotics? Swiss Med Wkly 147:w14553

Brüssow H, Desiere F (2006) Evolution of tailed phages-insights from comparative phage genomics. In: Calendar R, Abedon ST (eds) The bacteriophages. Oxford University Press, Oxford, pp 26–36

Bryan D, El-Shibiny A, Hobbs Z, Porter J, Kutter EM (2016) Bacteriophage T4 infection of stationary phase *E. coli*: life after log from a phage perspective. Front Microbiol 7:1391

Buckling A, Rainey PB (2002) Antagonistic coevolution between a bacterium and a bacteriophage. Proc R Soc Lond B Biol Sci 269:931–936

Bull JJ, Christensen KA, Scott C, Jack BR, Crandall CJ, Krone SM (2018) Phage-bacterial dynamics with spatial structure: self organization around phage sinks can promote increased cell densities. Antibiotics (Basel) 7(1):pii: E8

Burrowes B, Harper DR (2012) Phage therapy of non-wound infections. In: Hyman P, Abedon ST (eds) Bacteriophages in health and disease. CABI Press, Wallingford

Burrowes BH, Molineux IJ, Fralick JA (2019) Directed *in vitro* evolution of therapeutic bacteriophages: the Appelmans protocol. Viruses 11:241

Campbell A (1994) Comparative molecular biology of lambdoid phages. Annu Rev Microbiol 48:193–222

Campbell AM (2006) General aspects of lysogeny. In: Calendar R, Abedon ST (eds) The bacteriophages. Oxford University Press, Oxford, pp 66–73

Ceri H, Olson ME, Stremick C, Read RR, Morck D, Buret A (1999) The Calgary Biofilm Device: new technology for rapid determination of antibiotic susceptibilities of bacterial biofilms. J Clin Microbiol 37:1771–1776

Chakravorty P (2018) What is a signal? IEEE Signal Process Mag 35:175–177

Chan BK, Abedon ST (2012) Bacteriophage adaptation, with particular attention to issues of phage host range. In: Quiberoni A, Reinheimer J (eds) Bacteriophages in dairy processing. Nova Science Publishers, Hauppauge/New York, pp 25–52

Chan BK, Abedon ST (2015) Bacteriophages and their enzymes in biofilm control. Curr Pharm Des 21:85–99

Chan BK, Turner PE, Kim S, Mojibian HR, Elefteriades JA, Narayan D (2018) Phage treatment of an aortic graft infected with *Pseudomonas aeruginosa*. Evol Med Public Health 1:60–66

Chang RYK, Wallin M, Lin Y, Leung SSY, Wang H, Morales S, Chan HK (2018) Phage therapy for respiratory infections. Adv Drug Deliv Rev 133:76–86

Chattopadhyay D, Puls RW (2000) Forces dictating colloidal interactions between viruses and soil. Chemosphere 41:1279–1286
Chen Y, Golding I, Sawai S, Guo L, Cox EC (2005) Population fitness and the regulation of *Escherichia coli* genes by bacterial viruses. PLoS Biol 3:e229
Christie GE, Allison HA, Kuzio J, McShan M, Waldor MK, Kropinski AM (2012) Prophage-induced changes in cellular cytochemistry and virulence. In: Hyman P, Abedon ST (eds) Bacteriophages in health and disease. CABI Press, Wallingford, pp 33–60
Chróst RJ (1991) Aquatic microbial ecology. Springer, New York
Clark JM (2005) Microbe-laden aerosols. Microbiology Today, November, pp 172–173
Cochran PK, Kellogg CA, Paul JH (1998) Prophage induction of indigenous marine lysogenic bacteria by environmental pollutants. Mar Ecol Prog Ser 164:125–133
Cohan FM, Roberts MS, King EC (1991) The potential for genetic exchange by transformation within a natural-population of *Bacillus subtilis*. Evol Int J Org Evol 45:1393–1421
Colavecchio A, Cadieux B, Lo A, Goodridge LD (2017) Bacteriophages contribute to the spread of antibiotic resistance genes among foodborne pathogens of the *Enterobacteriaceae* family – a review. Front Microbiol 8:1108
Colegrave N (2002) Sex releases the speed limit on evolution. Nature (London) 420:664–666
Curtright AJ, Abedon ST (2011) Phage therapy: emergent property pharmacology. J Bioanalyt Biomed S3:010
Darch SE, Kragh KN, Abbott EA, Bjarnsholt T, Bull JJ, Whiteley M (2017) Phage inhibit pathogen dissemination by targeting bacterial migrants in a chronic infection model. MBio 8(2):pii: e00240-17
David HL, Clavel S, Clement F, Moniz-Pereira J (1980) Effects of antituberculosis and antileprosy drugs on mycobacteriophage D29 growth. Antimicrob Agents Chemother 18:357–359
Davies CM, Logan MR, Rothwell VJ, Krogh M, Ferguson CM, Charles K, Deere DA, Ashbolt NJ (2006) Soil inactivation of DNA viruses in septic seepage. J Appl Microbiol 100:365–374
Davis JA, Farrah SR, Wilkie AC (2006) Adsorption of viruses to soil: impact of anaerobic treatment. Water Sci Technol 54:161–167
Day MJ (2004) Transformation. In: Miller RV, Day MJ (eds) Microbial evolution: gene establishment, survival, and exchange. ASM Press, Washington, DC, pp 158–172
Day MJ, Miller RV (2008) Phage ecology of terrestrial environments. In: Abedon ST (ed) Bacteriophage ecology. Cambridge University Press, Cambridge, pp 281–301
de Jonge PA, Nobrega FL, Brouns SJJ, Dutilh BE (2019) Molecular and evolutionary determinants of bacteriophage host range. Trends Microbiol 27:51–63
De Sordi L, Lourenco M, Debarbieux L (2019) "I will survive": a tale of bacteriophage-bacteria coevolution in the gut. Gut Microbes 10:92–99
Dedrick RM, Guerrero-Bustamante CA, Garlena RA, Russell DA, Ford K, Harris K, Gilmour KC, Soothill J, Jacobs-Sera D, Schooley RT, Hatfull GF, Spencer H (2019) Engineered bacteriophages for treatment of a patient with a disseminated drug-resistant *Mycobacterium abscessus*. Nat Med 25:730–733
Dennehy JJ, Friedenberg NA, Yang YW, Turner PE (2006) Bacteriophage migration via nematode vectors: host-parasite-consumer interactions in laboratory microcosms. Appl Environ Microbiol 72:1974–1979
Diaz-Munoz SL, Koskella B (2014) Bacteria-phage interactions in natural environments. Adv Appl Microbiol 89:135–183
Diggle SP, Gardner A, West SA, Griffin AS (2007) Evolutionary theory of bacterial quorum sensing: when is a signal not a signal? Philos Trans R Soc Lond Ser B Biol Sci 362:1241–1249
Doermann AH (1948) Lysis and lysis inhibition with Escherichia coli bacteriophage. J Bacteriol 55:257–275
Doolittle WF (1998) You are what you eat: a gene transfer ratchet could account for bacterial genes in eukaryotic nuclear genomes. Trends Genet 14:307–311
Doron S, Melamed S, Ofir G, Leavitt A, Lopatina A, Keren M, Amitai G, Sorek R (2018) Systematic discovery of antiphage defense systems in the microbial pangenome. Science 359(6379):pii: eaar4120

Dou C, Xiong J, Gu Y, Yin K, Wang J, Hu Y, Zhou D, Fu X, Qi S, Zhu X, Yao S, Xu H, Nie C, Liang Z, Yang S, Wei Y, Cheng W (2018) Structural and functional insights into the regulation of the lysis-lysogeny decision in viral communities. Nat Microbiol 3:1285–1294

Drake JW, Ripley LS (1994) Induced mutagenesis and isolation of T4 mutants. In: Karam JD (ed) Molecular biology of bacteriophage T4. ASM Press, Washington, pp 447–451

Duboise SM, Moore BE, Sorber CA, Sagik BP (1979) Viruses in soil systems. In: Isenberg HD (ed) CRC critical reviews in microbiology. CRC Press, Boca Raton, pp 245–285

Duffy S, Turner PE (2008) Introduction to phage evolutionary biology. In: Abedon ST (ed) Bacteriophage ecology. Cambridge University Press, Cambridge, pp 147–176

Duffy S, Turner PE, Burch CL (2006) Pleiotropic costs of niche expansion in the RNA bacteriophage ϕ6. Genetics 172:751–757

Dy RL, Richter C, Salmond GP, Fineran PC (2014) Remarkable mechanisms in microbes to resist phage infections. Annu Rev Virol 1:307–331

Elena SF, Sanjuán R (2003) Climb every mountain? Science (New York, NY) 302:2074–2075

Erez Z, Steinberger-Levy I, Shamir M, Doron S, Stokar-Avihail A, Peleg Y, Melamed S, Leavitt A, Savidor A, Albeck S, Amitai G, Sorek R (2017) Communication between viruses guides lysis-lysogeny decisions. Nature (London) 541:488–493

Eriksen RS, Svenningsen SL, Sneppen K, Mitarai N (2018) A growing microcolony can survive and support persistent propagation of virulent phages. Proc Natl Acad Sci U S A 115:337–342

Errington J (2003) Regulation of endospore formation in *Bacillus subtilis*. Nat Rev Microbiol 1:117–126

Espeland EM, Lipp EK, Huq A, Colwell RR (2004) Polylysogeny and prophage induction by secondary infection in *Vibrio cholerae*. Environ Microbiol 6:760–763

Evans KJ, Hobley L, Lambert C, Sockett RE (2007) *Bdellovibrio*: lone hunter "cousin" of the "pack hunting" myxobacteria. In: Whitworth DE (ed) Myxobacteria: multicellularity and differentiation. ASM Press, Washington, DC, pp 351–362

Fellous S, Salvaudon L (2009) How can your parasites become your allies? Trends Parasitol 25:62–66

Ferguson CM, Davies CM, Kaucner C, Krogh M, Rodehutskors J, Deere DA, Ashbolt NJ (2007) Field scale quantification of microbial transport from bovine faeces under simulated rainfall events. J Water Health 5:83–95

Fernández L, Rodríguez A, García P (2018) Phage or foe: an insight into the impact of viral predation on microbial communities. ISME J 12:1171–1179

Ferry T, Leboucher G, Fevre C, Herry Y, Conrad A, Josse J, Batailler C, Chidiac C, Medina M, Lustig S, Laurent F (2018) Salvage debridement, antibiotics and implant retention ("DAIR") with local injection of a selected cocktail of bacteriophages: is it an option for an elderly patient with relapsing *Staphylococcus aureus* prosthetic-joint infection? Open Forum Infect Dis 5:ofy269

Fish R, Kutter E, Wheat G, Blasdel B, Kutateladze M, Kuhl S (2016) Bacteriophage treatment of intransigent diabetic toe ulcers: a case series. J Wound Care 25(Suppl 7):S27–S33

Fisher MB, Love DC, Schuech R, Nelson KL (2011) Simulated sunlight action spectra for inactivation of MS2 and PRD1 bacteriophages in clear water. Environ Sci Technol 45:9249–9255

Fisher RA, Gollan B, Helaine S (2017) Persistent bacterial infections and persister cells. Nat Rev Microbiol 15:453–464

Fitch WM (2000) Homology a personal view on some of the problems. Trends Genet 16:227–231

Flannery J, Rajko-Nenow P, Keaveney S, O'Flaherty V, Dore W (2013) Simulated sunlight inactivation of norovirus and FRNA bacteriophage in seawater. J Appl Microbiol 115(3):915–922

Forde SE, Thompson JN, Bohannan BJM (2004) Adaptation varies through space and time in a coevolving host–parasitoid interaction. Nature (London) 431:841–844

Fry BA (1959) Conditions for the infection of *Escherichia coli* with lambda phage and for the establishment of lysogeny. J Gen Microbiol 21:676–684

Fuhrman JA (1999) Marine viruses and their biogeochemical and ecological effects. Nature (London) 399:541–548

Fukuyo M, Sasaki A, Kobayashi I (2012) Success of a suicidal defense strategy against infection in a structured habitat. Sci Rep 2:238

Gallego Del SF, Penades JR, Marina A (2019) Deciphering the molecular mechanism underpinning phage arbitrium communication systems. Mol Cell 74:59–72

Gallet R, Shao Y, Wang I-N (2009) High adsorption rate is detrimental to bacteriophage fitness in a biofilm-like environment. BMC Evol Biol 9:241

Gama JA, Reis AM, Domingues I, Mendes-Soares H, Matos AM, Dionisio F (2013) Temperate bacterial viruses as double-edged swords in bacterial warfare. PLoS One 8:e59043

Ghosh D, Roy K, Williamson KE, Srinivasiah S, Wommack KE, Radosevich M (2009) Acyl-homoserine lactones can induce virus production in lysogenic bacteria: an alternative paradigm for prophage induction. Appl Environ Microbiol 75:7142–7152

Ghosh D, Roy K, Williamson KE, White DC, Wommack KE, Sublette KL, Radosevich M (2008) Prevalence of lysogeny among soil bacteria and presence of 16S rRNA and trzN genes in viral-community DNA. Appl Environ Microbiol 74:495–502

Gill JJ, Abedon ST (2003) Bacteriophage ecology and plants. APSnet feature. http://www.apsnet.org/online/feature/phages/

Gogarten JP (1994) Which is the most conserved group of proteins? Homology-orthology, paralogy, xenology, and the fusion of independent lineages. J Mol Evol 39:541–543

González JM, Suttle CA (1993) Grazing by marine nanoflagellates on viruses and virus-sized particles: ingestion and digestion. Mar Ecol Prog Ser 94:1–10

Gupta P, Singh HS, Shukla VK, Nath G, Bhartiya SK (2019) Bacteriophage therapy of chronic nonhealing wound: clinical study. Int J Low Extrem Wounds 18(2):171–175. https://doi.org/10.1177/1534734619835115

Haaber J, Leisner JJ, Cohn MT, Catalan-Moreno A, Nielsen JB, Westh H, Penades JR, Ingmer H (2016) Bacterial viruses enable their host to acquire antibiotic resistance genes from neighbouring cells. Nat Commun 7:13333

Hadas H, Einav M, Fishov I, Zaritsky A (1997) Bacteriophage T4 development depends on the physiology of its host *Escherichia coli*. Microbiology 143:179–185

Hagens S, Habel A, Blasi U (2006) Augmentation of the antimicrobial efficacy of antibiotics by filamentous phage. Microb Drug Resist 12:164–168

Hannan S, Ready D, Jasni AS, Rogers M, Pratten J, Roberts AP (2010) Transfer of antibiotic resistance by transformation with eDNA within oral biofilms. FEMS Immunol Med Microbiol 59:345–349

Hansen MF, Svenningsen SL, Roder HL, Middelboe M, Burmolle M (2019) Big impact of the tiny: bacteriophage-bacteria interactions in biofilms. Trends Microbiol 27:739

Hargreaves KR, Kropinski AM, Clokie MR (2014) What does the talking?: quorum sensing signalling genes discovered in a bacteriophage genome. PLoS One 9:e85131

Harms A, Diard M (2019) Crowd controlled-host quorum sensing drives phage decision. Cell Host Microbe 25:179–181

Harper DR (2013) Biological control by microorganisms. eLS. Wiley, Chichester. https://doi.org/10.1002/9780470015902.a0000344.pub3

Hassen A, Jamoussi F, Saidi N, Mabrouki Z, Fakhfakh E (2003) Microbial and copper adsorption by smectitic clay – an experimental study. Environ Technol 24:1117–1127

Häusler T (2006) Viruses vs. superbugs: a solution to the antibiotics crisis. Macmillan, New York

Hay ID, Lithgow T (2019) Filamentous phages: masters of a microbial sharing economy. EMBO Rep 20:e47427

Heineman RH, Springman R, Bull JJ (2008) Optimal foraging by bacteriophages through host avoidance. Am Nat 171:E150–E157

Hendrickx L, Hausner M, Wuertz S (2003) Natural genetic transformation in monoculture *Acinetobacter* sp. strain BD413 biofilms. Appl Environ Microbiol 69:1721–1727

Hendrix RW (2002) Bacteriophages: evolution of the majority. Theor Popul Biol 61:471–480

Hendrix RW (2008) Phage evolution. In: Abedon ST (ed) Bacteriophage ecology. Cambridge University Press, Cambridge, pp 177–194

Hendrix RW, Hatfull GF, Smith MCM (2003) Bacteriophages with tails: chasing their origins and evolution. Res Microbiol 154:253–257

Hendrix RW, Lawrence JG, Hatfull GF, Casjens S (2000) The origins and ongoing evolution of viruses. Trends Microbiol 8:504–508

Hendrix RW, Smith MCM, Burns RN, Ford ME, Hatfull GF (1999) Evolutionary relationships among diverse bacteriophages and prophages: all the world's a phage. Proc Natl Acad Sci U S A 96:2192–2197

Hendrix RW, Smith MCM, Burns RN, Ford ME, Hatfull GF (2002) Evolutionary relationships among diverse bacteriophages and prophages: all the world's a phage. In: Syvanen M, Kado CI (eds) Horizontal gene transfer. Academic, San Diego, pp 133–140

Herron MD, Hackett JD, Aylward FO, Michod RE (2009) Triassic origin and early radiation of multicellular volvocine algae. Proc Natl Acad Sci U S A 106:3254–3258

Herskowitz I, Banuett F (1984) Interaction of phage, host, and environmental factors in governing the λ lysis-lysogeny decision. In: Chopra VL, Joshi BC, Sharma RP, Bansal HC (eds) Genetics, new frontiers: proceedings of the XV international congress of genetics, vol I. Oxford and I.B.H, New Delhi, pp 59–73

Hewson I, Fuhrman JA (2003) Viriobenthos production and virioplankton sorptive scavenging. Microb Ecol 46:337–347

Hobbs Z, Abedon ST (2016) Diversity of phage infection types and associated terminology: the problem with 'Lytic or lysogenic'. FEMS Microbiol Lett 363:fnw047

Hoffman DB Jr, Rubenstein I (1968) Physical studies of lysogeny. I. Properties of intracellular parental bacteriophage DNA from λ-infected sensitive bacteria. J Mol Biol 35:375–399

Holder KK, Bull JJ (2001) Profiles of adaptation in two similar viruses. Genetics 159:1393–1404

Howard-Varona C, Hargreaves KR, Abedon ST, Sullivan MB (2017) Lysogeny in nature: mechanisms, impact and ecology of temperate phages. ISME J 11:1511–1520

Hoyle N, Zhvaniya P, Balarjishvili N, Bolkvadze D, Nadareishvili L, Nizharadze D, Wittmann J, Rohde C, Kutateladze M (2018) Phage therapy against *Achromobacter xylosoxidans* lung infection in a patient with cystic fibrosis: a case report. Res Microbiol 169:540–542

Hu J, Miyanaga K, Tanji Y (2010) Diffusion properties of bacteriophages through agarose gel membrane. Biotechnol Prog 26:1213–1221

Hughes KA, Sutherland IW, Jones MV (1998) Biofilm susceptibility to bacteriophage attack: the role of phage-borne polysaccharide depolymerase. Microbiology 144:3039–3047

Hyman P (2019) Phages for phage therapy: isolation, characterization, and host range breadth. Pharmaceuticals (Basel) 12(1):pii: E35

Hyman P, Abedon ST (2018) Viruses of microorganisms. Caister Academic Press, Norwich

Hyman P, Abedon ST (2008) Phage ecology of bacterial pathogenesis. In: Abedon ST (ed) Bacteriophage ecology. Cambridge University Press, Cambridge, pp 353–385

Hyman P, Abedon ST (2010) Bacteriophage host range and bacterial resistance. Adv Appl Microbiol 70:217–248

Hyman P, Abedon ST (2012) Smaller fleas: viruses of microorganisms. Scientifica 2012:734023

Igler C, Abedon ST (2019) Commentary: a host-produced quorum-sensing autoinducer controls a phage lysis-lysogeny decision. Front Microbiol 10:1171

Iranzo J, Lobkovsky AE, Wolf YI, Koonin EV (2015) Immunity, suicide or both? Ecological determinants for the combined evolution of anti-pathogen defense systems. BMC Evol Biol 15:43

Ito S-I, Nishimune T, Abe M, Kimoto M, Hayashi R (1986) Bacteriocinlike killing action of a temperate bacteriophage ϕBA1 of *Bacillus aneurinolyticus*. J Virol 59:103–111

Jacquet S, Zhong X, Peduzzi P, Thingstad TF, Parikka KJ, Weinbauer MG (2018) Virus interactions in the aquatic world. In: Hyman P, Abedon ST (eds) Viruses of microorganisms. Caister Academic Press, Norwich, pp 115–141

Jawetz E, Gunnison JB, Speck RS, Coleman VR (1951) Studies on antibiotic synergism and antagonism; the interference of chloramphenicol with the action of penicillin. AMA Arch Intern Med 87:349–359

Jiang SC, Paul JH (1996) Occurence of lysogenic bacteria in marine microbial communities as determined by prophage induction. Mar Ecol Prog Ser 142:27–38

Jiang SC, Paul JH (1998) Gene transfer by transduction in the marine environment. Appl Environ Microbiol 64:2780–2787

Joh RI, Weitz JS (2011) To lyse or not to lyse: transient-mediated stochastic fate determination in cells infected by bacteriophages. PLoS Comput Biol 7:e1002006

Johnson LJ, Koufopanou V, Goddard MR, Hetherington R, Schafer SM, Burt A (2004) Population genetics of the wild yeast *Saccharomyces paradoxus*. Genetics 166:43–52

Jolivet-Gougeon A, Bonnaure-Mallet M (2014) Biofilms as a mechanism of bacterial resistance. Drug Discov Today Technol 11:49–56

Karlovsky P (2008) Secondary metabolites in soil ecology. In: Karlovsky P (ed) Secondary metabolites in soil ecology. Springer, Berlin, pp 1–19

Knezevic P, Sabo VA (2019) Combining bacteriophages with other antibacterial agents to combat bacteria. In: Miedzybrodzki R, Borysowski J (eds) Górski A. Phage Therapy, A Practical Approach. Springer, pp 257–293

Keeling PJ, Palmer JD (2008) Horizontal gene transfer in eukaryotic evolution. Nat Rev Genet 9:605–618

Kerr B, West J, Bohannan BJM (2008) Bacteriophage: models for exploring basic principles of ecology. In: Abedon ST (ed) Bacteriophage ecology. Cambridge University Press, Cambridge, pp 31–63

Koch AL (1960) Encounter efficiency of coliphage-bacterium interaction. Biochim Biophys Acta 39:311–318

Korona R, Levin BR (1993) Phage-mediated selection and the evolution and maintenance of restriction-modification. Evol Int J Org Evol 47:556–575

Kourilsky P (1973) Lysogenization by bacteriophage lambda. I. Multiple infection and the lysogenic response. Mol Gen Genet 122:183–195

Kouzel N, Oldewurtel ER, Maier B (2015) Gene transfer efficiency in gonococcal biofilms: role of biofilm age, architecture, and pilin antigenic variation. J Bacteriol 197:2422–2431

Krisch HM (2003) The view from Les Treilles on the origins, evolution and diversity of viruses. Res Microbiol 154:227–229

Krone SM, Abedon ST (2008) Modeling phage plaque growth. In: Abedon ST (ed) Bacteriophage ecology. Cambridge University Press, Cambridge, pp 415–438

Kutter E (2005) Phage therapy: bacteriophages as natural, self-limiting antibiotics. In: Pizzorno W (ed) Textbook of natural medicine, 3rd edn. Churchill Livingston, St. Louis, pp 1147–1161

Kutter E, De Vos D, Gvasalia G, Alavidze Z, Gogokhia L, Kuhl S, Abedon ST (2010) Phage therapy in clinical practice: treatment of human infections. Curr Pharm Biotechnol 11:69–86

Kutter E, Kellenberger E, Carlson K, Eddy S, Neitzel J, Messinger L, North J, Guttman B (1994) Effects of bacterial growth conditions and physiology on T4 infection. In: Karam JD, Kutter E, Carlson K, Guttman B (eds) The molecular biology of bacteriophage T4. ASM Press, Washington, DC, pp 406–418

Kutter EM, Kuhl SJ, Abedon ST (2015) Re-establishing a place for phage therapy in western medicine. Future Microbiol 10:685–688

Labrie SJ, Samson JE, Moineau S (2010) Bacteriophage resistance mechanisms. Nat Rev Microbiol 8:317–327

Lacroix-Gueu P, Briandet R, Lévêque-Fort S, Bellon-Fontaine MN, Fountaine-Aupart MP (2005) In situ measurements of viral particles diffusion inside mucoid biofilms. C R Biol 328:1065–1072

Law N, Logan C, Yung G, CLL F, Lehman SM, Morales S, Rosas F, Gaidamaka A, Bilinsky I, Grint P (2019) Successful adjunctive use of bacteriophage therapy for treatment of multidrug-resistant *Pseudomonas aeruginosa* infection in a cystic fibrosis patient. Infection 47:665–668

Lawrence JG, Hendrickson H (2003) Lateral gene transfer: when will adolescence end? Mol Microbiol 50:739–749

Lawrence JG, Hendrix RW, Casjens S (2001) Where are the pseudogenes in bacterial genomes? Trends Microbiol 9:535–540

Lawrence JG, Ochman H (1997) Amelioration of bacterial genomes: rates of change and exchange. J Mol Evol 44:383–397

Lehman SM (2018) Bacteriophage diversity. In: Hyman P, Abedon ST (eds) Viruses of microorganisms. Caister Academic Press, Norwich, pp 145–165

Lemire S, Figueroa-Bossi N, Bossi L (2011) Bacteriophage crosstalk: coordination of prophage induction by trans-acting antirepressors. PLoS Genet 7:e1002149

Levin BR (1988) Frequency-dependent selection in bacterial populations. Philos Trans R Soc Lond Ser B Biol Sci 319:459–472

Levin BR, Bergstrom CT (2000) Bacteria are different: observations, interpretations, speculations, and opinions about the mechanisms of adaptive evolution in prokaryotes. Proc Natl Acad Sci U S A 97:6981–6985

Levin BR, Bull JJ (2004) Population and evolutionary dynamics of phage therapy. Nat Rev Microbiol 2:166–173

Levine M (1957) Mutations in the temperate phage P22 and lysogeny in *Salmonella*. Virology 3:22–41

Li YH, Lau PC, Lee JH, Ellen RP, Cvitkovitch DG (2001) Natural genetic transformation of *Streptococcus mutans* growing in biofilms. J Bacteriol 183:897–908

Lieb M (1953) The establishment of lysogenicity in Escherichia coli. J Bacteriol 65:642–651

Lin H, Paff ML, Molineux IJ, Bull JJ (2018) Antibiotic therapy using phage depolymerases: robustness across a range of conditions. Viruses 10:pii: E622

Little JW (2005) Lysogeny, prophage induction, and lysogenic conversion. In: Waldor MK, Friedman DI, Adhya SL (eds) Phages: their role in bacterial pathogenesis and biotechnology. ASM Press, Washington, DC, pp 37–54

Loc-Carrillo C, Abedon ST (2011) Pros and cons of phage therapy. Bacteriophage 1:111–114

Loc-Carrillo C, Wu S, Beck JP (2012) Phage therapy of wounds and related purulent infections. In: Hyman P, Abedon ST (eds) Bacteriophages in health and disease. CABI Press, Wallingford, pp 185–202

Lorenz MG, Wackernagel W (1994) Bacterial gene transfer by natural genetic transformation in the environment. Microbiol Rev 58:563–602

Los J, Zielinka S, Krajewska A, Filipiak M, Malachowska A, Kwasnicka K, Lobocka MB, Szkop U, Los M (2019) Temperate phages, prophages and lysogeny. In: Harper DR, Abedon ST, Burrowes BH, McConville M (eds) Bacteriophages: biology, technology, therapy. Springer, Cham

Los M, Kuzio J, McConnell MR, Kropinski AM, Wegrzyn G, Christie GE (2010) Lysogenic conversion in bacteria of importance to the food industry. In: Sabour PM, Griffiths MW (eds) Bacteriophages in the control of food- and waterborne pathogens. ASM Press, Washington, DC, pp 157–198

Los M, Wegrzyn G (2012) Pseudolysogeny. Adv Virus Res 82:339–349

Lwoff A (1953) Lysogeny. Bacteriol Rev 17:269–337

Macia MD, Rojo-Molinero E, Oliver A (2014) Antimicrobial susceptibility testing in biofilm-growing bacteria. Clin Microbiol Infect 20:981–990

Martens E, Demain AL (2017) The antibiotic resistance crisis, with a focus on the United States. J Antibiot (Tokyo) 70:520–526

Mavrich TN, Hatfull GF (2017) Bacteriophage evolution differs by host, lifestyle and genome. Nat Microbiol 2:17112

Mavrich TN, Hatfull GF (2019) Evolution of superinfection immunity in Cluster A mycobacteriophages. MBio 10:e00971–e00919

Maxwell KL (2019) Phages tune in to host cell quorum sensing. Cell 176:7–8

McKay LD, Harton AD, Wilson GV (2002) Influence of flow rate on transport of bacteriophage in shale saprolite. J Environ Qual 31:1095–1105

Miller RV, Day M (2008) Contribution of lysogeny, pseudolysogeny, and starvation to phage ecology. In: Abedon ST (ed) Bacteriophage ecology. Cambridge University Press, Cambridge, pp 114–143

Molin S, Tolker-Nielsen T (2003) Gene transfer occurs with enhanced efficiency in biofilms and induces enhanced stabilisation of the biofilm structure. Curr Opin Biotechnol 14:255–261

Morita RY (1997) Bacteria in oligotrophic environments: starvation-survival lifestyle. Chapman and Hall Ltd, London

Morrisette T, Kebriaei R, Lev KL, Morales S, Rybak MJ (2020) Bacteriophage Therapeutics: A Primer for Clinicians on Phage-Antibiotic Combinations. Pharmacotherapy: The Journal ofHuman Pharmacology and Drug Therapy 40 (2):153–168
Murray AG, Jackson GA (1992) Viral dynamics: a model of the effects of size, shape, motion, and abundance of single-celled planktonic organisms and other particles. Mar Ecol Prog Ser 89:103–116
Murray AG, Jackson GA (1993) Viral dynamics II: a model of the interaction of ultraviolet light and mixing processes on virus survival in seawater. Mar Ecol Prog Ser 102:105–114
Nobrega FL, Costa AR, Kluskens LD, Azeredo J (2015) Revisiting phage therapy: new applications for old resources. Trends Microbiol 23:185–191
Ocampo PS, Lazar V, Papp B, Arnoldini M, Abel Zur WP, Busa-Fekete R, Fekete G, Pal C, Ackermann M, Bonhoeffer S (2014) Antagonism between bacteriostatic and bactericidal antibiotics is prevalent. Antimicrob Agents Chemother 58:4573–4582
Ogunseitan OA (2008) Genetic transduction in freshwater ecosystems. Freshw Biol 53:1228–1239
Olsen I (2015) Biofilm-specific antibiotic tolerance and resistance. Eur J Clin Microbiol Infect Dis 34:877–886
Oppenheim VA, Willsky AS (1997) Signals & systems. Prentice Hall, Engelwood Cliffs
Pantastico-Caldas M, Duncan KE, Istock CA, Bell JA (1992) Population dynamics of bacteriophage and *Bacillus subtilis* in soil. Ecology 73:1888–1902
Paolozzi L, Ghelardini P (2006) The bacteriophage Mu. In: Calendar R, Abedon ST (eds) The bacteriophages. Oxford University Press, Oxford, pp 469–496
Paul JH (2008) Prophages in marine bacteria: dangerous molecular time bombs or the key to survival in the seas? ISME J 2:579–589
Pedulla ML, Ford ME, Houtz JM, Karthikeyan T, Wadsworth C, Lewis JA, Jacobs-Sera D, Falbo J, Gross J, Pannunzio NR, Brucker W, Kumar V, Kandasamy J, Keenan L, Bardarov S Jr, Kriakov J, Lawrence JG, Jacobs WR Jr, Hendrix RW, Hatfull GF (2003) Origins of highly mosaic mycobacteriophage genomes. Cell 113:171–182
Pietramellara G, Ascher J, Borgogni F, Ceccherini MT, Guerri G, Nannipieri P (2009) Extracellular DNA in soil and sediment: fate and ecological relevance. Biol Fertil Soils 45:219–235
Pires DP, Oliveira H, Melo LD, Sillankorva S, Azeredo J (2016) Bacteriophage-encoded depolymerases: their diversity and biotechnological applications. Appl Microbiol Biotechnol 100:2141–2151
Popa O, Dagan T (2011) Trends and barriers to lateral gene transfer in prokaryotes. Curr Opin Microbiol 14:615–623
Popa O, Landan G, Dagan T (2017) Phylogenomic networks reveal limited phylogenetic range of lateral gene transfer by transduction. ISME J 11:543–554
Presloid JB, Ebendick-Corp ZS, Novella IS (2008) Antagonistic pleiotropy involving promoter sequences in a virus. J Mol Biol 382:342–352
Rabinovitch A, Aviram I, Zaritsky A (2003) Bacterial debris – an ecological mechanism for coexistence of bacteria and their viruses. J Theor Biol 224:377–383
Reche I, D'Orta G, Mladenov N, Winget DM, Suttle CA (2018) Deposition rates of viruses and bacteria above the atmospheric boundary layer. ISME J 12:1154–1162
Reteno DGI, Bajrai LH, Aherfi S, Colson P, La Scola B (2018) Protozoal giant viruses. In: Hyman P, Abedon ST (eds) Viruses of microorganisms. Caister Academic Press, Norwich, pp 245–269
Reyes-Robles T, Dillard RS, Cairns LS, Silva-Valenzuela CA, Housman M, Ali A, Wright ER, Camilli A (2018) *Vibrio cholerae* outer membrane vesicles inhibit bacteriophage infection. J Bacteriol 200:e00792–e00717
Rodriguez-Brito B, Li L, Wegley L, Furlan M, Angly F, Breitbart M, Buchanan J, Desnues C, Dinsdale E, Edwards R, Felts B, Haynes M, Liu H, Lipson D, Mahaffy J, Martin-Cuadrado AB, Mira A, Nulton J, Pasic L, Rayhawk S, Rodriguez-Mueller J, Rodriguez-Valera F, Salamon P, Srinagesh S, Thingstad TF, Tran T, Thurber RV, Willner D, Youle M, Rohwer F (2010) Viral and microbial community dynamics in four aquatic environments. ISME J 4:739–751
Rodriguez-Verdugo A, Carrillo-Cisneros D, Gonzalez-Gonzalez A, Gaut BS, Bennett AF (2014) Different tradeoffs result from alternate genetic adaptations to a common environment. Proc Natl Acad Sci U S A 111:12121–12126

Ross A, Ward S, Hyman P (2016) More is better: selecting for broad host range bacteriophages. Front Microbiol 7:1352

Samson JE, Magadan AH, Sabri M, Moineau S (2013) Revenge of the phages: defeating bacterial defences. Nat Rev Microbiol 11:675–687

Scanlan PD, Buckling A, Hall AR (2015) Experimental evolution and bacterial resistance: (co) evolutionary costs and trade-offs as opportunities in phage therapy research. Bacteriophage 5:e1050153

Schneider CL (2017) Bacteriophage-mediated horizontal gene transfer: transduction. In: Harper DR, Abedon ST, Burrowes B, McConville M (eds) Bacteriophages: biology, technology, therapy. Springer, New York City. https://link.springer.com/referenceworkentry/10.1007/978-3-319-40598-8_4-1

Schooley RT, Biswas B, Gill JJ, Hernandez-Morales A, Lancaster J, Lessor L, Barr JJ, Reed SL, Rohwer F, Benler S, Segall AM, Taplitz R, Smith DM, Kerr K, Kumaraswamy M, Nizet V, Lin L, McCauley MD, Strathdee SA, Benson CA, Pope RK, Leroux BM, Picel AC, Mateczun AJ, Cilwa KE, Regeimbal JM, Estrella LA, Wolfe DM, Henry MS, Quinones J, Salka S, Bishop-Lilly KA, Young R, Hamilton T (2017) Development and use of personalized bacteriophage-based therapeutic cocktails to treat a patient with a disseminated resistant *Acinetobacter baumannii* infection. Antimicrob Agents Chemother 61:e00954–e00917

Segall AM, Roach DR, Strathdee SA (2019) Stronger together? Perspectives on phage-antibiotic synergy in clinical applications of phage therapy. Current Opinion in Microbiology 51:46–50

Short SM, Staniewski MA, Chaban YV, Long AM, Wang D (2018) Diversity of viruses infecting eukaryotic algae. In: Hyman P, Abedon ST (eds) Viruses of microorganisms. Caister Academic Press, Norwich, pp 211–243

Sillankorva S, Azeredo J (2014) The use of bacteriophages and bacteriophage-derived enzymes for clinically relevant biofilm control. In: Borysowski J, Miedzybrodzki R, Górski A (eds) Phage therapy: current research and applications. Caister Academic Press, Norfolk

Silpe JE, Bassler BL (2019) A host-produced quorum-sensing autoinducer controls a phage lysis-lysogeny decision. Cell 176:268–280

Silver-Mysliwiec T, Bramucci MG (1990) Bacteriophage-enhanced sporulation: comparison of the spore converting bacteriophages PMB12 and SP10. J Bacteriol 172:1948–1953

Simmons M, Drescher K, Nadell CD, Bucci V (2017) Phage mobility is a core determinant of phage-bacteria coexistence in biofilms. ISME J 12:531–543

Sisler FD (1940) The transmission of bacteriophage by mosquitoes. University of Maryland, College Park

Six E (1961) Inheritance of prophage P2 in superinfection experiments. Virology 14:220–233

Sonenshein AL (2006) Bacteriophages: how bacterial spores capture and protect phage DNA. Curr Biol 16:R14–R16

Song I, Choi CY, O'Shaughnessy S, Gerba CP (2005) Effects of temperature and moisture on coliphage PRD-1 survival in soil. J Food Prot 68:2118–2122

Stanley SY, Maxwell KL (2018) Phage-encoded anti-CRISPR defenses. Annu Rev Genet 52:445–464

Stent GS, Wollman EL (1952) On the two step nature of bacteriophage adsorption. Biochim Biophys Acta 8:260–269

Stewart FM, Levin BR (1984) The population biology of bacterial viruses: why be temperate. Theor Popul Biol 26:93–117

Sumby P, Smith MCM (2002) Genetics of the phage growth limitation (Pgl) system of *Streptomyces coelicolor* A3(2). Mol Microbiol 44:489–500

Sutherland IW, Hughes KA, Skillman LC, Tait K (2004) The interaction of phage and biofilms. FEMS Microbiol Lett 232:1–6

Suttle CA (2007) Marine viruses – major players in the global ecosystem. Nat Rev Microbiol 5:801–812

Tagliaferri TL, Jansen M, Horz H (2019) Fighting Pathogenic Bacteria on Two Fronts: Phages and Antibiotics as Combined Strategy. Frontiers in Cellular and Infection Microbiology 9:22

Tan JSH, Reanney DC (1976) Interactions between bacteriophages and bacteria in soil. Soil Biol Biochem 8:145–150

Thingstad TF, Bratbak G, Heldal M (2008) Aquatic phage ecology. In: Abedon ST (ed) Bacteriophage ecology. Cambridge University Press, Cambridge, pp 251–280
Turner PE, Duffy S (2008) Evolutionary ecology of multi-phage infections. In: Abedon ST (ed) Bacteriophage ecology. Cambridge University Press, Cambridge, pp 195–216
Vainio EJ, Hantula J (2018) Fungal viruses. In: Hyman P, Abedon ST (eds) Viruses of microorganisms. Caister Academic Press, Norwich, pp 193–209
Van Cuyk S, Siegrist RL (2007) Virus removal within a soil infiltration zone as affected by effluent composition, application rate, and soil type. Water Res 41:699–709
Wang BY, Chi B, Kuramitsu HK (2002) Genetic exchange between *Treponema denticola* and *Streptococcus gordonii* in biofilms. Oral Microbiol Immunol 17:108–112
Waters CM, Bassler BL (2005) Quorum sensing: cell-to-cell communication in bacteria. Annu Rev Cell Dev Biol 21:319–346
Webb V, Leduc E, Spiegelman GB (1982) Burst size of bacteriophage SP82 as a function of growth rate of its host *Bacillus subtilis*. Can J Microbiol 28:1277–1280
Weinbauer MG (2004) Ecology of prokaryotic viruses. FEMS Microbiol Rev 28:127–181
Weinbauer MG, Suttle CA (1999) Lysogeny and prophage induction in coastal and offshore bacterial communities. Aquat Microb Ecol 18:217–225
Weledji EP, Weledji EK, Assob JC, Nsagha DS (2017) Pros, cons and future of antibiotics. New Horiz Transl Med 4:9–14
Wienhold SM, Lienau J, Witzenrath M (2019) Towards inhaled phage therapy in Western Europe. Viruses 11:295
Wilhelm SW, Suttle CA (1999) Viruses and nutrient cycles in the sea: viruses play critical roles in the structure and function of aquatic food webs. Bioscience 49:781–788
Wilkinson JF (1958) The extracellular polysaccharides of bacteria. Bacteriol Rev 22:46–73
Williams GC (1957) Pleiotropy, natural selection, and the evolution of senescence. Evol Int J Org Evol 11:398–411
Williams ST, Lanning S (1984) Studies of the ecology of streptomycete phage in soil. In: Ortiz-Ortiz L, Bojalil LF, Yakoleff V (eds) Biological, biochemical and biomedical aspects of actinomycetes. Academic, London, pp 473–483
Williams ST, Mortimer AM, Manchester L (1987) Ecology of soil bacteriophages. In: Goyal SM, Gerba CP, Bitton G (eds) Phage ecology. Wiley, New York, pp 157–179
Winter C, Bouvier T, Weinbauer MG, Thingstad TF (2010) Trade-offs between competition and defense specialists among unicellular planktonic organisms: the "killing the winner" hypothesis revisited. Microbiol Mol Biol Rev 74:42–57
Witzany G (2017) Key levels of biocommunication. In: Biocommunication: sign-mediated interactions between cells and organisms. World Scientific, London, pp 37–61
Witzany G (2019) Communication as the main characteristic of life. In: Handbook of astrobiology. CRC Press, Boca Raton, pp 91–105
Wommack KE, Colwell RR (2000) Virioplankton: viruses in aquatic ecosystems. Microbiol Mol Biol Rev 64:69–114
Wommack KE, Hill RT, Muller TA, Colwell RR (1996) Effects of sunlight on bacteriophage viability and structure. Appl Environ Microbiol 62:1336–1341
Wong TP, Byappanahalli M, Yoneyama B, Ray C (2008) An evaluation of the mobility of pathogen indicators, *Escherichia coli* and bacteriophage MS-2, in a highly weathered tropical soil under unsaturated conditions. J Water Health 6:131–140
Wright A, Hawkins CH, Anggård EE, Harper DR (2009) A controlled clinical trial of a therapeutic bacteriophage preparation in chronic otitis due to antibiotic-resistant *Pseudomonas aeruginosa*; a preliminary report of efficacy. Clin Otolaryngol 34:349–357
Wuertz S, Hendrickx L, Kuehn M, Rodenacker K, Hausner M (2001) In situ quantification of gene transfer in biofilms. Methods Enzymol 336:129–143
Yin J, McCaskill JS (1992) Replication of viruses in a growing plaque: a reaction-diffusion model. Biophys J 61:1540–1549
Zeph LR, Casida LJ (1986) Gram-negative versus gram-positive (actinomycete) nonobligate bacterial predators of bacteria in soil. Appl Environ Microbiol 52:819–823

Chapter 3
Phage Communication and the Ecological Implications on Microbial Interactions, Diversity, and Function

Xiaolong Liang and Mark Radosevich

Abstract Viruses are extraordinarily abundant on Earth, containing a vast reservoir of genetic diversity. Despite the numerical abundance of soil viruses, even the most basic aspects of their ecology are poorly investigated. We know very little concerning the spatial and temporal variability of viral abundance and diversity, rates of viral turnover/production, and host specificity. Viruses in other environments are known to modulate microbial communities and their host processes through viral-mediated cell lysis and/or lysogenic interactions. Thus, the ecological impact of viruses on the community level can depend on whether their reproduction is predominantly lytic or lysogenic. Evidence has emerged suggesting temperate viruses can modulate their reproductive effort (either lytic or lysogenic pathways) based on cell-cell, phage-cell, and phage-phage communication. In this review, recent progress in aspects of phage ecology are discussed with emphasis on phage-bacterial interactions at both molecular and community levels.

3.1 Introduction

In the microbial world, bacteriophages (also known as phages, viruses that infect and replicate within prokaryotes) are ubiquitous and extremely abundant in the environment. Considerable research has revealed that viruses greatly influence microbial diversity and functions (Williamson et al. 2005, 2017; Breitbart et al. 2018; Emerson et al. 2018; Liang et al. 2019c). Phages may impose top-down control over host populations through predation and drive evolutionary processes of microbial communities. Phages have diverse life histories and can be classified into two general categories, i.e., virulent and temperate. While in the extracellular stage, phages are metabolically inactive and act as colloidal particles passively transported

X. Liang · M. Radosevich (✉)
Department of Biosystems Engineering and Soil Science, The University of Tennessee, Knoxville, TN, USA
e-mail: mrad@utk.edu

G. Witzany (ed.), *Biocommunication of Phages*,
https://doi.org/10.1007/978-3-030-45885-0_3

via hydrodynamic forces in the environment. When a phage encounters a suitable host cell and initiates viral infection, the phage genome is injected into the host cell entering an intracellular stage, in which virulent phages take control of the transcription and translation machinery in the cell for virion production ultimately leading to cell lysis. In contrast, temperate phages have two general replication-cycle options (either lytic or lysogenic) in the intracellular stage, and the phage can enter the lytic cycle and eventually destroy its host, releasing progeny virions, or follow the lysogenic pathway as the phage genome integrates into its host genome and stably maintains as a prophage and replicates as the host cell divides (Feiner et al. 2015). Temperate phages have the ability to select replication strategies pursuing either the lytic or lysogenic replication pathways depending on environmental and physiological conditions of the host cell (Goldberg et al. 2014; Brum et al. 2016; Howard-Varona et al. 2017). Analysis of the mechanisms underlying the lysis-lysogeny decisions in phage communities is the basis for formative work on phage-host interactions and phage ecological roles (Weitz et al. 2008; Liang and Radosevich 2019).

The switch between lysis and lysogeny of temperate phages is critical for potential influence of phage infection on host populations and environmental processes and has aroused interest and exploration on this topic. Several hypothesized models addressing the correlations of phage abundances and replication strategies with host abundances have been previously reported. The Kill-the-Winner (KtW) model proposed that lytic infections lead to steep collapses of the most dominant bacterial populations, while lysogeny is promoted at low host density, and the paradigm is supported by a wide range of studies (Thomas et al. 2011; Weitz and Wilhelm 2012; Payet and Suttle 2013; Silpe and Bassler 2019). However, a recent study by Knowles et al. (2016) observed low viral abundance at high host densities and attributed the nonlinear relationship to suppressed lytic phage-host dynamics. The increased prevalence of lysogeny at increasing host abundance led to proposal of the Piggyback-the-Winner (PtW) model wherein temperate dynamics dominate over lytic cycles in systems with high host cell densities leading to fewer viruses and more microbes. This work subsequently sparked a debate on viral lysis and lysogeny in natural systems concerning the analysis and interpretation of largely empirical measurements (Weitz et al. 2017). Knowles et al. (2017) took a further step by exploring the correlation of chemical agent based induction estimates of environmental lysogeny with host density and proposed that the fraction of chemically inducible bacterial cells is generally insensitive to host cell densities, and concluded that more innovative methods are needed to investigate viral lysis-lysogeny decisions.

Microbial communication is important in linking microbial membership to microbially-mediated processes such as biofilm formation, pathogenicity, antibiotic production, symbiosis, conjugation, and elemental cycling. Hall et al. (2018) proposed a framework for understanding the mechanisms of microbiomes influencing the habitat in which the microbial characteristics (i.e., abundance, diversity and functional processes) are linked to the system-level processes. The cell-to-cell communication among bacteria through chemical signal molecules is well known as quorum sensing in which bacteria can regulate gene expression coordinating group behaviors based on cell population density (Waters and Bassler 2005). Quorum sensing signals may also play a role in lytic-lysogenic decisions (Ghosh et al. 2009; Patterson et al. 2016; Høyland-Kroghsbo et al. 2017). Diverse microbially released molecules serve as the language of microbial communication and are crucial in regulating group behaviors and ecological function. Phages are extremely small (normally with 20–70 nm diameter and up-to-200 nm length) and must depend on their host cell system for replication, thus phage communication if/when it occurs, most likely occurs during the infection state in host cells (Abedon 2017). Despite the tremendous efforts in exploring environmental virology, phage communication and the mechanisms of phage functioning remain underexplored making phages "dark matter" in microbial ecology (Roux et al. 2016). In this review, recent progress in phage ecology including phage distribution, diversity, molecular basis of phage communication, and influence of phage infection on microbial communities are discussed with emphasis directed on phage-bacterial interactions at both molecular and community levels.

3.2 Approaches for Ecological Investigation of Phages

The ecological importance of phages had long been overlooked, but the discovery of the exceedingly high phage abundances in aquatic environments and more recently soils has propelled a significant increase in viral ecology research. Indeed, viruses are recognized as the most abundant lifeforms on earth with an estimated global abundance or 10^{31} virions in the biosphere (Breitbart and Rohwer 2005; Williamson et al. 2017). Restricted by the inefficiency of phage extraction and counting methods, investigations of phage abundance and distribution in many terrestrial ecosystems (e.g., soil and sediment environments) has lagged behind those of aquatic environments but steady recent progress suggests that Earth's soils harbor an even greater abundance and diversity of viruses than the world's oceans (Williamson et al. 2017). Assessment of virus abundances in soils is complicated by soil heterogeneity and difficulties optimizing efficient extraction protocols applicable to a wide variety of soils. This has made assessment of geographical distributions and inter-study comparisons difficult. (Williamson et al. 2013, 2017).

Beyond direct microscopic counting, molecular methods based on structural or functional marker genes and metagenomics/metatranscriptomics have been employeed for characterizing phage alpha- and beta-diversity, functional potential, and evolutionary history. Virome sequencing has been used for assessing viral community diversity and potential ecological functions (Brum and Sullivan 2015; Paez-Espino et al. 2016; Enault et al. 2017; Segobola et al. 2018; Gregory et al. 2019; Daly et al. 2019). These efforts have revealed that the viral communities are highly diverse across all types of ecosystems and have critical roles in microbial mortality, gene transfer, auxiliary metabolic genes (AMGs) mediated metabolic reprogramming and global elemental cycling. Besides directly sequencing viral DNA samples, mining viral sequences from publicly available microbial genomes and metagenomes is also valuable (Roux et al. 2015a, b). Roux et al. (2015b) tried to resolve phage-host interactions from publicly available microbial genomic data, and their work led to identification of 12,498 microbial host-linked high-confidence viral genomes and augmentation of public virus genome databases. The authors' efforts also identified viral sequences from 13 bacterial phyla in which no viral sequences were previously reported. Notably, mining viral genomes from microbial genomes was also important in taxonomically identifying the 'unknown' sequence space in viromes. Roux et al. (2016) identified 7–38% of such virome sequences with the recovered viral sequences from microbial genomes.

While metagenomics and classical phage isolation studies have contributed to better understanding of viral ecology, identifying the hosts of most viruses remains a challenge (Tadmor et al. 2011; Allers et al. 2013). Tadmor et al. (2011) employed microfluidic digital PCR targeting a phage and bacterial gene to explore single-cell level phage-host interactions in the termite hindgut. With this method, the authors were able to find the diversity of specific viral marker alleles (terminase gene) co-localized with SSU rRNA genes and revealed genus-wide infection patterns (Tadmor et al. 2011).

For better understanding of phage-host interactions, Allers et al. (2013) developed a technique known as "phageFISH" to detect and visualize intra- and extra-cellular phage DNA. PhageFISH was optimized from the geneFISH technique that consists of gene detection (based on in situ hybridization of double-stranded DNA probes) and rRNA detection (based on hybridization of horseradish peroxidase-labelled oligonucleotide probes) (Allers et al. 2013). The phageFISH protocol can identify and quantify both the intracellular replicating phages and extracellular free phages, along with the host cells and also has a high gene detection efficiency of more than 92%. Moreover, phageFISH can show the relative abundance of infected cells and the single-cell relative extent of phage infection by measuring the area of phage signal.

Substantial progress has been made in virology studies with diverse model systems of virus-host cultivation, experimental and bioinformatic methods, and modelling incorporating viruses. However, challenges remain, and future efforts are in earnest need of better understanding of virus-host ecology. Conceptual models describing the functions of viruses in nutrient and elemental cycling require empirical studies of characterizing viral lysis released dissolved organic matter and corresponding recycling by microorganisms. Taxonomic assignment to viral sequences and annotation of virome contigs require expansion of viral genome database. Viral ecology studies in soil ecosystems, deep terrestrial biosphere and other understudied environments are needed to catalogue viral diversity and examine virus-host interaction dynamics.

3.3 Phage Abundance and Distribution

Phages contacting and subsequently infecting host cells is the basis for the propagation of phages. The opportunity and/or probability of phage-host contact varies depending on the nature of the environment and the abundance of phages and host cells. For example, phages and host cells may be freely dispersed in aquatic environments for virus-host contact and interactions. In contrast, the heterogeneous porous nature of soils can impose physical barriers to phage-host encounters. Here, attachment/detachment to/from mineral and organic surfaces of phages and hosts can dictate phage-host contact rates. Further, phages due to their much smaller size may become spatially separated and sequestered from their host cells in small pores that restrict entry to relatively larger host cells. Therefore, the pore-scale hydrology and moisture content of soil become extremely important factors affecting phage-host contact rates. The phage-host contact rate is assumed to be proportional to virus and microbial cell abundances and, thus estimates of microbial cell abundance may be used to predict phage-host contact rate and phage production (Wigington et al. 2016). Microbial host cell densities also influence phages' reproduction strategy decisions. The relationship between phages and their microbial hosts and the influence of phages on microbial community dynamics and biogeochemical processes are primarily based on the relative abundance of phages and microbial cells. Virus-to-bacterium ratio (VBR) has been used in many studies to represent the contact rates between phages and host cells and infer the strength of phages' influence in prokaryotic communities (Williamson et al. 2005, 2007; Våge et al. 2016; Wigington et al. 2016; Parikka et al. 2017; Emerson et al. 2018). Phage abundances were reported to be positively correlated with host cell densities (Brum et al. 2016; Edwards et al. 2016; Pan et al. 2017; Liang et al. 2019a); however a broad range of VBR implies a loose connection between phage abundances and microbial cell abundances, and the sublinear relationship showing a lower VBR at increasing microbial densities is attributed to a combination of exogeneous (e.g., temperature, eutrophic conditions and radiation) and endogenous factors (e.g., life history traits of viruses and susceptibility of microbes) (Knowles et al. 2016; Weitz et al. 2017; Wigington et al. 2016).

Phage abundances in the environments are influenced by a wide range of factors, and the drivers of phage abundance and distribution in ecosystems are summarized in Table 3.1. Previous studies have shown that the most common factors influencing phage abundance and distribution generally include nutrient availability, temperature, pH, season, host cell density, and human or other biological activities. The conditions in the living environment of phages change continually imposing threats on phage survival. It is important for phages to modulate their population sizes in response to changes in environmental conditions through specific mechanisms. Great efforts have been made in characterizing the mechanism of phages modulating their population dynamics, and the function of phage communication in this process has been demonstrated (Erez et al. 2017; Silpe and Bassler 2019; Liang et al. 2019b).

Table 3.1 Factors influencing phage abundance and distribution

Habitat	Factors
Ocean	Nutrient concentrations (Finke et al. 2017)
	Salinity and temperature (Junger et al. 2017)
	Prokaryote abundances (Gainer et al. 2017)
	Organic/inorganic particles (Mojica and Brussaard 2014)
	Season (Brum et al. 2016; Gainer et al. 2017)
	Ultraviolet radiation (Mojica and Brussaard 2014)
River and lake	Dissolved inorganic carbon (Keshri et al. 2017)
	Hydrodynamics (turbidity, water transparency, partial pressure of carbon dioxide), dissolved organic carbon (Almeida et al. 2015)
Sediment	Suspended solids, inorganic nutrients (Hewson et al. 2001)
	Depth in sediment cores (Danovaro et al. 2005)
	Bacterial densities (Danovaro and Serresi 2000)
	Physical-chemical conditions (e.g., temperature; Manini et al. 2008)
Soil	pH (Narr et al. 2017)
	Total nitrogen content (Narr et al. 2017)
	Moisture content (Williamson et al. 2005, 2013)
	Land use (Williamson et al. 2005)
	Season, water-table height and dissolved oxygen (Ballaud et al. 2016)
	Organic matter (Amossé et al. 2013)
	Biological activity (Amossé et al. 2013)
Aquifer & ground water	Trophic conditions (Wilhartitz et al. 2013)
	Geochemical conditions (e.g., groundwater uranium and DOC; Pan et al. 2017)
	Abundance of prokaryotic cells (Kyle et al. 2008)
Human body	Sites (Letarov and Kulikov 2009)

3.4 Collective Decisions on Phage Lysis-Lysogeny Switching

3.4.1 Phage-Phage Communication

Cell-cell communication among bacteria using chemical signal molecules (termed "autoinducers"), known as quorum sensing, enables bacteria to monitor the environmental conditions and coordinate a wide range of group behaviors (Waters and Bassler 2005; Ng and Bassler 2009). In quorum sensing, bacteria produce, secrete, detect, and respond to autoinducers that carry chemical information, and the communication ultimately synchronizes the gene expression of the group for collective behaviors. In comparison, phage communication associated with infection of bacteria has been much less studied (Abedon 2017). Phage-phage interactions are diverse and extensively present in nature (Díaz-Muñoz and Koskella 2014; Díaz-Muñoz et al. 2017; Ofir and Sorek 2018). The reproduction strategy selection is an important feature for phages, and there has been a lot of debate on the mechanism(s) of phage lysis-lysogeny decision. However, all the models proposed for predicting phage reproductive strategy and their interactions with host community (e.g., KtW and PtW) suggest that phages can sense their host cell density and make decisions accordingly (Knowles et al. 2016; Weitz et al. 2017; Silpe and Bassler 2019). Meanwhile, lysis-lysogeny decisions occur at both the individual and group levels (Weitz et al. 2008; Zeng et al. 2010). As reported in Zeng et al. (2010), each individual phage, following the infection of host cells by multiple phages, makes a decision between lysis and lysogeny based on the phage particle abundance per cell volume. Then, a precise integration of the decisions by all phages infecting the same cell leads to a group decision, in which lysogeny is only reached by a unanimous "vote" by all phage members (Zeng et al. 2010). These studies suggest that phages should have specific communication systems for making group decisions.

The first report of a phage-to-phage based communication system by Erez et al. (2017) showed that *Bacillus* phage phi3T encoded hexapeptides are produced during infection of its host cell, and the signal peptides are secreted by host *Bacillus* and accumulate in the local environment. The progeny phages can sense the population level of nearby relatives through the host *Bacillus*-secreted extracellular signals and make lysis-lysogeny decisions accordingly. The small-molecule communication system, termed "arbitrium" system, includes three phage genes (i.e., *aimP*, *aimR*, and *aimX*) that are involved in the production of the peptide signals (AimP), intracellular receptor (AimR) of the peptide, and a negative regulator (AimX) of lysogeny (Fig. 3.1a). The expression of the *aimP* and *aimR* starts immediately upon infection of *Bacillus* host cell, while the expression of *aimX* needs to be activated by AimR. AimX inhibits lysogeny of the phage leading to a lytic cycle. However, AimR, as the arbitrium receptor, can bind to the mature arbitrium molecule preventing activation of *aimX* expression. The *aimP*-encoded peptides are recognized, cleaved, and secreted outside the cell, and the released molecules are further processed into six-amino-acid-long mature arbitrium peptides by host extracellular proteases. The molecular mechanism of phage communication systems was also described in detail by Wang et al. (2018) and del Sol et al. (2019). Upon infection, phages detect the concentration of the signal peptides using the arbitrium receptor AimR and decide on whether to lyse or lysogenize the host cell (Fig. 3.1b). High concentrations of arbitrium peptides lead to lysogenic cycles in phage-infected cells, and low concentrations of arbitrium peptides direct phages to lytic reproduction pathways. Thus, the arbitrium communication between phages enables infecting phages to lysogenize and extend the infection period as prophages when potential host cells are mostly phage infected.

Erez et al. (2017) also found that the communication peptide in arbitrium systems is phage-specific and can only influence the closely related phages. Two different phage-encoded communication peptides (SAIRGA and GMPRGA) that are effective in driving infection kinetics of specific.

Bacillus phages in a different fashion have also been reported (Dou et al. 2018). These studies demonstrated that there are multiple molecular mechanisms in determining phage lysis-lysogeny decisions and subtle structural changes of phage-encoded communication peptides can lead to distinct lysis-lysogeny decision pathways. The viral arbitrium communication mechanism has important ecological

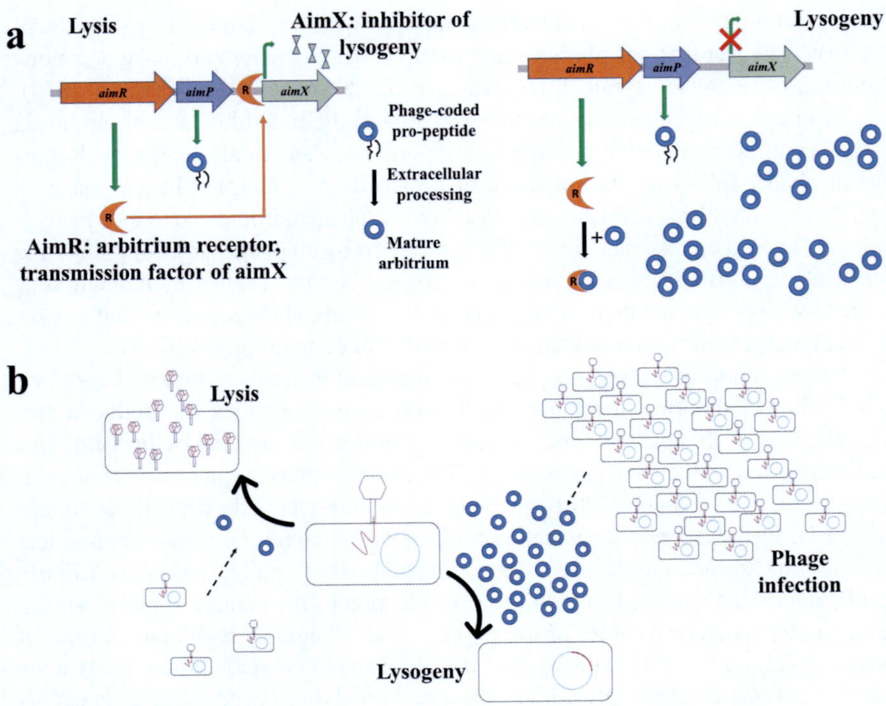

Fig. 3.1 Mechanistic model of phage-phage communication for lysis-lysogeny decisions

implications. Previous research showed that lysogeny is favored under conditions of low host cell density and nutrient availability (Ghosh et al. 2008; Thomas et al. 2011; Brum et al. 2016; Touchon et al. 2016), however the arbitrium communication mechanism suggests that high abundances of infected host cells cause high concentrations of arbitrium signal and guide phages into lysogenic cycles. The small-molecule phage communication mechanism may provide indirect evidence and theoretical support for the recently proposed Piggyback-the-Winner model (explained in Sect. 3.1 Introduction) (Knowles et al. 2016; Silveira and Rohwer 2016; Coutinho et al. 2017).

3.4.2 Phage-Host Communication

The long-term existence of phages depends on prokaryotic hosts, and preliminary efforts have been made to characterize the complicated interactions between bacteria and phages, e.g., parasitic infection, coevolution of arms race dynamics, and symbiosis through metabolic reprograming (Roucourt and Lavigne 2009; Díaz-Muñoz and Koskella 2014; De Sordi et al. 2019; Moreno-Gallego et al. 2019). Furthermore, phage reproductive cycles have significant impacts on host populations, physiology, and also phage-host dynamics. Quorum sensing systems, indicative of bacterial density as discussed above, were reported to activate prophage induction in soil and groundwater bacteria and a model bacterium *E. coli* (Ghosh et al. 2009). The authors also showed that the mechanism of homoserine lactone-based prophage induction was SOS (i.e. *recA*) independent. This study offered the first experimental evidence of cell-density dependent prophage induction indicating that prophage induction mechanisms can also be activated during times of high host cell density when the physiological conditions of potential host cells was favorable for growth.

The new molecular mechanism of quorum sensing-based lysis-lysogeny decisions was recently reported (Silpe and Bassler 2019). Specifically, the *Vibrio* quorum-sensing signal molecules 3,5-dimethylpyrazin-2-ol (DPO) are used as a que for vibriophage VP882 to determine the host cell density. The mechanistic model for host quorum sensing-based phage reproduction strategy selection, in which phage genes gp55, gp56, gp59, gp62, gp69, gp70, and gp71 are included, is shown in Fig. 3.2. The vibriophage genes gp69, gp70, and gp71 are necessary to execute lysis of the infected host cell. In the lysogeny circuit, Gp59 (shown as cl repressor, Fig. 3.2a) can directly repress the expression of gene 62 which codes the lysis genes-launching protein Q anti-terminator. Host cell lysis is promoted when the concentration of host-produced quorum-sensing DPO is high, and large amounts of DPO-binding $VqmA_{phage}$ protein of vibriophage VP882 bind to DPO and are

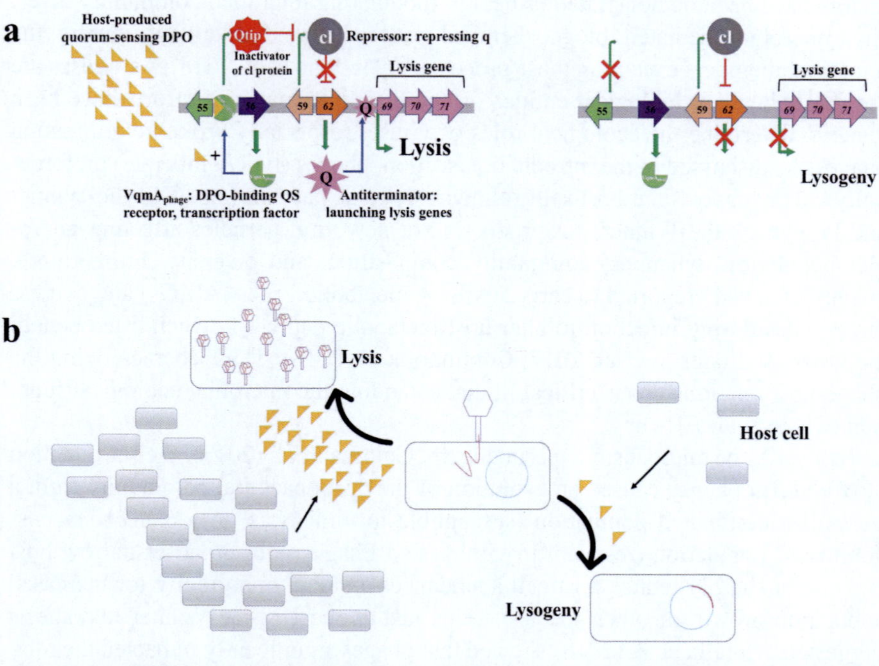

Fig. 3.2 Mechanistic model for quorum sensing-mediated lysis-lysogeny decisions. (Modified from Liang and Radosevich 2019)

activated. The DPO-activated VqmA$_{phage}$ binds upstream of gene gp55 activating the expression of Q-tip anti-repressor. Q-tip directly aggregates the cl repressor preventing cl from binding to the promoter of gene 62 which allows normal production of Q anti-terminator. Q activates lysis genes inducing lysis of the host cell. In summary, the host quorum-sensing autoinducer concentration indicating host cell density can guide the lysis-lysogeny decisions of phages (Fig. 3.2b).

The communication between phages and bacteria has significance for microbial ecology (Liang et al. 2019b). As Silpe and Bassler (2019) showed in the model of quorum sensing-mediated viral reproductive strategy selection, lysogeny is favored under conditions of low host cell density, while high host cell density tend to induce lytic cycles of phages. These reports provide strong support for the Kill-the-Winner hypothesis.

3.5 Phage Top-Down Controls over Host Community Dynamics

3.5.1 Microbial Activity and Community Diversity

Microbes play substantial roles in biogeochemical cycling on a global scale, and many ecosystem processes are predominately microbially driven. Environmental factors have been demonstrated critical in modulating microbial community diversity, microbial-mediated biogeochemical processes, and interplays among the microbial members executing these processes (Coutinho et al. 2018). As viruses are the most abundant biological entities on the planet, tremendous efforts have been made to determine the ecological roles of viruses, especially prokaryote-infecting phages. As discussed in the introduction section, phages (lytic/temperate) preferentially kill or support their host cells following specific patterns (e.g., Kill-the-Winner and Piggyback-the-Winner) for production of new viral particles affecting microbial population dynamics, community composition, and diversity. Furthermore, phages have been reported to carry auxiliary metabolic genes (AMGs) and express these genes during infection to alter host metabolic pathways which often benefit the hosts (Williamson et al. 2017; Coutinho et al. 2018). Thus, characterizing the phage-host relationships is critical in understanding the microbial ecological function (Chow et al. 2014).

In a series of microcosm experiments by Coloma et al. (2017), the introduction of *Nodularia* phages caused an evolution of the cyanobacterial community from a *Nodularia*-dominated population (susceptible to viral lysis) to a *Synechococcus*-dominated population (resistant to viral lysis). Phages with broad or narrow host ranges can sharply reduce host cell abundances creating opportunity for increased population growth of other species/strains just as the Kill-the-Winner hypothesis suggested. Morella et al. (2018) showed that phages significantly impacted the relative abundance of dominant bacterial community members, and the overall bacterial abundance decreased due to phage-mediated lysis of most abundant and/or fastest-growing bacterial species during initial colonization of a new plant. Phage infections as an important factor of shaping microbial community structure may cause large variation in microbial taxonomic composition (Steffen et al. 2015; Storesund et al. 2015; Louca and Doebeli 2018). Previous efforts have also demonstrated the importance of phages in microbial respiration (Allen et al. 2010), carbon processing (Trubl et al. 2018), Fe(III)-bioreduction (Liang et al. 2019a), and nutrient cycling (Weitz and Wilhelm 2012), yet it is still largely unknown how phage infections influence microbial community function.

3.5.2 Phage-Host Coevolution

Close interactions between phages and host prokaryotes lead to their dynamic coexistence and coevolution. Phages can mediate horizontal gene transfer and continually contribute to the appearance of new genotypes facilitating the evolution of prokaryotic hosts (Thomas and Nielsen 2005). Phages as viral predators of prokaryotes constantly attack their hosts to reproduce. In response, the prokaryotes have evolved diverse defense strategies against viral infections for better survival (Dy et al. 2014; Seed 2015). In turn, phages may rapidly evolve effective strategies to overcome the infection barriers, resulting in constant cycles of evolutionary arms race between the interacting predator and prey (Koskella and Brockhurst 2014; De Sordi et al. 2019). The parasite-host coevolution battles over every stage of the phage life cycle, i.e., phage attachment to host cell surface, phage DNA entry, and phage replication and release.

The coevolution of phages and prokaryotes influences the population dynamics of both sides and have essential roles in driving and preserving microbial diversity. The frequency of virus-resistant hosts, that substantially influences the community dynamics, may fluctuate as the phages and prokaryotes coevolve (Coloma et al. 2019). Arms race dynamics (ARD, featuring increasing host resistance and phage infectivity with time) and fluctuating selection dynamics (FSD, characteristic of frequency-based selection of host resistance alleles by phage evolution) are two most commonly considered models for phage-host antagonistic coevolution (Gómez and Buckling 2011; Scanlan 2017; De Sordi et al. 2019).

Quorum sensing system can coordinate bacterial group behaviors based on population density and also play an important role in phage-host interactions via influencing the lysis-lysogeny decisions of phages (discussed in Sect. 3.2 Phage-host communication). The impacts of quorum sensing on bacterial susceptibility to phage infection had been overlooked until Høyland-Kroghsbo et al. (2013) reported that the quorum-sensing signals N-acyl-L-homoserine lactones (AHLs) significantly reduced the phage adsorption due to cell-density-dependent down-regulation of the surface phage receptors and drastically increased the frequency of uninfected cells. Though not reviewed here, the mechanism that quorum sensing affect host resistance against phages also includes strengthening CRISPR-Cas adaptive immune systems in host cells (Patterson et al. 2016; Høyland-Kroghsbo et al. 2017).

3.6 Conclusion

Bacteriophages are ubiquitous and the most abundant biological entities on earth, typically outnumbering their coexisting microbial host cells by 10- to 1000-fold. Phage infections have been demonstrated as a key determinant of microbial population size, composition, structure, and evolution. The impacts of phages on their microbial hosts are closely related with the life properties of phages, infection strategies, morphology, host range, etc. The biology of phages is fundamental to elucidate phage-host relationship and further understanding microbial systems (Clokie et al. 2011). Phages were once believed too small to have molecular systems for group communication and coordinated behaviors. However, some recent studies have revealed the phage-phage (Erez et al. 2017; Dou et al. 2018; Wang et al. 2018; del Sol et al. 2019) phage-bacteria (Silpe and Bassler 2019) communication mechanisms that govern phage lysis-lysogeny decisions which is a key determinant of phage-host interactions and coevolution.

References

Abedon ST (2017) Commentary: communication between viruses guides lysis–lysogeny decisions. Front Microbiol 8:983

Allen B, Willner D, Oechel WC, Lipson D (2010) Top-down control of microbial activity and biomass in an Arctic soil ecosystem. Environ Microbiol 12:642–648

Allers E, Moraru C, Duhaime MB, Beneze E, Solonenko N, Barrero-Canosa J, Amann R, Sullivan MB (2013) Single-cell and population level viral infection dynamics revealed by phage FISH, a method to visualize intracellular and free viruses. Environ Microbiol 15:2306–2318

Almeida RM, Roland F, Cardoso SJ, Farjalla VF, Bozelli RL, Barros NO (2015) Viruses and bacteria in floodplain lakes along a major Amazon tributary respond to distance to the Amazon River. Front Microbiol 6:158

Amossé J, Bettarel Y, Bouvier C, Bouvier T, Duc TT, Thu TD, Jouquet P (2013) The flows of nitrogen, bacteria and viruses from the soil to water compartments are influenced by earthworm activity and organic fertilization (compost vs. vermicompost). Soil Biol Biochem 66:197–203

Ballaud F, Dufresne A, Francez AJ, Colombet J, Sime-Ngando T, Quaiser A (2016) Dynamics of viral abundance and diversity in a sphagnum-dominated peatland: temporal fluctuations prevail over habitat. Front Microbiol 6:1494

Breitbart M, Rohwer F (2005) Here a virus, there a virus, everywhere the same virus? Trends Microbiol 13:278–284

Breitbart M, Bonnain C, Malki K, Sawaya NA (2018) Phage puppet masters of the marine microbial realm. Nat Microbiol 3:754–766

Brum JR, Sullivan MB (2015) Rising to the challenge: accelerated pace of discovery transforms marine virology. Nat Rev Microbiol 13:147–159

Brum JR, Hurwitz BL, Schofield O, Ducklow HW, Sullivan MB (2016) Seasonal time bombs: dominant temperate viruses affect Southern Ocean microbial dynamics. ISME J 10:437–449

Chow CET, Kim DY, Sachdeva R et al (2014) Top-down controls on bacterial community structure: microbial network analysis of bacteria, T4-like viruses and protists. ISME J 8:816–829

Clokie MR, Millard AD, Letarov AV, Heaphy S (2011) Phages in nature. Bacteriophage 1:31–45

Coloma SE, Dienstbier A, Bamford DH, Sivonen K, Roine E, Hiltunen T (2017) Newly isolated Nodularia phage influences cyanobacterial community dynamics. Environ Microbiol 19:273–286

Coloma S, Gaedke U, Sivonen K, Hiltunen T (2019) Frequency of virus-resistant hosts determines experimental community dynamics. Ecology 100:e02554

Coutinho FH, Silveira CB, Gregoracci GB, Thompson CC, Edwards RA, Brussaard CP et al (2017) Marine viruses discovered via metagenomics shed light on viral strategies throughout the oceans. Nat Commun 8:15955

Coutinho FH, Gregoracci GB, Walter JM, Thompson CC, Thompson FL (2018) Metagenomics sheds light on the ecology of marine microbes and their viruses. Trends Microbiol 11:955–965

Daly RA, Roux S, Borton MA, Morgan DM, Johnston MD, Booker AE et al (2019) Viruses control dominant bacteria colonizing the terrestrial deep biosphere after hydraulic fracturing. Nat Microbiol 4:352–361

Danovaro R, Serresi M (2000) Viral density and virus-to-bacterium ratio in deep-sea sediments of the eastern Mediterranean. Appl Environ Microbiol 66:1857–1861

Danovaro R, Corinaldesi C, Dell'Anno A, Fabiano M, Corselli C (2005) Viruses, prokaryotes and DNA in the sediments of a deep-hypersaline anoxic basin (DHAB) of the Mediterranean Sea. Environ Microbiol 7:586–592

De Sordi L, Lourenço M, Debarbieux L (2019) The battle within: interactions of bacteriophages and Bacteria in the gastrointestinal tract. Cell Host Microbe 25:210–218

del Sol FG, Penadés JR, Marina A (2019) Deciphering the molecular mechanism underpinning phage arbitrium communication systems. Mol Cell 74:59–72

Díaz-Muñoz SL, Koskella B (2014) Bacteria–phage interactions in natural environments. In: Gadd G, Sariaslani S (eds) Advances in applied microbiology, vol 89. Academic, San Diego, pp 135–183

Díaz-Muñoz SL, Sanjuán R, West S (2017) Conflict, cooperation, and communication among viruses. Cell Host Microbe 22:437–441

Dou C, Xiong J, Gu Y, Yin K, Wang J, Hu Y et al (2018) Structural and functional insights into the regulation of the lysis–lysogeny decision in viral communities. Nat Microbiol 3:1285–1294

Dy RL, Richter C, Salmond GP, Fineran PC (2014) Remarkable mechanisms in microbes to resist phage infections. Annu Rev Virol 1:307–331

Edwards RA, McNair K, Faust K, Raes J, Dutilh BE (2016) Computational approaches to predict bacteriophage–host relationships. FEMS Microbiol Rev 40:258–272

Emerson JB, Roux S, Brum JR, Bolduc B, Woodcroft BJ, Jang HB et al (2018) Host-linked soil viral ecology along a permafrost thaw gradient. Nat Microbiol 3:870–880

Enault F, Briet A, Bouteille L, Roux S, Sullivan MB, Petit MA (2017) Phages rarely encode antibiotic resistance genes: a cautionary tale for virome analyses. ISME J 11:237–247

Erez Z, Steinberger-Levy I, Shamir M, Doron S, Stokar-Avihail A, Peleg Y et al (2017) Communication between viruses guides lysis–lysogeny decisions. Nature 541:488–593

Feiner R, Argov T, Rabinovich L, Sigal N, Borovok I, Herskovits AA (2015) A new perspective on lysogeny: prophages as active regulatory switches of bacteria. Nat Rev Microbiol 13:641–650

Finke JF, Hunt BP, Winter C, Carmack EC, Suttle CA (2017) Nutrients and other environmental factors influence virus abundances across oxic and hypoxic marine environments. Viruses 9:152

Gainer PJ, Pound HL, Larkin AA, LeCleir GR, DeBruyn JM, Zinser ER, Johnson ZI, Wilhelm SW (2017) Contrasting seasonal drivers of virus abundance and production in the North Pacific Ocean. PLoS One 12:e0184371

Ghosh D, Roy K, Williamson KE, White DC, Wommack KE, Sublette KL et al (2008) Prevalence of lysogeny among soil bacteria and presence of 16S rRNA and trzN genes in viral-community DNA. Appl Environ Microbiol 74:495–502

Ghosh D, Roy K, Williamson KE, Srinivasiah S, Wommack KE, Radosevich M (2009) Acyl-homoserine lactones can induce virus production in lysogenic bacteria: an alternative paradigm for prophage induction. Appl Environ Microbiol 75:7142–7152

Goldberg GW, Jiang W, Bikard D, Marraffini LA (2014) Conditional tolerance of temperate phages via transcription-dependent CRISPR-Cas targeting. Nature 514:633–637

Gómez P, Buckling A (2011) Bacteria-phage antagonistic coevolution in soil. Science 332:106–109

Gregory AC, Zayed AA, Conceição-Neto N, Temperton B, Bolduc B, Alberti A et al (2019) Marine DNA viral macro-and microdiversity from pole to pole. Cell 177:1109–1123

Hall EK, Bernhardt ES, Bier RL, Bradford MA, Boot CM, Cotner JB et al (2018) Understanding how microbiomes influence the systems they inhabit. Nat Microbiol 3:977–982
Hewson I, O'Neil JM, Fuhrman JA, Dennison WC (2001) Virus-like particle distribution and abundance in sediments and overlying waters along eutrophication gradients in two subtropical estuaries. Limnol Oceanogr 46:1734–1746
Howard-Varona C, Hargreaves KR, Abedon ST, Sullivan MB (2017) Lysogeny in nature: mechanisms, impact and ecology of temperate phages. ISME J 11:1511–1520
Høyland-Kroghsbo NM, Mærkedahl RB, Svenningsen SL (2013) A quorum-sensing-induced bacteriophage defense mechanism. MBio 4:e00362–e00312
Høyland-Kroghsbo NM, Paczkowski J, Mukherjee S, Broniewski J, Westra E, Bondy-Denomy J, Bassler BL (2017) Quorum sensing controls the *Pseudomonas aeruginosa* CRISPR-Cas adaptive immune system. Proc Natl Acad Sci U S A 114:131–135
Junger PC, Amando AM et al (2017) Salinity drives the virioplankton abundance but not production in tropical coastal lagoons. Microbial Ecol 75:52–63
Keshri J, Ram ASP, Colombet J, Perriere F, Thouvenot A, Sime-Ngando T (2017) Differential impact of lytic viruses on the taxonomical resolution of freshwater bacterioplankton community structure. Water Res 124:129–138
Knowles B, Silveira CB, Bailey BA, Barott K, Cantu VA, Cobián-Güemes AG et al (2016) Lytic to temperate switching of viral communities. Nature 531:466–470
Knowles B, Bailey B, Boling L, Breitbart M, Cobián-Güemes A, del Campo J et al (2017) Variability and host density independence in inductions-based estimates of environmental lysogeny. Nat Microbiol 2:17064
Koskella B, Brockhurst MA (2014) Bacteria–phage coevolution as a driver of ecological and evolutionary processes in microbial communities. FEMS Microbiol Rev 38:916–931
Kyle JE, Eydal HS, Ferris FG, Pedersen K (2008) Viruses in granitic groundwater from 69 to 450 m depth of the Äspö hard rock laboratory, Sweden. ISME J 2:571–574
Letarov A, Kulikov E (2009) The bacteriophages in human-and animal body-associated microbial communities. J Appl Microbiol 107:1–13
Liang X, Radosevich M (2019) Commentary: a host-produced quorum-sensing autoinducer controls a phage lysis-Lysogeny decision. Front Microbiol 10:1201
Liang X, Zhuang J, Löffler FE, Zhang Y, DeBruyn J, Wilhelm S, Schaeffer SM, Radosevich M (2019a) Viral and bacterial community responses to stimulated Fe(III)-bioreduction during simulated subsurface bioremediation. Environ Microbiol 21:2043–2055
Liang X, Wagner RE, Li B, Zhang N, Radosevich M (2019b) Prophage induction mediated by quorum sensing signals alters soil bacterial community structure. bioRxiv:805069
Liang X, Wagner RE, Zhuang J, DeBruyn JM, Wilhelm SW, Liu F, Yang L, Staton ME, Sherfy AC, Radosevich M (2019c) Viral abundance and diversity vary with depth in a southeastern United States agricultural ultisol. Soil Biol Biochem 137:107546
Louca S, Doebeli M (2018) Taxonomic variability and functional stability in microbial communities infected by phages. Environ Microbiol 19:3863–3878
Manini E, Luna GM, Corinaldesi C, Zeppilli D, Bortoluzzi G, Caramanna G, Raffa F, Danovaro R (2008) Prokaryote diversity and virus abundance in shallow hydrothermal vents of the Mediterranean Sea (Panarea Island) and the Pacific Ocean (North Sulawesi-Indonesia). Microb Ecol 55:626–639
Mojica KD, Brussaard CP (2014) Factors affecting virus dynamics and microbial host–virus interactions in marine environments. FEMS Microbiol Ecol 89:495–515
Morella NM, Gomez AL, Wang G, Leung MS, Koskella B (2018) The impact of bacteriophages on phyllosphere bacterial abundance and composition. Mol Ecol 27:2025–2038
Moreno-Gallego JL, Chou SP, Di Rienzi SC, Goodrich JK, Spector TD, Bell JT, Youngblut ND, Hewson I, Reyes A, Ley RE (2019) Virome diversity correlates with intestinal microbiome diversity in adult monozygotic twins. Cell Host Microbe 25:261–272
Narr A, Nawaz A, Lukus LY, et al (2017) Soil viral vommunities vary temporally and along a land use transect as revealed by virus-like particle counting and a modified community fingerprinting approach (fRAPD). Frontiers Terrestrial Microbiol 8:Article 175

Ng WL, Bassler BL (2009) Bacterial quorum-sensing network architectures. Annu Rev Genet 43:197–222

Ofir G, Sorek R (2018) Contemporary phage biology: from classic models to new insights. Cell 172:1260–1270

Paez-Espino D, Eloe-Fadrosh EA, Pavlopoulos GA, Thomas AD, Huntemann M, Mikhailova N, Rubin E, Ivanova NN, Kyrpides NC (2016) Uncovering Earth's virome. Nature 536:425–430

Pan D, Nolan J, Williams KH, Robbins MJ, Weber KA (2017) Abundance and distribution of microbial cells and viruses in an alluvial aquifer. Front Microbiol 8:1199

Parikka KJ, Le Romancer M, Wauters N, Jacquet S (2017) Deciphering the virus-to-prokaryote ratio (VPR): insights into virus–host relationships in a variety of ecosystems. Biol Rev 92:1081–1100

Patterson AG, Jackson SA, Taylor C, Evans GB, Salmond GP, Przybilski R, Staals RH, Fineran PC (2016) Quorum sensing controls adaptive immunity through the regulation of multiple CRISPR-Cas systems. Mol Cell 64:1102–1108

Payet JP, Suttle CA (2013) To kill or not to kill: the balance between lytic and lysogenic viral infection is driven by trophic status. Limnol Oceanogr 58:465–474

Roucourt B, Lavigne R (2009) The role of interactions between phage and bacterial proteins within the infected cell: a diverse and puzzling interactome. Environ Microbiol 11:2789–2805

Roux S, Enault F, Hurwitz BL, Sullivan MB (2015a) VirSorter: mining viral signal from microbial genomic data. PeerJ 3:e985

Roux S, Hallam SJ, Woyke T, Sullivan MB (2015b) Viral dark matter and virus–host interactions resolved from publicly available microbial genomes. elife 4:e08490

Roux S, Brum JR, Dutilh BE, Sunagawa S, Duhaime MB, Loy A et al (2016) Ecogenomics and potential biogeochemical impacts of globally abundant ocean viruses. Nature 537:689–693

Scanlan PD (2017) Bacteria–bacteriophage coevolution in the human gut: implications for microbial diversity and functionality. Trends Microbiol 25:614–623

Seed KD (2015) Battling phages: how bacteria defend against viral attack. PLoS Pathog 11:e1004847

Segobola J, Adriaenssens E, Tsekoa T, Rashamuse K, Cowan D (2018) Exploring viral diversity in a unique south African soil habitat. Sci Rep 8:111

Silpe JE, Bassler BL (2019) A host-produced quorum-sensing autoinducer controls a phage lysis-Lysogeny decision. Cell 176:1–13

Silveira CB, Rohwer FL (2016) Piggyback-the-winner in host-associated microbial communities. NPJ Biofilms Microbiol 2:16010

Steffen MM, Belisle BS, Watson SB, Boyer GL, Bourbonniere RA, Wilhelm SW (2015) Metatranscriptomic evidence for co-occurring top-down and bottom-up controls on toxic cyanobacterial communities. Appl Environ Microbiol 81:3268–3276

Storesund JE, Erga SR, Ray JL, Thingstad TF, Sandaa RA (2015) Top-down and bottom-up control on bacterial diversity in a western Norwegian deep-silled fjord. FEMS Microbiol Ecol 91:fiv076

Tadmor AD, Ottesen EA, Leadbetter JR, Phillips R (2011) Probing individual environmental bacteria for viruses by using microfluidic digital PCR. Science 333:58–62

Thomas CM, Nielsen KM (2005) Mechanisms of, and barriers to, horizontal gene transfer between bacteria. Nat Rev Microbiol 3:711–721

Thomas R, Berdjeb L, Sime-Ngando T, Jacquet S (2011) Viral abundance, production, decay rates and life strategies (lysogeny versus lysis) in Lake Bourget (France). Environ Microbiol 13:616–630

Touchon M, Bernheim A, Rocha EP (2016) Genetic and life-history traits associated with the distribution of prophages in bacteria. ISME J 10:2744–2754

Trubl G, Jang HB, Roux S, Emerson JB, Solonenko N, Vik DR, Solden L, Ellenbogen J, Runyon AT, Bolduc B, Woodcroft BJ (2018) Soil viruses are underexplored players in ecosystem carbon processing. mSystems 3:e00076–e00018

Våge S, Pree B, Thingstad TF (2016) Linking internal and external bacterial community control gives mechanistic framework for pelagic virus-to-bacteria ratios. Environ Microbiol 18:3932–3948

Wang Q, Guan Z, Pei K, Wang J, Liu Z, Yin P, Peng D, Zou T (2018) Structural basis of the arbitrium peptide–AimR communication system in the phage lysis–lysogeny decision. Nat Microbiol 3:1266–1273

Waters CM, Bassler BL (2005) Quorum sensing: cell-to-cell communication in bacteria. Annu. Rev Cell Dev Biol 21:319–346

Weitz JS, Wilhelm SW (2012) Ocean viruses and their effects on microbial communities and biogeochemical cycles. F1000 Biol Rep 4:17

Weitz JS, Mileyko Y, Joh RI, Voit EO (2008) Collective decision making in bacterial viruses. Biophys J 95:2673–2680

Weitz JS, Beckett SJ, Brum JR, Cael BB, Dushoff J (2017) Lysis, lysogeny and virus–microbe ratios. Nature 549:E1–E3

Wigington CH, Sonderegger D, Brussaard CP, Buchan A, Finke JF, Fuhrman JA et al (2016) Re-examination of the relationship between marine virus and microbial cell abundances. Nat Microbiol 1:15024

Wilhartitz IC, Kirschner AK, Brussaard CP, Fischer UR, Wieltschnig C, Stadler H, Farnleitner AH (2013) Dynamics of natural prokaryotes, viruses, and heterotrophic nanoflagellates in alpine karstic groundwater. Microbiology 2:633–643

Williamson KE, Radosevich M, Wommack KE (2005) Abundance and diversity of viruses in six Delaware soils. Appl Environ Microbiol 71:3119–3125

Williamson KE, Radosevich M, Smith DW, Wommack KE (2007) Incidence of lysogeny within temperate and extreme soil environments. Environ Microbiol 9:2563–2574

Williamson KE, Corzo KA, Drissi CL, Buckingham JM, Thompson CP, Helton RR (2013) Estimates of viral abundance in soils are strongly influenced by extraction and enumeration methods. Biol Fertil Soils 49:857–869

Williamson KE, Fuhrmann JJ, Wommack KE, Radosevich M (2017) Viruses in soil ecosystems: an unknown quantity within an unexplored territory. Annu Rev Virol 4:201–219

Zeng L, Skinner SO, Zong C, Sippy J, Feiss M, Golding I (2010) Decision making at a subcellular level determines the outcome of bacteriophage infection. Cell 141:682–691

Chapter 4
Phage-Phage Interactions

Jimmy T. Trinh and Lanying Zeng

Abstract Often considered to be the most numerous and most diverse biological entities, phages are relatively simple forms of life. Despite this, phages have remarkably similar behaviors to those characterized in more complex organisms. When considering the biological development of phages from this perspective, new ideas and narratives can be explored to achieve a better mechanistic understanding of these biological systems. The interactions of phages underpin their development, so by leveraging the proper tools to focus on these processes, fascinating details can be discovered in these tiny lives.

4.1 Introduction

The interactions of individuals guide how life develops. Beyond the interactions among complex lifeforms, we can also look at interactions in the world of single cells. Indeed, single-cell organisms and the single cells comprising organisms undergo constant interactions, for example, in the form of communication, community, and killing to influence how one or many lifeforms develop. Apart from traditional forms of life, we can also see that viruses engage with cells in the situations described above. The presence of viral proteins can communicate the threat of invasion to immune cells, triggering a response. The genetic content of a virus can integrate with a cell and lie dormant or exchange genetic material with its host to mutually provide gain of function. In addition, viruses can infect and kill cells, fielding changes across an entire organism or community. Extending this logic, viruses can also interact with other viruses. Continuing detailed studies have been showing that these viral interactions occur, and that they play important roles in the biology of viruses.

We will focus on the interactions of bacteriophages (or simply phages), viruses that infect bacteria, within cells. Inside a cell, the phage exists as nucleic acid, most often as double-stranded DNA. The phage must interact with other phages that

J. T. Trinh · L. Zeng (✉)
Department of Biochemistry and Biophysics, Center for Phage Technology, Texas A&M University, College Station, TX, USA
e-mail: lzeng@tamu.edu

G. Witzany (ed.), *Biocommunication of Phages*,
https://doi.org/10.1007/978-3-030-45885-0_4

infect the same cell, which may be identical, similar, or very dissimilar genetically. To be clear, descriptions of phage interactions need to be understood as interpretations of viral behaviors. For example, there is molecular communication between phages in a cell. When the phage communicates, it can be thought of as a language where proteins are the "words." Other phage DNAs can "listen" with specific sequences in their genome that specific proteins bind to, triggering programmed responses. This is a molecular description of a process such as gene regulation, but these specific events occur between related and unrelated phages inside cells to achieve varied outcomes. These fundamental biological mechanisms are responsible for many divergent forms of interaction and consequent behaviors of phages.

4.2 Phage Decisions Made by Individual Phages

It is through high-resolution techniques that researchers can disentangle the biological mechanisms occurring within single cells. Previously uncharacterized phenomena and deeper comprehension of viral machinations emerge from careful analysis of the data that had eluded researchers before, but such data are now accessible with current technology (Shao et al. 2019; Golding 2016). This can potentially revitalize the paradigms of phage biology, provoking interesting perspectives to understand the world of phages. The abundance of data, tools and techniques make phage lambda ideal for developing and verifying new technologies, as well as applying modern technologies to the discovery of new details about phage biology.

Lambda is a temperate phage, meaning that it enacts a post-infection decision to either kill or integrate with its host, *E. coli*. If lambda integrates with the cell, the resulting lysogen harbors the lambda prophage and remains dormant unless the virus switches itself into active growth to kill the cell. Lambda has been among the most popular models for cellular decision-making and bistable switch. Regarding lambda decision-making, it was long understood that phage interactions in the cell contribute to the outcome because the frequency of lysogeny increases with multiplicity of infection (MOI), meaning that as cells are infected by more individual phages, there is a corresponding shift in decision-making development, although the details of this interplay were not clear before (Kourilsky 1974; Kourilsky and Knapp 1974; Kourilsky 1973). With the implementation of high-resolution studies, new models have elaborated on how a phage interacts with another phage to make decisions. In one line of experimentation, the capsid decoration protein of lambda, gpD, was genetically modified with a fluorescent protein fusion, which produced fluorescent virions (Alvarez et al. 2007). By using these particles, precise enumeration of phages adsorbed to cells was achieved. This modification reported the lytic decision because fluorescence would be produced as a consequence of the decision to make new progeny virions. A separate lysogenic reporter was constructed on a plasmid harbored by the host cells. Using single-cell, single-phage experimentation, it was revealed that different lambda phages infecting the same cell would separately commit to decisions, termed as "voting" for cellular outcomes by viruses

(Zeng et al. 2010). For this model to work, there must be interaction between phages within the same cell.

To successfully establish lysogeny, lambda's master repressor, a protein named CI, must inhibit gene expression for all copies of phage DNA in the cell (Ptashne 1967). The basic process of establishing lysogeny requires interaction between lambda clones. Consider that a CI-producing phage has made a decision to project a message to other phages in the cell advocating the lysogenic lifestyle. Other copies of lambda DNA can both "hear" and "accept" this message, because they have operators for CI to bind to and repress gene expression, which reinforces this interaction. If there are non-related phages in the cell, then the message will not get through. The "words" of CI may find their way to other phages via non-specific DNA binding and search mechanisms but will ultimately be ignored (Loverdo et al. 2009). As for lambda, the phage is programmed to mutually communicate because CI is under positive-feedback regulation, so other lambda copies can voice agreement upon receiving the lysogeny message.

During the lytic decision, lambda commits to expressing its lysis and structural genes via anti-termination of the pR' transcript, the late operon, by a protein named Q (Grayhack et al. 1985; Grayhack and Roberts 1982; Roberts 1975). There are reports that this lytic process is the default pathway of development and that there is cis-preferential action involved in this decision (Burt and Brammar 1982; Echols et al. 1976). In other words, different phages will all work towards this lytic development eventually and somewhat separately, unless a lysogenic decision diverts them away.

One interesting consequence of the voting model is that phages can also apparently "refuse" to be repressed, since different phages have the ability to vote differently in the model. This mixed voting predominately results in the culmination of the lytic decision (Zeng et al. 2010; Shao et al. 2016). It was suggested that a unanimous vote for lysogeny was required to successfully establish this process, although the lysogenic reporter was a shared plasmid, incapable of providing information on individual lysogenic votes. Any dissenting lytic vote appeared to veto the lysogenic decision. It is not entirely clear how this process happens, but it might involve limited numbers of CI relative to the amount of phage DNA (Reichardt and Kaiser 1971). During the decision-making process, phage DNAs replicate, producing more individuals that need to receive the message to lysogenize. In other words, the message might get lost in the crowd of phages, and because the endpoint of the lytic decision is cell death, there is a time limit on enacting lysogeny (Wang et al. 2000). Diving deeper into the decision-making process, there is more in play than just CI and cell lysis, which are just the final events. Part of the process to produce phage progeny involves the assembly of phage particles and packaging of phage DNA into the capsids. Regarding how phages may override the lysogenic message, there could be an interaction with the packaging machinery, or the terminase. DNA packaging proteins are effectively a message to move out of the cell. Lambda DNAs, which can replicate into concatemers, can accept this message as the terminase complex of lambda packages lambda DNA into maturing capsids (Bastia and Sueoka 1975; Gold and Becker 1983). This interaction would inactivate any phage DNAs that may be sending their own messages, silencing them.

4.3 Interactions of Phage Decision-Making in Higher Resolution

Building upon the voting model, the concept of phage individuality was explored in more detail by constructing fluorescent reporters for different pathways. Specifically, by engineering two populations of lambda, each with different fluorescent lytic and lysogenic reporters directly on the genomes, the concept of individual decision-making behavior was probed. The lytic reporters were gpD fluorescent fusions as described above, allowing discrimination of different infecting phages and the lytic voting behavior of different phages in a single cell. The lysogenic reporters were different fluorescent transcriptional fusions to CI in each phage, which was appropriate because CI is downstream of a lysogenic vote. With this setup, different phages had different "voices" to express either lytic, lysogenic, or both votes, that could be directly observed using microscopy (Trinh et al. 2017).

There was an expectation that cells infected by both phages would result in both phages voting for decisions, which was partially fulfilled, as there was a subpopulation where both phages expressed themselves. Surprisingly, there was also a subpopulation where only one phage voted despite both phages being present. From the analysis of the voting patterns, it became clear that in lytic cells, a single phage typically dominates the cellular discourse, but in lysogenic cells, phages mutually enact their decisions. This competition during lysis and cooperation during lysogeny was an interesting view into the lives of individual phages during different lifestyles and prompted ideas about how this affects their fitness (Trinh and Zeng 2017). It is generally thought that conditions with plentiful, rapidly growing host cells favor the lytic decision. The selfishness during lytic development actually positions the winning phage for reproductive success by dominating the share of progeny within a particular niche. Conversely, poorer conditions with less hosts are known to favor lysogeny (Kourilsky 1973). More phages will infect single cells when there are less hosts, and the subsequent cooperation between phages in a lysogen encourages genetic diversity by allowing multiple phages to propagate in a single cell, without selecting for the genes maximizing lytic development. Regardless of whether the integrated genotypes are necessarily advantageous in the existing condition, lysogens are capable of moving to unknown environments and niches, spreading the genes through genetic drift. Because the lysogens can switch into lytic development in the future, the genetic diversity is maintained, giving the phage opportunities to find an advantage elsewhere. In combination, differential interactions in separate lifestyles might optimize evolutionary fitness over time. What different sets of interactions might reinforce these divergent pathways?

Earlier discussion about phage interactions during lysis or lysogeny can be expanded. For lysogeny, CI represents the final step of the process (Oppenheim et al. 2005). CI is produced from two different promoters, pRE and pRM. The pRE promoter is activated by a protein named CII and is where CI is initially produced. The pRM promoter is activated by CI itself via auto-positive and negative feedback regulation. The pRE promoter is strong, which is important because the initial lysogenic message must be forceful. The phage must not only divert itself away from lytic development, but also other phages in the cell. CII, the initial words that support lysogeny, delivers a multifaceted message for lysogeny. CII activates pRE, overriding the default lysis-supporting promoter, pR. CII also activates a promoter that antagonizes the expression of Q, the protein controlling the transcription of the late promoter for lysis. CII also promotes the expression of a protein named Integrase (Int), which acts to integrate phage DNA into the bacterial chromosome. From this, it should be clear why lysogeny is a cooperative effort. The action of CII and the proteins it promotes, CI and Int, act on any phage DNAs they encounter. It is a message that, even when announced by one phage, is far-reaching. During decision-making, lambda replicates its DNA, producing more phages that can interact in the cell. CI will inhibit the phage proteins needed to further replicate lambda DNA, limiting the extent that the lysogenic message must spread to be effective. The level of CI in the cell is maintained in the low hundreds in an established lysogen by the action of pRM, so the balance of CI numbers relative to phage DNA copies during infection definitely plays a role in whether lysogeny establishes successfully. Thus, DNA replication is a key aspect of phage decision-making and development, with important roles in both lysogeny and lysis.

The divergence in interactions between lysis and lysogeny can be explained by the phage's biology. To properly undergo lytic development, specific proteins must accumulate, and large amounts of phage DNA need to be produced to assemble progeny. Therefore, the competition during lysis could likely occur the DNA replication step because there is a positive feedback mechanism to replicate one phage over another. Simply put, it is a numbers game. If one phage can build up larger numbers initially, then that one phage will dominate DNA replication because it has more templates which are more likely to be replicated. By using single-DNA sensitive fluorescent reporters, it was shown that phages dominated during lysis were actually not able to replicate their DNA extensively (Trinh et al. 2017). This means that there is a selfish interaction where one phage precludes another from DNA replication. For lambda, DNA replication requires certain host factors as resources. Using computational simulations, it was shown that one phage could dominate a limited number of resources to out-compete another co-infecting phage. This means that there must be a mechanism for the phage to gather the resources to itself, which is an interaction that can be explored.

Regarding the remainder of the lytic pathway, DNA replication itself can be considered as a mechanism for a phage to decide to ignore attempts to lysogenize by simply titrating the message. Furthermore, there is a protein named Cro which is expressed from the early promoter pR. Cro binds to the same operators as CI, but with different binding affinities. In this manner, lambda can bias the decision-making process by making it harder for CI to bind to its targets. Cro is not a gene product specific to the lytic decision since it is produced early in the infection cycle, before commitment to a decision. Cro can be considered to support the lytic pathway because it antagonizes CI, so when new copies of phage DNA are produced during DNA replication, Cro can dampen the effect of CI by competing with CI for binding sites on the new templates. The late genes are under the control of Q, a transcriptional anti-terminator. When Q accumulates beyond a threshold, it will permit the expression of genes that will lyse the cell and assemble new phages. These processes occur within finite amounts of time, so the ability of phages to change their fate depends on when they enter the cell and express their own genes (Cortes et al. 2017). When it comes to the stage of cell lysis, there does not appear to be any communication, because the lysis genes do not interact with phage particles or DNA. If DNA replication is blocked, such as in a P^{-} mutant, lysis will not occur, which is credited to a lack of sufficient gene expression (Shao et al. 2018). To clarify these interactions, it would seem that there are many attempts to communicate and interact among phages during lysogeny to prevent lysis, but less direct mechanisms during lytic development to inhibit lysogeny. These data refine the voting model by demonstrating that infecting phages are not beholden to actually voting, since sometimes a phage will be silenced due to a lack of substantial replication. This silencing plays a major role in lytic development as phages jostle for advantages during propagation. Regarding lysogeny, different phages typically do voice their decisions, and this is due to the ability of one phage to produce CI and enforce the lysogenic decision upon the whole cell.

This idea of interactivity during voting actually allows for both lytic and lysogenic decisions to proceed concurrently. From the reporter systems of different phages, it is actually clear that different phages can have mixed voting patterns, such as when one phage does not vote for lysogeny, but another one does. This represents a confusion state where CI is inferred to exist in the cell, but phages are also being actively assembled and preparing for cell lysis. In general, because of the large numbers of phage DNA during lysis, the lytic decision overcomes attempts to lysogenize, dooming the cell. Notably, this is not a strict rule as, rarely, individual cells during mixed voting are capable of preventing lysis, subsequently dividing as apparent lysogens despite the production of phage particles (Trinh et al. 2017). Regarding the unanimous voting rule discussed earlier, this higher resolution study provides additional details that clarify some of the details. Lysogeny still requires unanimity as a general rule, as the phage's biology favors the abrogation of lysogeny during lytic development, but there are exceptions at the single-cell level. Though strange, this interaction might represent a mechanism to halt lysis if phage DNA replication stops during development, possibly due to non-ideal cellular conditions for lysis. The phage must sense something in the environment to initiate its lysogenic decision and does not stop attempting this development. Even if there is a lytic decision in the same cell, it does not specifically prevent the continued message by the lysogenic DNAs, but the concurrent lysogenic development does specifically inhibit lytic development. In mixed-voting cells, it was observed that the lysis time was significantly delayed compared to cells without it. Because the phage has evolved these functions, it appears that there is a mechanism to lysogenize even at the last minute, possibly providing some marginal benefit in niche scenarios.

4.4 Phage Personal Space

There is a sense of communication and community that develops during lysogeny versus the default selfishness of individual phages during lysis, but there is room to explore how this dichotomy is established. It does appear that DNA replication is one of the triggers. One of the ultimate goals of a lysogenizing phage is to prevent further active DNA replication by the phage, and this is a message that is attempted via multiple channels even in the moments before cell lysis. Recent work has been done to dissect this process by labeling phage DNAs as individuals and as a population. Regarding the mechanism of selfishness, we discussed DNA replication resources as something for phages to compete over. Following this, one essential, low-abundance resource was targeted, DnaB, the helicase in *E. coli* (Mallory et al. 1990; Klein et al. 1980). By fluorescently labeling DnaB, in conjunction with the phage DNA and lytic reporters, multiple steps of the infection process were monitored (Trinh et al. 2020).

The subject of phage lambda biology within cellular space has been largely overlooked, along with the details of lambda DNA replication, so a study combining the two concepts has major potential for uncovering new insights. After infection, individual phage DNAs were tracked with a specific single-DNA reporter, through the SeqA-FP (FP: fluorescent protein) system (Shao et al. 2015; Babic et al. 2008). It was observed that DnaB, initially diffused in the cell in absence of infection, coalesced at the position of the single phage DNAs. This revealed an important detail about lambda DNA replication, that the individual phage gathered essential resources to its own location. This accumulation of resources persisted over time, explaining the domination during lysis and providing a mechanism for the models described earlier. Originating at the sites of the individual phage DNAs and their DnaB resources, clusters of phage DNA labeled by a *tetO*/TetR-FP system appeared and grew (Wang et al. 2011). The data characterized hitherto unknown behaviors that supported phage individuality, the spatial separation of cellular resources by individual phages and the conglomeration of their replicated kin. Also surprising were the behaviors of multiple phages in the same cell, where they would gather their own resources, and subsequently replicate in separated locations without fusing together. This demonstrated the maintenance of individuality by different phages via spatial separation observed in single cells. Upon further study, it was characterized that the host DNA formed a barrier between phage individuals. From these data, it became increasingly clear that the phages interacted in a selfish manner to replicate themselves.

Extending this study, the transcription profile of intracellular phage groups was characterized, finding that different phage transcripts coincided as individual clusters with different phage DNA groups (Trinh et al. 2020). The individual phages construct their individual messages in separate parts of the cell, in the form of transcripts attached to phage DNAs. The proteins translated from these transcripts, however, are capable of moving throughout the cell to spread the message of one group of phages to any other phages in the cell. There is individual assertiveness at the level of resource gathering to facilitate domination by one phage. Cell-wide communication and cooperation arise via the diffusion of proteins across the barriers that separate phages. Different messages by different phages in the same cell are crafted perhaps due to varying conditions in the same cell, and the interplay between these can promote conflict that needs to be resolved by the phages. In this manner, the interactions and communication of individual phages has been traced from individual phage DNAs asserting themselves by gathering resources, to separate groups of individual phages replicating competitively, to their gene expression profiles, which differ across the cell for different phage groups. The spatial dynamics of phage biology represents an exciting area to target for study, promising more information about how the viruses interact as they develop.

4.5 Phage Communication Through Space and Time

Besides the interactions between phages inside a cell during development, there is also the case of communication among phages between cells. When the temperate phage phi3T infects *B. subtilis* grown in conditioned medium, which is the medium from a previous infection cycle of phi3T with cellular debris and phages removed, lysogeny was the preferred outcome of infection (Erez et al. 2017). This simple observation belies the interesting mechanisms and implications of this phage's decision-making. The proposed model is that phi3T produces a peptide during infection which is matured and secreted to be imported into neighboring cells. Once inside cells, these "arbitrium" peptides bind to phage DNA and exert transcriptional regulation to bias new infections toward lysogeny. This effect is concentration dependent, where cells are more likely to be inundated with peptides if there were many nearby infections taking place. That situation would signal host scarcity, ostensibly making lysogeny a better option. This appears to be a message that is meant for phages outside of the infecting phage's cell, and this mechanism could be a means for the phages to interact through time and space. When thinking about phages, intercellular communication is a fairly radical concept. The phage's world is the cell. The phage DNA is only ever active within the confines of this cellular world, so all of the interactions and related narratives generally only play out inside or on the cell. This idea, that the proteins produced in the cell can move on into other cell and act on other phages, fundamentally shifts how we can conceptualize phage interactions.

Not only can a phage use its own quorum sensing peptide to communicate, it also takes advantage of the host quorum sensing system to regulate its lysis-lysogeny decision-making. It was reported that a *Vibro cholerae* phage utilizes the host quorum sensing autoinducer DPO to sense host density for its lysis-lysogeny decision, in which the phage lyses the cell more often when there is more DPO around, indicating higher cell density (Silpe and Bassler 2019). The authors also found that phages can be designed to activate lysis in response to user-defined cues. In addition, phage lambda was reported to respond to a quorum sensing chemical,

N-acyl-L-homoserine lactone (AHL) to increase its prophage induction in its lysogen form (Ghosh et al. 2009). However, a later study suggested that the increased lambda concentration via prophage induction was due to AHL reducing the number of lambda receptors on *E. coli* surface rather than the increased prophage induction rate (Hoyland-Kroghsbo et al. 2013). Therefore, this AHL chemical provides *E. coli* a quorum-sensing-regulated anti-phage defense system protecting it from high risk of lambda infection (Hoyland-Kroghsbo et al. 2013). Nevertheless, quorum sensing seems to play a significant role in phage lambda infection process.

4.6 Self Versus Non-self among Phages

Broadly, many phage systems undergo viral interactions during lysogeny. One basic feature of a lysogen is that it is homo-immune, or that the lysogen is not susceptible to superinfection by like phages. In general, prophages interact with incoming phages of the same type by having some level of repressor expressed in the cell. Because repressor proteins act to silence lytic gene expression, newly infecting phages will immediately be inhibited. This is a message from the prophage that the cell is occupied and not available as a host. In the case of lambda, no integrase or phage recombinases are expressed during lysogeny, discouraging recombination, so the prophage does not welcome like invaders once lysogeny has been established, in contrast to its cooperative behavior during the decision-making process (Lin et al. 1977). If lysogens could simply be infected by other like phages, it would defeat the point of this dormant state of propagation. As such, when the phage chooses this alternate form of replicating, it must interact in various ways to protect itself.

The process of superinfection immunity does not protect the lysogen from different phages. If an unrelated phage infects the lysogen, the prophage may die with the cell in the case of a lytic phage infection or find itself with a new prophage neighbor in the case of a lysogenic phage infection (Casjens 2003). Another interaction that a prophage might use to defend itself is via restriction systems. Phage P1 can lysogenize *E. coli* and encodes a type III restriction modification system (Humbelin et al. 1988). Similar systems are found in many bacteria and can act as defenses against foreign DNA. In the case of the P1 EcoP1 system, it has been shown to cut lambda DNA during infection, which will lower lambda's infection capability. In this interaction, the phage must distinguish self from non-self, and P1 encodes a methyltransferase that prevents the restriction system from acting on its own DNA. The expression of this system presents a hostile environment to incoming phages, although it is not without its flaws. It has been reported that when lambda is plated on P1 lysogens, lambda will be recovered at low efficiencies, but these escapees will freely kill P1 lysogens (Arber and Dussoix 1962; Rao et al. 2014). This is because the restriction system was evaded by the invaders, and the modification system transformed the non-self phage into a "self" phage. DNA modification in general represents a method for phages to distinguish themselves. Another example of a phage that can target other phages is phage T4, which is a lytic phage that replaces all of its cytosine bases with hydroxymethylcytosine, which are then glucosylated (Warren 1980). T4 also encodes nucleases whose primary job is likely to destroy the host chromosome in order to utilize the nucleotides for DNA replication (Miller et al. 2003). This would also result in the destruction of any prophages or incoming phages that enter the cell at the wrong time. Such nuclease systems are commonplace in virulent phages, representing hostility to other phages, external DNA elements, and hosts alike.

If a virulent phage infects a lysogen, then the only outcome of successful infection is cell death. Furthermore, the resulting progeny of lytic infection can kill other lysogens as well. What capability does a prophage have to avert this? In the case of a virulent infection, survival of the cell is unlikely once the virulent phage ejects its DNA into the cell. In some phages, there is an interaction to limit further damage to lysogens at the population level. There are abortive infection (ABI) systems that inhibit the cell from producing phage progeny in the case of superinfection. Take the case of phage lambda lysogens, where lambda produces RexA and RexB as a prophage (Matz et al. 1982; Gussin and Peterson 1972). These two gene products are under the control of pRM, and are expressed at the same time as CI. Lambda lysogens prevent productive infection by certain genotypes of phage T4, specifically, *rII* mutants of T4. The Rex system of lambda is a toxin/antitoxin system, and the actions of *rII* T4 abrogate the antitoxin, and kill the cell (Engelberg-Kulka et al. 1998). In other words, upon lysogenizing the cell, lambda primes a trap or failsafe. Although the outcome is not necessarily good for the specific lambda copy in the cell, this type of system is often interpreted as altruistic, benefiting the population of lambda lysogens (Shub 1994). These systems only make sense in the bigger picture, between antagonistic phages, implying evolutionary battles. The Rex system is actually named after the fact that it is an *r**II* excluding set of genes, meaning that phage T4 evolved a non-essential gene to combat the ABI system. So, where lambda lays traps, which are unseen until another phage invades, T4 preemptively deploys a defense against a potential ambush. It is unclear if the expression of this proactive defense is a burden for T4 in terms of maximizing progeny, but the defense system has likely been selected for due to the many interactions these and similar phages have had.

4.7 Phage Keeping Sole Possession of Its Host

As a lytic phage, whose only ostensible goal is to make more phages, it is interesting that T4 "considers" interacting with other phages in a variety of manners. T4's normal infection cycle produces ~200 progeny and lyses the cell in about 25 min (Chen and Young 2016). This is quite rapid, allowing one T4 infection to spread to many hosts quickly. Interestingly, if there are many T4 phages relative to hosts, and T4 superinfects an ongoing infection within ~5 min, the cell will not lyse in 25 min (Bode 1967; Paddison et al. 1998). Instead, lysis will be delayed depending on the number of superinfecting phages, and the burst size will increase accordingly. In other words, T4 senses that there are not enough hosts to go around and adapts its lytic lifestyle to produce more phages with less bacteria. The first infecting T4 monitors the periplasm for new T4 DNA in an uncharacterized manner and actually locks out superinfections. In this manner, the early T4 exploits the tardy T4 for its own gain by boosting its own replication as hosts are dwindling at the expense of another phage. This lysis inhibition can also be triggered by related, non-T4 phages, so T4 evolved to sense and sabotage some virulent phage competitors (Abedon 1990). There is some collateral damage to be expected from this interaction, since T4 will defeat other T4 infections, but perhaps the increased progeny gain and competitor denial favors T4.

There are advantages in preventing superinfection even by like phages, as it ensures more progeny are direct descendants of the infecting phage. Unlike in the case of T4, there are other systems that do not send related phages to a dead end in the event of superinfection. Phage T5 has a strategy to deter superinfection via expression of a lipoprotein at the beginning of infection to modify its receptor, FhuA, preventing adsorption by other T5 phages (Pedruzzi et al. 1998). Because the protein is expressed immediately after infection, the phage quickly marks the cell as its own, communicating that the cell is occupied before other phages enter the cell. This strategy could also maximize the amount of free phages in a local environment since related phages will not eject their DNA spuriously, and instead remain active even if hosts die off. Furthermore, after a host cell dies, the remaining cellular debris will often still contain the receptors, which progeny phages could bind and become inactivated (Braun et al. 1994). The strategy of T5 actually prevents this occurrence as well since the receptors are marked as "used." To expound on this interaction, the modification of the actual cellular receptor does not make the cell invisible to other phages. T5 will still be able to reversibly bind to occupied cells via its tail fiber interaction with LPS, even if there is an ongoing infection, but the message that the cell has other T5 phages inside will be clear since FhuA will not be accessible (Heller and Braun 1979). In such cases, extracellular T5 can simply move on in the environment.

4.8 Phages Breed Diversity

Despite the variety of ways that phages can safeguard their host cell as their exclusive phage factory, different phages do often infect the same cell. This is quite evident when bacterial genomes are sequenced, as many strains harbor multiple complete prophages, as well as prophage remnants. Phage genomes are also quite variegated, with genetic bits of one phage here and of other phages elsewhere. The prevalence of multiple phage infections was also suggested in the studies that roughly estimated the number of phages in ocean water being ten-fold higher than bacteria, meaning that phages are constantly encountering the same cells (Suttle 2005). One of the most common phage interactions, therefore, is recombination events between phage DNAs. Homologous recombination is something that bacteria have in common, so naturally, co-infecting phages can alter their DNA based on their hosts recombination functions (Didelot and Maiden 2010). However, phages can also encode their own recombination systems. For example, lambda expresses the red system, a homologous recombination system that has been a coopted as a widespread molecular tool (Datsenko and Wanner 2000; Yu et al. 2000). Since the phage evolved this function and expresses it during vegetative growth, it suggests that the phage searches for DNA to recombine with during its infection cycle. The phage does not "know" if it will encounter non-self DNA to recombine with whenever it infects or induces, so the expression of the recombination system is a general interaction to exchange genetic information. Such a strategy would promote genetic diversity over time, and as previously established for lambda, competition and cooperation will allow different genotypes to spread. Beyond lambda, it is clear that phages are generally recombining with other phages and their hosts, and this interplay contributes to the astounding diversity of phages.

4.9 Phages Exploiting Other Phages

Competition between phages can be fierce, and as such, varied strategies are concocted to optimize fitness. An interesting interaction is found in helper/satellite phage systems, sometimes described as pirate phage systems. One example is the P2/P4 system, where P2 is the helper and P4 is the satellite (Six and Klug 1973). Phage P4 contains genes responsible for replication and lysogeny, but conspicuously lacks lysis and morphogenesis genes. On the surface, this suggests that P4 is a mobile genetic element without an extracellular physical form, not quite a phage. However, P4 virions exist, and this is because P4 pirates the structural elements from phage P2 (Christie and Dokland 2012). The only situations in which P4 forms virions and escapes the cell is when it encounters P2. This can occur when P2 infects a P4 lysogen, when P4 infects a P2 lysogen, or when P2 and P4 infect the same cell concurrently. When P4 infects a P2 lysogen, P4 has specific proteins that derepress P2 to force it into the lytic pathway. The terminology of helper phage invokes some partnership where both helper and satellite propagate productively, but the outcome of P4's derepression is ~100 progeny per cell, whereas P2 produces ~0.001 progeny per cell. In this case, the pirating narrative captures the interaction more accurately. In the case where P2 infects a P4 lysogen, P4 senses this and is able to propagate in the dying cell, although less efficiently at ~1 or less progeny per cell. During co-infection, P4 gene expression is similar to a P4 solo infection, and its efficiency of propagation is likely influenced by relative timing of P2 and P4 infections. During P4 dominated infections, it is clear that P4 interferes with P2. Because P4 has a smaller genome size than P2, it requires a smaller capsid. P4 redirects P2's capsid assembly to produce smaller capsids almost exclusively, so P2 DNA cannot be packaged (Marvik et al. 1995). The late genes of P4 are activated earlier in the presence of P2, meaning that P4 evolved binding sites for P2 proteins to accelerate its own development. P4 also produces proteins that trans-activate P2's late gene products, completely bypassing P2's gene regulation and DNA replication (Lindqvist et al. 1993). In this way, P4 is always surveilling its environment to exploit a helper phage in case one shows up.

Though not optimal, P4 lysogens also have a backup plan to survive P2 infection. An antirepressor of P2 can activate P4 transcription, but this activity occurs after P2 has replicated its DNA and is undergoing development, so P4 prophages would lag behind in development (Christie and Dokland 2012). Still, the fact that P4 has a binding site to sense an invading helper phage while dormant suggests that it remains vigilant even during lysogeny. P4 also has what could be described as a decision-making process to either replicate as plasmid copies or become an integrated lysogen, and these pathways are not mutually exclusive (Briani et al. 2001; Goldstein et al. 1982). In a sense, this is similar to the mixed voting or confusion interaction described in lambda decision-making, particularly because host cells with P4 plasmids grow slowly and often filament. This suggests that the replicating plasmids are not optimal for host growth and may cause cells to die more frequently, which may not be ideal for the phage.

4.10 Perspectives

The unparalleled diversity of phages offers limitless opportunity for biological research. The interactions of phages are one aspect of phage biology that is potentially vast owing to the sheer number of phages. This sampling of phage interactions only represents a small amount of the work that has been done and does not begin to broach the potential of this area of research. In a world abound with phages, future progress will require creativity in thinking to discover the myriad unknown interactions.

References

Abedon ST (1990) Selection for lysis inhibition in bacteriophage. J Theor Biol 146(4):501–511. https://doi.org/10.1016/s0022-5193(05)80375-3

Alvarez LJ, Thomen P, Makushok T, Chatenay D (2007) Propagation of fluorescent viruses in growing plaques. Biotechnol Bioeng 96(3):615–621. https://doi.org/10.1002/bit.21110

Arber W, Dussoix D (1962) Host specificity of DNA produced by Escherichia coli. I Host controlled modification of bacteriophage lambda. J Mol Biol 5:18–36. https://doi.org/10.1016/s0022-2836(62)80058-8

Babic A, Lindner AB, Vulic M, Stewart EJ, Radman M (2008) Direct visualization of horizontal gene transfer. Science 319(5869):1533–1536. https://doi.org/10.1126/science.1153498

Bastia D, Sueoka N (1975) Studies on the late replication of phage lambda: rolling-circle replication of the wild type and a partially suppressed strain, Oam29 Pam80. J Mol Biol 98(2):305–320

Bode W (1967) Lysis inhibition in Escherichia coli infected with bacteriophage T4. J Virol 1(5):948–955

Braun V, Killmann H, Herrmann C (1994) Inactivation of FhuA at the cell surface of Escherichia coli K-12 by a phage T5 lipoprotein at the periplasmic face of the outer membrane. J Bacteriol 176(15):4710–4717. https://doi.org/10.1128/jb.176.15.4710-4717

Briani F, Deho G, Forti F, Ghisotti D (2001) The plasmid status of satellite bacteriophage P4. Plasmid 45(1):1–17. https://doi.org/10.1006/plas.2000.1497

Burt D, Brammar W (1982) The cis-specificity of the Q-gene product of bacteriophage lambda. Mol Gen Genet 185:468–472

Casjens S (2003) Prophages and bacterial genomics: what have we learned so far? Mol Microbiol 49(2):277–300. https://doi.org/10.1046/j.1365-2958.2003.03580.x

Chen Y, Young R (2016) The last r locus unveiled: T4 RIII is a cytoplasmic Antiholin. J Bacteriol 198(18):2448–2457. https://doi.org/10.1128/JB.00294-16

Christie GE, Dokland T (2012) Pirates of the Caudovirales. Virology 434(2):210–221. https://doi.org/10.1016/j.virol.2012.10.028

Cortes MG, Trinh JT, Zeng L, Balazsi G (2017) Late-arriving signals contribute less to cell-fate decisions. Biophys J 113(9):2110–2120. https://doi.org/10.1016/j.bpj.2017.09.012

Datsenko KA, Wanner BL (2000) One-step inactivation of chromosomal genes in Escherichia coli K-12 using PCR products. Proc Natl Acad Sci U S A 97(12):6640–6645. https://doi.org/10.1073/pnas.120163297

Didelot X, Maiden MC (2010) Impact of recombination on bacterial evolution. Trends Microbiol 18(7):315–322. https://doi.org/10.1016/j.tim.2010.04.002

Echols H, Court DL, Green L (1976) On the nature of cis-acting regulatory proteins and genetic organization in bacteriophage: the example of gene Q of bacteriophage lambda. Genetics 33:5–10

Engelberg-Kulka H, Reches M, Narasimhan S, Schoulaker-Schwarz R, Klemes Y, Aizenman E (1998) rexB of bacteriophage lambda is an anti-cell death gene. Proc Natl Acad Sci U S A 95(26):15481–15486. https://doi.org/10.1073/pnas.95.26.15481

Erez Z, Steinberger-Levy I, Shamir M, Doron S, Stokar-Avihail A, Peleg Y (2017) Communication between viruses guides lysis-lysogeny decisions. Nature 541(7638):488–493. https://doi.org/10.1038/nature21049

Ghosh D, Roy K, Williamson KE, Srinivasiah S, Wommack KE, Radosevich M (2009) Acyl-homoserine lactones can induce virus production in lysogenic bacteria: an alternative paradigm for prophage induction. Appl Environ Microbiol 75(22):7142–7152. https://doi.org/10.1128/AEM.00950-09

Gold M, Becker A (1983) The bacteriophage lambda terminase. Partial purification and preliminary characterization of properties. J Biol Chem 258(23):14619–14625

Golding I (2016) Single-cell studies of phage λ: hidden treasures under Occam's Rug. Annu Rev Virol 3(7):1–20. https://doi.org/10.1146/annurev-virology-110615-042127

Goldstein R, Sedivy J, Ljungquist E (1982) Propagation of satellite phage P4 as a plasmid. Proc Natl Acad Sci U S A 79(2):515–519. https://doi.org/10.1073/pnas.79.2.515

Grayhack EJ, Roberts JW (1982) The phage-lambda Q gene-product – activity of a transcription anti-terminator Invitro. *Cell* 30(2):637–648. https://doi.org/10.1016/0092-8674(82)90260-4

Grayhack EJ, Yang XJ, Lau LF, Roberts JW (1985) Phage lambda gene-Q Antiterminator recognizes Rna-polymerase near the promoter and accelerates it through a pause site. Cell 42(1):259–269. https://doi.org/10.1016/S0092-8674(85)80121-5

Gussin G, Peterson V (1972) Isolation and properties of rex- mutants of bacteriophage lambda. Virology 10(4):760–765

Heller K, Braun V (1979) Accelerated adsorption of bacteriophage T5 to Escherichia coli F, resulting from reversible tail fiber-lipopolysaccharide binding. J Bacteriol 139(1):32–38

Hoyland-Kroghsbo NM, Maerkedahl RB, Svenningsen SL (2013) A quorum-sensing-induced bacteriophage defense mechanism. MBio 4(1):e00362–e00312. https://doi.org/10.1128/mBio.00362-12

Humbelin M, Suri B, Rao DN, Hornby DP, Eberle H, Pripfl T (1988) Type III DNA restriction and modification systems EcoP1 and EcoP15. Nucleotide sequence of the EcoP1 operon, the EcoP15 mod gene and some EcoP1 mod mutants. J Mol Biol 200(1):23–29. https://doi.org/10.1016/0022-2836(88)90330-0

Klein A, Lanka E, Schuster H (1980) Isolation of a complex between the P-protein of phage-lambda and the Dnab-protein of Escherichia-Coli. Euro J Biochem 105(1):1–6. https://doi.org/10.1111/j.1432-1033.1980.tb04467.x

Kourilsky P (1973) Lysogenization by bacteriophage lambda-I. multiple infection and the lysogenic response. Mol Gen Genet 122:183–195

Kourilsky P (1974) Lysogenization by bacteriophage lambda-II. Identification of genes involved in the multiplicity dependent processes. *Biochimie* 56:1511–1515

Kourilsky P, Knapp A (1974) Lysogenization by bacteriophage lambda. III Multiplicity dependent phenomena occuring upon infection by lambda. *Biochimie* 56(11–12):1517–1523

Lin PF, Bardwell E, Howard-Flanders P (1977) Initiation of genetic exchanges in lambda phage--prophage crosses. Proc Natl Acad Sci U S A 74(1):291–295. https://doi.org/10.1073/pnas.74.1.291

Lindqvist BH, Deho G, Calendar R (1993) Mechanisms of genome propagation and helper exploitation by satellite phage P4. Microbiol Rev 57(3):683–702

Loverdo C, Benichou O, Voituriez R, Biebricher A, Bonnet I, Desbiolles P (2009) Quantifying hopping and jumping in facilitated diffusion of DNA-binding proteins. Phys Rev Lett 102(18). https://doi.org/10.1103/Physrevlett.102.188101

Mallory JB, Alfano C, McMacken R (1990) Host virus interactions in the initiation of bacteriophage lambda DNA replication. Recruitment of Escherichia coli DnaB helicase by lambda P replication protein. J Biol Chem 265(22):13297–13307

Marvik OJ, Dokland T, Nokling RH, Jacobsen E, Larsen T, Lindqvist BH (1995) The capsid size-determining protein Sid forms an external scaffold on phage P4 procapsids. J Mol Biol 251(1):59–75. https://doi.org/10.1006/jmbi.1995.0416

Matz K, Schmandt M, Gussin GN (1982) The rex gene of bacteriophage lambda is really two genes. Genetics 102(3):319–327

Miller ES, Kutter E, Mosig G, Arisaka F, Kunisawa T, Ruger W (2003) Bacteriophage T4 genome. Microbiol Mol Biol Rev 67(1):86–156

Oppenheim AB, Kobiler O, Stavans J, Court DL, Adhya SL (2005) Switches in lambda development. Annu Rev Genet 39:409–429

Paddison P, Abedon ST, Dressman HK, Gailbreath K, Tracy J, Mosser E (1998) The roles of the bacteriophage T4 r genes in lysis inhibition and fine-structure genetics: a new perspective. Genetics 148(4):1539–1550

Pedruzzi I, Rosenbusch JP, Locher KP (1998) Inactivation in vitro of the Escherichia coli outer membrane protein FhuA by a phage T5-encoded lipoprotein. FEMS Microbiol Lett 168(1):119–125

Ptashne M (1967) Isolation of the lambda phage repressor. PNAS 57:306–313

Rao DN, Dryden DT, Bheemanaik S (2014) Type III restriction-modification enzymes: a historical perspective. Nucleic Acids Res 42(1):45–55. https://doi.org/10.1093/nar/gkt616

Reichardt L, Kaiser A (1971) Control of lambda repressor synthesis. PNAS 68(9):2185–2189

Roberts J (1975) Transcription termination and late control in phage lambda. PNAS 72(9):3300–3304

Shao Q, Hawkins A, Zeng L (2015) Phage DNA dynamics in cells with different fates. Biophys J 108(8):2048–2060. https://doi.org/10.1016/j.bpj.2015.03.027

Shao Q, Trinh JT, McIntosh CS, Christenson B, Balázsi G, Zeng L (2016) Lysis-lysogeny coexistence: prophage integration during lytic development. *MicrobiologyOpen* 6:e00395. https://doi.org/10.1002/mbo3.395

Shao Q, Cortes MG, Trinh JT, Guan J, Balazsi G, Zeng L (2018) Coupling of DNA replication and negative feedback controls gene expression for cell-fate decisions. iScience 6:1–12. https://doi.org/10.1016/j.isci.2018.07.006

Shao Q, Trinh JT, Zeng L (2019) High-resolution studies of lysis-lysogeny decision-making in bacteriophage lambda. J Biol Chem 294(10):3343–3349. https://doi.org/10.1074/jbc.TM118.003209

Shub DA (1994) Bacterial viruses. Bacterial altruism? Curr Biol 4(6):555–556. https://doi.org/10.1016/s0960-9822(00)00124-x

Silpe JE, Bassler BL (2019) A host-produced quorum-sensing autoinducer controls a phage lysis-Lysogeny decision. Cell 176(1–2):268–280. e13. https://doi.org/10.1016/j.cell.2018.10.059

Six EW, Klug CA (1973) Bacteriophage P4: a satellite virus depending on a helper such as prophage P2. Virology 51(2):327–344

Suttle CA (2005) Viruses in the sea. Nature 437(7057):356–361. https://doi.org/10.1038/nature04160

Trinh JT, Zeng L (2017) Virus interactions: cooperation or competition? Future Microbiol 12:561–564. https://doi.org/10.2217/fmb-2017-0048

Trinh JT, Szekely T, Shao Q, Balazsi G, Zeng L (2017) Cell fate decisions emerge as phages cooperate or compete inside their host. Nat Commun 8:14341. https://doi.org/10.1038/ncomms14341

Trinh JT, Shao Q, Guan J, Zeng L (2020) Emerging heterogeneous compartments by viruses in single bacterial cells. Under Review

Wang IN, Smith DL, Young R (2000) Holins: the protein clocks of bacteriophage infections. Annu Rev Microbiol 54:799–825. https://doi.org/10.1146/annurev.micro.54.1.799

Wang W, Li GW, Chen C, Xie XS, Zhuang X (2011) Chromosome organization by a nucleoid-associated protein in live bacteria. Science 333(6048):1445–1449. https://doi.org/10.1126/science.1204697

Warren RA (1980) Modified bases in bacteriophage DNAs. Annu Rev Microbiol 34:137–158. https://doi.org/10.1146/annurev.mi.34.100180.001033

Yu D, Ellis HM, Lee EC, Jenkins NA, Copeland NG, Court DL (2000) An efficient recombination system for chromosome engineering in Escherichia coli. Proc Natl Acad Sci U S A 97(11):5978–5983. https://doi.org/10.1073/pnas.100127597

Zeng L, Skinner SO, Zong C, Sippy J, Feiss M, Golding I (2010) Decision making at a subcellular level determines the outcome of bacteriophage infection. Cell 141(4):682–691. https://doi.org/10.1016/j.cell.2010.03.034

Chapter 5
Social Interactions Among Bacteriophages

Pilar Domingo-Calap and Rafael Sanjuán

Abstract Although viruses lack many of the social adaptations shown by more complex organisms, different types of social interactions have been unraveled in viruses. Phage research has contributed significantly to the development of this field, called sociovirology, with the discovery of processes such as intracellular and extracellular public good production, prudent host exploitation, cheating, and inter-phage communication. We here review and discuss these processes from a social evolution approach. Similar to other organisms, the origin and maintenance of phage-phage interactions can be explained using kin selection, group selection and game theory approaches. Key determinants of phage social evolution include genetic relatedness, spatial population structure, and frequency-dependent selection, among others. Finally, we discuss possible applications of phage social interactions.

5.1 Social Interactions Go Viral

Social interactions have been investigated by evolutionary biologists in many organisms, from bacteria to mammals. Social interactions arise when the traits of one individual influence the fitness of other individuals in a population, and have been classified according to the direct fitness effects experienced by the actor expressing a given social trait and the recipient of this action. Mutualism takes place when the interaction benefits both the actor and the recipient. A selfish trait benefits the actor at the expense of the recipient whereas, conversely, altruism takes place when the recipient obtains a fitness benefit but the actor experiences a fitness loss. Finally, spiteful interactions are detrimental to both actor and recipient (West et al. 2007).

Traits that are detrimental to the actor according to the above classification are difficult to explain from an evolutionary point of view. Indeed, a major puzzle for

P. Domingo-Calap (✉) · R. Sanjuán
Institute for Integrative Systems Biology, Universitat de València-Consejo Superior de Investigaciones Científicas, Paterna, Spain
e-mail: domingocalap@gmail.com

G. Witzany (ed.), *Biocommunication of Phages*,
https://doi.org/10.1007/978-3-030-45885-0_5

the evolution of cooperation (including mutualistic and altruistic interactions) is that, in principle, any population of cooperators is susceptible of being invaded by cheaters. Cheaters are selfish individuals who benefit from pre-existing cooperators without reciprocating. If the cooperative action entails some cost, then any interaction between a cheater and a cooperator will increase the fitness of the former at the expense of the latter. As a result, cheaters will increase in frequency, jeopardizing the maintenance of cooperation.

One way of solving this apparent conundrum is to define fitness as a compound of direct and indirect effects (inclusive fitness). Direct effects are those felt by the actor as a result of the actor's trait, whereas indirect effects are experienced by others. An altruistic trait may be favored by natural selection if it provides indirect benefits to the actor despite the direct costs. Such indirect effects can be obtained by directing altruism towards genetically related individuals who tend to share the altruistic trait (or by directing a spiteful behavior specifically towards unrelated individuals). By increasing the fitness of individuals who are also altruistic, altruism could be indirectly selected. Generally speaking, for cooperation to be maintained in a population, cooperators should preferentially interact with other cooperators. These concepts have been formalized within the frameworks of kin selection, group selection, and game theory (West et al. 2006; Nowak 2006; Lion et al. 2011; Marshall 2011; Birch 2018).

Social evolution has been applied to viruses recently (Díaz-Muñoz et al. 2017). For this "sociovirology" approach to provide useful explanations of certain aspects of viral biology, it is important that the mechanisms involved in virus-virus interactions are unraveled. The mechanistically-informed bottom-up approach allows us to understand which evolutionary outcomes are possible. For instance, in vertebrates, learning and memory play an important role in social interactions since they allow individuals to identify cooperators, but this is obviously not an expected scenario in viruses. However, viruses have been recently shown to possess communication capabilities that allow demographical information to be transmitted among members of the viral population (Erez et al. 2017). Another well-known factor allowing for the evolution of cooperation is kin recognition, which is displayed by many organisms including bacteria (Wall 2016), but this process is unknown in viruses. Yet, a major driver of cooperation at all levels of biological organization and which applies fully to viruses is population spatial structure. By producing local groups of genetically related individuals, spatial structure can promote cooperation. Recent work has indeed shown that spatial structure may determine the social evolution of important viral traits such as innate immunity evasion (Domingo-Calap et al. 2019).

In addition to unraveling new virus-virus interactions, the social evolution perspective has cast light on processes that have been known for decades in the virology field. For instance, a classical observation in virology is that, when a virus is cultured at high population densities such that on average each cell receives more than one infectious particle (high multiplicity of infection, MOI), defective interfering particles (DIPs) can invade the viral population (Rezelj et al. 2018). DIPs are viruses lacking large and essential portions of the genome and which can replicate and package their genomes only in cells coinfected with the normal virus. By competing with normal viruses, DIPs can severely reduce the production of functional genomes and lower mean population fitness dramatically. DIPs constitute an extreme form of social cheating which, interestingly, reveal cooperation, since the pre-existence of cooperation is needed for cheating to evolve. Social evolution theory predicts that DIP evolution should depend inversely on levels of genetic relatedness in the population, which in turn are determined by a number of factors such as mutation rate, population bottlenecks, and spatial structure.

Another recent progress in sociovirology has been the realization that levels of coinfection in a given population do not depend solely on the probability that two independent viral particles reach the same cell or host since, in many viruses, viral spread involves collective infectious units (Sanjuán 2017). These are structures that contain multiple copies of a virus genome, which are transmitted jointly. Examples of collective infectious units include virion aggregates, pools of virions encapsulated in extracellular lipid vesicles, polyploid virions, and occlusion bodies, among other structures. Collective infectious units have been reported in widely different types of animal viruses including paramyxovirurses, rhabdoviruses, enteroviruses, rotaviruses, noroviruses, polyomaviruses, and baculoviruses, among others. This transmission mode should promote virus-virus interactions at the intracellular level. For instance, it has been shown that two variants of a given virus can interact in a mutualistic manner by complementing genetic defects, or by combining their gene products in a way that allows new functions to be performed (Shirogane et al. 2012). However, collective spread can also favor the emergence of DIPs (Andreu-Moreno and Sanjuán 2019).

As outlined above, social processes have been demonstrated in different types of viruses. Here, though, we will focus on the sociobiology of bacteriophages. Phages have contributed decisively to our understanding of social interactions among viruses thanks to the experimental demonstration of prudent host exploitation, cheating, public good production, and viral communication. In some cases, our current understanding of these interactions allows linking specific molecular mechanisms to social processes at the population level.

5.2 Cheating Among Phages

As outlined above, DIPs have been reported in a wide number of animal viruses. In contrast, surprisingly few studies have described DIPs in phages. One study showed rapid accumulation of defective mutants lacking more than 70% of the genome in the small DNA phage f1 after serial passaging under a high MOI regime (Enea et al. 1977). There are several possible reasons for such unbalance in DIP reporting between animal viruses and phages. First, most DIPs have been found in RNA viruses, possibly because their replication mechanism is more prone to generating large deletions via template jumps of the viral polymerase during replication. Whereas RNA viruses are highly abundant among eukaryotic viruses, use of RNA as genetic material seems less frequent among phages. Second, the maintenance of DIPs in a population critically requires high MOIs. Since the virus/cell size ratio is lower for eukaryotic viruses than for phages, high MOIs might be more easily achieved in the former. Also, many animal cells are static and cultured as adherent monolayers, as opposed to bacteria. In a population of static cells, viral infections usually propagate as foci showing high local virus densities and hence high MOIs, particularly if the medium is viscous. Third, it has been hypothesized that DIPs might play a role as regulators of innate immunity in some animal viruses (Yang et al. 2019), leading to the speculation that these viruses might have evolved a propensity to produce DIPs.

Less extreme forms of cheating have been investigated in phages, such as for instance in *Pseudomonas sp.* phage Ø6 (Turner and Chao 1999). In this study, the evolutionary outcome of the interaction between a cooperator and a cheater was conceptualized using game theory. For this, a simple two-by-two payoff matrix was used to define the fitness of cooperators when they interact with other cooperators ($f_{A|A}$), of cooperators when they interact with cheaters ($f_{A|C}$), of cheaters when they interact with cooperators ($f_{C|A}$), and of cheaters when they interact with cheaters ($f_{C|C}$). For DIPs, the following holds true: $f_{C|A} > f_{A|A} > f_{A|C} > f_{C|C}$. In a deterministic scenario, this configuration leads to coexistence of cheaters and cooperators, because cheaters are fitter when cooperators are highly abundant, but cooperators become fitter when cheaters are highly abundant (frequency-dependent selection). Indeed, DIPs cannot replicate alone and hence cannot fully outcompete cooperators ($f_{C|C} = 0$). Yet, in the above study (Turner and Chao 1999), the following configuration was obtained after measuring the fitness of each variant: $f_{C|A} > f_{A|A} > f_{C|C} > f_{A|C}$. This leads to deterministic extinction of cooperators, despite the fact that population fitness would be highest if cheaters went extinct. This paradoxical situation is akin to the well-known Prisoner's dilemma game, in which defecting is always the optimal strategy (Fig. 5.1).

Hence, simple game theory models show that, under some situations, natural selection can reduce mean population fitness. Social evolution theory has investigated under which conditions invasion of the population by cheaters can be avoided. As explained above, this is achieved by ensuring that cooperators interact preferentially with other cooperators and, in viruses, spatial structure is probably the most important factor allowing for such assortment. Confirming this, a follow-up study with phage Ø6 showed that the Prisoner's dilemma could be avoided by propagating the virus under conditions in which population structure promoted high levels of genetic relatedness (clonality) (Turner and Chao 2003).

5.3 Intracellular Public Goods in Phages

Public goods are resources produced by some or all members of a population and available to other members of the population. Because the resource is public, it is cheatable by individuals who benefit from the goods without contributing to the common pool. If producing the goods is costly, cheaters will benefit from a fitness advantage over producers (cooperators). This will lead to an increase in the

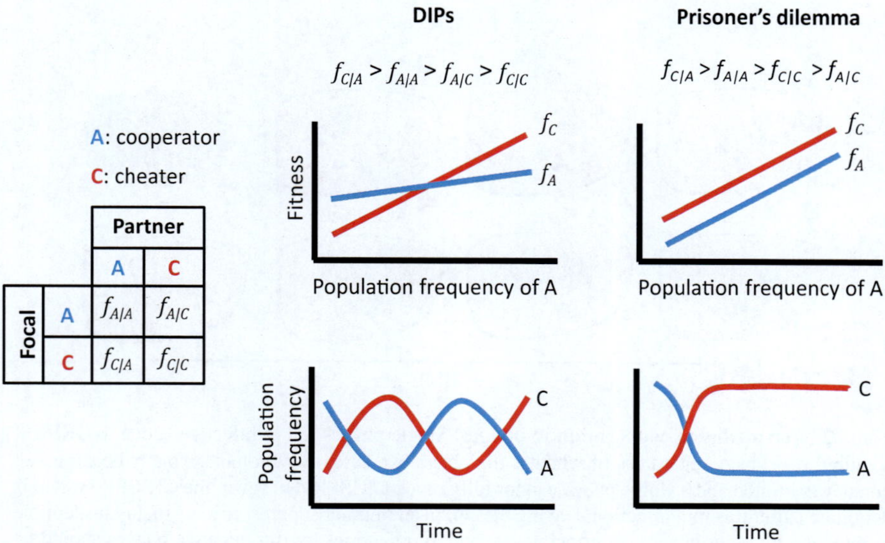

Fig. 5.1 Frequency-dependent selection in populations of cooperators and cheaters. Simple two-by-two payoff matrix defining the fitness of cooperators and cheaters as a function of virus-virus interactions. Two possible scenarios are represented, DIPs and Prisoner's dilemma, showing frequency-dependency and the resulting population dynamics

population frequency of cheaters and, ultimately, to extinction of the public good, a situation known and the tragedy of the commons. Again, for public goods to be preserved in a population there has to be some mechanism preventing cheater invasion, which typically involves directing the goods preferentially to cooperators.

The very existence of cheaters (particularly DIPs) under high MOI regimes shows that viral proteins can function as public goods at the intracellular level. Yet, not all viral proteins should be necessarily sharable among viruses coinfecting a cell. This depends on whether the involved proteins diffuse passively in the cytoplasm or remain associated to the genomes encoding them by means of intracellular ultrastructures or DNA-protein interactions. This again stresses the importance of identifying the molecular basis of virus-virus interactions when studying social evolution processes.

Although intracellular phage public goods are still poorly understood, anti-CRISPR proteins provide an interesting example. CRISPR (clustered regularly interspaced short palindromic repeats) and their associated proteins have evolved as a major system for preventing phage infection. Immunity against future infections is acquired by incorporating small sequences from the phage as spacers into the CRISPR array located in the bacterial chromosome. Subsequent transcription of small CRISPR RNAs guides the specific destruction of re-infecting phages by cellular nucleases, based on sequence complementarity between the phage and spacers (Brouns et al. 2008). Among other escape mechanisms, phages have evolved proteins capable of blocking the CRISPR response, called anti-CRISPR (Acr) proteins (Bondy-Denomy et al. 2013). Importantly, though, the action of Acr proteins is often imperfect, leading only to a partial suppression of CRISPR that results in

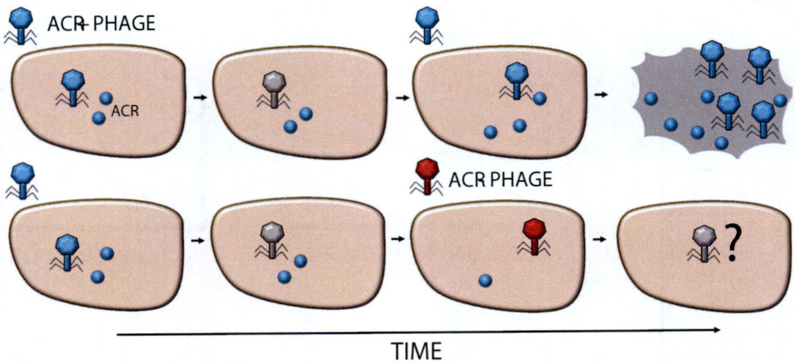

Fig. 5.2 Acr-mediated cooperation in phages. Single phages fail to infect cells due to CRISPR acquired resistance, regardless of whether the phage expresses Acr proteins. This is because Acr potency is limited and a single phage cannot fully block CRISPR. However, the CRISPR system is debilitated after this first infection, leaving the cell in an immunosuppressed state that is susceptible to infection by other phages. In principle, secondary infections by Acr-negative phages should be less successful than those by Acr-positive phages, although this remains to be experimentally tested

failure to infect and degradation of the incoming virus. However, a sufficiently high intracellular concentration of Acr proteins can succeed in blocking CRISPR, and this can be achieved by combining the Acr proteins of several members of the phage population, as shown for *Pseudomonas aeruginosa* phage DMS3*m* (Landsberger et al. 2018). Interestingly, this type of cooperation does not necessarily require that multiple phages coinfect a cell simultaneously. A first infection might fail due to insufficient Acr potency, but might nevertheless reduce CRISPR activity and leave the cell in an immunosuppressed state for a certain amount of time, during which a second phage might enter. The added Acr activity of the second phage would then achieve a more efficient CRISPR inhibition, producing a successful infection (Fig. 5.2). This cooperation should be relatively robust against cheating, because Acr-negative variants in the phage population might not achieve a sufficient level of CRISPR inhibition even in previously immunosuppressed cells.

Although work on the social nature of Acr proteins is still restricted to *Pseudomonas* phages, new Acr proteins have been described recently in phages infecting other bacterial genera such as *Streptococcus* (Hynes et al. 2018), as well as in archaeal viruses (He et al. 2018).

5.4 Extracellular Public Goods in Phages

There is a vast literature on the social evolution of secreted factors that function as public goods in bacteria and other microorganisms (West et al. 2006; Nadell et al. 2008; Özkaya et al. 2017). In contrast, extracellular public goods remain largely understudied in viruses, probably because they are rare. A potentially interesting system, though, is constituted by phage depolymerases. Many bacteria are encapsulated in a protective layer of external exopolysaccharides (EPS). For instance, biomedically relevant bacteria such as *Klebsiella pneumoniae*, *Escherichia coli*, and *Acinetobacter baumanii*, are major EPS producers. The EPS layer functions as a barrier against phage infection, hampers immune recognition, and limits entry of antibiotics (Labrie et al. 2010). In their continual arms race against hosts, some phages have acquired hydrolases capable of digesting EPS (depolymerases), allowing the phage to cross the EPS barrier (Pires et al. 2016).

Most phage depolymerases are anchored to the virion tail-spikes, but some are secreted as free enzymes (Latka et al. 2017). In principle, this defines free depolymerases as diffusible public goods, whereas tail-spike linked depolymerases ought to be considered as privatized goods as long as they are only available to the phage encoding the depolymerase. Whether depolymerase privatization occurred throughout evolutionary history as a means of preventing cheater invasion or whether depolymerase anchoring to the tail-spikes obeys other evolutionary processes (e.g. increased local efficiency) remains to be assessed. It should be noted, though, that tail-spike depolymerases might also be used by other members of the population if lysis releases tail-spike-depolymerase complexes that have not been incorporated into virions, or if depolymerized EPS become accessible to other phages.

Empirical support for the use of phage depolymerases as public goods was provided in a study with coliphages, albeit in this case the interaction was established between two phages belonging to different species (Schmerer et al. 2014). One of the phages was strongly lytic but was poorly infectious due to its inability to break the EPS barrier, whereas the other phage was capable of digesting EPS and infect cells, but had less lytic power. The combination of both phages resulted in a synergistic interaction whereby cells were more efficiently killed when inoculated with both phages as compared to the effects of single-phage infections. In addition to this cooperative effect, other types of interaction mediated by depolymerases are possible, including antagonistic interactions (Fig. 5.3).

5.5 Evolution of Prudent Phages

In general, "prudent" use of resources can be viewed as a cooperative action aimed at increasing fitness at the population level over the long term at the expense of reducing short-term fitness. Specifically, if a phage spends more time inside the host before triggering lysis (reduced virulence) or spends more time in the extracellular milieu before entering a host (lower infectivity), nearby uninfected bacteria will have more time to replicate before being infected and lysed (Fig. 5.4). This should increase the number of susceptible hosts accessible to the phage in subsequent generations, providing a long-term fitness benefit to the phage. In contrast to this prudent strategy, "rapacious" phages benefit from a rapid release of their progeny but this could exhaust the host population more rapidly, lowering fitness over the long

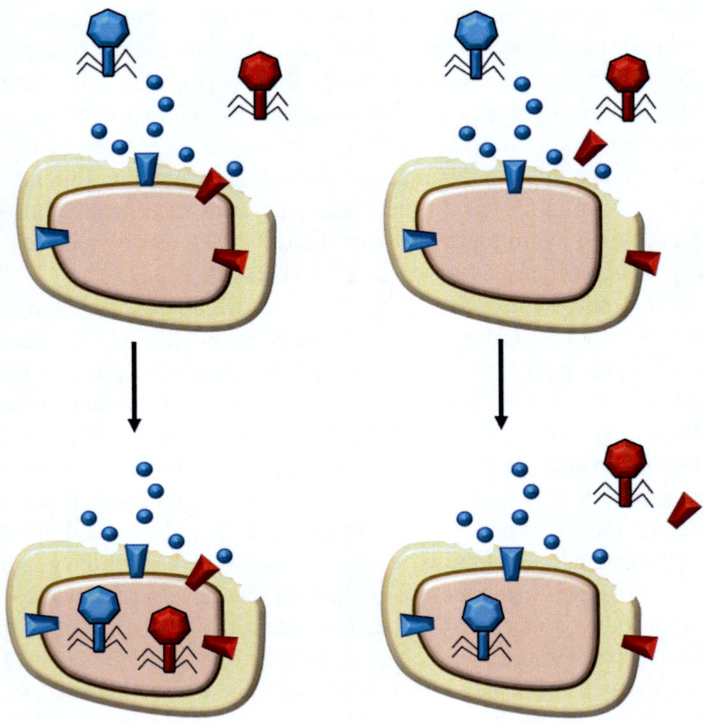

Fig. 5.3 Soluble phage depolymerases as public goods. Left: phage receptors are buried inside the EPS capsule. Depolymerases encoded by the blue phage digest EPS and allow entry of the blue phage, but also of another phage (red) that does not produce depolymerases. This can lead to synergistic effects of phage combinations on lysis, but also to competition and/or cheating. Right: another possible type of interaction. The red phage uses a receptor anchored to the EPS. Depolymerases encoded by the blue phage block entry of the red phage because receptors are lost following EPS digestion

term. How this tradeoff is solved depends on the spatial structure of the population (Boots and Mealor 2007). In viscous populations (low dispersal) where interactions take place preferentially among genealogically related phages, the benefits of prudent exploitation are felt preferentially by similar (prudent) phages. This assortment between cooperators should favor the evolution of prudent host exploitation. In contrast, in populations experiencing high dispersal rates, prudent phages (cooperators) will often compete with rapacious phages (cheaters). The latter will be favored because they exploit available resources more rapidly and, hence, benefit from a short-competitive advantage over prudent counterparts. This leads to a tragedy of the commons (the common goods being susceptible bacteria) whereby virulence increases and mean population fitness decreases (Kerr et al. 2006).

The molecular mechanism underlying prudency was investigated in *Escherichia coli* phage ID11 after passaging the virus for 500 generations in the presence of spatial structure. An adaptive mutation was found in the evolved lines that

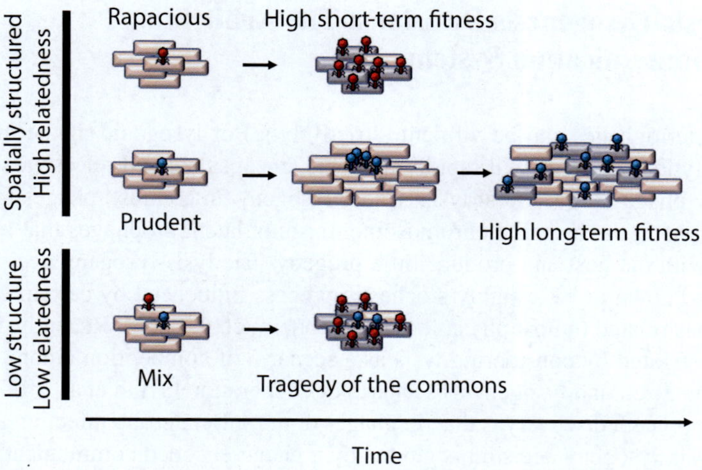

Fig. 5.4 Prudent versus rapacious host exploitation. Rapacious phages infect and kill cells more rapidly than prudent phages and hence have a short-term fitness benefit. However, rapacious exploitation exhausts the host population rapidly and reduces long-term population fitness. In spatially structured populations (high genetic relatedness), prudency can evolve because prudent and rapacious phages do not compete directly, but only indirectly at the inter-deme level. In contrast, in the absence of this structure (low relatedness) rapacious phages outcompete prudent phages, leading to suboptimal long-term fitness (tragedy of the commons)

substituted a single amino acid in the major capsid protein. This change reduced adsorption efficiency without altering lysis time or burst size. By increasing the time required to complete the infection cycle, prudent phages gave more time for hosts to reproduce, allowing phages to infect more cells in subsequent generations. The observable consequence was an increased plaque size compared to the wild-type phage (Roychoudhury et al. 2014).

However, the relationship between genetic relatedness (or spatial structure) and virulence evolution is not straightforward. A work with *Pseudomonas fluorescens* phage Ø2 showed that this relationship was dependent on resource availability. Under low host density (limited resource availability), fast-killing variants were selected regardless of relatedness. In contrast, under non-limiting host availability and high relatedness, time to kill was significantly slowered. Thus, prudency evolved in spatially structured environments, but only when resources were not limiting (Leggett et al. 2017). Here, the key factor determining evolutionary outcome was whether intra-plaque (intra-deme) competition prevented the evolution of prudent exploitation. More generally, whether prudency can evolve depends on the relative importance of local (within group) versus global (inter group) competition.

5.6 Lysis-Lysogenesis Decisions: The Arbitrium Communication System

Many bacteriophages can be virulent (strictly lytic) or lysogenic. If the infection becomes lytic, the phage will rapidly replicate, encapsidate and release assembled virions by phage-induced host lysis. Under lysogeny, in contrast, phages integrate their genome in the bacterial chromosome, forming latent prophages that replicate together with the host and produce little progeny. The lysis-lysogeny decision has been long known to be a highly stochastic process influenced by certain environmental cues related to host physiology (Oppenheim et al. 2005). Recently, this has been investigated by considering lysis as a scenario of competition among phages coinfecting a cell and lysogeny as scenario of cooperation (Trinh et al. 2017).

A major recent discovery is that, in phages of the spBeta group infecting *Bacillus sp.*, the lysis-lysogeny decision is guided by a phage-encoded communication system called "arbitrium", which is based on secretion of a small peptide (AimP) that allows the phage to sense population density (Erez et al. 2017). Bacteria internalize the AimP peptide via the oligopeptide permease transporter, and AimP is then recognized by phage receptors. Once internalized, this peptide suppresses the transcription of a negative regulator of lysogeny called AimX. When phage population density is high, the AimP concentration in the medium becomes high and, consequently, AimX expression diminishes, de-repressing lysogeny. In contrast, when phage population density is low, lysogeny is repressed and lysis takes place (Fig. 5.5).

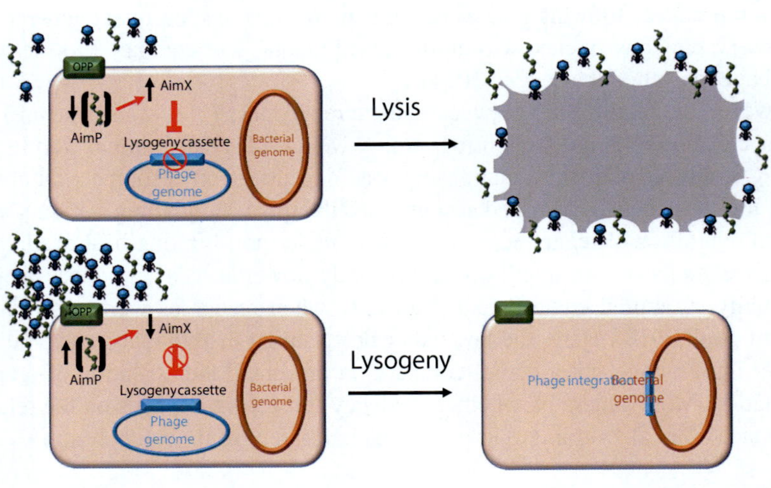

Fig. 5.5 Schematic representation of the arbitrium system. Viral lysogeny is regulated in a density-dependent manner. At high phage population densities, the concentration of the phage-encoded peptide AimP is high, and signal transduction then leads to lysogeny. In contrast, when phage population density is low, lysis is triggered. *OPP:* oligopeptide permease transporter. AimP: peptide arbitrium. AimX: negative regulator of lysogeny

The arbitrium system is clearly social, since it influences basic life-history traits (reproduction and dormancy) in other members of the phage population. An interesting aspect of this system is that it could be potentially used for suppressing the reproduction of competitor phages. It appears that, as a result, the system has rapidly evolved under diversifying selection to be species-specific. AimP peptide sequences differ among phage species and the mechanisms involved in lysis-lysogeny decision making also vary. Modulation of conformational changes in peptide receptors has been proposed to determine specificity in the mechanism of decision making (Dou et al. 2018).

5.7 Superinfection Exclusion

Many viruses have evolved mechanisms to avoid secondary infections by the same or a closely related virus (superinfection exclusion). This phenomenon usually prevents virion attachment to an already infected cell (Karam et al. 1994), although exclusion has been also observed at other levels including viral entry, replication, and transcription (Biryukov and Meyers 2018). Superinfection exclusion was first demonstrated using coliphages, where a role of early gene expression was proved to be key to avoid secondary infections (Abedon 2015). Subsequently, superinfection exclusion was described in many others viruses infecting bacteria, plants, and animals, including human viruses (Broecker and Moelling 2019).

Infection of bacteria containing prophages can result in prophage loss if the incoming phage is virulent. In order to avoid this scenario, prophages have also evolved mechanisms to modify the host cell and prevent superinfection. Studies with *Pseudomonas aeruginosa* have shown that superinfection exclusion can be mediated by different mechanisms, including induction of surface changes in the type IV pilus and O-antigen, but also through internal changes in the cell (Bondy-Denomy et al. 2016). In *Streptococcus thermophilus* prophages, superinfection exclusion is mediated by the lipoprotein Ltp encoded within the lysogeny module. After transcription, the lipoprotein is tethered to the outer membrane of the cell, preventing DNA injection by future phages (Sun et al. 2006).

From a social evolution perspective, superinfection exclusion can be interpreted in different ways. First, exclusion might not have an adaptive significance per se, but instead be a passive process whereby an infected cell becomes refractory to secondary infections simply as a byproduct of physiological changes associated to the infection process per se. However, this seems unlikely given that phages, and also prophages, contain specific genes controlling superinfection exclusion. Second, superinfection exclusion might be an adaptive trait favoring the resident virus by preventing competition from a secondary virus. Third, and less intuitively, superinfection exclusion might be a cooperative trait whereby the resident virus favors other members of the population by preventing them from initiating secondary infections that would be poorly productive, allowing them instead to reach alternate phage-free hosts containing more available resources (Fig. 5.6). Although the

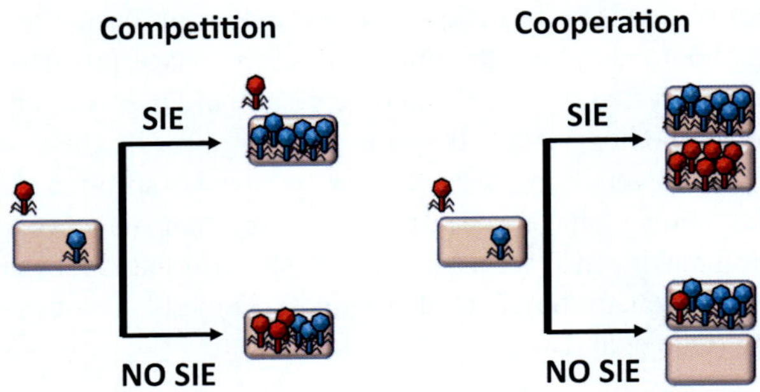

Fig. 5.6 Two possible evolutionary explanations for superinfection exclusion. Superinfection exclusion (SIE) could have evolved as a trait allowing the resident phage (blue) to avoid competition exerted by secondary infections (red phage). Alternatively, SIE could have evolved as a trait allowing secondary phages to avoid entering already-occupied cells, which would increase their chances of finding cells with more available resources. A key observation allowing us to distinguish between these two hypotheses would be whether, in the absence of SIE, the secondary phage can produce progeny efficiently and interferes with progeny production of the resident phage. Another key observation would be whether, once excluded from a pre-infected cell, the secondary phage is able to find other susceptible hosts (for this, pre-entry exclusion is required)

question whether superinfection exclusion is competitive or cooperative remains largely unexplored, a study with vaccinia virus supported the cooperation hypothesis (Doceul et al. 2010).Vaccinia-infected cells express surface proteins induced by the virus that promote repulsion of superinfecting virions. This is achieved by the expression of actin tails that propel virions away from the cell surface and allows them to reach uninfected cells, accelerating viral spread. Whether a similar process takes place on other viruses, including phages, remains unknown.

5.8 Collective Infection Units

Viruses are usually found in the environment as free viral particles. However, it is being increasingly recognized that viruses can also form collective infectious units, defined as multi-virion structures that allow for the co-spread of multiple viral genomes and cellular coinfection. Different types of collective infectious units have been described in animal viruses, including polyploid capsids, virion aggregates, virion-containing extracellular vesicles, occlusion bodies, virions attached to the surface of microbiota, and structures allowing for the transfer of pools of viral particles directly from cell to cell (Sanjuán 2017). The joint transmission of multiple particles to the same cell sets the stage for social virus interactions even when the overall ratio of virions to cells in a given population is low. These interactions can be cooperative (e.g. sharing viral products at the intracellular level) or selfish (e.g. DIP emergence), as detailed above. For instance, recent work with vesicular stomatitis virus has shown that virion aggregation accelerates the early stages of the cell infection cycle and increases the per-capita short-term fitness of the virus, defining an Allee effect at the level of cell invasion (Andreu-Moreno and Sanjuán 2018).

Collective infectious units remain largely unexplored in phages compared to animal viruses. Early work showed that f1 phage can form polyploid capsids, called "polyphage", but the implications for virus social evolution were not investigated (López and Webster 1983). In contrast, phage aggregation has been studied in some more detail. Phage aggregation can be determined by external factors such as changes in pH, temperature, ionic strength, and organic matter content (Gerba and Betancourt 2017). For example, free coliphage MS2 virions tend to aggregate near or at their isoelectric point where virion surface charge is close to neutral and repulsive electrostatic forces are minimal. In addition, it has been proposed that phosphate linking of the amino acid lysine to viral coat proteins could result in the formation of aggregates (Yuan et al. 2008). Phage aggregation was also reported in coliphage ØX174, although the presence of the spikes resulted in less compact aggregates (Kazumori 1981). Recent work has demonstrated that aggregation in T4 is mediated by environmental ion availability (Szermer-Olearnik et al. 2017). In this work, it was speculated that ion concentration functions as a regulator of phage aggregation/dispersal. In the neighborhood of lysed cells, locally high sodium concentrations tend to promote de-aggregation of phage particles, whereas as distance to cells increases sodium concentration decreases and aggregation is promoted. When aggregates of phage particles approach new healthy cells, local elevation of sodium levels would again promote de-aggregation (Fig. 5.7).

Aggregation might influence social processes by allowing invasion of cells with multiple particles at once. This might be the case of cooperation among Acr-encoding phages to overcome CRISPR immunity, although this remains to be investigated.

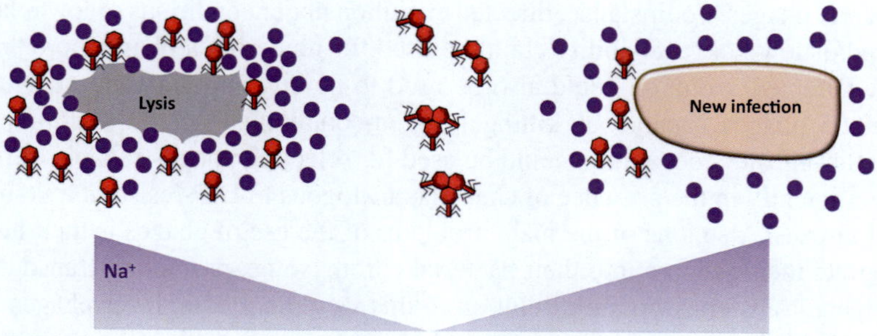

Fig. 5.7 Ion-dependent phage aggregation. Ionic strength might function as a regulator of phage aggregation/dispersal. After cell lysis, locally high concentration of Na^+ promote virions de-aggregation. As phages diffuse away from cells, the decrease in Na^+ concentration promotes aggregation. Once the phage approaches new healthy bacteria, local elevation of ionic levels might again promote de-aggregation. Whether these Na-dependent changes have an effect on phage fitness/infectivity remains to be addressed

5.9 Conclusions and Possible Applications of Phage Social Evolution

Whereas social evolution has been extensively applied to microbes, sociovirology is still an incipient field of research. We have nevertheless shown that phages display a variety of social traits such as prudency, public good production, cheating, communication, codispersal, and so on. Recent work has led to the discovery of exciting social features in phages, such as the arbitrium communication system regulating lysis-lysogeny decissions, and cooperation among Acr-producing phages to overcome CRISPR-based resistance. Some phage social processes have also been demonstrated in eukaryotic viruses, whereas others remain phage-exclusive, particularly phage communication. On the other hand, other processes, such as DIP emergence, have been extensively studied in animal viruses but remain less investigated in phages. Understanding the reasons for these differences as well as using current phage knowledge to guide research with eukaryotic viruses and vice versa may open new interesting avenues of research.

Besides allowing us to better understand phage biology, research on social evolution might also have some practical implications. Phage therapy is a promising alternative to current antibiotics, and use of phages cocktails has been suggested as an interesting strategy to increase the range of action of phage-based therapies and to limit the emergence of resistances (Ho 2001). Extensive phage-phage interactions might occur within phage cocktails, but these remain largely unexplored. Characterizing and exploiting these interactions might allow us to increase the efficacy of phage therapy. For instance, as outlined above, the fact that depolymerases can function as public goods makes it possible to make synergistic combinations of phages against a given bacterial strain.

Social evolution suggests additional interventions for increasing the efficacy of therapeutic phages. For instance, directed evolution under conditions of low relatedness and/or low host availability could be used for rendering a phage more "rapacious". Directed evolution could also be used to modify superinfection exclusion and create phages capable of killing lysogen-containing bacteria. Additionally, cooperator-cheater coevolution could be used for selecting phages capable of replicating efficiently in the presence of cheaters, analogous to DIP-resistant mutants in animal viruses. Also, one of the main problems of the use of phages is their ability to integrate their genome into their bacterial chromosome, promoting transduction lateral gene transfer. A possible solution to this would be to engineer phages with modified arbitrium systems that no longer undergo lysogeny. Finally, aggregation could be manipulated to render phages more stable in the environment, or to increase lysis potency in case Allee effects take place at the cellular level. This could be achieved, for instance, by constructing smart nanocapsules to control virion delivery (Drab 2018).

References

Abedon ST (2015) Bacteriophage secondary infection. Virol Sin 30:3–10. https://doi.org/10.1007/s12250-014-3547-2

Andreu-Moreno I, Sanjuán R (2018) Collective infection of cells by viral aggregates promotes early viral proliferation and reveals a cellular-level Allee effect. Curr Biol 28:3212–3219.e4. https://doi.org/10.1016/j.cub.2018.08.028

Andreu-Moreno I, Sanjuán R (2019) Collective viral spread mediated by virion aggregates promotes the evolution of defective interfering particles. MBio. (in press)

Birch J (2018) Kin selection, group selection, and the varieties of population structure. Br J Philos Sci. axx028. https://doi.org/10.1093/bjps/axx028

Biryukov J, Meyers C (2018) Superinfection exclusion between two high-risk human papillomavirus types during a coinfection. J Virol 92:e01993–e01917. https://doi.org/10.1128/JVI.01993-17

Bondy-Denomy J, Pawluk A, Maxwell KL, Davidson AR (2013) Bacteriophage genes that inactivate the CRISPR/Cas bacterial immune system. Nature 493:429–432. https://doi.org/10.1038/nature11723

Bondy-Denomy J, Qian J, Westra ER, Buckling A, Guttman DS, Davidson AR, Maxwell KL (2016) Prophages mediate defense against phage infection through diverse mechanisms. ISME J 10:2854–2866. https://doi.org/10.1038/ismej.2016.79

Boots M, Mealor M (2007) Local interactions select for lower pathogen infectivity. Science 315:1284–1286. https://doi.org/10.1126/science.1137126

Broecker F, Moelling K (2019) Evolution of immune systems from viruses and transposable elements. Front Microbiol 10. https://doi.org/10.3389/fmicb.2019.00051

Brouns SJJ, Jore MM, Lundgren M, Westra ER, Slijkhuis RJH, Snijders APL, Dickman MJ, Makarova KS, Koonin EV, van der Oost J (2008) Small CRISPR RNAs guide antiviral defense in prokaryotes. Science 321:960–964. https://doi.org/10.1126/science.1159689

Díaz-Muñoz SL, Sanjuán R, West S (2017) Sociovirology: conflict, cooperation, and communication among viruses. Cell Host Microbe 22:437–441. https://doi.org/10.1016/j.chom.2017.09.012

Doceul V, Hollinshead M, van der Linden L, Smith GL (2010) Repulsion of superinfecting virions: a mechanism for rapid virus spread. Science 327:873–876. https://doi.org/10.1126/science.1183173

Domingo-Calap P, Segredo-Otero E, Durán-Moreno M, Sanjuán R (2019) Social evolution of innate immunity evasion in a virus. Nat Microbiol 4:1006–1013. https://doi.org/10.1038/s41564-019-0379-8

Dou C, Xiong J, Gu Y, Yin K, Wang J, Hu Y, Zhou D, Fu X, Qi S, Zhu X, Yao S, Xu H, Nie C, Liang Z, Yang S, Wei Y, Cheng W (2018) Structural and functional insights into the regulation of the lysis-lysogeny decision in viral communities. Nat Microbiol 3:1285–1294. https://doi.org/10.1038/s41564-018-0259-7

Drab M (2018) Phage aggregation-dispersion by ions: striving beyond antibacterial therapy. Trends Biotechnol 36:875–881. https://doi.org/10.1016/j.tibtech.2018.03.002

Enea V, Horiuchi K, Turgeon BG, Zinder ND (1977) Physical map of defective interfering particles of bacteriophage f1. J Mol Biol 111:395–414. https://doi.org/10.1016/S0022-2836(77)80061-2

Erez Z, Steinberger-Levy I, Shamir M, Doron S, Stokar-Avihail A, Peleg Y, Melamed S, Leavitt A, Savidor A, Albeck S, Amitai G, Sorek R (2017) Communication between viruses guides lysis-lysogeny decisions. Nature 541:488–493. https://doi.org/10.1038/nature21049

Gerba CP, Betancourt WQ (2017) Viral aggregation: impact on virus behavior in the environment. Environ Sci Technol 51:7318–7325. https://doi.org/10.1021/acs.est.6b05835

He F, Bhoobalan-Chitty Y, Van LB, Kjeldsen AL, Dedola M, Makarova KS, Koonin EV, Brodersen DE, Peng X (2018) Anti-CRISPR proteins encoded by archaeal lytic viruses inhibit subtype I-D immunity. Nat Microbiol 3:461–469. https://doi.org/10.1038/s41564-018-0120-z

Ho K (2001) Bacteriophage therapy for bacterial infections. Rekindling a memory from the pre-antibiotics era. Perspect Biol Med 44:1–16. https://doi.org/10.1353/pbm.2001.0006

Hynes AP, Rousseau GM, Agudelo D, Goulet A, Amigues B, Loehr J, Romero DA, Fremaux C, Horvath P, Doyon Y, Cambillau C, Moineau S (2018) Widespread anti-CRISPR proteins in virulent bacteriophages inhibit a range of Cas9 proteins. Nat Commun 9:2919. https://doi.org/10.1038/s41467-018-05092-w

Karam JD, Drake JW, Kreuzer KN (1994) Molecular biology of bacteriophage T4, 1st edn. American Society for Microbiology, Washington, DC

Kazumori Y (1981) Electron microscopic studies of bacteriophage φ X174 intact and 'eclipsing' particles, and the genome by the staining and shadowing method. J Virol Methods 2:159–167. https://doi.org/10.1016/0166-0934(81)90034-3

Kerr B, Neuhauser C, Bohannan BJM, Dean AM (2006) Local migration promotes competitive restraint in a host-pathogen "tragedy of the commons". Nature 442:75–78. https://doi.org/10.1038/nature04864

Labrie SJ, Samson JE, Moineau S (2010) Bacteriophage resistance mechanisms. Nat Rev Microbiol 8:317–327. https://doi.org/10.1038/nrmicro2315

Landsberger M, Gandon S, Meaden S, Rollie C, Chevallereau A, Chabas H, Buckling A, Westra ER, van Houte S (2018) Anti-CRISPR phages cooperate to overcome CRISPR-Cas immunity. Cell 174:908–916.e12. https://doi.org/10.1016/j.cell.2018.05.058

Latka A, Maciejewska B, Majkowska-Skrobek G, Briers Y, Drulis-Kawa Z (2017) Bacteriophage-encoded virion-associated enzymes to overcome the carbohydrate barriers during the infection process. Appl Microbiol Biotechnol 101:3103–3119. https://doi.org/10.1007/s00253-017-8224-6

Leggett HC, Wild G, West SA, Buckling A (2017) Fast-killing parasites can be favoured in spatially structured populations. Philos Trans R Soc B Biol Sci 372:20160096. https://doi.org/10.1098/rstb.2016.0096

Lion S, Jansen VAA, Day T (2011) Evolution in structured populations: beyond the kin versus group debate. Trends Ecol Evol 26:193–201. https://doi.org/10.1016/j.tree.2011.01.006

Lopez J, Webster RE (1983) Morphogenesis of filamentous bacteriophage f1: orientation of extrusion and production of polyphage. Virology 127:177–193. https://doi.org/10.1016/0042-6822(83)90382-3

Marshall JAR (2011) Group selection and kin selection: formally equivalent approaches. Trends Ecol Evol 26:325–332. https://doi.org/10.1016/j.tree.2011.04.008

Nadell CD, Bassler BL, Levin SA (2008) Observing bacteria through the lens of social evolution. J Biol 7:27. https://doi.org/10.1186/jbiol87

Nowak MA (2006) Five rules for the evolution of cooperation. Science 314:1560–1563. https://doi.org/10.1126/science.1133755

Oppenheim AB, Kobiler O, Stavans J, Court DL, Adhya S (2005) Switches in bacteriophage lambda development. Annu Rev Genet 39:409–429. https://doi.org/10.1146/annurev.genet.39.073003.113656

Özkaya Ö, Xavier KB, Dionisio F, Balbontín R (2017) Maintenance of microbial cooperation mediated by public goods in single- and multiple-trait scenarios. J Bacteriol 199:22. https://doi.org/10.1128/JB.00297-17

Pires DP, Oliveira H, Melo LDR, Sillankorva S, Azeredo J (2016) Bacteriophage-encoded depolymerases: their diversity and biotechnological applications. Appl Microbiol Biotechnol 100:2141–2151. https://doi.org/10.1007/s00253-015-7247-0

Rezelj VV, Levi LI, Vignuzzi M (2018) The defective component of viral populations. Curr Opin Virol 33:74–80. https://doi.org/10.1016/j.coviro.2018.07.014

Roychoudhury P, Shrestha N, Wiss VR, Krone SM (2014) Fitness benefits of low infectivity in a spatially structured population of bacteriophages. Proc R Soc B Biol Sci 281:20132563. https://doi.org/10.1098/rspb.2013.2563

Sanjuán R (2017) Collective infectious units in viruses. Trends Microbiol 25:402–412. https://doi.org/10.1016/j.tim.2017.02.003

Schmerer M, Molineux IJ, Bull JJ (2014) Synergy as a rationale for phage therapy using phage cocktails. PeerJ 2:e590. https://doi.org/10.7717/peerj.590

Shirogane Y, Watanabe S, Yanagi Y (2012) Cooperation between different RNA virus genomes produces a new phenotype. Nat Commun 3:1235. https://doi.org/10.1038/ncomms2252

Sun X, Göhler A, Heller KJ, Neve H (2006) The ltp gene of temperate Streptococcus thermophilus phage TP-J34 confers superinfection exclusion to Streptococcus thermophilus and Lactococcus lactis. Virology 350:146–157. https://doi.org/10.1016/j.virol.2006.03.001

Szermer-Olearnik B, Drab M, Mąkosa M, Zembala M, Barbasz J, Dąbrowska K, Boratyński J (2017) Aggregation/dispersion transitions of T4 phage triggered by environmental ion availability. J Nanobiotechnol 15:32. https://doi.org/10.1186/s12951-017-0266-5

Trinh JT, Székely T, Shao Q, Balázsi G, Zeng L (2017) Cell fate decisions emerge as phages cooperate or compete inside their host. Nat Commun 8:14341. https://doi.org/10.1038/ncomms14341

Turner PE, Chao L (1999) Prisoner's dilemma in an RNA virus. Nature 398:441–443. https://doi.org/10.1038/18913

Turner PE, Chao L (2003) Escape from prisoner's dilemma in RNA phage Φ6. Am Nat 161:497–505. https://doi.org/10.1086/367880

Wall D (2016) Kin recognition in bacteria. Annu Rev Microbiol 70:143–160. https://doi.org/10.1146/annurev-micro-102215-095325

West SA, Griffin AS, Gardner A, Diggle SP (2006) Social evolution theory for microorganisms. Nat Rev Microbiol 4:597–607. https://doi.org/10.1038/nrmicro1461

West SA, Griffin AS, Gardner A (2007) Social semantics: altruism, cooperation, mutualism, strong reciprocity and group selection. J Evol Biol 20:415–432. https://doi.org/10.1111/j.1420-9101.2006.01258.x

Yang Y, Lyu T, Zhou R, He X, Ye K, Xie Q, Zhu L, Chen T, Shen C, Wu Q, Zhang B, Zhao W (2019) The antiviral and antitumor effects of defective interfering particles/genomes and their mechanisms. Front Microbiol 10:1852. https://doi.org/10.3389/fmicb.2019.01852

Yuan B, Pham M, Nguyen TH (2008) Deposition kinetics of bacteriophage MS2 on a silica surface coated with natural organic matter in a radial stagnation point flow cell. Environ Sci Technol 42:7628–7633. https://doi.org/10.1021/es801003s

Chapter 6
Phage Protein Interactions in the Inhibition Mechanism of Bacterial Cell

Chandrabose Selvaraj and Sanjeev Kumar Singh

Abstract Bacteriophages are one of the most diversified microorganisms and are play an important role in designing effective antibiotics and also acts as the potential antimicrobial agents against the harmful pathogens in number of applications including food processing, biotechnology, and medicine. Due to its inherent ability and specificity, phage-host interactions have been received more attention in the field of research to develop potential antibiotics. The initial attachment of the phages is mediated by several numbers of receptor binding proteins, which significantly recognizes and binds several number of proteins localized in bacterial cell wall. The co-evolution of receptor binding proteins has forced endless adoption in host-phage interaction. Hence, phage may recognize as an alternative drug to prevent the growth of the antibiotic-resistant bacterial infection. This chapter mainly focused on the various receptor proteins of both host and phage and their recognition and proteins involved in phage adsorption and penetration into the host cell.

6.1 Introduction

Phages are viral particles that effectively infect the specific species of prokaryotic cells or certain strains of the same species. Like all virus, bacteriophages also absolute parasites and their genome encode the protein, which relies heavily on the host cell for their replication. Generally, the infection of phages involved a well-organized, specific program such as host cell wall adsorption and internalization of genome and transition of phage-directed metabolism, phage genome replication, and morphogenesis and host cell lysis (Guttman and Raya 2005). It can adhere to the microbial cell surface and intermediate protein and other protein molecules finally inject their genetic material and lead different cycles of infection such as a lytic and lysogenic cycle (Muñoz and Koskella 2014; Jassim and Limoges 2014).

C. Selvaraj · S. K. Singh (✉)
Computer Aided Drug Design and Molecular Modelling Lab, Department of Bioinformatics, Alagappa University, Karaikudi, Tamil Nadu, India

G. Witzany (ed.), *Biocommunication of Phages*,
https://doi.org/10.1007/978-3-030-45885-0_6

The genetic material of all the bacteriophage is widely varied species to species and this feature can be used for the prediction of protein synthesis and polynucleotide sequence has been used for the taxonomic classification including order, families, genus, and species (Klumpp et al. 2012; Sanger et al. 1977). The receptor proteins present in the bacterial cell surface plays a significant role in interaction with host and also its surrounding environment. Generally, the diversity of the bacterial cell envelops viewed as dichotomic (gram-positive and gram-negative bacteria). The microscopic observation revealed that the structural divergence in bacterial envelop, gram-positive bacteria shows the thick cell wall sedentary on the top of the cytoplasmic membrane (Desvaux et al. 2004; Silhavy et al. 2010).

The host-pathogen interaction mechanism is one of the major targets for microbiologists who seek to understand the impact of phage on their host. Compared to gram-positive bacteria, phage significantly targets the both gram-positive and negative bacteria; hence it has been widely used to study the host-pathogen interaction mechanism (Mahony and van Sinderen 2012). Receptor binding proteins and tail fiber proteins are the main key factor for all types of phages that determine the phage specificity. Stability of these proteins and their binding nature and affinity with the cellular receptor of host cells has been exploited as a therapeutic target to reduce bacterial colonization (Simpson et al. 2016; Waseh et al. 2010). Basic three diagnostic techniques such as phage amplification assay which effectively detect the live cells, it may act as recognition agent for the identification of intracellular components after phage induced lysis. Due to the specific cell surface receptors interaction with phage proteins, most of the bacteriophages infect the limited number of species or strain of a single species (Sulakvelidze and Alavidze 2001). Several cellular receptor proteins and cell surface components like cell surface proteins, peptidoglycans, teichoic acid, lipopolysaccharides, capsule proteins, flagellum, and oligosaccharides are highly recognized by phages (Kutter et al. 2005; Calendar et al. 2006). This limited host range of phage is considered a significant advantage by which the highly pathogenic bacteria can be targeted without affecting the normal flora of the colonies. Recently, the accessibility of synthetic biology consents the usage of engineered viruses with desired attributes. Lu et al. (Lu and Collins 2007) designed a genetically modified lytic *Enterobacteria* phage T7 that shows significant biofilm degrading enzymes mechanism. It highly expresses the biofilm-degrading enzyme that enables to hydrolyze exopolysaccharide compounds in bacterial biofilm during its propagation. Such modified phages are widely used in the eradication of monoculture or phage cocktail to treat mix-culture biofilm. However, the utilization of genetically engineered viruses may affect the balance between natural viruses and bacterial hosts in the environment. The genetically modified phages are additionally carrying several reporter genes including LRP, GFP, β-Gal or tagging peptides for staining and detection. Based on the bioluminescence, fluorescence and enzymatic reactions, electrochemical nature, measurement of ATP or bacterial enzymes and changes in media composition phage peptides are tagged with modified viruses for easy and accurate detection. Besides, characteristic features and the possibility of the phage derived proteins also used in diagnosis for bacterial eradication (Drulis-Kawa et al. 2015).

6.2 Structural Organization of Phage

Phages are generally had three major components such as a capsid where the genetic materials are compactly arranged, a tail that serves as pipe and used to transfer the viral genome into the host cell during infection and a unique adsorption apparatus acts as adhesive system and at the end it has tail that will significantly recognize the cell surface and penetrate the cell wall of the host cell. The function of typical phage is a consequence of the multistep process starts with a cellular component of both host and phages including capsid, tail scaffolding, etc. The icosahedral symmetries and fivefold structure are the important characteristic feature of dsDNA which are broken at the one end of the fivefold axis by the heat-to-tail interface (Tavares 2018). Dodecameric portal protein is the main component of the head to the tail interface (HTI) that represents the DNA-packing motor, which acts as an important part of the nanomachines. The oligomeric rings of HIT significantly involved in the making of the additional interface to ATP molecules that provide energy for DNA packing and also play a key role in portal protein and tail binding mechanism. In some species, it also severs as valves that make a close-open system by which it prevents the genome leakage from capsid at the same time it may easily open during the phage attachment (Orlova et al. 1999; Cingolani et al. 2002; Cerritelli et al. 2003). Typically, most of the phage proteins have conservative folds and have different additional domains (Hendrix et al. 1999; Bamford et al. 2005). Another important structural component is a tail which plays a key role in infection. Mainly the adsorption apparatus located on its distal end significantly recognize host cell receptors and the envelope chemistry of the host to ensure the genome delivery. The phage belongs to the family *Myoviridae* and *Siphoviridae*, have a series of stacked rings in their tail region along with a host recognition device at the end. The phages belong to the family *Podoviridae* possess the adsorption apparatus along with HTI which is surrounded in many phages by fibrils which may ensure the close-fitting attachment to the host cells. Negative stain electron microscopy was used in earlier time to elucidate the typical structural organization of the T4 phage head, it clearly shows distinct parts such as procapsid, capsid, and poly head with the position of dispensable Hoc and Soc proteins which is similar to the structure predicted by cryo-electron microscopic structure determination of isometric capsids. However, the copy number of inferred protein and the dimension of the capsid have been slightly changed based on the higher resolution (Fokine et al. 2004). The resolution (~0.3 nm) used in the cryo-electron microscope permits the rationalization of the head length mutation in capsid protein and also the vertex protein. It also showed the bypass mutants that substitute the other parameters of the capsid proteins; the wild type capsid does not have the Soc decoration protein subunits of the gp23 molecules in the interface of the gp23-gp24 (Fokine et al. 2006; Fokine et al. 2005). Due to this rationalization the major capsid protein significantly diverts the small fraction of the capsid to alter the size variables and may affect the head size mutations (Black et al. 1994).

The precursor of the phage capsid named as procapsid formed during the assembly process. This process was driven by scaffolding proteins (SPs) along with chaperoning major capsid protein (MCP) subunits for the development of dsDNA. The SPs are mainly found to the proto complex during the formation of the precursor with scaffolding inside. The conformational changes in sequence from a pre capsid to capsid leads the maturation process where the genome has packed and go through the series of intermediates (Wikoff et al. 2000; Wikoff et al. 2006; Huet et al. 2010; Veesler et al. 2012). In some phages, SPs is not present separately, for example, the capsid protein of HHK97, and T5, fused with a scaffolding domain at N –terminal region, which immediately cleaved off from scaffolding domain and assembled like the separate SP will be removed from the capsid to make space for compact assembly of genome (Wikoff et al. 2006; Huet et al. 2010). During the maturation, the spherical shell becomes thinner due to the alteration of inter-intra-subunit contacts. Most of the tailed phage has icosahedral capsid mainly formed by several numbers of proteins. It is characterized by about 60 copies of the major independent parts (Baker et al. 1999). The presence of Oligomers of the proteins in fivefold axes are referred as pentons, while the same complex present on the faces of the icosahedrons are formed six subunits and are named as hexons (Lander et al. 2008). In λ phage, the mature capsid is stabilized with decoration protein gpD which is significantly attached to the threefold vertex. The structure of the pre capsid was determined with 13.3 and 14.5 Å resolution and the matured capsid was at 6.8 Å (Lander et al. 2008).

6.3 Genome Organization of Phage

The complete genome of phage (bacteriophage øX174) is first sequenced in 1997; the information obtained from the genome analysis indicates that the genetic material varies in different bacteriophages that can predict the protein synthesis by polynucleotide sequence. Hence it also used to taxonomically classify the various phages in orders, families and genus and species (Klumpp et al. 2012). The size of the phage genome was varied in various species ranging from ~3300 nucleotide ssRNA (virus of *Escherichia coli*) to 500 kbp (Bacillus megaterium phage G). The smallest genome structure of dsDNA tailed phages are range from ~11.5 kbp, ~21 kbp, ~ 30 kbp of Mycoplasma phage P1, *Lactococcus* phage c2, and *Pasteurella* phage F108 respectively (Friedman et al. 2009; Tu et al. 2001; Lubbers et al. 1995). Generally, these genomes are compactly packed in similar densities into the capsids and the size of the capsid also varies depends on the genome size. The amount of the packed DNA in the capsid significantly influenced the viron infectivity, either a small amount or too much leads to the loss of viron stability. Besides, evolutionary pressure may also influence the viron stability and loss or gain of DNA to accommodate the package. Hence, the size of the genome has played a crucial role in understanding of evolutionary relationship between the bacteriophage and providing a gain and loss of DNA mechanism that is independent of gene function. The newly formed DNA provides the reservoir of genetic information for future use. A comparative genome analysis of bacteriophage reveals that there are enormously diverse (Hatfull 2008). For example, the phage genome has known close relatives that infect the same specific host or the close relative prophage. These results suggested that the host organisms represent a significant genetic barrier to exchange, while phages can switch the host cells more easily to the close relative host cell by a variety of mechanisms, still it is unclear the rate host interaction occurs in natural populations. In addition to the genetic barriers, it also exhibits other berries to genetic exchange (Pope et al. 2011; Kwan et al. 2005, 2006). Some temperate phages possess the peculiar lifestyle with alternating the lytic phase and prophage existence makes phages fascinating objects for biologists for the understanding of the evolution of parasitic DNA. In the case of selfish and mutualistic aspects of prophage for bacteria emerge from prophage genomics (Desiere et al. 2001). For example, *Vibrio cholera* is an interesting organisms shows the contribution of multiple phages in the bacterial pathogenicity. This indicates most of the bacteria are plays a key role in inter-strain genetic variability due to the presence of horizontally transferred DNA. About 3–10% DNA of the bacterial genome is contributed by prophages. In the case of *Siphoviridae* genomes fall into two size classes such as 121 ± 134 kb and 22 ± 56 kb. The range of SPb phage genome belongs to the class 121 ± 134 kb shows no similarity to the smaller size genomes of *Siphoviridae* and might be polyphyletic. The most putative structural genes of SPb are significantly unrelated to the entries in the database, whereas the tail fiber genes share some similarities with the host organisms (Brüssow and Desiere 2001; Brüssow et al. 2004; Stalin and Srinivasan 2016a, b).

6.4 Mechanism of Host-Phage Interaction

Host-phage interactions initiate by the basic steps such as adsorption and penetration into the host outer membrane and insert their genetic material by which leads the infection. Phage penetration in the higher organisms leads the direct contact with host cells, therefore it is a very important mechanism to know whether phage can interact with or infect host cells. During the infection, several elements of the phage tail structure can bind to the specific receptors on the surface of the host cells. It also recognized that phages cannot infect the host cells, because of the major difference between eukaryotes and prokaryotes in penetration mechanisms that are essential for translation and replication (Sharp 2011). After insertion of the phage genome into the host cell, the cell is termed as a virocell that carrying virus auxiliary metabolic gene which responsible for metabolic potential during infection (Rosenwasser et al. 2016). Based on the recognition of the location and nature of the host cell receptors by phages are varies on the host. Phages also have significant binding site to receptors that are in the cell wall of both gram-negative and gram-positive bacteria (Xia et al. 2011; Marti et al. 2013), and also in slime layers and in appendages (Fehmel et al. 1975; Guerrero-Ferreira et al. 2011). This diversity in receptors structure of the host cells are involved in multiplicity mechanism developed by phages and host to overcome the evolutionary strategies adopted by their counterparts (Clokie et al. 2011). The host cell envelop is a complex multilayered structure that maintains the minimum level of integrity to assurance cell survival. It protects the cell from various environmental factors at the same time allowing the influx of essential nutrients and the efflux of waste products (Silhavy et al. 2010; Dufresne and Paradis-Bleau 2015). Most of the phages are have dsDNA as a genome and are packed within a capsid connected to a tail hence all of them have come under the order *Caudovirales* (Ackermann 2009). Based on the tail morphology is divided into three families such as *Podoviridae* (have a short stub-like tail), *Myoviridae* (have contractile tails) and *Siphoviridae* (have long non-contractile tails) (Bertozzi Silva et al. 2016). Most of the phages belong to the family *Siphoviridae* are characterized as gram-positive targeting phages. Distal tail protein (Dit) is one of the highly conserved proteins of gram-positive and gram-negative targeting phages including T5 and lambda (Flayhan et al. 2014; Pell et al. 2009). Dit protein of all phage significantly enhances the arrangement of the tail and adsorption apparatus and making a central hexameric bub with distal tail-tube and baseplate components such as Tal and RBPs (Veesler et al. 2010). For example, the N-terminal domain Dit protein of the Lactococcal phage forms a central channel with 40 Å wide and the C-terminal domain forms a galectin-like β-sandwich with an arm (consist of 60 residues) which acts a connector to a single RBP timer (Bebeacua et al. 2013). Recently evolved Dit protein identified from *Lactobacillus casei* phage J-1 and PLE3 shows the presence of two carbohydrate-binding modules which is not present in classical types. The carbohydrate-binding modules (CBM) of the phage are protruded outwards from the tail baseplate and significantly interact with host receptors (Dieterle et al. 2017). Tail-Associated Lysin is another often used protein for attachment, most of the phages have the feature C-terminal peptidoglycan hydrolase (PGH) activity and indirectly play role in host recognition and degradation of a thick layer of host cells. The terminal peptidoglycan hydrolase may ensure penetration of phage into the outer membrane and leads the infection. The enzymatic activity of peptidoglycan hydrolase enhances the cell infection by enabling cleavage of cross-linked peptidoglycan. Some phages do not have hydrolase function, but these phages infect the host cells during the exponential phase (Kenny et al. 2004; Mahony et al. 2013). Upper baseplate protein also presents in most of the phages which act as a connector between the Dit and peripheral RBPs. For example, the upper baseplate protein of TP901–1 assembles as six asymmetric trimmers forming a ring structure that is used for the attachment of Dit hub through N-terminal domains while the C-terminal domain attaches the RBPs complex (Veesler et al. 2012; Li et al. 2016). The structural architecture and morphology of baseplate and RBP are greatly varied in different family members, depending on the nature of infection and host receptor recognition. Protein-based receptors like SPP1 of *B. subtilis* have typical features consist of central dodecameric organization of two Dit hexamer attached with a single receptor binding site (Fig. 6.1). The mechanism behind the host recognition by phage SPP1 is processed in two basic steps; irreversible attachment of inner membrane protein YueB and reversible binding of

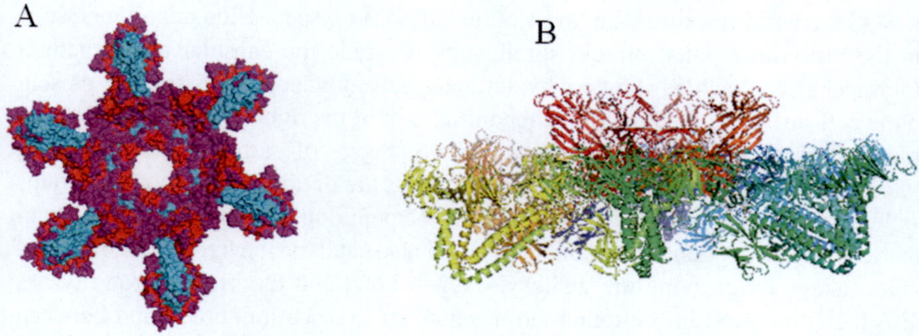

Fig. 6.1 (**a**) Top view of the baseplate. Each trimer of BppU shows an asymmetric organization. (**b**) The six BppU timers arranged like a circlet around the central part of Dit hexamer

glucosylated wall teichoic acid (WTA) (Mahony and van Sinderen 2012; Baptista et al. 2008).

6.5 Cell Surface Proteins of Phage

Most of the bacteria eating phages only infect the particular strain and the ability to affect the host cell is significantly influenced by highly specific interaction between host cell receptors and phage baseplate components (Weinbauer 2004; Gill and Hyman 2010). Phages are effectively-recognized the number of cell surface proteins including outer membrane proteins, teichoic acids, lipopolysaccharides, flagellum, sex pillus, oligosaccharides and fimbriae of the host cell (Kutter et al. 2005; Calendar et al. 2006). The virulence capacity of the phages is depended on the two aspects; the first aspects including the duration of efficient adhesion, latent period and progeny release, and the second aspect is the growth rate. The adsorption rate in certain strain and short phage generation is the main factor that determines the strong antibacterial efficacy of phage. Variations in environmental conditions also influence viral efficacy. Predators are effectively propagated thoroughly on the bacterial population, then the second aspect multiplicity of infection is fundamental for phage titer-dependent killing (Gill and Hyman 2010; Abedon 2009; Ryan et al. 2011). Several studies have been reported that the applications of both natural and modified phages like *Enterobacteria* phage M13, Ike and fd and *Pseudomonas* phage Pf3. Hagen and Blasi (Hagens and Blasi 2003) reported that the utilization of modified M13 and Pf3 filamentous phages to enhance anti-bacterial efficacy (Hagens and Blasi 2003; Hagens et al. 2004). In the modified phage, researchers have designed with additional genes of restriction endonuclease system that causing bacterial degradation or alter the inner membrane potential. The genetically modified phage Pf3 was designed in two different aspects; replicating, non-lytic and non-replicating. Later Lu and Collins have reported that the antibiotic-enhancing phage with standard drug treatment in phage M13mp18 carrying the lexA3 gene a potent repressor of SOS bacterial response (Lu and Collins 2009). Polysaccharide depolymerase is on the virion-associated protein significantly degrade the capsular and structural polysaccharides such as alginate, hyaluronan, polysialic acid, LPS and PG present in the cell surface of the host cell at the initial step of the tightly programmed phage infection process. Based on the degradation capacity of carbohydrate-containing polymers of host cells, phage associated enzymes are divided into two main groups such as hydrolases (EC3.2.1) which degrade the peptidoglycan or capsular polysaccharides and O-antigen side-chain of LPS and also catalyze the hydrolysis of glycosidic linkage by cleaving the glycosyl-oxygen bond and the second group lyases (EC 4.2.2) involved in β-elimination mechanism to leads a double bond between fourth carbon and fifth carbon atom of the uronic acid and also cleave the glycosidic linkage between a monosaccharide and fourth carbon atom of uronic acid (Sutherland 1999). Viron-associated peptidoglycan hydrolases (VAPGHs) is a phage-encoded lytic enzyme that specifically catalyzes the breakdown of important bonds in peptidoglycan by its degradation. Based on this catalytic enzymatic activity, phages are classified into four subgroups; lysozymes, lytic transglycosylases, glucosaminidases and endopeptidases and are noted in the current International Union of Biochemistry and Molecular biology (IUBMB) enzyme nomenclature (Stockdale et al. 2013; Caldentey and Bamford 1992; Rydman and Bamford 2000; Arisaka et al. 2003; Moak and Molineux 2000). Except for lytic transglycosylase, all the VAPGHs are hydrolases that cleave the β-1,4-glycosidic bond.

6.6 Phage Receptor Binding Proteins

Phage can infect the host cells through binding of either cell surface proteins or carbohydrate receptors which effectively distinguishes both primary and secondary receptors of the host cell. Phage receptor binding proteins (RBPs) are located at the tails and are responsible for the binding of specific receptors on the host cell (Garcia-Doval and van Raaij 2013). RBPs are effective acts as host recognition factors and are appropriately positioning the phage on the cell surface of the host organism before a successful infection. Depending on the phage the RBPs are termed as tail spikes, tail fibers or spike proteins. RBP of phage P22 is often rich in β-structure and intertwined beta-helix architecture which serves as a classical example of RBP spatial arrangement (Singh et al. 2012). Due to this, it has more stability and specificity and ease of overexpression, hence it may be used as potent alternative drugbility molecule instead of currently available antibodies for the development of effective diagnostic methods. Tail spike protein is another important receptor protein present in the *Podoviridae* family specifically binds to the repeating α-D-mannose-(1,4)-α-L-rhamnise-(1,3)-α-D-galactose O-antigen of the host cell lipopolysaccharides. It also can cleave the polysaccharides to assist in phage binding to the bacterial surface (Iwashita and Kanegasaki 1976; Steinbacher et al. 1996). Phage endolysin contains a C-terminal cell surface binding domain that specifically recognizes the surface glycans including N-acetylglucosamine- or choline-modified teichoic acids

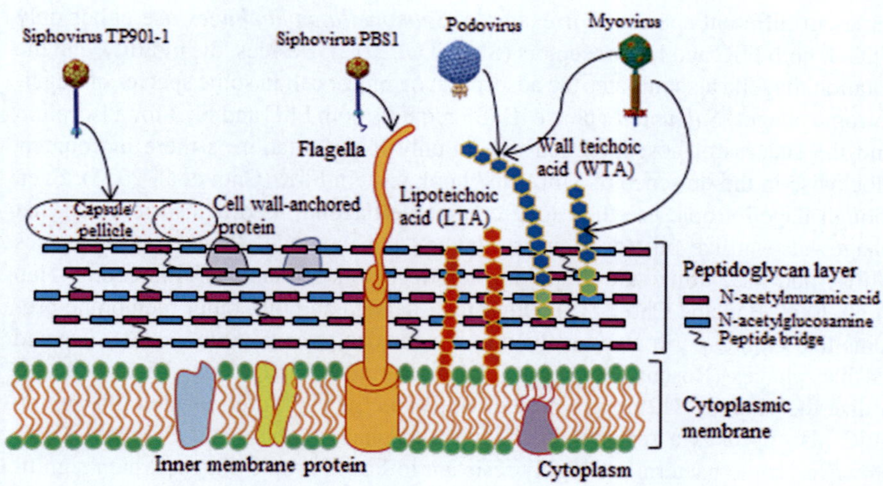

Fig. 6.2 Various mode of host-phage interaction mechanism of different phages

which possesses endopeptidase, N-actyle-β-D-muranidase activity (Fig. 6.2) (Schmelcher et al. 2012). The cell wall binding domains of the endolysins are effective binds with the gram-positive cell wall. Cell wall binding domain has a significant binding activity than antibodies due to the nature of magnetic beads hence it selectively captures the Listeria strain (Loessner et al. 2002; Schmelcher et al. 2010).

6.7 Cell Surface Proteins in Gram-Positive Bacteria

Cell protein of host cell also consider as an important key factors and effectively involved in phage infection, the proteins such as peptidoglycan or murein are important component consist of several amino acids subunits and sugar derivatives like N-acetylglucosamine and N-acetylmuramic acid and are associated with glycosidic bonds and form sheet of glycan tetrapeptide that are bound together through cross-linking of amino acids and also via peptide bonds among diaminopimelic acid and D-alanine and short peptide inner bridges that are numerous in gram-positive bacteria which leads characteristic features of the thick cell wall. Teichoic acid is another important polysaccharide significantly direct the process of phage adsorption in gram-positive bacteria. In some cases, the teichoic acids are bind with lipids of the plasma membrane and are known as lipoteichoic acids (LTA). The majority of the cell surface receptor of host cells are connected with peptidoglycan or teichoic acid. The interactions with the residues of teichoic acids and peptidoglycan are plays a key role in the adsorption of phage. The Interaction between the tail fiber and the bacterial flagella of the host cell is specific and strong, hence it can be easily adsorb to the cell wall via flagellin proteins such as phase 1 antigen FliC, phase 2 antigens FljB (Shin et al. 2012; Choi et al. 2013). The preference of these flagella receptors varies in different species. For example, *Salmonella siphophages* use either only FliC or both FliC and FljB receptors (Shin et al. 2012). Besides, the motility, and the rotation, flagella also initiates the adsorption og phage cell in some species of flagellotropic phages. *Salmonella* phage iEPS5 requires both FliC and FljB for adsorption and the successful host infection occurs only if the rotation is there in counterclockwise in the presence of MotA and hook protein FliK (Choi et al. 2013). Even though flagellotropic use their long contractile flagella, reports also revealed that phages also utilize pili for their attachment. For example, *P. aeruginosa* phages MPK7 and M22 utilize the IV pili as there receptors for successful infection (Heo et al. 2007; Bae and Cho 2013). Other than flagella and pili, outer membrane proteins like OmpC porin also involved in phage adsorption in *Salmonella Gifsy* and T4-like phages (Rosenwasser et al. 2016; Ho and Slauch 2001). T5-like phages utilize the vitamin B12 uptake protein BtuB and the conserved innate efflux pump TolC also used as a receptor by *Salmonella* phages (Ricci and Piddock 2010). However, some bacteria are highly resistant to flagellatropic phages which significantly targeting the BtuB and LPS. The possible interaction between the BtuB and LPS leads to the resistant of a bacterial strain to BtuB-targeting and LPS-targeting phages (Shin et al. 2012). In some organisms like *S. Typhimurium*, TolC has dual advantages it has been shown to colonization factor. Ferrichrome transport protein FhuA are recognize as potential receptors for the host-pathogenic interaction of Coliphages like *Salmonella* sp. (Casjens et al. 2005).

6.7.1 Protein Receptor

In gram-positive bacteria, the outer membrane proteins are classified into 5 classes; (i) structural protein that is significantly intermingling with peptidoglycan layer; (ii) specific and non-specific porins channel proteins; (iii) enzymes; (iv) transport proteins which play a key role in secretion and (v) substrate receptors those have high affinity. Most of the structural proteins act as a receptor for phage adsorption, the transmembrane protein OmpA composed of 8 antiparallel β-sheets that deeply fit into the outer membrane by the non-covalent bond interaction between the peptidoglycan layers with the free C-terminal vertex (Koebnik 1999). The complex of OmpA-LPS can inhibit the phage Tulb. Binding of bacteriophages with protein-LPS forms the precipitation of the complex with Mg^{2+} leads the irreversible phage attachment. OmpA protein potentially inhibits the bacteriophage K3 and the mutant form lacks the virulence and becomes resistant to phages (Datta et al. 1977). Porins such as OmpC and OmpF are the types of proteins present in gram negative bacteria and consist of major 3 subunits that form the channel in the outer membrane of the bacterial cell. OmpC serves as a significant receptor for phages like Hy2, ss4, and T4. T4 phages are effectively utilized the porin proteins in the combination with cell wall LPS. Another protein gp37 of phage T4 shaping the tail fibers and governs receptor recognition. This region is significantly recognizing the OmpC due the presence of important amino residues and has a large number of histidine residues.

Table. 6.1 shows the various outer membrane receptor proteins in both gram-positive and negative bacteria and their infective phages

Receptor	Host organisms	Family	Infective phages	References
Paracrystalline surface (S) layer protein	*Caulobacter crescentus*	*Myoviridae*	φCr30	Selvaraj et al. (2014b)
Protein *d* or 3A (OmpA) with LPS	*Escherichia coli*	*Myoviridae*	K3	Van Alphen et al. (1977)
Protein *c* (OmpC)	*Escherichia coli*	*Myoviridae*	Me1	Verhoef et al. (1977)
Antibiotic efflux protein TolC and the inner core of LPS	*Escherichia coli*	*Siphoviridae*	TLS	German and Misra (2001)
Protein O-8 (OmpC) with LPS	*Escherichia coli* K-12	Myoviridae	T4	Goldberg et al. (1994)
Proteins FhuA and TonB	*Siphoviridae*	*Siphoviridae*	φ80	Hancock and Braun (1976)
Protein OmpC	*Siphoviridae*	*Salmonella*	Gifsy-1Gifsy-2	Ho and Slauch (2001)
LPS	*Myoviridae*	*Pseudomonas aeruginosa*	JG004	Garbe et al. (2011)
Truncated O-chain of LPS	*Cystoviridae*	*Pseudomonas syringae*	φ13	Daugelavicius and Cvirkaite (2005)
Poly(glycerophosphate) moiety of LTA	*Siphoviridae*	*Staphylococcus aureus*	φSLT	Kaneko et al. (2009)
N-acetylglucosamine (GlcNAc) glycoepitope on WTA	*Siphoviridae*	*Staphylococcus aureus*	Wφ13, φ47, φ77, φSa2m	Xia et al. (2011)
Cell wall saccharides and pellicle phosphohexasaccharide motifs	*Siphoviridae*	*Lactococcus lactis*	P2	Bebeacua et al. (2013)
N-acetyl-muramic acid (MurNAc) of peptidoglycan in the cell wall	*Tectiviridae*	*Bacillus thuringiensis*	Bam35	Gaidelyte et al. (2006)
Glucosyl residues of poly(glycerophosphate) on WTA	*Siphoviridae*	*Bacillus subtilis*	P1	Baptista et al. (2008)
Membrane surface-anchored protein gamma phage receptor (GamR)	*Siphoviridae*	*Bacillus anthracis*	λ	Davison et al. (2005)

The OmpF receptor of T2 phages are contrasted to the T4 phage, the receptor binding site is located in the regions of gp38 and attached to the terminal part of gp37. Instead of histidine bases, gp38 possesses glycine residues in the terminal region (Selvaraj et al. 2014a). Transport protein like LamB is acted as a receptor in λ phage, contrasting, the non-selective porins such as OmpC and OmpF proteins form a narrow channel that allows specific molecules to pass through it (Table 6.1). The λ phages recognize the LamB protein via gbJ which acts as the important factor that defining the host range of the phages (Charbit et al. 1998). Enzymes present in the outer membrane are proteases such as OmpT and OmpX and serve as a potential receptor for T-like phages (Hashemolhossieni et al. 1994).

6.7.2 Lipopolysaccharide (LPS) Receptors

LPS is one of the important constituents present on the outer membrane in gram-positive bacterial along with different types of proteins. LPS also recognized as a potential receptor for phage adsorption. LPS is a complex structure composed of several units of monosaccharides and fatty acids. Structurally, it has three major portions such as lipid A, core and O-chain. Lipid A consists of disaccharides (D-glycosamine) linked together by β-1, 6-bond with attached fatty acids. Two types of LPS have been characterized in gram-positive bacteria such as smooth and rough LPS. Smooth type comprising lipid A core and O-chain, but Rough type lack O-chain. Generally, the phages are fixing to LPS O-chain for their adsorption through specific enzymatic cleavage of the polysaccharide chain. For example, phage P22 possesses endorhamnosidase activity to lyse the bond between Rha-1 and 3-Gal in O-chain of *Salmonella anatum*. Phage Ω8 significantly adsorbed on the *E. coli* outer membrane and shows endomannosidase activity by which it breaks the bond between Man-1 and 3-Man and releases prevailing levels if Hexa and monosaccharides (Reske et al. 1973).

6.8 Cell Surface Proteins in Gram-Negative Bacteria

Both structurally and chemically gram-native bacteria are different from gram-positive ones. The main constituent of the gram-negative bacteria is peptidoglycan composed of disaccharide monomers such as N-acetyl glucosamine and N-acetylmuramic acid. Teichoic acids are the major component present in the gram-negative bacteria contains a large ratio of D-alanine binds with hydroxyl groups and other substitutes. It has the bulk of bacterial antigen on the surface, the phages specific to the *Staphylococcus aureus* including 3C, 71, 79, 77, and 80 are irreversibly inactivated the complex of peptidoglycan and teichoic acid. The attachment of glycan fibers with teichoic acids are is the major event for reversible adsorption. On another hand, the presence of N-acetylglucosamine in the teichoic acid and O-actyle groups in muramic acids are playing a crucial role in the phage adsorption (Shaw and Chatterjee 1971). Due to this structural organization and resemblance phages specific for *S. aureus* significantly influence the phage adsorption in *B. subtilis* (Rakieten and Rakieten 1937). A mutant form of teichoic acids lacks D-Glucose moiety that potentially inhibits the adsorption of phages like SP3, SP10, and SP02 on the surface of *B. substilis* (Lindberg 1973).

6.9 Phage Penetration and Adsorption

Membrane penetration of prokaryotic infective phages is still unclear than the penetration of eukaryotic phages. Most of the prokaryotic infecting phages are possess long contractile tails that composed of distinct capsid head with tail and viral genome. Compared to the eukaryotic phages, prokaryotic phages use a unique mechanism for its genome delivery. Prokaryotic infective phages inject their genome into the cytoplasm of the host cell, while it leaves their outer empty capsid on the surface. In case of eukaryotic phages, the viral genome is packed in capsid via genome packing motor, this process is accompanied by accumulated pressure up to 60 atm pressure in the capsid and this is the major force which induces the injection of viral genome in the host cells during the infection (Molineux and Panja 2013; Rickgauer et al. 2008; Smith et al. 2001). Both gram-positive and negative bacteria are surrounded by an outer membrane with several numbers of receptors, phage tails are responsible for the host cell recognition, cell wall penetration, signal transmission and also trigger genome injection (Leiman and Shneider 2012). The rate of adsorption values and concentration play a key role in the host pair attachment. In T4 phages the initial step of adsorption consists of the reversible attachment of long-tail fiber which significantly involved in adsorption and triggers the conformational changes for DNA penetration into the cell. Also, the peptidoglycan layer and basal plate protein gp5 facilitate the penetration (Kanamaru et al. 2002). In the case of T5 and similar phages, protein FhuA located in the cellular outer membrane acts as a potential receptor for phage penetration and genome release in the absence of other factors. Phage T1 and φ 80 also use transport protein FhuA as a receptor, though it requires energy (Letellier et al. 2004).

6.10 Phage as Biocontrol Agents and Therapeutic Application

Generally, bacteriophages are acts as a highly promising agent for the treatment of various types of bacterial diseases. Due to the unique properties like self-limiting and self-replication they can effectively inhibit the colonization of pathogenic bacteria that no longer have a place to replicate (Summers 2001; Clark and March 2006). In recent years phages are used as potential biocontrol agents to reduce the colonization of pathogenic microorganisms. For example, phages are effectively used as biocontrol agents for the reduction of *Campylobacter* and *Salmonella* in both broiler and chicken products (Atterbury et al. 2005). Another study, Raya et al. (Raya et al. 2006) reported the effective use of phages to reduce the growth of *E. coli* in sheep pre-slaughter intervention. Listeria is one of the severe and important pathogenic microbes and often vital consequences to the susceptible individual and also can survive in the food products under cool conditions (Hudson et al. 2005). Another study, Vongkamjan et al. (Vongkamjan et al. 2012) revealed the various applications of listeriaphages as biocontrol agents to kill the harmful bacteria in smoked fish. In the year 2007, Atterbury and co-workers isolated the over 200 phages and are effective to inhibit the broadest host range (Atterbury et al. 2007). Mostly the pathogenic bacteria contaminate the food products during the steps of harvesting, processing, and packing and also becoming resistant to available antibiotics, hence phages are used as potent less toxic antimicrobial agents against most of the pathogenic bacteria and can ensure the safety of food products from production to consumption. Due to their unique properties such as easy handling, self-limiting and high species-specific nature make them a potentially auspicious substitute to antibiotics (Martínez et al. 2008; Bueno et al. 2012; Atterbury et al. 2003; Oliver et al. 2005). The increasing evidence of antimicrobial resistance and limited antibiotic treatment and development leads to another alternative research and led to a recovery of phages as agents to inhibit the pathogenic microorganism. The first use of phages and their antibacterial effect was discovered and published in the year 1921. The various pharmacological concepts are required for the application of phage therapy (Abedon 2017). It mainly utilizes the lytic phages to inhibit the population of their respective bacterial hosts, resulted in the case of life-saving therapeutic use and multiple clinical trials. For example, Forti et al. (Forti et al. 2018) use a cocktail of six phages to treat the respiratory *P. aeruginosa* infection in mice. The ability to penetrate the biofilms of *P. aeruginosa* is a major advantage over conventional phage treatments (Fong et al. 2017; Waters et al. 2017). Besides, the co-administration of phages along with antibiotics has been reported as one of the important mechanisms for restoring antibiotic sensitivity (Chan et al. 2016). Currently, most of the phages are effectively used for the elimination of multiple drug-resistant pathogens like *P. aerugnosa* in patients with burns. Several study reports revealed that the combination therapy (bacteriophage and antibiotics) shows better clinical dynamics in patients infected burns compared to the antibiotics alone treated patients (Lazareva et al. 2001).

6.11 Nanotechnology in Phage Therapy

In recent years, phage nanoparticle generation received more attention for the development of potential nanomaterial scaffolds for tissue engineering. For example, M13 nanofibers show several hallmark features and that makes them captivating building blocks for in tissue engineering and a large number of blocks can be produced via application. Phage does not have the intrinsic tropism in mammalian cells, hence it can be used as a remarkable safety nanofibers phages profile for clinical applications (Farr et al. 2014). The generation of phage nanoparticles is performed through phage display which provides the possibility for using various controls on the phage particles for other applications in medicine (Zamit et al. 2010). Phage peptide libraries are the most important prospects which offered the neuronal generation. These libraries provide several numbers of peptide ligands and are contributing to the development of potential drug molecule and therapeutic approaches in neural regeneration. One of the most significant prospects of bacteriophage nanoparticles for neural regeneration is offered by phage peptide libraries. These libraries through providing access to an innumerable number of peptide ligands may contribute to the development of novel therapeutic approaches for the field of neural regeneration. The affinity screening of phage peptide libraries over the cells belonging to different parts of the nervous system can lead to the identification of peptides that are capable of specific binding to the desired cells. These nerve cell-binding peptides can be used for the selective delivery of therapeutic cargoes into damaged cells of the nervous system. It screens all the ligand molecules based on the binding affinity selection-based methodology called panning. In these methods biomolecules such as protein or antigen or tissue types are used as target (Bakhshinejad et al. 2014).

6.12 Concluding Remarks

The understanding of Phage- host interactions at the molecular level is more important for hastening the exploitation of phages in the drug development and diagnostic process. The species-specific inherent ability and the capable of infecting the other bacterial genera phages are highly used as potential antibiotics. The molecular interaction mechanism of host-phage interaction and the structural composition of pathogenic microorganism is the great interest of the research community and are applied in various fields such as foodborne pathogen detection, phage therapy, and agricultural applications. The therapeutic application of phage becomes an efficient alternative treatment with available antibiotics against various bacterial infections. Adsorption and penetration of phages are the main and initial steps in phage infection hence the understanding the fundamental receptors is imperative to allow a rational structure-based approach to words the modification of receptor binding proteins of both host cells. Future research should focus on the nature and structure of both host and phage receptor and molecular mechanism of phage-host communications.

Acknowledgement The authors Chandrabose Selvaraj and Sanjeev Kumar Singh thankfully acknowledge MHRD – Rashtriya Uchchatar Shiksha Abhiyan (RUSA) Phase 2.0 (grant sanctioned wide letter no. F. 24-51/ 2014-U, Policy (TN Multi-Gen), Dept. of Edn. Govt. of India, Dt. 09.10.2018).

References

Abedon ST (2009) Kinetics of phage-mediated biocontrol of Bacteria. Foodborne Pathog Dis 6:807–815

Abedon ST (2017) Information phage therapy research should report. Pharmaceuticals 10:43

Ackermann HW (2009) Phage classification and characterization. Methods Mol Biol 501:127–140

Arisaka F, Kanamaru S, Leiman P, Rossmann MG (2003) The tail lysozyme complex of bacteriophage T4. Int J Biochem Cell Biol 35:16–21

Atterbury RJ, Connerton PL, Dodd CER, Rees CED, Connerton IF (2003) Isolation and characterization of campylobacter bacteriophages from retail poultry. Appl Environ Microbiol 69:4511–4518

Atterbury RJ, Dillon E, Swift C, Connerton PL, Frost JA, Dodd CER, Rees CED, Connerton IF (2005) Correlation of campylobacter bacteriophage with reduced presence of hosts in broiler chicken ceca. Appl Environ Microbiol 71:4885–4887

Atterbury RJ, Van Bergen MAP, Ortiz F, Lovell MA, Harris JA, De Boer A, Wagenaar JA, Allen VM, Barrow PA (2007) Bacteriophage therapy to reduce Salmonella colonization of broiler chickens. Appl Environ Microbiol 73:4543–4549

Bae H, Cho Y (2013) Complete genome sequence of *Pseudomonas aeruginosa* podophage MPK7, which requires type IV pili for infection. Genome Announc 1:1

Baker TS, Olson NH, Fuller SD (1999) Adding the third dimension to virus life cycles: three-dimensional reconstruction of icosahedral viruses from cryo-electron micrographs. Microbiol Mol Biol Rev 63:862–922

Bakhshinejad B, Karimi M, Sadeghizadeh M (2014) Bacteriophages and medical oncology: targeted gene therapy of cancer. Med Oncol 31:110

Bamford DH, Grimes JM, Stuart DI (2005) What does structure tell us about virus evolution? Curr Opin Struct Biol 15:655–663

Baptista C, Santos MA, São-José C (2008) Phage SPP1 reversible adsorption to Bacillus subtilis cell wall teichoic acids accelerates virus recognition of membrane receptor YueB. J Bacteriol 190:4989–4996

Bebeacua C, Tremblay D, Farenc C, Chapot-Chartier MP, Sadovskaya I, van Heel M, Veesler D, Moineau S, Cambillau C (2013) Structure, adsorption to host, and infection mechanism of virulent Lactococcal phage p2. J Virol 87:12302–12312

Bertozzi Silva J, Storms Z, Sauvageau D (2016) Host receptors for bacteriophage adsorption. FEMS Microbiol Lett 363

Black LW, Showe MK, Steven AC (1994) Morphogenesis of the T4 head. In: Karam JM (ed) Molecular biology of bacteriophage T4. ASM Press, Washington, DC, pp 218–258

Brüssow H, Desiere F (2001) Comparative phage genomics and the evolution of Siphoviridae: insights from dairy phages. Mol Microbiol 39:213–222

Brüssow H, Canchaya C, Hardt WD (2004) Phages and the evolution of bacterial pathogens: from genomic rearrangements to lysogenic conversion. Microbiol Mol Biol Rev 68:560–602

Bueno E, García P, Martínez B, Rodríguez A (2012) Phage inactivation of staphylococcus aureus in fresh and hard type cheese. Int J Food Microbiol 158:23–27

Caldentey J, Bamford DH (1992) The lytic enzyme of the Pseudomonas phage f6. Purification and biochemical characterization. Biochim Biophys. Acta Protein Struct Mol Enzymol 1159:44–50

Calendar R, Abedon ST, Young R, Wang IN (2006) Phage lysis the bacteriophages. Oxford University Press, New York, pp 104–125

Casjens SR, Gilcrease EB, Winn-Stapley DA, Schicklmaier P, Schmieger H, Pedulla ML (2005) The generalized transducing *Salmonella* bacteriophage es18: complete genome sequence and DNA packaging strategy. J Bacteriol 187:1091–1104

Cerritelli ME, Trus BL, Smith CS, Cheng N, Conway JF, Steven AC (2003) A second symmetry mismatch at the portal vertex of bacteriophage T7: 8-fold symmetry in the procapsid core. J Mol Biol 327:1–6

Chan BK, Sistrom M, Wertz JE, Kortright KE, Narayan D, Turner PE (2016) Phage selection restores antibiotic sensitivity in MDR *Pseudomonas aeruginosa*. Sci Rep 6:26717

Charbit A, Wang J, Michel V, Hofnung M (1998) A cluster of charged and aromatic residues in the C-terminal portion of maltoporin participates in sugar binding and uptake. Mol Gen Genet 260:185–192

Choi Y, Shin H, Lee JH, Ryu S (2013) *Identification* and characterization of a novel flagellum-dependent *Salmonella*-infecting bacteriophage, iEPS5. Appl Environ Microbiol 79:4829–4837

Cingolani G, Moore SD, Prevelige PE Jr, Johnson JE (2002) Preliminary crystallographic analysis of the bacteriophage P22 portal protein. J Struct Biol 139:46–54

Clark JR, March JB (2006) Bacteriophages and biotechnology: vaccines, gene therapy and antibacterial. Trends Biotechnol 24:212–218

Clokie MR, Millard AD, Letarov AV, Heaphy S (2011) Phages in nature. Bacteriophage 1:31–45

Datta DB, Arden B, Henning U (1977) Majorprotein of the Escherichia Coli outer cell envelope membrane as bacteriophage receptor. J Bacteriol 131:821–829

Daugelavicius R, Cvirkaite V, Gaidelyte A, Bakienė E, Gabrėnaitė-Verkhovskaya R (2005) Bamford DHPenetration of enveloped double-stranded RNA bacteriophages ϕ13 and ϕ6 into Pseudomonas syringae cells. J Virol 79:5017–5026

Davison S, Couture-Tosi E, Candela T, Mock M, Fouet A (2005) Identification of the *Bacillus anthracis* γ phage receptor. J Bacteriol 187:6742–6749

Desiere F, McShan WM, van Sinderen D, Ferretti JJ, Brussow H (2001) Comparative genomics reveals close genetic relationships between phages from dairy bacteria and pathogenic streptococci: evolutionary implications for prophage-host interactions. Virology 288:325–341

Desvaux M, Parham NJ, Henderson IR (2004) The autotransporter secretion system. Res Microbiol 155:53–60

Dieterle ME, Spinelli S, Sadovskaya I, Piuri M, Cambillau C (2017) Evolved distal tail carbohydrate binding modules of Lactobacillus phage J-1: a novel type of anti-receptor widespread among lactic acid bacteria phages. Mol Microbiol 104:608–620

Drulis-Kawa Z, Majkowska-Skrobek G, Maciejewska B (2015) Bacteriophages and phage-derived proteins--application approaches. Curr Med Chem 22:1757–1773

Dufresne K, Paradis-Bleau C (2015) Biology and assembly of the bacterial envelope. Adv Exp Med Biol 883:41–76

Farr R, Choi DS, Lee SW (2014) Phage-based nanomaterials for biomedical applications. Acta Biomater 10:1741–1175

Fehmel F, Feige U, Niemann H, Stirm S (1975) *Escherichia coli* capsule bacteriophages VII. Bacteriophage 29-host capsular polysaccharide interactions. J Virol 16:591–601

Flayhan A, Vellieux FMD, Lurz R, Maury O, Contreras-Martel C, Girard E, Boulanger P, Breyton C (2014) Crystal structure of pb9, the distal tail protein of bacteriophage T5: a conserved structural motif among all Siphophages. J Virol 88:820–828

Fokine A, Chipman PR, Leiman PG, Mesyanzhinov VV, Rao VB, Rossmann MG (2004) Molecular architecture of the prolate head of bacteriophage T4. Proc Natl Acad Sci U S A 101:6003–6008

Fokine A, Leiman PG, Shneider MM, Ahvazi B, Boeshans KM, Steven AC, Black LW, Mesyanzhinov VV, Rossmann MG (2005) Structural and functional similarities between the capsid proteins of bacteriophages T4 and HK97 point to a common ancestry. Proc Natl Acad Sci U S A 102:7163–7168

Fokine A, Battisti AJ, Kostyuchenko VA, Black LW, Rossmann MG (2006) Cryo-EM structure of a bacteriophage T4 gp24 bypass mutant: the evolution of pentameric vertex proteins in icosahedral viruses. J Struct Biol 154:255–259

Fong SA, Drilling A, Morales S, Cornet ME, Woodworth BA, Fokkens WJ, Psaltis AJ, Vreugde S, Wormald PJ (2017) Activity of bacteriophages in removing biofilms of *Pseudomonas aeruginosa* isolates from chronic rhinosinusitis patients. Front Cell Infect Microbiol 7:418

Forti F, Roach DR, Cafora M, Pasini ME, Horner DS, Fiscarelli EV, Rossitto M, Cariani L, Briani F, Debarbieux L, Ghisotti D (2018) Design of a broad-range bacteriophage cocktail that reduces *Pseudomonas aeruginosa* biofilms and treats acute infections in two animal models. Antimicrob Agents Chemother 62:e02573–e02517

Friedman SD, Genthner FJ, Gentry J, Sobsey MD, Vinje J (2009) Gene mapping and phylogenetic analysis of the complete genome from 30 single-stranded RNA male-specific coliphages (family Leviviridae). J Virol 83:11233–11243

Gaidelyte A, Cvirkaite-Krupovic V, Daugelavicius R, Bamford JK, Bamford DH (2006) The entry mechanism of membrane-containing phage Bam35 infecting *Bacillus thuringiensis*. J Bacteriol 188:5925–5934

Garbe J, Bunk B, Rohde M, Schobert M (2011) Sequencing and characterization of Pseudomonas aeruginosa phage JG004. BMC Microbiol 11:102

Garcia-Doval C, van Raaij MJ (2013) Bacteriophage receptor recognition and nucleic acid transfer. Subcell Biochem 68:489–518

German GJ, Misra R (2001) The TolC protein of Escherichia coli serves as a cell-surface receptor for the newly characterized TLS bacteriophage. J Mol Biol 308:579–585

Gill JJ, Hyman P (2010) Phage choice, isolation, and preparation for phage therapy. Curr Pharm Biotechnol 11:2–14

Goldberg E, Grinius L, Letellier L (1994) Recognition, attachment and injection. In: Karam JD (ed) Molecular biology of bacteriophage T4. American Society for Microbiology, Washington, pp 347–356

Guerrero-Ferreira RC, Viollier PH, Ely B, Poindexter JS, Georgieva M, Jensen GJ, Wright ER (2011) Alternative mechanism for bacteriophage adsorption to the motile bacterium *Caulobacter crescentus*. Proc Natl Acad Sci U S A 108:9963–9968

Guttman B, R Raya, and E. Kutter. 2005. Basic phage biology, p. 29–66. *In*: E. Kutter and A. Sulakvelidze (eds.), Bacteriophages: biology and application. CRC Press, Boca Raton

Hagens S, Blasi U (2003) Genetically modified filamentous phage as bactericidal agents a pilot study. Lett Appl Microbiol 37:318–323

Hagens S, Ahsen U, Gabain A (2004) Therapy of experimental pseudomonas infections with a nonreplicating genetically modified phage. Antimicrob Agents Chemother 48:3817–3822

Hancock REW, Braun V (1976) Nature of the energy requirement for the irreversible adsorption of bacteriophages T1 and ϕ80 to Escherichia coli. J Bacteriol 125:409–415

Hashemolhossieni S, Holmes Z, Mutschler B, Henning U (1994) Alterations of receptor specificities of coliphage of the T2 family. J Mol Biol 240:105–110

Hatfull GF (2008) Bacteriophage genomics. Curr Opin Microbiol 11:447–453

Hendrix RW, Smith MCM, Burns RN, Ford ME, Hatfull GF (1999) Evolutionary relationships among diverse bacteriophages and prophages: all the world's a phage. Proc Natl Acad Sci U S A 96:2192–2197

Heo YJ, Chung IY, Choi KB, Lau GW, Cho YH (2007) Genome sequence comparison and superinfection between two related *Pseudomonas aeruginosa* phages, D3112 and MP22. Microbiology 153:2885–2895

Ho TD, Slauch JM (2001) OmpC is the receptor for Gifsy-1 and Gifsy-2 bacteriophages of Salmonella. J Bacteriol 183:1495–1498

Hudson JA, Billington C, Carey-Smith G, Greening G (2005) Bacteriophages as biocontrol agents in food. J Food Prot 68:426–437

Huet A, Conway JF, Letellier L, Boulanger P (2010) In vitro assembly of the T-13 procapsid of bacteriophage T5 with its scaffolding domain. J Virol 84:9350–9358

Iwashita S, Kanegasaki S (1976) Deacetylation reaction catalyzed by *Salmonella* phage c341 and its baseplate parts. J Biol Chem 251:5361–5365

Jassim SAA, Limoges RG (2014) Natural solution to antibiotic resistance: bacteriophages: the l p iving drugs. World J Microbiol Biotechnol 30:2153–2170

Kanamaru S, Leiman PG, Kostyuchenko VA, Chipman PR, Mesyanzhinov VV, Arisaka F, Rossmann MG (2002) Structure of the cell-puncturing device of bacteriophage T4. Nature 415:553–557

Kaneko J, Narita-Yamada S, Wakabayashi Y, Kamio Y (2009) Identification of ORF636 in phage ϕSLT carrying panton-valentine leukocidin genes, acting as an adhesion protein for a poly(glycerophosphate) chain of lipoteichoic acid on the cell surface of Staphylococcus aureus. J Bacteriol 191:4674–4680

Kenny JG, McGrath S, Fitzgerald GF, van Sinderen D (2004) Bacteriophage Tuc2009 Encodes a Tail-Associated Cell Wall-Degrading Activity. J Bacteriol 186:3480–3491

Klumpp J, Fouts DE, Sozhamannan S (2012) Next generation sequencing technologies and the changing landscape of phage genomics. Bacteriophage 2:190–199

Koebnik R (1999) Structural and functional roles of the surface exposed loops of the beta-barrel membrane protein OmpA from Escherichia coli. J Bacteriol 181:3688–3694

Kutter E, Sulakvelidze A, Guttman B, Raya R, Kutter E (2005) Bacteriophages biology and application bacteriophages biology and application. CRC Press, Boca Raton, pp 29–66

Kwan T, Liu J, DuBow M, Gros P, Pelletier J (2005) The complete genomes and proteomes of 27 *Staphylococcus aureus* bacteriophages. Proc Natl Acad Sci U S A 102:5174–5179

Kwan T, Liu J, Dubow M, Gros P, Pelletier J (2006) Comparative genomic analysis of 18 *Pseudomonas aeruginosa* bacteriophages. J Bacteriol 188:1184–1187

Lander GC, Evilevitch A, Jeembaeva M, Potter CS, Carragher B, Johnson JE (2008) Bacteriophage lambda stabilization by auxiliary protein gpD:timing, location, and mechanism of attachment determined by cryoEM. Structure 16:1399–1406

Lazareva EB, Smirnov SV, Khvatov VB, Spiridonova TG, Bitkova EE, Darbeeva OS, Maĭskaia LM, Parfeniuk RL, Men'shikov DD (2001) Efficacy of bacteriophages in complex treatment of patients with burn wounds. Antibiot Khimioter 46:10–14

Leiman PG, Shneider MM (2012) Contractile tail machines of bacteriophages. Adv Exp Med Biol 726:93–114

Letellier L, Boulanger P, Plancon L, Jacquot P, Santamaria M (2004) Main features on tailed phage, host recognition and DNA uptake. Front Biosci 9:1228–1339

Li X, Koç C, Kühner P, Stierhof YD, Krismer B, Enright MC, Penadés JR, Wolz C, Stehle T, Cambillau C (2016) An essential role for the baseplate protein Gp45 in phage adsorption to *Staphylococcus aureus*. Sci Rep 6:26455

Lindberg AA (1973) Bacteriophage receptors. Annu Rev Microbiol 27:205–241

Loessner MJ, Kramer K, Ebel F, Scherer S (2002) C−terminal domains of *Listeria monocytogenes* bacteriophage murein hydrolases determine specific recognition and high−affinity binding to bacterial cell wall carbohydrates. Mol Microbiol 44:335–349

Lu TK, Collins JJ (2007) Dispersing biofilms with engineered enzymatic bacteriophage. Proc Natl Acad Sci U S A 104:11197–11202

Lu TK, Collins JJ (2009) Engineered bacteriophage targeting gene networks as adjuvants for antibiotic therapy. Proc Natl Acad Sci U S A 106:4629–4634

Lubbers MW, Waterfield NR, Beresford TP, Le Page RW, Jarvis AW (1995) Sequencing and analysis of the prolate-headed lactococcal bacteriophage c2 genome and identification of the structural genes. Appl Environ Microbiol 61:4348–4356

Mahony J, van Sinderen D (2012) Structural aspects of the interaction of dairy phages with their host Bacteria. Viruses 4:1410–1424

Mahony J, Martel B, Tremblay DM, Neve H, Heller KJ, Moineau S, van Sinderen D (2013) Identification of a new P335 subgroup through molecular analysis of lactococcal phages Q33 and BM13. Appl Environ Microbiol 79:4401–4409

Marti R, Zurfluh K, Hagens S, Pianezzi J, Klumpp J, Loessner MJ (2013) Long tail fibres of the novel broad-host-range T-even bacteriophage S16 specifically recognize *Salmonella* OmpC. Mol Microbiol 87:818–834

Martínez B, Obeso MJ, Ana Rodríguez A, García P (2008) Nisinbacteriophages cross-resistance in Staphylococcus aureus. Int J Food Microbiol 122:253–258

Moak M, Molineux IJ (2000) Role of the Gp16 lytic Transglycosylase motif in bacteriophage T7 Virions at the initiation of infection. Mol Microbiol 37:345–355

Molineux IJ, Panja D (2013) Popping the cork: mechanisms of phage genome ejection. Nat Rev Microbiol 11:194–204

Muñoz SLD, Koskella B (2014) Bacteria-phage interactions in natural environments. Adv Appl Microbiol 89:135–183

Oliver SP, Jayarao BM, Almeida RA (2005) Foodborne pathogen in milk and the dairy farm environment: food safety and public health implications. Foodborne Pathog Dis 2:115–129

Orlova EV, Dube P, Beckmann E, Zemlin F, Lurz R, Trautner TA, Tavares P, van Heel M (1999) Structure of the 13-fold symmetric portal protein of bacteriophage SPP1. Nat Struct Biol 6:842–846

Pell LG, Liu A, Edmonds L, Donaldson LW, Howell PL, Davidson AR (2009) The X-ray crystal structure of the phage λ tail terminator protein reveals the biologically relevant Hexameric ring structure and demonstrates a conserved mechanism of tail termination among diverse long-tailed phages. J Mol Biol 389:938–951

Pope WH, Jacobs-Sera D, Russell DA, Peebles CL, Al-Atrache Z, Alcoser TA et al (2011) Expanding the diversity of mycobacteriophages: insights into genome architecture and evolution. PLoS One 6:e16329

Rakieten ML, Rakieten TL (1937) Relationships between Staphylococci and Bacilli belonging to the subtilis group as shown by bacteriophage absorption. J Bacteriol 34:285–300

Raya RR, Varey P, Oot RA, Dyen MR, Callaway TR, Edrington TS, Kutter EM, Brabban AD (2006) Isolation and characterization of a new T-even bacteriophage, CEV1, and determination of its potential to reduce Escherichia coli 0157:H7 levels in sheep. Appl Environ Microbiol 72:6405–6410

Reske K, Wallenfels B, Jann K (1973) Enzymatic degradation of O-antigenic lipopolysaccharides by coliphage Ω8. Eur J Biochem 36:167–171

Ricci V, Piddock LVJ (2010) Exploiting the role of TolC in pathogenicity; identification of a bacteriophage for eradication of Salmonella serovars form poultry. Appl Environ Microbiol 76:1704–1706

Rickgauer JP, Fuller DN, Grimes S, Jardine PJ, Anderson DL, Smith DE (2008) Portal motor velocity and internal force resisting viral DNA packaging in bacteriophage phi29. Biophys J 94:159–167

Rosenwasser S, Ziv C, Creveld SG, Vardi A (2016) Virocell metabolism: metabolic innovations during host-virus interactions in the ocean. Trends Microbiol 24:821–832

Ryan EM, Gorman SP, Donnelly RF, Gilmore BF (2011) Recent advances in bacteriophage therapy how delivery routes, formulation, concentration and timing influence the success of phage therapy. J Pharm Pharmacol 63:1253–1264

Rydman PS, Bamford DH (2000) Bacteriophage PRD1 DNA entry uses a viral membrane-associated Transglycosylase activity. Mol Microbiol 37:356–363

Sanger F, Air GM, Barrell BG, Brown NL, Coulson AR, Fiddes CA, Hutchison CA, Slocombe PM, Smith M (1977) Nucleotide sequence of bacteriophage fX174 DNA. Nature 265:687–695

Schmelcher M, Shabarova T, Eugster MR, Eichenseher F, Tchang VS, Banz M, Loessner MJ (2010) Rapid multiplex detection and differentiation of *Listeria* cells by use of fluorescent phage endolysin cell wall binding domains. Appl Environ Microbiol 76:5745–5756

Schmelcher M, Donovan DM, Loessner MJ (2012) Bacteriophage Endolysins as novel antimicrobials. Future Microbiol 7:1147–1171

Selvaraj C, Singh P, Singh SK (2014a) Investigations on the interactions of λPhage-derived peptides against the SrtA mechanism in bacillus anthracis. Appl Biochem Biotechnol 172(4):1790–1806

Selvaraj C, Bharathi Priya R, Singh SK (2014b) Communication of γ phage Lysin plyG enzymes binding toward SrtA for inhibition of bacillus anthracis: protein–protein interaction and molecular dynamics study. Cell Commun Adhes 21:257–265

Sharp R (2011) Bacteriophages: biology and history. J Chem Technol Biotechnol 76:667–672

Shaw DRD, Chatterjee AN (1971) O-acetyl groups as a component of the bacteriophage receptor on *Staphylococcus aureus* cell walls. J Bacteriol 1085:84–85

Shin H, Lee J-H, Kim H, Choi Y, Heu S, Ryu S (2012) Receptor diversity and host interaction of bacteriophages infecting *Salmonella enterica* Serovar typhimurium. PLoS One 7(8):e43392

Silhavy TJ, Kahne D, Walker S (2010) The bacterial cell envelope. Cold Spring Harb Perspect Biol 2:a000414

Simpson DJ, Sacher JC, Szymanski CM (2016) Development of an assay for the identification of receptor binding proteins from bacteriophages. Viruses 8:17

Singh A, Arutyunov D, Szymanski CM, Evoy S (2012) Bacteriophage based probes for pathogen detection. Analyst 137:3405–3421

Smith DE, Tans SJ, Smith SB, Grimes S, Anderson DL, Bustamante C (2001) The bacteriophage straight phi29 portal motor can package DNA against a large internal force. Nature 413:748–752

Stalin N, Srinivasan P (2016a) Efficacy of potential phage cocktails against Vibrio harveyi and closely related Vibrio species isolated from shrimp aquaculture environment in the south east coast of India. Vet Microbiol 207:83–96

Stalin N, Srinivasan P (2016b) Molecular characterization of antibiotic resistant Vibrio harveyi isolated from shrimp aquaculture environment in the south east coast of India. Microb Pathog 97:110–118

Steinbacher S, Baxa U, Miller S, Weintraub A, Seckler R, Huber R (1996) Crystal structure of phage P22 tailspike protein complexed with Salmonella sp. O-antigen receptors. Proc Natl Acad Sci U S A 93:10584–10588

Stockdale SR, Mahony J, Courtin P, Chapot-Chartier MP, Pijkeren JP, van Britton RA, Neve H, Heller KJ, Aideh B, Vogensen FK (2013) The lactococcal phages Tuc2009 and TP901-1 incorporate two alternate forms of their tail fiber into their virions for infection specialization. J Biol Chem 288:5581–5590

Sulakvelidze A, Alavidze Z (2001) MINIREVIEW bacteriophage therapy. Antimicrob Agents Chemother 45:649–659

Summers WC (2001) Bacteriophage therapy. Annu Rev Microbiol 55:437–451

Sutherland IW (1999) Polysaccharases for microbial exopoly- saccharides. Carbohydr Polym 38:319–328

Tavares P (2018) The bacteriophage head-to-tail interface. Subcell Biochem 88:305–328

Tu AH, Voelker LL, Shen X, Dybvig K (2001) Complete nucleotide sequence of the mycoplasma virus P1 genome. Plasmid 45:122–126

Van Alphen L, Havekes L, Lugtenberg B (1977) Major outer membrane protein d of Escherichia coli K12. Purification and in vitro activity of bacteriophage k3 and f-pilus mediated conjugation. FEBS Lett 75:285–290

Veesler D, Robin G, Lichière J, Auzat I, Tavares P, Bron P, Campanacci V, Cambillau C (2010) Crystal structure of bacteriophage SPP1 distal tail protein (gp19.1) a baseplate hub paradigm in gram-positive infecting phages. J Biol Chem 285:36666–36673

Veesler D, Spinelli S, Mahony J, Lichière J, Blangy S, Bricogne G, Legrand P, Ortiz-Lombardia M, Campanacci V, van Sinderen D (2012) Structure of the phage TP901-1 1.8 MDa baseplate suggests an alternative host adhesion mechanism. Proc Natl Acad Sci U S A 109:8954–8958

Verhoef C, de Graaff PJ, Lugtenberg EJJ (1977) Mapping of a gene for a major outer membrane protein of Escherichia coli K12 with the aid of a newly isolated bacteriophage. Mol Gen Genet 150:103–105

Vongkamjan K, Switt AM, de Bakker CH, Fortes ED, Wiedmann M (2012) Silage collected from dairy farms harbours an abundance of Listeriaphages with considerable host range and genome size diversity. Appl Environ Microbiol 78:8666–8675

Waseh S, Hanifi-Moghaddam P, Coleman R, Masotti M, Ryan S, Foss M, MacKenzie R, Henry M, Szymanski CM, Tanha J (2010) Orally administered P22 phage Tailspike protein reduces Salmonella colonization in chickens: prospects of a novel therapy against bacterial infections. PLoS One 5:17

Waters EM, Neill DR, Kaman B, Sahota JS, Clokie MRJ, Winstanley C, Kadioglu A (2017) Phage therapy is highly effective against chronic lung infections with *Pseudomonas aeruginosa*. Thorax 72:666–667

Weinbauer MG (2004) Ecology of prokaryotic viruses. FEMS Microbiol Rev 28:127–181

Wikoff WR, Liljas L, Duda RL, Tsuruta H, Hendrix RW, Johnson JE (2000) Topologically linked protein rings in the bacteriophage HK97 capsid. Science 289:2129–2133

Wikoff WR, Conway JF, Tang J, Lee KK, Gan L, Cheng N (2006) Timeresolved molecular dynamics of bacteriophage HK97 capsid maturation interpreted by electron cryo-microscopy and X-ray crystallography. J Struct Biol 153:300–306

Xia G, Corrigan RM, Winstel V, Goerke C, Gründling A, Peschel A (2011) Wall teichoic acid-dependent adsorption of staphylococcal siphovirus and myovirus. J Bacteriol 193:4006–4009
Zamit AL, Ostrowski M, Fondevila N, Wigdorovitz A, Romera A, Bratanich AC (2010) Use of phage display peptides libraries for epitope mapping of bovine viral diarrhea virus E2 protein. Res J Immunol 3:31–36

Chapter 7
Are Phages Parasites or Symbionts of Bacteria?

Lorena Rodríguez-Rubio, Pedro Blanco-Picazo, and Maite Muniesa

Abstract Bacteria and phages have co-existed for several billions of years. The direct observation of their relationship, and the evidence that phages cause bacterial lysis and death, has led researchers to believe that phages and bacteria are natural enemies and that phages can be applied as antimicrobial agents. However, phages are also known to provide various benefits for their bacterial hosts, and on many occasions the phage-host interaction resembles a symbiotic relationship. In this chapter, we evaluate findings in different bacterial genera and their associated phages to assess if phages should be considered as bacterial parasites or symbiotic organisms.

7.1 Background

Bacteriophages or phages, viruses that infect bacteria, are the most abundant and ubiquitous organisms in the biosphere, and are found in all the environments inhabited by their hosts (Utter et al. 2014). The phage life cycle is mandatorily linked to the bacterial (or archaeal) cells and as such phages are considered obligate parasites of the cells they depend on to propagate. Bacteria for which no phages have been identified are exceptional cases. The ubiquity of phages can be seen as the consequence of a successful evolutionary adaptation, which has converted them into highly efficient predators of all living prokaryotic cells, despite the efforts of the latter to avoid infection. An alternative view, however, is that prokaryotic cells tolerate phages, at least certain phage groups, and even promote their interference, because they bestow several advantages on the bacterial population. These contentions are among the questions dealt with in this chapter, whose main purpose is to present arguments that may help the reader evaluate if phages can be considered as obligate parasites or symbiotic organisms. In other words, it is examined if phages are pathogenic entities harmful for bacteria, or if, after billions of years of

L. Rodríguez-Rubio · P. Blanco-Picazo · M. Muniesa (✉)
Department of Genetics, Microbiology and Statistics, University of Barcelona, Barcelona, Spain
e-mail: mmuniesa@ub.edu

G. Witzany (ed.), *Biocommunication of Phages*,
https://doi.org/10.1007/978-3-030-45885-0_7

co-evolution, phages and bacteria have reached a win-win equilibrium that confers mutual benefits.

7.1.1 Structure, Specificity and Persistence of Phages

The bacteriophage in its extracellular form is called a virion, which generally consists of one nucleic acid molecule, the phage genome, surrounded by a protein coating, the capsid (Adams 1959). The capsid is responsible for the phage specificity for a given bacterial host, as it allows the virion to recognize the receptor required for adsorption to the bacterial surface at the start of the infection process (Bertozzi Silva et al. 2016). Host cell components acting as receptors include surface proteins, polysaccharides, lipopolysaccharides and carbohydrate moieties and even other structures such as the sexual pili (Rajala-Mustonen and Heinonen-Tanski 1994; Bertozzi Silva et al. 2016). The ability to use widely distributed molecules as receptors is characteristic of polyvalent phages with a broad host range, allowing them to infect and propagate in diverse species or even genera (Souza et al. 1972; Jensen et al. 1998; Evans et al. 2010; Yu et al. 2016). In contrast, the use of cell wall molecules found only in a few bacterial species or even strains (Beumer and Robinson 2005) confers a high specificity to the infective phages for the given host.

The relatively simple structure of the virion endows the phage particles with an extracellular persistence (Jończyk et al. 2011; Lee and Sobsey 2011; Calero-Cáceres and Muniesa 2016) and an enduring infectivity. In certain environments some phages can persist longer than their bacterial hosts (Durán et al. 2002; Mocé-llivina et al. 2003; Allué-Guardia et al. 2014), and this staying power indicates that phages can act as environmental reservoirs of the genes contained and protected within their capsids (Subirats et al. 2016). Bacteriophage persistence has relevant implications, as will be explained in the following sections.

7.1.2 Life Cycle

Bacteriophages can be divided into those able to propagate only through the lytic cycle, which results in lysis of the host cells, or through the lysogenic pathway, in which phage DNA is integrated in the bacterial genome. Virulent phages can only follow the lytic cycle, which begins with the attachment of the phage to a specific receptor, followed by injection and replication of the nucleic acid, synthesis of the capsid proteins, assembly of the capsids, packaging of the nucleic acid, cell lysis and release of the virions. In the lysogenic cycle, which depends on integrase activity (Groth and Calos 2004), temperate phage DNA is integrated into the host chromosome, and is maintained in a "dormant" state as a prophage until a stressing factor activates a response in the bacteria leading to phage induction.

Although some differences can be found among phages, phage λ provides a valid model for the induction of many temperate phages. Phage λ induction begins with a stressor activating the SOS response, which leads to the degradation of the protein that promotes the lysogenic cycle (protein CI) and derepresses the protein responsible for the activation of the lytic cycle (protein Cro) (Oppenheim et al. 2005). After induction, the prophage excises from the bacterial chromosome, replicates and expresses the phage proteins. Capsid proteins then assemble, phage DNA is packaged, and phage virions are released by cell lysis.

Both virulent and temperate phages can transfer bacterial DNA from one bacterium to another through a transduction process. The released virion transfers this bacterial DNA from the donor cell to a recipient cell through the extracellular environment. The size of the bacterial DNA packaged into a phage particle ranges from a short fragment, such as a single gene, with the rest of the capsid content being phage DNA, to a sequence of up to 100 kb (chromosomal or plasmidic), with an absence of phage genes.

7.1.3 Transduction

Transduction was originally thought to be the result of errors in the phage packaging machinery, and therefore a rare event (Thierauf et al. 2009; Penades et al. 2015). However, in nature, any error that proves prejudicial for the individual will be negatively selected. So the ubiquity of transduction is an indication that in some way it must benefit either the phages or the bacteria. Moreover, recent data suggest that transduction is far more frequent than previously believed (García-Aljaro et al. 2017) and can be induced by different agents and conditions, including antibiotics such as β-lactams and above all quinolones (Kimmitt et al. 2000; Miller et al. 2004; Johnson et al. 2017; Chen et al. 2018).

The influence of transduction in bacterial evolution is unquestionable, although the extent of its contribution to bacterial genetic variability depends on the DNA-packaging mechanism and the amount of bacterial DNA packaged. Transduction has traditionally been classified as either generalized or specialized but recently a third mechanism has surfaced, that of lateral transduction.

7.1.3.1 Generalized Transduction

Generalized transducing particles can be produced during the lytic cycle of both virulent and temperate phages (Thierauf et al. 2009). This process occurs during the packaging step of phage propagation when the phage machinery apparently fails to recognize the *pac* site where the phage DNA should be cut and instead cuts into homologous *pac*-like sites located in the bacterial chromosome. As a result, a fragment of bacterial DNA is cleaved and packaged inside the capsid (Doolittle 2002).

Particles derived from generalized transduction cannot propagate because they do not transport phage DNA and therefore lack the phage genes necessary for replication. However, their capacity for infection is not diminished, as this depends on the capsid (Muniesa et al. 2013). After infection of a new host cell, the DNA mobilized in the phage particle can be incorporated into the host genome by recombination mechanisms (Moat et al. 2003). The capacity for generalized transduction is greater in certain phages, such as T4, P22 and P1; approximately 30% of the particles present in a lysate of these phages were reported to be derived from generalized transduction (Sternberg and Coulby 1990; Thierauf et al. 2009).

7.1.3.2 Specialized Transduction

Specialized transduction involves an inaccurate excision of a prophage from a specific integration site on the bacterial chromosome. The released fragment comprises the phage genome plus an adjacent DNA fragment of the bacterial genome, both of which are packaged into a single phage particle (Miller et al. 2004; Thierauf et al. 2009). Phages that carry a specific bacterial gene, not necessarily adjacent to the site of integration, are also sometimes considered as specialized transducing phages. Once the phage infects a new host, and the phage genome integrates into a specific site of the host bacterium by means of the phage integrase, the bacterial gene is also incorporated in the bacterial genetic pool. As the specialized prophage possesses all the genes necessary for its replication and propagation, it can be induced and is able to produce new progeny (carrying the extra bacterial fragment) (Fortier and Sekulovic 2013).

Specialized transduction is probably only a minor contributor to genetic transfer in the environment compared to generalized transduction. However, the frequency of transduction of specific bacterial DNA fragments is high, as all the phage particles produced after the prophage induction carry this particular fragment.

7.1.3.3 Lateral Transduction

Recently described in staphylococci, lateral transduction is highly efficient and able to mobilize large fragments of the bacterial genome. Moreover, the frequency of this transduction mechanism is 1000-fold higher than previously observed for the other types of transduction.

Staphylococcus aureus contains several prophages that can be induced and enter the lytic cycle, as indicated above. However, a form of lateral transduction has been described in this species that differs from the typical model in that after induction the prophages do not excise from the bacterial chromosome until late in the virus life cycle. As a result, the phage DNA is replicated while still incorporated within the bacterial chromosome (Chen et al. 2018). DNA packaging initiates *in situ* from the integrated prophages by a head-full mechanism. Once the capsid is full, another generated capsid continues packaging the adjacent fragment belonging to the host bacterial genome. In this way, large segments of the *S. aureus* genome of several hundred kb are packaged in phage heads, generating large quantities of lateral-transducing particles that can transduce bacterial DNA at a very high frequency. It is expected that this mechanism will be identified in species other than *S. aureus*.

Considering the number of prophages present in *S. aureus*, as well as in many other bacterial species (Casjens 2003; Paul 2008; Fortier and Sekulovic 2013; Chen et al. 2015; Brown-Jaque et al. 2018), it is plausible that bacteria can mobilize almost all, if not their entire genome by phage transduction.

7.2 Benefits of Phages for Bacteria: Phages as Symbionts

7.2.1 Spread of Genes as Virions: Horizontal Gene Transfer

Phage-mediated horizontal gene transfer (HGT) plays a key role in the evolution of bacteria by supplying hosts with new functional genes that allow access to new environmental niches, including pathogenicity. Phage-mediated HGT takes place through two main mechanisms: transduction (explained in Sect. 7.1.3) and lysogenic conversion. The latter occurs when a phage is inserted in the bacterial chromosome and the expression of prophage genes leads to phenotypic changes in the host, affecting virulence, motility and bacterial competition (Touchon et al. 2017). The success of these processes depends on the efficiency of the bacterial defense system (Labrie et al. 2010) and the ability of phage tools to overcome these defenses (Samson et al. 2013). HGT can endow bacteria with a range of beneficial genes responsible for virulence, metabolism and evolution.

7.2.1.1 Genes Related to Virulence

Many prophages encode toxins and other proteins, such as antigens and effector proteins, which are responsible for a number of well-known human diseases. This is the case in the Shiga toxin-producing *Escherichia coli* strains (STEC) (Morabito 2014). The principal virulence factor produced by these pathogens is a Stx toxin, which is encoded in a prophage. The most studied STEC strain is O157:H7, which has become a serious global public health concern since its discovery in 1982. Genetic evidence suggests that the strain evolved from *E. coli* strain O55:H7 by a small number of genetic events, notably the sequential acquisition of Stx2 and then Stx1 phage (Shaikh and Tarr 2003), which encode Stx toxins.

Found in marine environments, *Vibrio cholerae* enters the human body by an oral route, colonizes the small intestine and secretes toxins that cause severe diarrhea. The two human-pathogenic serogroups, O1 and O139, contain two key factors for their pathogenicity: the cholera toxin and the toxin-coregulated pilus, encoded by phages or phage-like elements, respectively (Waldor and Mekalanos 1996; Karaolis et al. 1999).

Human botulism is caused by the consumption of food contaminated with botulinum toxins. Produced by *Clostridium botulinum*, these toxins are classified in 8 groups: A, B, C1, D, E, F and G, and C1 and D are encoded by prophages (Inoue and Iida 1970; Eklund et al. 1972). Toxinotypes C and D are interconvertible, depending on the infecting phage (Eklund and Poysky 1974).

Corynebacterium diphtheriae is the leading causative agent of diphtheriae, although only toxigenic strains are responsible for the disease. The toxigenic phenotype of *C. diphtheriae* is determined by temperate corynephages whose genomes carry the *tox* gene encoding the diphtheria toxin (Parsons and Frobisher 1951).

Bacteriophage φCTX from *Pseudomonas aeruginosa* PA158 carries a cytotoxin gene (*ctx*), and is able to convert non-CTX-producing *P. aeruginosa* strains into CTX producers (Hayashi et al. 1990).

Many *S. aureus* virulence factors are also encoded by phages and phage-related mobile elements, such as the shock syndrome toxin 1, which is carried by pathogenicity island SapI-1 (Lindsay et al. 1998), and the latter is in turn mobilized at high frequency by the generalized staphylococcal transducing phage 80α (Ruzin et al. 2001). More examples are the Panton-Valentine leukocidin (Kaneko et al. 1997), enterotoxin A (Betley and Mekalanos 1985) and the exfoliative toxin (Yamaguchi et al. 2000), all of them encoded by lysogenized phages. In addition to the phage-encoded virulence factors, it was shown that the genomic island νSaβ, encoding staphylococcal superantigens, proteases, and leukotoxins, can be transferred between *S. aureus* strains by a resident prophage. This transfer is achieved by transducing particles carrying overlapping segments of the island (Moon et al. 2015).

Besides exotoxins, bacteriophages also encode other virulence factors responsible for the adhesion, colonization and invasion capacity of their hosts (reviewed in Wagner and Waldor, 2002).

All these examples show the intimate relationship between phages and bacterial virulence factors. This association ensures a widespread dissemination of phages among bacterial populations and also their persistence, as phages survive longer than bacteria in the environment (Muniesa et al. 1999).

7.2.1.2 Other Beneficial Genes

Cyanobacteria and their bacteriophages offer a clear example of gene transfer between viruses and hosts. Cyanophages are key factors in the evolution of cyanobacteria populations, regulating their size and diversity and playing a role as genetic information stores and carriers (Shestakov and Karbysheva 2015). Cyanophage genes control metabolic processes in the host bacteria that allow the host cell to survive under adverse conditions, including transcription regulators that switch off unnecessary metabolic pathways when cells are growing slowly (Shestakov and Karbysheva 2015).

Cyanophages also carry genes encoding host-like proteins that enhance the photosynthetic process and prevent photo-inhibition in infected cells by expressing their own copies (Mann et al. 2003). This is the case of *psbA* and *psbD*, two photosystem II (PSII) reaction genes. These host-like genes are present in cyanophage genomes (Sullivan et al. 2006) probably because cyanophage production is optimal when photosynthesis is maintained during infection. At the same time, the expression of these auxiliary metabolic genes benefits the host by supplying extra energy and carbon, thereby boosting their photosynthetic activity.

7.2.1.3 Bacterial Evolution and Emergence of New Strains

Bacteriophage-mediated HGT also contributes to the emergence of new strains. This is the case of *E. coli* O104:H4, responsible for the large outbreak of diarrhea and hemolytic uremic syndrome (HUS) in Europe from May to July 2011 (Muniesa et al. 2012). Isolates obtained during the outbreak had an unusual combination of virulence factors attributed to Shiga toxin production by STEC and the aggregative-adherence fimbriae of enteroaggregative *E. coli* (EAEC) (Grad et al. 2013). A plausible explanation for the emergence of this new type of *E. coli* pathogen is a Stx_2-encoding bacteriophage, integrated in the chromosome of an enteroaggregative *E. coli*. The resulting new strain was designated as enteroaggregative hemorrhagic *E. coli* (EAHEC) (Brzuszkiewicz et al. 2011).

S. aureus is another species whose diversity is determined by mobile genetic elements, mainly prophages and pathogenicity islands, and its strain evolution is closely linked to phages. A frequent colonizer of human skin, *S. aureus* is present in 20–30% of the general population without any clinical manifestation, but it is also responsible for a wide spectrum of diseases in humans and animals. The versatility of this pathogen can be explained by different adaptive strategies and virulence properties, some of which are intrinsic to the bacterium, whereas others require adaptation and genomic evolution. *Staphylococcus* phages can mobilize pathogenicity islands, chromosomal markers and other extra-chromosomal elements such as plasmids (Xia and Wolz 2014). As mentioned in previous sections, many virulence factors important for the success of certain strains are encoded in phage genomes. The importance of phages in *S. aureus* evolution is illustrated by the case of ST398. This lineage was orginally described in livestock pigs and humans in contact with the animals (Graveland et al. 2011). Currently, the infection spectrum has widened to include, among others, humans in animal-free environments, and ST398 was isolated in a few cases of bloodstream infections (Verkade et al. 2012). A recent study has shown that, whereas the ancestral isolate associated with animals lacks prophages, emerging clades causing human infections possess φ3 prophage variants encoding two immune-modulating proteins, which alter or prevent chemotaxis, phagocytosis and killing of *S. aureus* by human neutrophils (Diene et al. 2017). The authors thus hypothesized that lysogeny has played an important role in the ability of ST398 to cause infections in humans.

7.2.2 Bacterial Survival and Adaptation to the Host

Besides conferring new features to bacteria through HGT, the presence of prophages in bacterial genomes has other advantages. For instance, prophages protect their hosts against secondary infections by preventing similar phages from attaching, in what is known as superinfection exclusion. In addition, prophages can affect the behavior and fitness of their hosts. Phages carry a variety of auxiliary metabolic genes encoding critical ecological functions, such as photosynthesis, nucleotide, carbon and phosphate metabolism, and stress response. Once again, examples can be found in STEC, where Stx-encoding bacteriophages were observed to modulate the metabolism of their *E. coli* host by increasing the rates of respiration and cell proliferation, which in turn alters resistance to osmotic stress and strengthens the host's antimicrobial tolerance (Holt et al. 2017).

The Shiga toxins encoded by Stx phages are also known to modulate the immune system in cattle during STEC infection by altering the cytokine expression pattern in macrophages (Stamm et al. 2008) and intraepithelial lymphocytes (Moussay et al. 2006) and by suppressing the activation and proliferation of mucosal (Menge et al. 2004) and peripheral lymphocytes *in vitro* (Menge et al. 1999). This modulation is believed to play a role in the persistence of *E. coli* O157:H7 in ruminants.

In STEC, the presence of two Stx prophages reduces the activation of the lytic cycle, as their repressor proteins mutually influence their spontaneous induction (Serra-Moreno et al. 2008). A reduction of lytic induction implies a lower toxin production, which in turns produces a less virulent strain, a logical strategy for the effective colonization of a new environment. The destruction of a fraction of the bacterial population by cell lysis maintains the virulence of the overall population. On the other hand, the fraction kept in a stable lysogenic state, with a reduced pathogenicity, ensures the population survives (Imamovic et al. 2016). This mechanism is covered in more depth in Sect. 7.3.3.

Analysis of the *E. coli* O157:H7 str. Sakai genome revealed the presence of 55 small RNA (sRNA) within bacteriophage-derived regions (Tree et al. 2014), one of which modulates the activity of the host core genome regulatory sRNA by mimicking its substrate. The loss of this anti-sRNA strongly reduced the competitiveness of the strain in bovine rectal mucus but not in laboratory media. These results suggest this pathogenicity island-associated sRNA is involved in the growth of the strain within its animal host reservoir at the specialized site colonized by the pathogen.

Bacteriophage Pf, a filamentous phage from *P. aeruginosa,* allows its host to evade the immune system by triggering an antiviral immune response that suppresses bacterial clearance from infected wounds (Sweere et al. 2019). After the uptake of the phage by human and murine leukocytes, Pf produces RNA that in turn triggers antiviral pattern recognition receptors, driving type I IFN production, inhibiting TNF production and suppressing phagocytosis. Conversely, vaccination and passive immunization with monoclonal antibodies against Pf prevented *P. aeruginosa* colonization (Sweere et al. 2019). This immune evasion was also described in *Neisseria meningitidis*, whose prophages express the TspB protein, which is present on the bacterial surface and binds to human IgG. The resulting bacterial aggregates can protect bacteria from the immune system (Müller et al. 2013).

Bacteriophages can also promote the establishment and persistence of their hosts by reducing their virulence. Prophage Rs551, for example, was found to negatively regulate the virulence of *Ralstonia solanacearum* UW551 strain (Ahmad et al. 2017). However, the absence of the prophage decreased the competitiveness of a mutant UW551 strain when it was co-inoculated with another strain in tomato plants. These results suggest that ϕRs551 may play an important ecological role by regulating the virulence of the UW551 strain and offering a fitness advantage for persistence in the environment.

7.2.3 Other Benefits for Bacteria

The presence of prophages can also be beneficial for other bacteria that compete with the host lysogenic strains. For example, in the nasopharynx the production of H_2O_2 by *Streptococcus pneumoniae* triggers the SOS response in staphylococci, including *S. aureus* lysogenic strains. As well as activating DNA repair, it induces resident prophages, resulting in cell lysis and ultimately in the displacement of the *S. aureus* population (Selva et al. 2009). Since most *S. aureus* strains are lysogenic, the production of H_2O_2 is a very effective anti-staphylococcal strategy used by *S. pneumoniae.*

Several studies have highlighted bacteriophages as a way for bacteria to spread their genetic content, focusing particularly on antibiotic resistant genes (ARGs). As described above, in phage-mediated transfer of bacterial DNA, the nucleic acid is protected in a protein capsid, which affords an extracellular persistence unavailable for naked DNA or RNA. The presence and abundance of ARGs has been confirmed in the phage fraction isolated from human and animal faeces, faecally polluted environments and food (Colomer-Lluch et al. 2011a, b; Quirós et al. 2014; Balcazar 2014; Marti et al. 2014; Calero-Cáceres and Muniesa 2016; Larrañaga et al. 2018; Gómez-Gómez et al. 2019). As generalized transducing particles do not contain phage DNA, they are not affected by phage superimmunity and can inject their genetic content in any susceptible cell. Although this lack of immunity means transductants are more susceptible to being destroyed by other lytic phages present in the environment, they constitute a significant threat to human health, because of their ability to mobilize antibiotic resistances, protect the ARG in the extracellular environment, and mobilize it in an environmentally persistent particle.

7.3 Disadvantages of Phages for Bacteria: Phages as Parasites

7.3.1 Phages Affect Fitness Cost and Bacterial Functions

The effect of carrying numerous prophages on bacterial fitness is similar to that of carrying and replicating several plasmids. During lysogeny, the bacterial host and the phage have a symbiotic relationship and it is advantageous for the prophage to provide genes that increase the fitness of the host cell. Despite this genetic contribution, the prophage can be costly for the cell, as large DNA insertions disrupt gene expression (Feiner et al. 2015). Another energetic burden is associated with the amount of DNA that must be replicated when the bacterial cell divides and the prophages are vertically transmitted.

It is in the interest of the prophage integrated in the bacterial genome to prevent further phage infection (superinfection) in an attempt to save its own genome from destruction. Prophages can defend their bacterial host from superinfection of phages of the same (homoimmunity) or different (heteroimmunity) kind. This protection can be an advantage for the cell if other infectious phages are present, but the defense mechanisms operate even in their absence. Defense strategies include superinfection exclusion proteins expressed from the resident prophages (Bondy-Denomy and Davidson 2014; Bondy-Denomy et al. 2016). The proteins block phage genome injection by interacting with the cytoplasmic membrane or prevent the binding of the new phage by interacting with the phage receptor on the bacterial outer membrane. In the process, some activities beneficial for the host cell might be repressed by the resident prophage. A notable example can be found in *Pseudomonas* phage D3112, which uses the type IV pilus (T4P) for infection. The Tip protein of phage D3112 inhibits the T4P, thereby preventing superinfection by phage MP22, which uses the same receptor (Chung et al. 2014). However, this protection against a potential enemy (in this case phage MP22) also disturbs the bacterial cell, as the T4P structure mediates twitching motility, which plays several roles in *P. aeruginosa* pathogenesis and biofilm formation.

Bacteria have developed different defense mechanisms against phage infection, and the presence of phages should result in the selection of a resistant population. Mechanisms inhibit phage adsorption by blocking or modifying the phage receptor, producing an extracellular matrix or by competitive inhibitors. Phage receptor modification by structural adaptation of surface receptors or the three-dimensional conformation has been studied in different Gram-negative bacteria. The resistance of *P. aeruginosa* to certain phages is based on conformational changes of the OprM protein used as a receptor (Chan et al. 2016), which alter the phage affinity to the receptor, preventing or reducing phage adsorption.

In *E. coli*, the outer membrane protein TraT masks or modifies the conformation of the outer membrane protein OmpA, which is the receptor on many T-even-like *E. coli* phages (Riede and Eschbach 1986). In *Salmonella,* changes in the innate efflux pump TolC (Ricci and Piddock 2010) or the porin OmplC in *Edwarsiella ictaluri* (Hossain et al. 2012) affect the degree of phage susceptibility.

As many of these proteins are also used as mechanisms to eliminate antimicrobial compounds, they indirectly confer resistance to certain antibiotics. *P. aeruginosa* OprD mediates antibiotic resistance through changes in cell permeability (Nikaido 2003). In *Klebsiella pneumoniae*, susceptibility to β-lactam antibiotics is associated with a shift from the expression of OmpK35 and OmpK36 to OmpK37 (Doménech-Sánchez et al. 2003). Changes in efflux pumps affect a wide range of antibiotics, including synthesis inhibitors, fluoroquinolones, β-lactams, carbapenems and polymyxins. One of the most studied is the *E. coli* AcrAB-TolC system, which functions as a proton antiporter and is able to transport a wide array of substrates, conferring resistance to tetracycline, chloramphenicol, novobiocin, some β-lactams, fusidic acid and fluoroquinolones (Du et al. 2015).

However, resistance to phage infection can lead to changes in the susceptibility of the strain to a given antibiotic. In *P. aeruginosa*, the number of bacterial cells is reduced by phage infection and lysis, but during the infectious process, spontaneous phage-resistant mutants quickly start to appear (Chan et al. 2016). The mutants show a modification in OprM that disables its function as an efflux pump, used by the cell to remove certain antibiotics (Ma et al. 1994; Poole 2002; Martinez et al. 2009; Delmar et al. 2014). As a consequence, *P. aeruginosa* becomes increasingly sensitive to several antibiotics, especially ceftazidime and tetracycline (Chan et al. 2016).

7.3.2 Disadvantages of Multiple Prophages

The frequency of multilysogeny, demonstrated in many bacterial genera, suggests it is beneficial for the phage and/or the bacteria. The benefit for the bacteria is not always entirely clear, as in the case of Shiga toxin-producing phages (Stx phages) infecting pathogenic Shiga toxin-producing *E. coli* (STEC). Stx phages can apparently avoid phage homoimmunity and hence two or even three identical or very similar Stx prophages can be found in the same strain (Serra-Moreno et al. 2008). Studies have shown that *E. coli* O157:H7 isolates harboring two Stx prophages produce less toxin than strains from the same clone carrying only one (Muniesa et al. 2003; Herold et al. 2005; Bielaszewska et al. 2006; Serra-Moreno et al. 2008). The opposite effect is also reported, when two prophages lead to two copies of the toxin gene and consequently a higher production of the Shiga toxin. The two possible outcomes might depend on the expression levels of phage late genes. If very high, the resources of an infected cell may be overwhelmed (Fogg et al. 2012), resulting in a decrease in toxin expression in multiple Stx phage lysogens. On the contrary, if the strain is able to handle the transcription of the proteins encoded in multiple phages, the result is an enhanced Stx production. Regardless, both scenarios can create problems for the bacteria. Stx is a powerful weapon for the bacteria, so a reduction of phage induction and Stx expression is a clear disadvantage for multilysogenic strains. In compensation, the strains will survive longer if the lytic cycle induction is reduced. On the other hand, when two *stx* copies lead to a higher Stx expression by means of Stx phage induction, the price to pay is cell lysis.

7.3.3 Phages Cause Bacterial Lysis and Death

Lysogeny is a ticking time bomb in which prophage induction represents the detonating mechanism. When encountering stress conditions, bacteria trigger the SOS response to survive in an unfavorable environment, but with the side-effect that the prophage switches from the lysogenic to the lytic cycle (Oppenheim et al. 2005; Ubeda et al. 2005; Waldor and Friedman 2005). Upon activation of the SOS response, autocleavage of the phage lytic repressor activates transcription of lytic genes and induces the lytic cycle, resulting in cell death and the release of free phages. Phage induction is therefore an extreme threat for bacterial survival, potentially leading to population extinction.

Yet despite the inherent risks, lysogeny has in many cases been positively selected and maintained (Bednarz et al. 2014). Some prophages, providing beneficial new genes for the host, are not rejected. However, there are paradoxical cases, such as that of the aforementioned Stx phages. These provide virulence genes, yet expression of the toxin at usable levels and its release outside the cell depend on the activation of the lytic cycle. It has been demonstrated in *E. coli* (and to a lesser extent in *S. aureus* and *S. enteritidis*) that the key to understanding this paradox is the heterogeneity of the lysogenic bacterial population, which enables it to circumvent extinction. After phage induction, a fraction of the cells maintain lysogeny, allowing population survival, while another fraction undergo lysis, increasing Stx production and spreading Stx phages (Imamovic et al. 2016).

Even if bacteria can subdivide their populations to take advantage of both lysis and lysogeny, phage-induced lysis remains a burden. As a consequence, bacteria have acquired defense mechanisms and phages themselves can control their lytic activities to avoid wiping out their bacterial hosts.

7.3.3.1 Bacterial Defenses against Phage Infection

Maintaining the integrity of a phage genome is necessary for successful infection after DNA injection. Bacteria defend themselves by using innate and adaptive immune systems, such as CRISPR-Cas (Barrangou et al. 2007; Mojica and Rodriguez-Valera 2016), or restriction/modification systems (Labrie et al. 2010; Dupuis et al. 2013). The mere existence of such mechanisms is indirect evidence that phages pose a threat for bacteria (because of lysis or uncontrolled gene acquisition), inducing bacteria to develop protection systems. In counteraction, some phages have incorporated genes encoding Acr proteins that overcome phage DNA cleavage by inactivating CRISPR/Cas systems (Bondy-Denomy et al. 2012).

However, CRISPR/Cas system inactivation relies on Acr reaching a certain concentration, which is only achieved after the cooperative contribution of multiple prophages within a single cell (Borges et al. 2018).

Paradoxically, multiple examples show bacteria carrying imperfect, incomplete and no longer immunogenic CRISPR-Cas systems (Perez-Rodriguez et al. 2011; Shariat et al. 2015; Crispim et al. 2018), suggesting that phage infection and prophage incorporation is not always an undesired event.

Although CRISPR-Cas systems are mechanisms used to fight against phage infection and lysis, and hence are contrary to phage interests, they are also used by phages to compete against other mobile genetic elements. For example, in *Clostridium difficile* phages incorporate new CRISPR loci to avoid superinfection by other phages (Soutourina et al. 2013). In *V. cholerae* phage ICP1 encodes an entire CRISPR-Cas system to compete against a phage-inducible chromosomal island. Upon phage infection, the island excises from the genome and circularizes to limit a productive phage infection and in response the phage-encoded CRISPR-Cas system neutralizes the activity of that island (Seed et al. 2013).

CRISPR/Cas systems can also be used by bacteria in a far more complicated strategy. Bacteria require previous phage infection to generate transducing particles containing their own DNA. However, the presence of wild phage particles is a threat because they lyse the recipient cells. CRISPR-Cas inactivates the lytic phage but is ineffective against transducing particles that only contain bacterial DNA. By activating this mechanism, bacteria can act specifically against wild-type phages, protecting the sensitive population while promoting the generation of transductants and HGT (Watson et al. 2018).

Similar observations have been reported in environmental α-proteobacteria such as *Rhodobacter* or *Bartonella* (Lang et al. 2012), which produce phage-like particles called gene transfer agents (GTAs) that package almost exclusively bacterial DNA and transfer it to other bacterial cells. GTAs resemble phage capsids except that their formation does not require previous phage infection, as the capsid genes are already present in the bacterial genome. The maintenance of the GTA genes in the bacterial chromosome indicates they benefit the host as HGT vehicles. It is reasonable to expect that these or similar mechanisms could play a role in the spread of bacterial DNA in other bacterial groups, including human pathogens.

7.3.3.2 Phage Control of the Lytic Cycle

Phages as well as bacteria have evolved to regulate their populations by mechanisms that control the induction of the lytic cycle, which can destroy the host cell population if it occurs excessively. Arbitrium communication is such a mechanism, involving the AimP peptide encoded in *Bacillus* phages. During infection, AimP is processed by the host and released in a mature form into the environment after lysis (Erez et al. 2017). AimP has an effect on neighboring cells and by binding to AimR controls the expression of AimX, which is a negative regulator of lysogeny. By blocking the expression of the negative regulator, AimP biases the cells toward lysogeny. The overall process can be interpreted as the viruses of one generation informing subsequent generations of host scarcity, blocking the lytic cycle and promoting their own stability through lysogeny, particularly at high levels of the peptide.

7.4 Final Remarks

Bacteriophages and bacteria have co-existed for over a billion years, in contrast with the short time we have studied their relationship or tried to harness it for our own purposes. Their coexistence has driven the evolution of diverse strategies, as bacteria and phages seek to stay ahead of each other in the arms race (Prof. Vincent A Fischetti @microbephage 1019 Twitter). As usual in nature, rather than a winner or loser, this contest has produced a finely tuned equilibrium of forces, in which each player selects the best options for survival. For this reason, before deciding which side to take, it would be wise to carefully observe and understand these balanced forces, in case the opponents turn out to be mutual allies.

Acknowledgements This work was supported by the Ministerio de Ciencia e Innovación (AGL2016-75536-P). P. B-P has a grant from the Spanish Ministry of Economy, Industry and Competitiveness (BES-2017-081296). L. R.-R is supported by the Beatriu de Pinós postdoctoral program of the Government of Catalonia.

References

Adams MH (1959) Bacteriophages. In, Bacteriophages, p 592

Ahmad AA, Stulberg MJ, Huang Q (2017) Prophage Rs551 and its repressor gene orf14 reduce virulence and increase competitive fitness of its Ralstonia solanacearum carrier strain UW551. Front Microbiol 8:2480. https://doi.org/10.3389/fmicb.2017.02480

Allué-Guardia A, Martínez-Castillo A, Muniesa M (2014) Persistence of infectious Shiga toxin-encoding bacteriophages after disinfection treatments. Appl Environ Microbiol 80:2142–2149. https://doi.org/10.1128/AEM.04006-13

Balcazar JL (2014) Bacteriophages as vehicles for antibiotic resistance genes in the environment. PLoS Pathog 10:e1004219. https://doi.org/10.1371/journal.ppat.1004219

Barrangou R, Fremaux C, Deveau H et al (2007) CRISPR provides acquired resistance against viruses in prokaryotes. Science 315:1709–1712. https://doi.org/10.1126/science.1138140

Bednarz M, Halliday JA, Herman C, Golding I (2014) Revisiting bistability in the lysis/lysogeny circuit of bacteriophage lambda. PLoS One 9:e100876. https://doi.org/10.1371/journal.pone.0100876

Bertozzi Silva J, Storms Z, Sauvageau D (2016) Host receptors for bacteriophage adsorption. FEMS Microbiol Lett 363:fnw002. https://doi.org/10.1093/femsle/fnw002

Betley MJ, Mekalanos JJ (1985) Staphylococcal enterotoxin a is encoded by phage. Science (80-) 229:185–187. https://doi.org/10.1126/science.3160112

Beumer A, Robinson JB (2005) A broad-host-range, generalized transducing phage (SN-T) acquires 16S rRNA genes from different genera of bacteria. Appl Environ Microbiol 71:8301–8304. https://doi.org/10.1128/AEM.71.12.8301-8304.2005

Bielaszewska M, Prager R, Zhang W et al (2006) Chromosomal dynamism in progeny of outbreak-related sorbitol-fermenting enterohemorrhagic Escherichia coli O157:NM. Appl Environ Microbiol 72:1900–1909. https://doi.org/10.1128/AEM.72.3.1900-1909.2006

Bondy-Denomy J, Davidson AR (2014) When a virus is not a parasite: the beneficial effects of prophages on bacterial fitness. J Microbiol 52:235–242. https://doi.org/10.1007/s12275-014-4083-3

Bondy-Denomy J, Pawluk A, Maxwell KL, Davidson AR (2012) Bacteriophage genes that inactivate the CRISPR/Cas bacterial immune system. Nature 493:429–432. https://doi.org/10.1038/nature11723

Bondy-Denomy J, Qian J, Westra ER et al (2016) Prophages mediate defense against phage infection through diverse mechanisms. ISME J 10:2854–2866. https://doi.org/10.1038/ismej.2016.79

Borges AL, Zhang JY, Rollins MF et al (2018) Bacteriophage cooperation suppresses CRISPR-Cas3 and Cas9 immunity. Cell 174:917–925.e10. https://doi.org/10.1016/j.cell.2018.06.013

Brown-Jaque M, Calero-Cáceres W, Espinal P et al (2018) Antibiotic resistance genes in phage particles isolated from human feces and induced from clinical bacterial isolates. Int J Antimicrob Agents 51:434–442

Brzuszkiewicz E, Thürmer A, Schuldes J et al (2011) Genome sequence analyses of two isolates from the recent Escherichia coli outbreak in Germany reveal the emergence of a new pathotype: Entero-aggregative-haemorrhagic Escherichia coli (EAHEC). Arch Microbiol 193:883–891. https://doi.org/10.1007/s00203-011-0725-6

Calero-Cáceres W, Muniesa M (2016) Persistence of naturally occurring antibiotic resistance genes in the bacteria and bacteriophage fractions of wastewater. Water Res 95:11–18. https://doi.org/10.1016/j.watres.2016.03.006

Casjens S (2003) Prophages and bacterial genomics: what have we learned so far? Mol Microbiol 49:277–300

Chan BK, Sistrom M, Wertz JE et al (2016) Phage selection restores antibiotic sensitivity in MDR Pseudomonas aeruginosa. Sci Rep 6:26717. https://doi.org/10.1038/srep26717

Chen J, Ram G, Penadés JR et al (2015) Pathogenicity Island-directed transfer of unlinked chromosomal virulence genes. Mol Cell 57:138–149. https://doi.org/10.1016/j.molcel.2014.11.011

Chen J, Quiles-Puchalt N, Chiang YN et al (2018) Genome hypermobility by lateral transduction. Science (80-) 362:207–212. https://doi.org/10.1126/science.aat5867

Chung I-Y, Jang H-J, Bae H-W, Cho Y-H (2014) A phage protein that inhibits the bacterial ATPase required for type IV pilus assembly. Proc Natl Acad Sci 111:11503–11508. https://doi.org/10.1073/pnas.1403537111

Colomer-Lluch M, Imamovic L, Jofre J, Muniesa M (2011a) Bacteriophages carrying antibiotic resistance genes in fecal waste from cattle, pigs, and poultry. Antimicrob Agents Chemother 55:4908–4911. https://doi.org/10.1128/AAC.00535-11

Colomer-Lluch M, Jofre J, Muniesa M (2011b) Antibiotic resistance genes in the bacteriophage DNA fraction of environmental samples. PLoS One 6:e17549

Crispim JS, Dias RS, Vidigal PMP et al (2018) Screening and characterization of prophages in Desulfovibrio genomes. Sci Rep 8:9273. https://doi.org/10.1038/s41598-018-27423-z

Delmar JA, Su C-C, Yu EW (2014) Bacterial Multidrug Efflux Transporters. Annu Rev Biophys 43:93–117. https://doi.org/10.1146/annurev-biophys-051013-022855

Diene SM, Corvaglia AR, François P et al (2017) Prophages and adaptation of Staphylococcus aureus ST398 to the human clinic. BMC Genomics 18:133. https://doi.org/10.1186/s12864-017-3516-x

Doménech-Sánchez A, Martínez-Martínez L, Hernández-Allés S et al (2003) Role of Klebsiella pneumoniae OmpK35 porin in antimicrobial resistance. Antimicrob Agents Chemother 47:3332–3335

Doolittle WF (2002) Lateral DNA transfer: mechanisms and consequences. CSHL Press

Du D, van Veen HW, Luisi BF (2015) Assembly and operation of bacterial tripartite multidrug efflux pumps. Trends Microbiol 23:311–319. https://doi.org/10.1016/j.tim.2015.01.010

Dupuis M-È, Villion M, Magadán AH, Moineau S (2013) CRISPR-Cas and restriction–modification systems are compatible and increase phage resistance. Nat Commun 4:2087. https://doi.org/10.1038/ncomms3087

Durán AE, Muniesa M, Méndez X et al (2002) Removal and inactivation of indicator bacteriophages in fresh waters. J Appl Microbiol 92:338–347

Eklund MW, Poysky FT (1974) Interconversion of type C and D strains of *Clostridium botulinum* by specific bacteriophages. Appl Microbiol 27:251–258

Eklund MW, Poysky FT, Reed SM (1972) Bacteriophage and the toxigenicity of clostridium botulinum type D. Nat New Biol 235:16–17. https://doi.org/10.1038/newbio235016a0

Erez Z, Steinberger-Levy I, Shamir M et al (2017) Communication between viruses guides lysis–lysogeny decisions. Nature 541:488–493. https://doi.org/10.1038/nature21049

Evans TJ, Crow MA, Williamson NR et al (2010) Characterization of a broad-host-range flagellum-dependent phage that mediates high-efficiency generalized transduction in, and between, Serratia and Pantoea. Microbiology 156:240–247. https://doi.org/10.1099/mic.0.032797-0

Feiner R, Argov T, Rabinovich L et al (2015) A new perspective on lysogeny: prophages as active regulatory switches of bacteria. Nat Rev Microbiol 13:641–650. https://doi.org/10.1038/nrmicro3527

Fogg PCM, Saunders JR, McCarthy AJ, Allison HE (2012) Cumulative effect of prophage burden on Shiga toxin production in Escherichia coli. Microbiology 158:488–497. https://doi.org/10.1099/mic.0.054981-0

Fortier L-C, Sekulovic O (2013) Importance of prophages to evolution and virulence of bacterial pathogens. Virulence 4:354–365. https://doi.org/10.4161/viru.24498

García-Aljaro C, Ballesté E, Muniesa M (2017) Beyond the canonical strategies of horizontal gene transfer in prokaryotes. Curr Opin Microbiol 38:95–105. https://doi.org/10.1016/j.mib.2017.04.011

Gómez-Gómez C, Blanco-Picazo P, Brown-Jaque M et al (2019) Infectious phage particles packaging antibiotic resistance genes found in meat products and chicken feces. Sci Rep 9:13281. https://doi.org/10.1038/s41598-019-49898-0

Grad YH, Godfrey P, Cerquiera GC et al (2013) Comparative genomics of recent Shiga toxin-producing Escherichia coli O104:H4: short-term evolution of an emerging pathogen. MBio 4:e00452–e00412. https://doi.org/10.1128/mBio.00452-12

Graveland H, Duim B, van Duijkeren E et al (2011) Livestock-associated methicillin-resistant Staphylococcus aureus in animals and humans. Int J Med Microbiol 301:630–634. https://doi.org/10.1016/j.ijmm.2011.09.004

Groth AC, Calos MP (2004) Phage integrases: biology and applications. J Mol Biol 335:667–678

Hayashi T, Baba T, Matsumoto H, Terawaki Y (1990) Phage-conversion of cytotoxin production in Pseudomonas aeruginosa. Mol Microbiol 4:1703–1709. https://doi.org/10.1111/j.1365-2958.1990.tb00547.x

Herold S, Siebert J, Huber A, Schmidt H (2005) Global expression of Prophage genes in Escherichia coli O157:H7 strain EDL933 in response to Norfloxacin. Antimicrob Agents Chemother 49:931–944. https://doi.org/10.1128/AAC.49.3.931-944.2005

Holt GS, Lodge JK, McCarthy AJ et al (2017) Shigatoxin encoding bacteriophage φ24B modulates bacterial metabolism to raise antimicrobial tolerance. Sci Rep 7:40424. https://doi.org/10.1038/srep40424

Hossain MJ, Rahman KS, Terhune JS, Liles MR (2012) An outer membrane porin protein modulates phage susceptibility in Edwardsiella ictaluri. Microbiology 158:474–487. https://doi.org/10.1099/mic.0.054866-0

Imamovic L, Ballesté E, Martínez-Castillo A et al (2016) Heterogeneity in phage induction enables the survival of the lysogenic population. Environ Microbiol 18:957–969. https://doi.org/10.1111/1462-2920.13151

Inoue K, Iida H (1970) Conversion of Toxigenicity in Clostridium botulinum type C. Jpn J Microbiol 14:87–89. https://doi.org/10.1111/j.1348-0421.1970.tb00495.x

Jensen EC, Schrader HS, Rieland B et al (1998) Prevalence of broad-host-range lytic bacteriophages of Sphaerotilus natans, Escherichia coli, and Pseudomonas aeruginosa. Appl Environ Microbiol 64:575–580

Johnson TA, Looft T, Severin AJ et al (2017) The in-feed antibiotic carbadox induces phage gene transcription in the swine gut microbiome. MBio 8. https://doi.org/10.1128/mBio.00709-17

Jończyk E, Kłak M, Międzybrodzki R, Górski A (2011) The influence of external factors on bacteriophages--review. Folia Microbiol (Praha) 56:191–200. https://doi.org/10.1007/s12223-011-0039-8

Kaneko J, Kimura T, Kawakami Y et al (1997) Panton-valentine leukocidin genes in a phage-like particle isolated from mitomycin C-treated Staphylococcus aureus V8 (ATCC 49775). Biosci Biotechnol Biochem 61:1960–1962

Karaolis DKR, Somara S, Maneval DR et al (1999) A bacteriophage encoding a pathogenicity island, a type-IV pilus and a phage receptor in cholera bacteria. Nature 399:375–379. https://doi.org/10.1038/20715

Kimmitt PT, Harwood CR, Barer MR (2000) Toxin gene expression by Shiga toxin-producing Escherichia coli: the role of antibiotics and the bacterial SOS response. Emerg Infect Dis 6:458–465

Labrie SJ, Samson JE, Moineau S (2010) Bacteriophage resistance mechanisms. Nat Rev Microbiol 8:317–327. https://doi.org/10.1038/nrmicro2315

Lang AS, Zhaxybayeva O, Beatty JT (2012) Gene transfer agents: phage-like elements of genetic exchange. Nat Rev Microbiol 10:472–482. https://doi.org/10.1038/nrmicro2802

Larrañaga O, Brown-Jaque M, Quirós P et al (2018) Phage particles harboring antibiotic resistance genes in fresh-cut vegetables and agricultural soil. Environ Int 115:133–141. https://doi.org/10.1016/j.envint.2018.03.019

Lee HS, Sobsey MD (2011) Survival of prototype strains of somatic coliphage families in environmental waters and when exposed to UV low-pressure monochromatic radiation or heat. Water Res 45:3723–3734. https://doi.org/10.1016/j.watres.2011.04.024

Lindsay JA, Ruzin A, Ross HF et al (1998) The gene for toxic shock toxin is carried by a family of mobile pathogenicity islands in Staphylococcus aureus. Mol Microbiol 29:527–543. https://doi.org/10.1046/j.1365-2958.1998.00947.x

Ma D, Cook DN, Hearst JE, Nikaido H (1994) Efflux pumps and drug resistance in gram-negative bacteria. Trends Microbiol 2:489–493. https://doi.org/10.1016/0966-842X(94)90654-8

Mann NH, Cook A, Millard A et al (2003) Bacterial photosynthesis genes in a virus. Nature 424:741. https://doi.org/10.1038/424741a

Marti E, Variatza E, Balcázar JL (2014) Bacteriophages as a reservoir of extended-spectrum β-lactamase and fluoroquinolone resistance genes in the environment. Clin Microbiol Infect 20:O456–O459. https://doi.org/10.1111/1469-0691.12446

Martinez JL, Sánchez MB, Martínez-Solano L et al (2009) Functional role of bacterial multidrug efflux pumps in microbial natural ecosystems. FEMS Microbiol Rev 33:430–449. https://doi.org/10.1111/j.1574-6976.2008.00157.x

Menge C, Wieler LH, Schlapp T, Baljer G (1999) Shiga toxin 1 from Escherichia coli blocks activation and proliferation of bovine lymphocyte subpopulations in vitro. Infect Immun 67:2209–2217

Menge C, Stamm I, Van Diemen PM et al (2004) Phenotypic and functional characterization of intraepithelial lymphocytes in a bovine ligated intestinal loop model of enterohaemorrhagic Escherichia coli infection. J Med Microbiol 53:573–579. https://doi.org/10.1099/jmm.0.45530-0

Miller C, Thomsen LE, Gaggero C et al (2004) SOS response induction by beta-lactams and bacterial defense against antibiotic lethality. Science 305:1629–1631. https://doi.org/10.1126/science.1101630

Moat AG, Foster JW, Spector MP (2003) Bacteriophage genetics. In: Microbial physiology. Fourth Edi. John Wiley & Sons, Inc., Hoboken, NJ, USA, pp 239–276

Mocé-llivina L, Muniesa M, Pimenta-vale H et al (2003) Survival of Bacterial Indicator Species and Bacteriophages after Thermal Treatment of Sludge and Sewage. Appl Environ Microbiol 69:1452–1456. https://doi.org/10.1128/AEM.69.3.1452

Mojica FJM, Rodriguez-Valera F (2016) The discovery of CRISPR in archaea and bacteria. FEBS J 283:3162–3169. https://doi.org/10.1111/febs.13766

Moon BY, Park JY, Hwang SY et al (2015) Phage-mediated horizontal transfer of a Staphylococcus aureus virulence-associated genomic island. Sci Rep 5:9784. https://doi.org/10.1038/srep09784

Morabito S (2014) Pathogenic Escherichia coli: molecular and cellular microbiology. Horizon Press, Linton

Moussay E, Stamm I, Taubert A et al (2006) Escherichia coli Shiga toxin 1 enhances il-4 transcripts in bovine ileal intraepithelial lymphocytes. Vet Immunol Immunopathol 113:367–382. https://doi.org/10.1016/j.vetimm.2006.06.007

Müller MG, Ing JY, Cheng MK-W et al (2013) Identification of a phage-encoded Ig-binding protein from invasive Neisseria meningitidis. J Immunol 191:3287–3296. https://doi.org/10.4049/jimmunol.1301153

Muniesa M, Lucena F, Jofre J (1999) Comparative survival of free Shiga toxin 2-encoding Phages and. Society 65:5615–5618

Muniesa M, De Simon M, Prats G et al (2003) Shiga toxin 2-converting bacteriophages associated with clonal variability in Escherichia coli O157:H7 strains of human origin isolated from a single outbreak. Infect Immun 71. https://doi.org/10.1128/IAI.71.8.4554-4562.2003

Muniesa M, Hammerl JA, Hertwig S et al (2012) Shiga toxin-producing Escherichia coli O104:H4: a new challenge for microbiology. Appl Environ Microbiol 78:4065–4073. https://doi.org/10.1128/AEM.00217-12

Muniesa M, Allué-Guardia A, Martínez-Castillo A (2013) Bacteriophage-driven emergence of novel pathogens. Futur Virol 8:323–325. https://doi.org/10.2217/fvl.13.13

Nikaido H (2003) Molecular basis of bacterial outer membrane permeability revisited. Microbiol Mol Biol Rev 67:593–656

Oppenheim AB, Kobiler O, Stavans J et al (2005) Switches in bacteriophage lambda development. Annu Rev Genet 39:409–429. https://doi.org/10.1146/annurev.genet.39.073003.113656

Parsons EI, Frobisher M (1951) Effect of bacteriophage on virulence of Corynebacterium diphtheriae. Exp Biol Med 78:746–747. https://doi.org/10.3181/00379727-78-19204

Paul JH (2008) Prophages in marine bacteria: dangerous molecular time bombs or the key to survival in the seas? ISME J 2:579–589. https://doi.org/10.1038/ismej.2008.35

Penades JR, Chen J, Quiles-Puchalt N et al (2015) Bacteriophage-mediated spread of bacterial virulence genes. Curr Opin Microbiol 23:171–178. https://doi.org/10.1016/j.mib.2014.11.019

Perez-Rodriguez R, Haitjema C, Huang Q et al (2011) Envelope stress is a trigger of CRISPR RNA-mediated DNA silencing in Escherichia coli. Mol Microbiol 79:584–599. https://doi.org/10.1111/j.1365-2958.2010.07482.x

Poole K (2002) Outer membranes and efflux: the path to multidrug resistance in gram-negative bacteria. Curr Pharm Biotechnol 3:77–98

Quirós P, Colomer-Lluch M, Martínez-Castillo A et al (2014) Antibiotic resistance genes in the bacteriophage DNA fraction of human fecal samples. Antimicrob Agents Chemother 58:606–609. https://doi.org/10.1128/AAC.01684-13

Rajala-Mustonen RL, Heinonen-Tanski H (1994) Sensitivity of host strains and host range of coliphages isolated from finnish and nicaraguan wastewater. Water Res 28:1811–1815. https://doi.org/10.1016/0043-1354(94)90254-2

Ricci V, Piddock LJV (2010) Exploiting the role of TolC in pathogenicity: identification of a bacteriophage for eradication of Salmonella Serovars from poultry. Appl Environ Microbiol 76:1704–1706. https://doi.org/10.1128/AEM.02681-09

Riede I, Eschbach ML (1986) Evidence that TraT interacts with OmpA of *Escherichia Coli*. FEBS Lett 205(2):241–5. https://doi.org/10.1016/0014-5793(86)80905-x

Ruzin A, Lindsay J, Novick RP (2001) Molecular genetics of SaPI1 - a mobile pathogenicity island in Ataphylococcus aureus. Mol Microbiol 41:365–377. https://doi.org/10.1046/j.1365-2958.2001.02488.x

Samson JE, Magadán AH, Sabri M, Moineau S (2013) Revenge of the phages: defeating bacterial defences. Nat Rev Microbiol 11:675–687. https://doi.org/10.1038/nrmicro3096

Seed KD, Lazinski DW, Calderwood SB, Camilli A (2013) A bacteriophage encodes its own CRISPR/Cas adaptive response to evade host innate immunity. Nature 494:489–491. https://doi.org/10.1038/nature11927

Selva L, Viana D, Regev-Yochay G et al (2009) Killing niche competitors by remote-control bacteriophage induction. PNAS 106:1234–1238

Serra-Moreno R, Jofre J, Muniesa M (2008) The CI repressors of Shiga toxin-converting prophages are involved in coinfection of Escherichia coli strains, which causes a down regulation in the production of Shiga toxin 2. J Bacteriol 190:4722–4735. https://doi.org/10.1128/JB.00069-08

Shaikh N, Tarr PI (2003) Escherichia coli O157:H7 Shiga toxin-encoding bacteriophages: integrations, excisions, truncations, and evolutionary implications. J Bacteriol 185:3596–3605

Shariat N, Timme RE, Pettengill JB et al (2015) Characterization and evolution of Salmonella CRISPR-Cas systems. Microbiology 161:374–386. https://doi.org/10.1099/mic.0.000005

Shestakov SV, Karbysheva EA (2015) The role of viruses in the evolution of cyanobacteria. Biol Bull Rev 5:527–537. https://doi.org/10.1134/s2079086415060079

Soutourina OA, Monot M, Boudry P et al (2013) Genome-wide identification of regulatory RNAs in the human pathogen Clostridium difficile. PLoS Genet 9:e1003493. https://doi.org/10.1371/journal.pgen.1003493

Souza KA, Ginoza HS, Haight RD (1972) Isolation of a polyvalent bacteriophage for Escherichia coli, Klebsiella pneumoniae and Aerobacter Aerogenes. J Virol 9:851–856

Stamm I, Mohr M, Bridger PS et al (2008) Epithelial and mesenchymal cells in the bovine colonic mucosa differ in their responsiveness to Escherichia coli Shiga toxin 1. Infect Immun 76:5381–5391. https://doi.org/10.1128/IAI.00553-08

Sternberg N, Coulby JN (1990) Cleavage of the bacteriophage P1 packaging site (pac) is regulated by adenine methylation. PNAS 87:8070–8074. https://doi.org/10.1016/0378-1119(88)90287-9

Subirats J, Sànchez-Melsió A, Borrego CM et al (2016) Metagenomic analysis reveals that bacteriophages are reservoirs of antibiotic resistance genes. Int J Antimicrob Agents 48:163–167. https://doi.org/10.1016/j.ijantimicag.2016.04.028

Sullivan MB, Lindell D, Lee JA et al (2006) Prevalence and evolution of core photosystem II genes in marine cyanobacterial viruses and their hosts. PLoS Biol 4:1344–1357. https://doi.org/10.1371/journal.pbio.0040234

Sweere JM, Van Belleghem JD, Ishak H et al (2019) Bacteriophage trigger antiviral immunity and prevent clearance of bacterial infection. Science (80-) 363:eaat9691. https://doi.org/10.1126/science.aat9691

Thierauf A, Perez G, Maloy AS (2009) Generalized transduction. Methods Mol Biol 501:267–286. https://doi.org/10.1007/978-1-60327-164-6_23

Touchon M, Moura de Sousa JA, Rocha EP (2017) Embracing the enemy: the diversification of microbial gene repertoires by phage-mediated horizontal gene transfer. Curr Opin Microbiol 38:66–73

Tree JJ, Granneman S, McAteer SP et al (2014) Identification of bacteriophage-encoded anti-sRNAs in pathogenic Escherichia coli. Mol Cell 55:199–213. https://doi.org/10.1016/j.molcel.2014.05.006

Ubeda C, Maiques E, Knecht E et al (2005) Antibiotic-induced SOS response promotes horizontal dissemination of pathogenicity island-encoded virulence factors in staphylococci. Mol Microbiol 56:836–844. https://doi.org/10.1111/j.1365-2958.2005.04584.x

Utter B, Deutsch DR, Schuch R et al (2014) Beyond the chromosome: the prevalence of unique extra-chromosomal bacteriophages with integrated virulence genes in pathogenic Staphylococcus aureus. PLoS One 9:e100502. https://doi.org/10.1371/journal.pone.0100502

Verkade E, Bergmans AMC, Budding AE et al (2012) Recent emergence of Staphylococcus aureus clonal complex 398 in human blood cultures. PLoS One 7:e41855. https://doi.org/10.1371/journal.pone.0041855

Wagner PL, Waldor MK (2002) Bacteriophage control of bacterial virulence. Infect Immun 70:3985–3993

Waldor MK, Friedman DI (2005) Phage regulatory circuits and virulence gene expression. Curr Opin Microbiol 8:459–465. https://doi.org/10.1016/j.mib.2005.06.001

Waldor MK, Mekalanos JJ (1996) Lysogenic conversion by a filamentous phage encoding cholera toxin. Science 272:1910–1914

Watson BNJ, Staals RHJ, Fineran PC (2018) CRISPR-Cas-mediated phage resistance enhances horizontal gene transfer by transduction. MBio 9. https://doi.org/10.1128/mBio.02406-17

Xia G, Wolz C (2014) Phages of Staphylococcus aureus and their impact on host evolution. Infect Genet Evol 21:593–601. https://doi.org/10.1016/j.meegid.2013.04.022

Yamaguchi T, Hayashi T, Takami H et al (2000) Phage conversion of exfoliative toxin a production in Staphylococcus aureus. Mol Microbiol 38:694–705. https://doi.org/10.1046/j.1365-2958.2000.02169.x

Yu P, Mathieu J, Li M et al (2016) Isolation of polyvalent bacteriophages by sequential multiple-host approaches. Appl Environ Microbiol 82:808–815. https://doi.org/10.1128/AEM.02382-15

Chapter 8
Microbial Communication Networks: Sketching a Method for Analyzing the Communication of Bacteriophages Inside Environmental Communities

Charles Bernard, Philippe Lopez, and Eric Bapteste

Abstract Recent functional studies have shed light into how the combinatorics of genes associated with quorum sensing (QS) – often described as an entity-entity communication mechanism – may support different communication modalities in bacteriophages. Specifically, only systems of QS genes for phage to phage communication and eavesdropping on bacterial communication molecules have been characterized so far, which represents only a fraction of the spectrum of all the possible communication modalities predicted by this combinatory logic. However, we argue that computational methods are already available to systematically mine the genomes of viruses and other microorganisms for QS genes, to compare these genes together across genomes, and to infer many novel links and types of communication between microbiological entities. All these putative communication links could be conveniently represented together in the form of a network, which would summarize which virus is suspected to interact with which microbiological entity, *via* which QS signaling molecule and foremost, under which communication modality. Besides, with the recent advents of metagenomics and metaviromics that enable accessing to genomes sequenced from the same environmental site, it would be theoretically possible to use this methodology to partially infer the big picture of communication inside a real community of viruses and cellular organisms. Finally, we discuss how the systematic analysis of such predicted microbial communication networks could provide insights into the many forms and "social" consequences that the biocommunication of viruses may imply.

C. Bernard
Institut de Systématique, Evolution, Biodiversité (ISYEB), Sorbonne Université, CNRS, Museum National d'Histoire Naturelle, Paris, France

Unité Molécules de Communication et Adaptation des Micro-organismes (MCAM), CNRS, Museum National d'Histoire Naturelle, Paris, France

P. Lopez · E. Bapteste (✉)
Institut de Systématique, Evolution, Biodiversité (ISYEB), Sorbonne Université, CNRS, Museum National d'Histoire Naturelle, Paris, France
e-mail: eric.bapteste@upmc.fr

G. Witzany (ed.), *Biocommunication of Phages*,
https://doi.org/10.1007/978-3-030-45885-0_8

8.1 Microbial Communication Via Quorum Sensing (QS) Involving Bacteriophages

Temperate bacteriophages are viruses that can insert all or part of their genetic information into their bacterial host genomes in a process named lysogeny, as an alternative to the host-destructive lytic cycle, through which they replicate and spread. When a temperate bacteriophage lysogenizes a bacterium, it becomes dormant under the form of a "prophage" and confers the lysogenized bacterium immunity to further infection by the bacteriophage free form (virion) (Oppenheim et al. 2005; Abedon 2015). Lysogeny can therefore be seen as a means to preserve the phage-host collective, notably when the lytic cycle starts becoming a threat for the survival of the host population, upon which the own phage population survival is conditioned. Consistently, when the number of host cells happens to be limited, fast growing fully-lytic phages may quickly end up eradicating their hosts (and therefore themselves) whereas temperate bacteriophages are able to maintain a slow but continuous growth through the back and forth switch between lysis and lysogeny, a strategy shown to be evolutionary stable (Wahl et al. 2018). Specifically, in order for this switch to be an optimal evolutionary strategy, it should ideally be synchronized within the viral population, and occur sufficiently late during the infection to ensure that the phages have produced a sufficiently high number of progeny at the expanse of the host population (maximization of the phage replication) but also sufficiently early to ensure that the population of hosts is still composed of enough individuals (protection of the phage-host collective). In other words, from the phage perspective, the switch from lysis to lysogeny should ideally intervene in a coordinated manner, and be dependent on both of the phage and the host population densities.

One well-known microbial mechanism that orchestrates the emergence of a collective behavior in a density dependent manner is called "quorum sensing" (QS) (Fuqua et al. 1994). *Via* this mechanism, initially identified in bacteria and further extended to all superkingdoms of evolving entities (bacteria (Miller and Bassler 2001), archaea (Paggi et al. 2003; Johnson et al. 2005; Zhang et al. 2012), eukaryota (Mehmood et al. 2019) and viruses (Erez et al. 2017)), individual microorganisms "signal" their presence by secreting a small diffusible molecule, the concentration of which indirectly reflects the density of the emitting population. The quorum, or the population density threshold above which the coordinated behavior is triggered, is next materialized by the concentration at which the signaling molecule starts to be effectively sensed by its cognate receptors within (or at the surface of) each individuals, and transduced to regulate the response at the basis of the emergent behavior. Mechanistically, a canonical QS mechanism then relies on a system of genes composed of three components: a synthase (or a cluster of synthases), a receptor and a response regulator, responsible for the signal production, perception, and downstream transcriptional response, respectively. Interestingly, because a shared complete QS system within a population of microorganisms allow, among other ends (Cornforth et al. 2014) -, one individual to produce a signal that can influence the behavior of another individual (and *vice et versa*), QS has often been described as a cell-cell communication mechanism (Waters and Bassler 2005). Besides, the specificity, or secrecy, of a QS signaling molecule may range from intra-species to inter-kingdom levels (Sztajer et al. 2014; Mukherjee and Bassler 2019) and may thus entail relationships of co-influence between distinct species of a same microbial community. Last but not least, the functional complementation of individual QS components of QS system(s) across different genomes may give rise to other forms of relationships than intra or inter-entity communication, such as, for example, eavesdropping (perception of exogenous communication molecules) (González and Venturi 2013; Karavolos et al. 2013; Silpe and Bassler 2019a) or manipulation (emission of a QS signaling molecule that happens to impair certain biological functions in another organism) (Hochstrasser and Hilbi 2017; Lee et al. 2018).

Surprisingly, although the density-dependent behavioral transitions mediated by QS could have appeared as an elegant candidate mechanism to control the lysis-lysogeny transition in temperate bacteriophages, 47 years passed between the first characterization of QS in the proteobacterium *Alivibrio fischerii* in 1970 (Nealson et al. 1970) and the first description of a functional QS system of genes inside viral genomes (in *Bacillus phages phi3T* and *SPbeta)*, which was furthermore guiding the lysis-lysogeny decision upon *Bacillus* infection (Erez et al. 2017)-. The groundbreaking discovery of this first phage-phage communication through QS possibly gave rise to emulation, because two years after its characterization, another form of communication modality has been characterized in bacteriophages, namely, eavesdropping on host QS signals. The next subsections will briefly describe all these fascinating discoveries.

8.1.1 Phage to Phage "Arbitrium" Communication System

Although, QS components had already been found in genomes of bacteriophages at that time (Hargreaves et al. 2014), the first experimental characterization of a phage-phage communication mediated by a viral QS system dates from 2017 (Erez et al. 2017). Much like the QS systems of firmicutes (Bhatt 2018), the uncovered viral QS system supports a highly secretive, intra-specific mode of communication, relying on the so-called "arbitrium" short signaling peptide (SAIRGA in *phi3T*, GMPRGA in *SPbeta*), rather than on a hormone-like signaling molecule of variable specificity, as observed for QS in proteobacteria. Upon Bacillus infection, the arbitrium peptide is released in the medium by cleavage of the AimP propeptide, accumulates as the phage replicates, and is sensed at high concentration by the AimR transcription factor, which then ceases to transactivate the AimX negative regulator of lysogeny. The arbitrium-based QS system thereby enables a population of infecting phages to synchronously lysogenize when it becomes critically large. Since then, hundreds of homologous QS systems have been found in the genomes of other phages and prophages and demonstrates that QS through intra-specific communication molecules is actually widespread in bacteriophages (Stokar-Avihail et al. 2019).

8.1.2 *Host Eavesdropping Enabled by the Externalization of a QS Receptor from the Host to the Phage*

Two years after the characterization of the arbitrium QS, a Vibrio phage encoded protein that is homologous to the VqmA sensor of the 3,5-dimethylpyrazin-2-ol (DPO) QS molecule secreted by *Vibrio cholerae* was uncovered (Silpe and Bassler 2019a). The $\text{Vqma}_{\text{phage}}$ encoding gene thus probably results from an externalization, *i.e*, the copy of genetic material from the host to the phage (Corel et al. 2018). Unlike the arbitrium peptide, the host-produced DPO signal is typical of the quorum sensing molecules secreted by proteobacteria, namely, small diffusible secondary metabolites biosynthesized by enzymes, with a spectrum of specificity than can range from intra-species to inter-kingdom levels of recognition (Papenfort and Bassler 2016). Consistently, the authors showed that DPO is produced in part by the threonine dehydrogenase (Tdh) of *V. cholerae*, and binds to both of the VqmA sensor of *V. cholerae* and of its homolog in *Vibrio phage VP882*. Thereby, Silpe and Bassler unraveled the first instance of a unidirectional communication link from a host bacterium towards a bacteriophage. Precisely, the phage-encoded VqmA transcription factor has been shown to confer the *Vibrio phage VP882* the ability to eavesdrop on the density status of its host *via* DPO sensing and to enact the lytic program accordingly.

Shortly after their discovery, Silpe and Bassler uncovered a putative analogous phenomenon of eavesdropping involving *Aeromonas phage phiARM81ld* and the *Aeromonas* bacterial host genus (Silpe and Bassler 2019b). The authors noticed that *phiARM81ld* encodes a QS receptor of the LuxR family, which detects Acyl-Homoserine Lactones (HSL), a type of QS molecule produced by many *Proteobacteria*, including *Aeromonas* species. Specifically, the viral LuxR protein of *phiARM81ld* presents 60% sequence identity with the bacterial LuxR protein of *Aeromonas popoffii CIP 105493* and exhibits high affinity towards C4-HSL, the only N-Acyl-Homoserine Lactone that is produced ubiquitously within the *Aeromonas* genus according to the Sigmol database of QS (Rajput et al. 2016). On the basis of these results, Silpe and Bassler speculated that the LuxR sensor of *Aeromonas phage phiARM81ld* could enable eavesdropping on the host-produced HSL QS signals, presumably to optimize the timing of execution of particular steps during the viral infection.

8.1.3 *State of the Art Knowledge of Communication Via QS in Bacteriophages*

In a nutshell, hundreds of homologous QS systems in bacteriophages have been so far characterized as a means for a phage to communicate on its own population density, and two distinct externalized QS receptors as a means of eavesdropping on the QS signals of bacterial hosts. So far, the cues gathered by the characterized phage QS components on either the phage or the host population densities have been shown to serve as basis to trigger behavioral transitions during a viral infection. These transitions initiated by QS have all been shown to sophistically balance the trade-off between phage replication (done at the expanse of the hosts) and the preservation of the phage-host collective (upon which the phage survival is conditioned), both essential for the evolutionary stability of the temperate lifestyle.

Finally, the espionage on host QS signals mediated by the externalization of the Vqma or of the LuxR QS sensors is particularly interesting from an evolutionary viewpoint, as it shows that there is not necessarily such thing as a phylogenetic limit to microbial communication. Indeed, it demonstrates that as long as two evolving entities happen to be in interaction together in a feedback loop and form thereby a collective (in such case, the phage-host collective), they might, in some cases, eventually develop a link of communication through the externalization of QS components, may these evolving entities have a common phylogenetic origin or not. This open many perspectives in the field of biocommunication of viruses as it allows speculating on many other communication links, involving many other species, from perhaps different superkingdoms. Moreover, beyond the externalization of QS sensors, it enables one to speculate on the potential externalization of host QS synthases or full QS systems that would support other types of communication modalities than phage-phage communication or eavesdropping on bacterial QS signals: for instance, host manipulation or even phage-host bi-directional communication (Table 8.1). In this chapter, we will propose a methodology to infer and analyze microbial communication network involving viruses, which could likely shed light into the many forms that biocommunication of viruses, including bacteriophages, may take.

Table 8.1 Type of edges in a microbial communication network

Communication modality	Node 1	Node 2	Genome 1	Genome 2
Canonical QS (intra-entity communication)	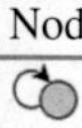		– QS synthase(s)	
			– QS receptor	
Inter-entity communication	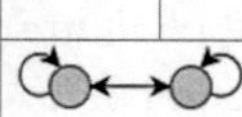		– QS synthase(s)	– QS synthase(s) homolog
			– QS receptor	– QS receptor homolog
Eavesdropping	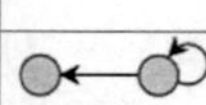		– QS receptor	– QS synthase(s)
				– QS receptor homolog
Signal amplification	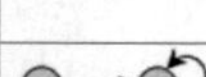		– QS synthase(s)	– QS synthase(s) homolog
				– QS receptor
Unilateral signaling by functional complementation	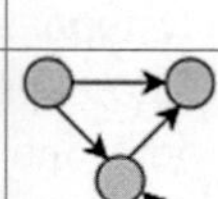		– QS synthase(s)	– QS receptor

8.2 Knowledge-Based Communication Network Involving Bacteriophages: An Introduction to Microbial Communication Networks

As we have seen in the previous section, the number of full or partial QS systems characterized inside complete genomes of bacteriophages has rapidly risen since 2017 and these recent discoveries will surely emulate many others in the near future. Because these viral QS systems involve multiple genes of different functions, and may serve as support for multi-partners, multi-kingdoms or even multi-systems communication, their qualitative descriptions are already, and will become more and more limiting to represent the whole picture of communication involving bacteriophages, other viruses and their hosts in a comprehensive way without computational methods. In this section, we will introduce the notion of "communication network", a framework particularly suited for algorithms, by drawing a simple graph from the previous qualitative descriptions of viral communication modalities. In the next section, we will take advantage of this reference knowledge-based network to attempt justifying each of its edges by an underlying rule of construction and imagine other rules for other types of edges. Indeed, a complete dictionary of formal rules may be used to substantially expand this first knowledge-based communication network and to provide thereby a rich and flexible analytical framework to study viral and microbial chemical communication *via* QS.

In the knowledge-based communication network, nodes could represent microbial genomes, and edges between nodes could be directed in order to capture the nature of the communication link between interacting entities. Basically, the nodes of this network would then correspond to *Bacillus phage phi3T*, *Bacillus virus SPbeta*, *Vibrio phage VP882*, *Vibrio cholerae*, *Aeromonas phage phiARM81ld* and *Aeromonas popoffii CIP 105493*. The first type of edges to be drawn could be loops, namely arrows that connect a node to itself, for entities that use QS to estimate their own population density and therefore "speak" and "listen" to their own kind. Hence, loops would be drawn for *phage phi3T* (SAIRGA arbitrium-based QS), *phage SPbeta* (GMPRGA arbitrium-based QS), *V. cholerae* (DPO-based QS) and *A. popoffii* (HSL-based QS). Additionally, the direction of an edge could depict the direction of the QS signal between two partners of communication. Hence, arrows that start from a host bacterium and points towards a phage would capture the concept of eavesdropping on host QS signals. Such arrows would then be drawn from *V. cholerae* towards *Vibrio phage VP882 (*DPO eavesdropping*)*, and from *A. popoffi* towards *Aeromonas phage phiARM81ld* (HSL eavesdropping). In the next sections, we will see that other types of edges could actually be drawn on such a network. Here, because we imposed that this communication network should strictly depict the actual knowledge of QS involving bacteriophages, these additional types of edges cannot yet be drawn because there is no experimental evidence to support them. Finally, from a visualization point of view, one could make this network even more analytically tractable by labelling the nodes according to qualitative attributes of genomes; for instance, blue for bacterial species and purple for viruses. Second,

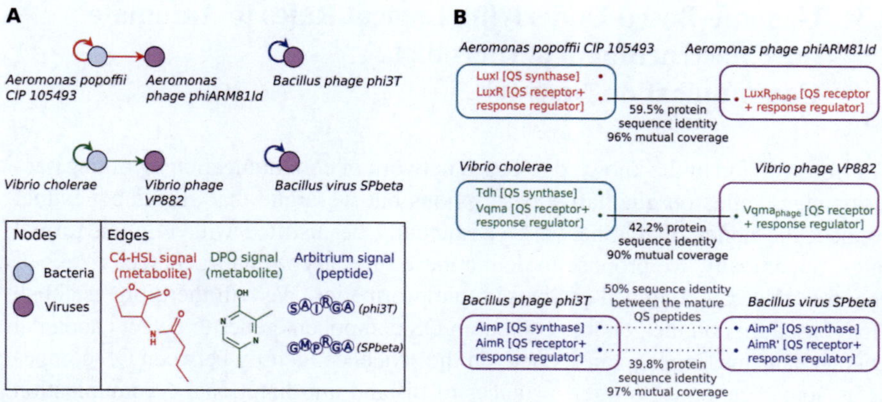

Fig. 8.1 Knowledge-based communication network involving bacteriophages. (a) Network depicting the state of the art knowledge of communication via QS molecules involving bacteriophages. (b) Genome content in QS genes of each node of the network and sequence identities between homologous QS components across genomes

one could color the edges according to the QS signal that support the interaction: blue for the arbitrium peptides, green for DPO and red for HSL. Ultimately, the resulting knowledge-based microbial communication network would look like as in Fig. 8.1a.

However, it should be said that *Bacillus phages*, *V. cholerae* and *A. popoffi CIP 105493* are only representative users of the arbitrium-based, DPO-based and HSL-based QS, respectively. As a result, the number of new nodes and new edges would considerably grow if one desires to picture by hand all the possible communication partners of the phage nodes. Moreover, it should be duly noted that edges of this network are limited exclusively to QS signals produced and/or recognized by bacteriophages. One should keep in mind, for instance, that *Vibrio cholerae* uses at least three other QS signals than DPO: Cholera Autoinducer-1 (CAI-1), Autoinducer-2 (AI-2) and Ethanolamine, responsible for intra-species, inter-species, and inter-kingdom communication, respectively (Jung et al. 2015; Watve et al. 2019). These observations introduce the issue that, if comprehensiveness is sought to describe communication in the microbial world, the manual construction of such a network based on the literature would become an overwhelming task. However, this expansion could be easily automated with computational methods, on the basis of formal rules. To facilitate their introduction, we purposely indicated the QS protein component(s) encoded by the genome of each interacting species in Fig. 8.1b. As we will see in greater details in Sect. 8.4, the advantages of inferring and analyzing a comprehensive communication network with computational methods are many, including the possibility to quantify the different types of communication modalities that a bacteriophage may support (intra-specific communication, inter-entities communication, eavesdropping, signal amplification …), to identify new putative phage-host associations, to assess the confidentiality of each phage QS signals, *etc.*

8.3 Genome-Based Underlying Logical Rules to Automate the Construction of a Microbial Communication Network

The construction of the knowledge-based network of communication involving bacteriophages relies on qualitative descriptions but we argue that all the edges that connect the nodes of this network can actually be justified with implicit formal rules. Specifically, we propose to define these rules according to characteristics of genomes that can be mined easily with bioinformatics. We will therefore establish them based first on the genome content in QS component genes to detect canonical QS mechanisms (loops), and second, on the sequence identity between QS components across genomes to infer instances of bi- and uni-directional communication between entities (other edges) (Fig. 8.1b). Before describing all these rules in detail, an overview of their basic principles as well as all the types of interactions that they might capture are given in Table 8.1.

One may notice that we took only into consideration QS synthases and receptors (including receptors with a response regulator activity (one-component systems)) but not response regulators. The reason for this is that their sequence similarity with transcription factors unrelated to QS would possibly lead to too many situations whereby a transcription factor inside a genome would be incorrectly detected as a QS response regulator.

8.3.1 Loops for Canonical QS (Intra-entity Communication)

- If the synthase(s) and the receptor of the same QS signal are together present inside a genome, then a loop is drawn for the node representing this genome (*e.g* presence of the AimP QS synthase and of the AimR QS receptor/response regulator in the genome of *Bacillus phage phi3T*).

8.3.2 Bi-directional Arrows for Inter-entities Communication

If two taxonomically different entities both encode a complete QS system relying on the same QS signal, they could theoretically communicate together. However, it is not trivial to predict whether two distinct entities will rely on the exact same QS signal(s) and could thus "speak" and "listen" likewise based on the presence of homologous QS systems inside their genomes. Such prediction would be pretty straightforward for different entities that share an identical QS system but this best case scenario is not to be expected quite often. Then, it feels like some thresholds of sequence identities between the components of a shared QS system across two distinct genomes could be set to decide whether the homologous QS systems are similar enough to likely produce at least one QS signal in common and support inter-genomes communication, or not. However, we advertise that such thresholds should be specific to the mode of production of the QS signal. Indeed, as we have seen earlier, the likelihood that a QS signal will serve as support for intra-entity or inter-entities communication is largely determined by the mode of production of the signal. When the QS signal is a peptide, and if this peptide (or its pro-peptide) is ribosomally synthesized as in the firmicutes phylum, its sequence is carried by a corresponding encoding region on a gene, and one non-synonymous mutation in this region is enough to alter the final conformation of the QS signal and therefore its recognition by the cognate receptor (*e.g* SAIRGA and GMPRGA in *Phi3T* and *SPbeta*). On the other hand, when the QS signal is a metabolite or a non-ribosomally synthesized peptide, whose synthesis are catalyzed by one or several enzymes as in the proteobacteria or actinobacteria phyla, there is less chance that one substitution in the sequence of this or these enzyme(s) will drastically alter the structure of the produced signaling molecule. Accordingly, one could set a condition on the nature of the QS signal to formalize the rules for drawing a bi-directional edge between two entities:

- If the QS signal is a peptide: if the two genomes encode an identical QS peptide and if they both share a homologous cognate receptor as defined by arbitrary stringent thresholds (*e.g.* sequence identity >70% over more than 90% of mutual coverage) then a bi-directional arrow between the nodes representing these two genomes is drawn.
- If the QS signal is a metabolite: if both of the synthase(s) and of the receptor of this signal are homologous across two genomes, as defined by more relaxed thresholds (*e.g.* identity >40% over more than 80% of mutual coverage) then a bi-directional arrow between the nodes representing these two genomes is drawn.

Here these thresholds are empirical but are nonetheless supported by the sequence identities between QS components that are known to give rise or not to an asserted communication link between two genomes (Fig. 8.1b).

8.3.3 Uni-Directional Arrow That Starts from a Node with a Loop for Eavesdropping

- If genome A has a full QS system (loop) and if genome B encodes a protein that is homologous to that of the QS receptor of A (see thresholds in 8.3.2) but does not possess the biosynthetic pathway of the associated QS signal, then an uni-directional arrow from the node of A to the node of B is drawn (*e.g.* homology between the QS receptors Vqma of *V. cholerae* and $\text{VqmA}_{\text{phage}}$ of *Vibrio phage VP882* but absence of homolog of the Tdh QS synthase of *V. cholerae* in *Vibrio phage VP882*).

8.3.4 Uni-Directional Arrow That Reaches a Node with a Loop for Signal Amplification

This case is the converse of the eavesdropping scenario, and could correspond to "signal amplification". It would imply the situation where entity A and entity B would conjointly produce the same QS signal but only B would perceive it, and would perhaps be led (or misled) to interpret the density of the collective as its own population density.

- If genome A possesses the QS synthase(s) but not the cognate receptor of the QS system of genome B (see thresholds in 8.3.2), then draw an uni-directional arrow from the node of A to the node of B

8.3.5 Uni-Directional Arrow That Reaches a Node Without a Loop for Unilateral Signaling

This case would be the one where neither genome A nor genome B would have a complete QS system but each of them would possess the QS component that would functionally complement the one of the other. From an evolutionary viewpoint, this situation can be envisaged only if the unit of selection of these entities would not correspond to the organismic level but to the collective level (*e.g.* relation of inter-dependence between A and B). Nevertheless, this case implies that A and B would have no QS genes in common to create a mean of comparison, and therefore a bridge between the two genomes. However, the presence in the network of a third genome, say C, could solve this issue as long as it would have the complete QS system. The rule for this second case would then be the following:

- If genome C has a complete QS system, and if genome A only possesses the QS synthase(s) of this system whereas genome B only the QS receptor, then draw uni-directional arrows from A to C, from C to B and from A to B

8.3.6 Pre and Post Processing of These Rules

All the rules of construction of edges between entities in a communication network rely on the comparison of QS components across genomes. Therefore, a tool to automatically detect full or partial QS systems of genes inside genomes and compare their sequence identities together is a prerequisite preprocessing step. Such a tool should therefore rely on a database of protein sequences corresponding to various QS systems of reference. On another note, in order to circumscribe the resulting microbial communication network to the study of biocommunication involving viruses, one could optionally set an ultimate filter: any node corresponding to a prokaryote or a eukaryote that is not connected to a viral node should be deleted from the network. Also, a last post processing operation could be to weigh each edge of a network according to the percentage of sequence identity between homologous QS components in order to quantify the likelihood of any inferred communication link.

8.3.7 *Should Genes of Prophages Belong to Viral or Bacterial Nodes?*

Besides, a tool to detect prophages inside bacterial genomes could be required to consider prophages as viral nodes or at least to distinguish between (i) genuine links of communication between phages and hosts due to externalization of QS genes and (ii) possible artefactual links of communication between phages and hosts originating from the sharing of QS genes between phages and prophages. However, one may also envisage that QS genes resulting from the internalization of a prophage inside a bacterial genome could possibly be expressed by the lysogenized bacterium and confer it a new functional QS system of genes. Indeed, it has been proposed that prophages may contribute to bacterial host phenotypes by providing adaptive genes (Schuch and Fischetti 2009; Oliver et al. 2009). Therefore, assigning a special annotation to prophages could be interesting if one would like to study the possible role of "social vector" exerted by temperate phages through the transfer of QS genes across bacterial genomes of the same community.

8.4 Microbial Communication Network: Examples of Perspectives That This Rich and Flexible Analytical Framework May Open

Now that we have introduced how to systematize the construction of a communication network from the protein sequences encoded by publicly available reference genomes, we will discuss which kind of perspectives its analysis may open. The goal of this section is of course not to be exhaustive but to give few examples of analyses that might be performed on such a network, and which kind of insights these might bring.

At the global level, the examination of the number of nodes and number of edges of this network could for instance allow to appreciate how much more viruses than the ones already characterized as QS users are suspected to communicate, and how much more microbial organisms are predicted to be involved in a communication link with viruses. Also, quantifying each type of edge involving a viral node (inter-entity communication, eavesdropping, signal amplification …) could tell us which communication modalities are predicted to be widespread in viruses and which are the ones that seem rather exceptional. Additionally, counting the number of edges that are supported by genes of prophages could allow assessing to which extent bacteriophages may contribute to the diffusion of QS genes across bacterial genomes.

On another note, filtering a whole communication network by the labels of edges would produce as many subnetworks as there are type of QS signals predicted to be used by viruses and could therefore allow quantifying how many and which different "languages" are suspected to be employed by viruses. Conveniently, the number of occurrences of a node in each of these subnetworks would also directly tell us how many distinct (complete or partial) putative QS systems are present inside its corresponding genome. Indeed, from a biological viewpoint, the identification of viruses that are suspected to be "multilingual" would be particularly interesting. Specifically, if a unique QS system allows a phage to assess its population density in a binary way (basically, below or above a given quorum), the accumulation of distinct QS systems could enable it to resolve its density with more refinement, through a succession of discrete measurements, corresponding to the distinct quorum associated with each QS signal. It has been demonstrated that these QS-based distinct steps could enable a population of growing proteobacteria to trigger different actions at different timing, much like a developmental program (Mehta et al. 2009). Hence, the prediction of plural QS systems in viral genomes (nodes with loops corresponding to distinct labels and thus distinct QS signals) could pave the way to the discovery of an analogous phenomenon in bacteriophages, namely that a hierarchy between distinct QS systems inside a same viral genome could control the transition between different stages of an infectious program. On another hand, even more exciting would be the discovery of plural QS systems with different levels of both completeness and specificity inside viral genomes. For instance, if a bacteriophage had a complete and intra-specific QS system to mediate phage-phage communication in addition to a QS receptor to eavesdrop on the density status of its host (node with a loop of one color and an edge directed towards it of another color), these QS components could theoretically enable it to contextualize its own density with respect to the one of its host. For instance, such a configuration could enable a bacteriophage not only to trigger lysis when it senses that its susceptible hosts are many (*e.g* DPO eavesdropping of *Vibrio phage VP882*) or trigger lysogeny when it senses that its own population becomes critically large (*e.g* arbitrium phage-phage communication system), but to switch between lysis and lysogeny at an optimal ratio of susceptible hosts available per virion.

Another example of analysis would be linked to the identification of connected components (CCs) inside these subnetworks. Basically, a CC designates a set of nodes within which each node is at least connected to another. Distinct CCs are then clearly identifiable as they have no connections between them. Specifically, the analysis of the composition of a CC may allow assessing the level of confidentiality of any communication molecule that a virus is suspected to rely on. For instance, small CCs, that may be composed of a single node, could depict an entity that rely on a very secretive QS signal (*e.g* the SAIRGA arbitrium peptide (Fig. 8.1)) whereas large CCs which harbor a substantial phylogenetic diversity could encompass entities that may "speak" and/or "listen to" a less confidential "language", such as, for example, the autoinducer-2 QS signal is suspected to be in bacteria (Sun et al. 2004). To sum up, the distribution of the viral nodes inside different classes of CCs (defined by a combination of their number of nodes, number of edges and intra phylogenetic diversity) may allow quantifying the confidentiality of each QS signal predicted to be used by viruses.

However, it should be said that a CC does not always depict a group of co-communicating species. For instance, two nodes may not be directly connected inside the same CC if their respective QS components (yet belonging to the same family of QS systems) were too divergent from each other but if "intermediate" QS components from other genomes nonetheless made the bridge between these distant nodes. This being said, a useful property of a network could be used to specifically detect preferential "discussion" groups inside CCs, namely, its modularity. The modularity of a CC or of a network is a scale that measures the density of edges inside communities of nodes to edges outside communities. Certain algorithms, such as the "Louvain method for community detection" aims at identifying set of nodes that optimize this value in order to identify community of nodes, or in other words, clusters of densely connected nodes (Blondel et al. 2008). Such a cluster of nodes, wherein most nodes are connected to each other, could therefore depict microbial and/or viral entities that may all inter-influence each other and form a preferential "discussion" group. From a broader point of view, identification of such groups could perhaps help discovering potential collective of entities. To follow up on this idea, the identification of a community of nodes that would include bacteriophages with many other bacteria (e.g *Vibrio phage VP882* and many other *Vibrio species*) could perhaps help identifying, in an unsupervised way, the spectrum of possible hosts for a given bacteriophage. Also, identifying "discussion" groups wherein the nodes are taxonomically different and the weight of edges are maximal (close to 100% sequence identity between QS components) would perhaps give more support to the hypothesis according to which bacteriophages may exert the role of "social vectors" in microbial community by facilitating the transfer of communication genes across phylogenetically diverse organisms. Besides, considering that many bacteria become virulent through QS, the identification of "eavesdropping" bacteriophages inside the same community of nodes as pathogenic bacteria may be of peculiar interest, especially if the viral function inferred to be triggered by QS corresponds to lysis. The *in-silico* identification of "eavesdropping" bacteriophages could indeed be of particular interest for phagotherapy as they could potentially prevent some pathogenic bacteria from spreading and switching to a virulent phenotype.

At a more local level, some analytical tools from the graph theory may help identifying influent microbial entities inside a communication network. Indeed, centrality indicators are measures that allow identifying important nodes within a network, according to different criteria. For instance, the "degree" centrality accounts for the number of edges that connects a node to another (loops are counted twice) (Shaw 1954; Freeman 1978) and may thus help identifying microbial or viral entity that tends to be more "social" than others, especially if their predicted "partners" of communication correspond to phylogenetically diverse entities. Again, the "betweenness" centrality accounts for the number of times that a node acts as a bridge along the shortest path between two other nodes (Freeman 1978; Newman 2005). Hence, the higher the betweenness of a node, the more it acts as a connector between distinct communities of nodes. A node with a high betweenness could therefore depict entities that could be adapted to plural communication contexts, or that could perhaps represent an intermediate state before the specialization in a given "discussion" group. Another example could arise from the "eigenvector" centrality, a measure of the influence that a node has in the network (Bonacich 1972). In a nutshell, a high eigenvector score means that a node is connected to many nodes who themselves have high scores. Specifically, the identification of a cluster of nodes with an associated average eigenvector centrality that is high could therefore highlight a potential "source" of externalization of QS components inside a microbial community. A final example could be given by the "eccentricity" centrality which attributes a high score to nodes that lie at the extremities of connected components (Bouttier et al. 2003). This indicator could precisely point at microbial entities that rely on one or several divergent quorum system(s) of genes that might thus be responsible for the production and/or the recognition of communication molecules that exhibit uncharacterized chemical structures.

Of course, all these examples are not exhaustive but illustrate nonetheless that a communication network is an analytical framework that is rich, flexible and suited for many analyses, both at the global and local levels.

8.5 On the Peculiar Interest of Inferring a Communication Network from Metaviromes and Metagenomes Sequenced in the Same Environmental Site

If building a network of communication based on reference genomes available on the public databases may open many interesting perspectives, it also exhibits limitations, inherent to the fact that most of the publicly available reference genomes have been sequenced independently from each other. Indeed, reference genomes are usually the results of the following protocol: isolation from their respective environments, cultivation in laboratory and sequencing. Hence, it is almost always impossible to access to the social context that these well-studied microorganisms experienced in their environment of extraction. It follows that the inferences made by our proposed method on the communication links between reference genomes would always suffer from the lack of proof that they are in contact with each other, in the very same environment. However, with the recent advent of metagenomics and metaviromics, which allow to reconstruct (more or less completely and confidently) genomes of unicellulars and viruses from environmental DNA sequences (metagenomics-assembled genomes) (Breitwieser et al. 2019), this barrier can nowadays be somewhat overcome. Using metagenomes and metaviromes from the same environmental site as input to our algorithm would indeed ensure that connected nodes in the resulting communication network always represent entities that are part of a same community.

Moreover, environmental DNA sequences that cannot be mapped to reference genomes, could capture the genetic information of viruses or unicellulars for which there is no isolated representatives, and may thus allow to access to a new diversity of biological systems (Albertsen et al. 2013). With this respect, it would be interesting to decipher the contribution of these uncharacterized genomes to the nodes of the communication network, as it may highlight how important these understudied microorganisms may be for the social dynamics of an environmental community. Finally, the information on the abundance of each metagenome and metavirome in a given environment could be mapped to the nodes of the corresponding communication network and could allow to study whether there is a correlation or not between the abundance of a given entity in its environment and its detected ability to communicate.

8.6 Key Perspectives from the Comparison of Several Environmental Networks

If the generation of a microbial communication network from one environment is already informative, we believe that the comparison of networks from different environments could be even more interesting.

For instance, the comparison of the abundances of each entity across distinct environments could be used to detect entities that are often found together, and at equivalent ratio of abundances in distinct communities. Indeed, the set of all co-occurrence observations between entities can be resumed in the form of a co-occurrence network (Yang et al. 2019) that can be analyzed conjointly with a communication network. For instance, deciphering whether there are areas of overlap between co-occurrence and communication networks could highlight how important communication links are to promote and/or reinforce interactions between entities (Fig. 8.2). With this respect, two genomes that would be linked by both a co-occurrence and a communication edges could further reinforce the prognosis of a suspected link of symbiosis. In this latter case, a metagenomics dataset that reflects the composition of a community at different time steps could be of particular interest, as tracking the abundances of genomes that vary likewise along time sequences could allow to infer symbiotic relationships with a higher confidence (Probst et al. 2018)

On another note, the comparison of different environmental metagenomics datasets could also unravel whether some microbial entities that harbor the same taxonomic label (Luo et al. 2015) but that thrive in distinct environments may possess different QS systems. In this regard, it could teach us to which extent the presence of QS systems inside the genome of a microbial species is determined by the habitat, and to which extent it is rather determined by their evolutionary paths. If the

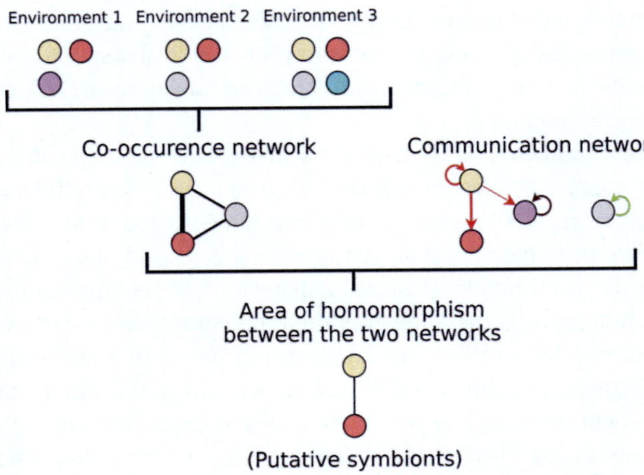

Fig. 8.2 Joint analysis of co-occurrence and communication networks to reinforce the prognosis of putative symbiotic relationships. Each node depicts a microbial entity, each directed edge a link of communication and each undirected edge a link of co-occurrence

habitat could determine the acquisition of QS systems, this would reinforce the hypothesis according to viruses and especially bacteriophages could play the role of "social" vectors within microbial communities, by transferring communication genes across genomes. With this respect, this putative role could perhaps be unraveled by a strong edge of communication between a virus and any genome that would harbor a QS system only in a given environment.

8.7 Concluding Remarks

In this chapter, we gave a brief overview of how quorum sensing genes mediate biocommunication in bacteriophages. We then argue that technical solutions are already available to mine genomes of viruses and other microorganisms for quorum sensing genes, upon which many entities rely to interact with others. We further proposed a methodology to use the results of this analysis to automatically construct networks of communication involving bacteriophages. Although such a network can be built from reference genomes sequenced independently from each other, we rather encourage inferring communication networks from metagenomes and metaviromes, sequenced altogether in the same environmental sites. Indeed, we speculate that the information that can be extracted from these networks could open many fascinating perspectives in the fields of biocommunication (*e.g.* discovery of new QS circuits in bacteriophages), ecology (*e.g.* discovery of new phage-host associations), evolution (*e.g.* examination of introgressive acquisitions of QS genes), medicine (*e.g.* discovery of bacteriophages that could antagonize pathogenic bacteria), and biochemistry (*e.g.* discovery of new communication molecules).

References

Abedon ST (2015) Bacteriophage secondary infection. Virol Sin 30(1):3–10. https://doi.org/10.1007/s12250-014-3547-2

Albertsen M, Hugenholtz P, Skarshewski A, Nielsen KL, Tyson GW, Nielsen PH (2013) Genome sequences of rare, uncultured bacteria obtained by differential coverage binning of multiple metagenomes. Nat Biotechnol 31(6):533–538. https://doi.org/10.1038/nbt.2579

Bhatt VS (2018) Quorum sensing mechanisms in gram positive bacteria, in implication of quorum sensing system in biofilm formation and virulence. Singapore, Springer Singapore, pp 297–311. https://doi.org/10.1007/978-981-13-2429-1_20

Blondel VD, Guillaume J-L, Lambiotte R, Lefebvre E (2008) Fast unfolding of communities in large networks. J Stat Mech: Theory Exp 2008(10):P10008. https://doi.org/10.1088/1742-5468/2008/10/P10008

Bonacich P (1972) Factoring and weighting approaches to status scores and clique identification. J Math Sociol 2(1):113–120. https://doi.org/10.1080/0022250X.1972.9989806

Bouttier J, Di Francesco P, Guitter E (2003) Geodesic distance in planar graphs. Nucl Phys B 663(3):535–567. https://doi.org/10.1016/S0550-3213(03)00355-9

Breitwieser FP, Lu J, Salzberg SL (2019) A review of methods and databases for metagenomic classification and assembly. Brief Bioinform 20(4):1125–1136. https://doi.org/10.1093/bib/bbx120

Corel E, Méheust R, Watson AK, McInerney JO, Lopez P, Bapteste E (2018) Bipartite network analysis of gene sharings in the microbial world. Mol Biol Evol 35(4):899–913. https://doi.org/10.1093/molbev/msy001

Cornforth DM, Popat R, McNally L, Gurney J, Scott-Phillips TC, Ivens A, Diggle SP, Brown SP (2014) Combinatorial quorum sensing allows bacteria to resolve their social and physical environment. Proc Natl Acad Sci U S A 111(11):4280–4284. https://doi.org/10.1073/pnas.1319175111

Erez Z, Steinberger-Levy I, Shamir M, Doron S, Stokar-Avihail A, Peleg Y, Melamed S et al (2017) Communication between viruses guides lysis-lysogeny decisions. Nature 541(7638):488–493. https://doi.org/10.1038/nature21049

Freeman LC (1978) Centrality in social networks conceptual clarification. Soc Networks 1(3):215–239. https://doi.org/10.1016/0378-8733(78)90021-7

Fuqua WC, Winans SC, Greenberg EP (1994) Quorum sensing in bacteria: the LuxR-LuxI family of cell density-responsive transcriptional regulators. J Bacteriol 176(2):269–275. https://doi.org/10.1128/jb.176.2.269-275.1994

González JF, Venturi V (2013) A novel widespread interkingdom signaling circuit. Trends Plant Sci 18(3):167–174. https://doi.org/10.1016/j.tplants.2012.09.007

Hargreaves KR, Kropinski AM, Clokie MRJ (2014) "What does the talking?: quorum sensing signalling genes discovered in a bacteriophage genome." edited by Gunnar F. Kaufmann. PLoS One 9(1):e85131. https://doi.org/10.1371/journal.pone.0085131

Hochstrasser R, Hilbi H (2017) Intra-species and inter-kingdom signaling of Legionella Pneumophila. Front Microbiol 8:79. https://doi.org/10.3389/FMICB.2017.00079

Johnson MR, Montero CI, Conners SB, Shockley KR, Bridger SL, Kelly RM (2005) Population density-dependent regulation of exopolysaccharide formation in the hyperthermophilic bacterium Thermotoga Maritima. Mol Microbiol 55(3):664–674. https://doi.org/10.1111/j.1365-2958.2004.04419.x

Jung SA, Chapman CA, Ng W-L (2015) Quadruple quorum-sensing inputs control vibrio cholerae virulence and maintain system robustness. PLoS Pathog 11(4):e1004837. https://doi.org/10.1371/journal.ppat.1004837

Karavolos MH, Winzer K, Williams P, Anjam Khan CM (2013) Pathogen espionage: multiple bacterial adrenergic sensors eavesdrop on host communication systems. Mol Microbiol 87(3):455–465. https://doi.org/10.1111/mmi.12110

Lee W, Lee S-H, Kim M, Moon J-S, Kim G-W, Jung H-G, Kim IH et al (2018) Vibrio Vulnificus quorum-sensing molecule cyclo(Phe-Pro) inhibits RIG-I-mediated antiviral innate immunity. Nat Commun 9(1):1606. https://doi.org/10.1038/s41467-018-04075-1

Luo C, Knight R, Siljander H, Knip M, Xavier RJ, Gevers D (2015) ConStrains identifies microbial strains in metagenomic datasets. Nat Biotechnol 33(10):1045–1052. https://doi.org/10.1038/nbt.3319

Mehmood A, Liu G, Wang X, Meng G, Wang C, Liu Y (2019) Fungal quorum-sensing molecules and inhibitors with potential antifungal activity: a review. Molecules (Basel, Switzerland) 24(10). https://doi.org/10.3390/molecules24101950

Mehta P, Goyal S, Long T, Bassler BL, Wingreen NS (2009) Information processing and signal integration in bacterial quorum sensing. Mol Syst Biol 5(1):325. https://doi.org/10.1038/msb.2009.79

Miller MB, Bassler BL (2001) Quorum sensing in bacteria. Annu Rev Microbiol 55(1):165–199. https://doi.org/10.1146/annurev.micro.55.1.165

Mukherjee S, Bassler BL (2019) Bacterial quorum sensing in complex and dynamically changing environments. Nat Rev Microbiol 17(6):371–382. https://doi.org/10.1038/s41579-019-0186-5

Nealson KH, Platt T, Hastings JW (1970) Cellular control of the synthesis and activity of the bacterial luminescent system. J Bacteriol 104(1):313–322

Newman MEJ (2005) A measure of betweenness centrality based on random walks. Soc Networks 27(1):39–54. https://doi.org/10.1016/J.SOCNET.2004.11.009

Oliver KM, Degnan PH, Hunter MS, Moran NA (2009) Bacteriophages encode factors required for protection in a symbiotic mutualism. Science (New York, NY) 325(5943):992–994. https://doi.org/10.1126/science.1174463

Oppenheim AB, Kobiler O, Stavans J, Court DL, Adhya S (2005) Switches in bacteriophage lambda development. Annu Rev Genet 39(1):409–429. https://doi.org/10.1146/annurev.genet.39.073003.113656

Paggi RA, Martone CB, Fuqua C, Castro RE (2003) Detection of quorum sensing signals in the haloalkaliphilic archaeon *Natronococcus Occultus*. FEMS Microbiol Lett 221(1):49–52. https://doi.org/10.1016/S0378-1097(03)00174-5

Papenfort K, Bassler BL (2016) Quorum sensing signal-response systems in gram-negative bacteria. Nat Rev Microbiol 14(9):576–588. https://doi.org/10.1038/nrmicro.2016.89

Probst AJ, Ladd B, Jarett JK, Geller-McGrath DE, Sieber CMK, Emerson JB, Anantharaman K et al (2018) Differential depth distribution of microbial function and putative symbionts through sediment-hosted aquifers in the deep terrestrial subsurface. Nat Microbiol 3(3):328–336. https://doi.org/10.1038/s41564-017-0098-y

Rajput A, Kaur K, Kumar M (2016) SigMol: repertoire of quorum sensing signaling molecules in prokaryotes. Nucleic Acids Res 44(D1):D634–D639. https://doi.org/10.1093/nar/gkv1076

Schuch R, Fischetti VA (2009) The secret life of the anthrax agent Bacillus Anthracis: bacteriophage-mediated ecological adaptations. PLoS One 4(8):e6532. https://doi.org/10.1371/journal.pone.0006532

Shaw ME (1954) Group structure and the behavior of individuals in small groups. J Psychol 38(1):139–149. https://doi.org/10.1080/00223980.1954.9712925

Silpe JE, Bassler BL (2019a) A host-produced quorum-sensing autoinducer controls a phage lysis-lysogeny decision. Cell 176(1–2):268–280.e13. https://doi.org/10.1016/J.CELL.2018.10.059

Silpe JE, Bassler BL (2019b) Phage-encoded LuxR-type receptors responsive to host-produced bacterial quorum-sensing autoinducers. BioRxiv March:577726. https://doi.org/10.1101/577726

Stokar-Avihail A, Tal N, Erez Z, Lopatina A, Sorek R (2019) Widespread utilization of peptide communication in phages infecting soil and pathogenic bacteria. Cell Host Microbe 25(5):746–755.e5. https://doi.org/10.1016/j.chom.2019.03.017

Sun J, Daniel R, Wagner-Döbler I, Zeng A-P (2004) Is autoinducer-2 a universal signal for interspecies communication: a comparative genomic and phylogenetic analysis of the synthesis and signal transduction pathways. BMC Evol Biol 4(1):36. https://doi.org/10.1186/1471-2148-4-36

Sztajer H, Szafranski SP, Tomasch J, Reck M, Nimtz M, Rohde M, Wagner-Döbler I (2014) Cross-feeding and interkingdom communication in dual-species biofilms of Streptococcus Mutans and Candida Albicans. ISME J 8(11):2256–2271. https://doi.org/10.1038/ismej.2014.73

Wahl LM, Betti MI, Dick DW, Pattenden T, Puccini AJ (2018) Evolutionary stability of the lysis-lysogeny decision: why be virulent? Evolution 73(1):evo.13648. https://doi.org/10.1111/evo.13648

Waters CM, Bassler BL (2005) QUORUM SENSING: cell-to-cell communication in bacteria. Annu Rev Cell Dev Biol 21(1):319–346. https://doi.org/10.1146/annurev.cellbio.21.012704.131001

Watve S, Barrasso K, Jung SA, Davis KJ, Hawver LA, Khataokar A, Palaganas RG, Neiditch MB, Perez LJ, Ng W-L (2019) Ethanolamine regulates CqsR quorum-sensing signaling in vibrio cholerae. BioRxiv March:589390. https://doi.org/10.1101/589390

Yang P, Yu S, Lin C, Ning K (2019) Meta-network: optimized species-species network analysis for microbial communities. BMC Genomics 20(S2):187. https://doi.org/10.1186/s12864-019-5471-1

Zhang G, Zhang F, Ding G, Li J, Guo X, Zhu J, Zhou L et al (2012) Acyl homoserine lactone-based quorum sensing in a methanogenic archaeon. ISME J 6(7):1336–1344. https://doi.org/10.1038/ismej.2011.203

Chapter 9
Information Stored in a Phage Particle: *Lactobacillus delbrueckii* Bacteriophage LL-H as a Case

Patricia Munsch-Alatossava and Tapani Alatossava

Abstract The bacteriophage LL-H, that infects *Lactobacillus delbrueckii* ssp. *lactis*, is a typical *pac*-type double-stranded DNA phage. The genome, of about 34.7 kbp, is packaged inside an icosahedral proteinous capsid of approximately 50 nm diameter; LL-H possesses a 170 nm long noncontractile tail and one tail fiber of 30–35 nm length. This chapter summarizes some of the research data obtained with LL-H since the late 1970s. Here, besides brief considerations of its genomic organization, the major discussion focuses on the biology of LL-H, on its special structural features of prime importance for its survival and infectivity during the period when its DNA is in an inactive state, more precisely during the time that elapses between the DNA packaging and injection steps; during that time interval, the information for the specific phage-host interactions is stored in the phage particle itself. Models on the calcium/magnesium dependent adsorption and injection steps are also discussed for LL-H.

9.1 Introduction

Viruses are obligatory parasites of living cells including both prokaryotes (bacteria and archaea) and eukaryotes (protists, fungi, plants and animals). Very often, viruses are host-specific and even cell-specific within a multicellular host species. Bacterial viruses, called bacteriophages or shortly phages, are capable of infecting vegetative bacterial cells. Viruses are considered as the predominant biological entities on earth with about 10^{31} virus particles (virions), most of which are phages (Wommack and Colwell 2000).

P. Munsch-Alatossava
Independent researcher, Helsinki, Finland

T. Alatossava (✉)
Postal Address: Department of Food and Nutrition, University of Helsinki, Helsinki, Finland
e-mail: tapani.alatossava@helsinki.fi

G. Witzany (ed.), *Biocommunication of Phages*,
https://doi.org/10.1007/978-3-030-45885-0_9

The host-specificity of phages is mostly very narrow (usually restricted to particular strains belonging to one bacterial species) and depends on the phage receptor structures and on the arsenal of various bacterial phage resistance mechanisms (like R/M and CRISPR/Cas systems) present and active in the considered target strain (Doron et al. 2018). It is widely accepted that every bacterial species has its infective phage(s): therefore, bacteria and phages are engaged in a coevolution since their appearance on earth, about 3.5 billion years ago (Ackermann and DuBow 1987).

A virus particle (or virion) consists of nucleic acid (DNA or RNA), constitutive of the genome, enclosed by a protein coat (capsid). Depending on the virus type, often other structural proteins and sometimes even lipid components can be present. The vast majority (about 95%) of all phages represents tailed phages having double stranded (ds)-DNA as a genome. Based on their tail morphologies, tailed phages are classified into three families: *Myoviridae* bear contractile tails, *Siphoviridae* possess long non-contractile tails, and *Podoviridae* harbor short non-contractile tails. Among all phages, *Siphoviridae* constitute the most frequently (about 60%) encountered phage group (Ackermann 2007).

9.2 *Lactobacillus delbrueckii* Phage LL-H and Its Genome

Often, bacteriophages cause quality problems during the manufacture of dairy products like fermented milks or cheeses, and accordingly promote economic losses. Phages of lactic acid bacteria (LAB) carry a bad reputation in dairies, since they are able to kill LAB starter strains and disturb or even prevent the starter promoted activities like the lactic acid fermentation for example.

Phage LL-H, one of the LAB phages, was originally isolated in 1972 in a local cheese factory situated in Hauho, Finland. This phage is able to infect some industrial *Lactobacillus delbrueckii* ssp. *lactis* starter strains used during the Emmental cheese manufacture in the dairy plant. Our group started to characterize this phage as a model of *Lactobacillus* phages in the late 1970s (Alatossava and Pyhtilä, 1980). The early electron microscopy (EM) studies on LL-H, by Alatossava and Pyhtilä (1980), revealed that this phage belonged to the most common taxonomic bacteriophage B1 group, which comprises phages having an isometric head, characterized by an icosahedral capsid symmetry and a long non-contractile tail (Ackermann et al. 1984). Later, more detailed EM studies on LL-H showed that the diameter of the capsid is approximately 45–50 nm, the length of the tail is 170 nm with 45–50 cross-striations, followed by a minor baseplate (of 15 nm diameter and 5–10 nm thickness) which is further connected to the tail fiber of 30–35 nm length (Alatossava 1987; Forsman and Alatossava 1991; Munsch-Alatossava and Alatossava 2013).

Phage LL-H was the first *Lactobacillus* phage, for which the ds-DNA genome was sequenced (Mikkonen 1996, Mikkonen et al. 1996). LL-H has a genome length, in the circular form, of 34,660 bp; it comprises 47 putative genes (ORFs) on the main coding strand and four small putative genes (ORFs) together with remains of an integrase gene on the opposite coding strand. The presence of a truncated (about

90% deleted) nonfunctional integrase gene suggests that LL-H is a virulent phage and possibly originates from a temperate phage or from a prophage, following a large deletion in the functional integrase (*int*) gene. LL-H presents a *pac*-type genome organization: the ds-DNA is packed inside the capsid in a linear form and includes 2.8 ± 0.2 kbp terminal repeats at the ends of the linear ds-DNA; at that stage, the total length of the genome is approximately 37.5 ± 0.2 kbp, which corresponds to an increase of 8.1 ± 0.6% compared to the size of the circular genome of LL-H. The minireview by Munsch-Alatossava and Alatossava (2013) included one minor nucleotide (nt) sequence correction in the region of gene *orf546*; compared to the original work by Mikkonen and Alatossava (1994), one additional G was observed after the nt position 13,888. From each concatemeric form of phage LL-H DNA replication intermediate, starting at the *pac*-site (which locates near the left end of the 3.7-kb *Sal*I fragment), at least six packaging rounds have been determined. In following, the LL-H DNA linear form is circularly permuted: this point was corroborated by electron microscopic observations of circular homoduplex LL-H DNA molecules (Forsman and Alatossava 1991). Moreover, the *pac*-type phage genome organization of LL-H enables the possibility to pack bacterial DNA from an infected *Lb. delbrueckii* cell, in addition to the concatemeric LL-H DNA: this feature was also confirmed for LL-H (Ravin et al. 2006). Consequently, phage LL-H is capable of generalized transduction promoting bacterial recombination among sensitive strains.

9.3 Phage LL-H Infection Cycle and Its Ca^{2+}/Mg^{2+} Dependency

Most of the LL-H biology studies were completed with MRS (De **M**an, **R**ogosa and **S**harpe) broth, supplemented with 20 mM $CaCl_2$, using *Lb. delbrueckii* ssp. *lactis* strain LL23 as the host strain, grown at 37 °C (Alatossava 1987). Calcium (Ca^{2+}) supplemented to the growth medium is essential for phage LL-H infectivity; among 10 different tested divalent cations, only magnesium (Mg^{2+}) was able to replace Ca^{2+}. An optimal efficiency of plating (e.o.p.) was obtained with rather high Ca^{2+}/Mg^{2+} (20–40 mM) concentrations (Alatossava et al. 1983).

The host strains LKT and ATCC 15808 (CNRZ 700 and CNRZ 326, respectively) have been later employed for phage LL-H studies because of the recovery of a higher relative e.o.p., indicating that more progeny phage LL-H particles were obtained with strains LKT and ATCC 15808, compared to the strain LL23 (Forsman 1994; Mikkonen 1996; Vasala 1998; Räisänen 2007; Riipinen 2011). However, regarding the kinetics of the major steps of the LL-H infection cycle, no differences have been found with these three strains, and consequently the data from these various phage LL-H infection studies are comparable.

Figure 9.1 presents a summary of the major steps of the *Lb. delbrueckii* ssp. *lactis* infection by LL-H, that adsorbs on the host cells irreversibly within a few

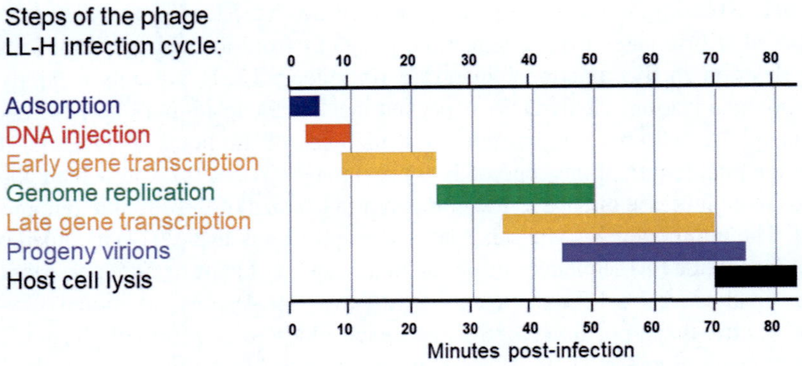

Fig. 9.1 Kinetics of phage LL-H infection of its *Lb. delbrueckii* subsp. *lactis* host cell. Data were compiled from Alatossava (1987) and Mikkonen et al. (1996)

minutes; after 5 min, about 95% of the phage particles are adsorbed, and the DNA injection is achieved within 10 min (Alatossava et al. 1987; Ravin et al. 2002). These steps are followed by RNA synthesis of phage LL-H early genes, with concomitant transient inhibition of total (host and phage) RNA synthesis after 10 min (Alatossava 1987, Mikkonen et al. 1996). The total DNA synthesis begins to increase strongly 25 min after the start of the infection, revealing the initiation of phage DNA replication (Alatossava 1987). At the same time, early gene expression of phage LL-H starts to quench, whereas the late gene expression of phage LL-H begins after about 35 min (Mikkonen et al. 1996). The rate of total protein synthesis is not affected by phage LL-H infection (Alatossava 1987). Progeny phage particles (virions) appear 40 min following the beginning of the infection (Alatossava 1987), and finally the lysis of the infected cells occurs after about 70 min, completing the infection cycle of phage LL-H, with a burst size of about 100 (Alatossava 1987, Mikkonen et al. 1996).

9.4 Phage LL-H Particle-Associated Interactions in Intra- and Extracellular Conditions

9.4.1 LL-H DNA Packaging and Assembly of Progeny Virions in Lb. delbrueckii *ssp.* lactis

The *pac*-type LL-H genome organization implies that the size of a packed DNA molecule is restricted by the volume of the phage LL-H precapsid according to the headful packaging mechanism (Rao and Feiss 2015). Among different LL-H virions (mature phage particles), the observed size variation of the packed LL-H DNA molecules is very small (S.D. is approximately ±0.6%, see paragraph 3.), which suggests that the structure of the packed LL-H DNA is very constant and accurate.

For the phage DNA packaging into the prohead, a phage-encoded terminase enzyme (terS and terL subunits in the case of phage LL-H) is required to cut the concatemeric DNA into the proper size during the DNA transfer process into the precapsid, through the particular phage-encoded connector (a portal protein located in one apex of the icosahedral precapsid). The DNA packaging process requests energy provided by ATP (Leforestier 2013; Rao and Feiss 2015; Bayfield et al. 2019). The negatively charged phosphate groups of the phage DNA need to be neutralized by positively charged counter components also at the level of the packaged DNA inside the capsid in order to eliminate the strong repulsive ionic forces between the phosphate groups and to allow the formation of a very condensed and solid-like state of DNA (Li et al. 2015). Major positively charged components present at molar concentrations inside a bacterial cell are potassium (K^+) and magnesium (Mg^{2+}) cations. As divalent Mg^{2+} cations have a much higher affinity for DNA than monovalent K^+ cations (Korolev et al. 1999), the negative charges on the packaged phage DNA are practically solely neutralized by Mg^{2+} ions; in addition, Mg^{2+} cations are capable of cross-linking two phosphate groups of the phage DNA coiled strands, and therefore promote a very high compactness of the DNA, like in the case of fibers or liquid crystals.

One could estimate that, with a stoichiometry of one Mg^{2+} cation per two phosphate groups, each LL-H phage DNA molecule could bind as many as 37,500 Mg^{2+} ions inside the phage capsid. This amount of Mg^{2+} ions inside one capsid (that presents an approximate volume of 6×10^4 nm^3) corresponds to a total Mg^{2+} concentration of nearly 1 M, which is 10- to 20-fold higher than the total (bound and free) intracellular Mg^{2+} concentration (50–100 mM) determined for a bacterial vegetative cell (Lusk et al. 1968; Moncany and Kellenberger 1981; Alatossava et al. 1985).

From the experiments with EDTA (a divalent cation chelating agent), we know that purified LL-H particles (in 5 mM Tris-HCl/10 mM $MgCl_2$ buffer, pH 6.6) were disrupted and released a portion of the phage packaged DNA after the capsid shell was broken (Alatossava 1982, 1987); this observation suggests that Mg^{2+} ions are essentially located inside the phage head. Later in this chapter (paragraph 9.4.4), we return to this point in the context of the LL-H DNA injection step.

Phage structural proteins for precapsids, tails and fibers are assembled by separate subpathways. For the control of the length of the tail, typically multifunctional, a large tape measure protein (TMP) is required to form a skeleton in the assembly of the major tail protein. The TMP is found to fill the core of the mature tail and is later ejected out from the tail core, preceding the phage DNA ejection through the tail core. The multifunctionality of TMP is related to its muralytic and fusogenic activities required for the successful phage DNA transfer through the cell wall and membrane during the DNA injection step (Boulanger et al. 2008). In the case of LL-H, after a nt sequence correction was made (Munsch-Alatossava and Alatossava 2013), ORF351 and ORF546 were considered as two TMP homologs. Phage LL-H ghost particles (LL-H particles without DNA in their capsids) have shown to lack the minor LL-H structural protein Gp54 (Trautwetter et al. 1986); Gp54 (equivalent to SP54) could correspond to *orf546* encoded TMP or its proteolytic fragment (Munsch-Alatossava and Alatossava 2013).

After packaging the DNA inside the precapsids, the mature phage heads are connected to mature phage tails with fibers, and finally the progeny phage particles (virions) are accumulated inside the infected bacterial cell. In following, the phage lysin with help of the phage holin protein causes local degradation of the bacterial peptidoglycan, promoting the lysis of the bacterial cell and finally leading to the release of the mature infective virions in the environment.

9.4.2 Stability Features of Phage LL-H in Extracellular Conditions

The unique metastability property of viruses ensures that the mature virions are stable enough to face various environmental challenges, but at the same time unstable enough to release the genome after irreversible adsorption (Li et al. 2015). The early studies on LL-H indicated that, after CsCl ultracentrifugation, the purified LL-H particles lost their infectivity during the dialysis step against Tris-HCl buffer (pH 6.6), contrarily to sodium phosphate or ammonium acetate buffers (pH 6.6): in other words, phage LL-H was particularly sensitive to Tris-HCl buffer that promoted its inactivation, which was prevented by $MgCl_2$ supplements at concentrations lower than 1 mM (Alatossava 1982).

The electron microscopy studies confirmed that the inactivation of LL-H was caused by the Tris-HCl –buffer promoted phage DNA ejection, which was prevented by Mg^{2+} or Ca^{2+} supplements, but not by monovalent cations like potassium (K^+) or sodium (Na^+). Like for the purified phage LL-H particles, the purified LL-H ghost particles contained all phage structural proteins except Gp54, which suggests that the phage LL-H DNA ejection *in vitro* resembles the phage LL-H DNA injection *in vivo*; furthermore, the energy required for phage DNA ejection "is stored "in the phage particle itself, obviously through the condensed state of the packaged LL-H DNA associated to 1 M Mg^{2+} counterions. However, the mode of action of the Tris-HCl buffer at molecular level is still unknown. In the absence of divalent cations, Tris (tris (hydroxymethyl) aminomethane) molecules may trigger conformational changes of the phage LL-H adsorption protein (antireceptor), similar to changes triggered by LL-H receptor molecules of the phage-susceptible *Lb. delbrueckii* strain in the presence of divalent cations.

Like in the case of many other widely studied tailed phages, the capsid of LL-H is permeable to small molecules like EDTA, ethidium bromide or uranyl acetate. However, according to the so-called macromolecular crowding effect, DNA in its condensed state (like packaged phage DNA inside the capsid) is independent or only slightly dependent on the ionic composition and concentrations of the bulk solution (Korolev et al. 1999). If this holds true for the packaged phage genome, the phage DNA- bound Mg^{2+} ions, inside the capsid, are not anymore exchanged with external cations (even Ca^{2+} or other divalent cations) present in the environment.

9.4.3 *Adsorption and LL-H-Specific Receptors at the Cell Surface of* Lb. delbrueckii

To deliver the phage genome inside the cell and achieve a successful infection of the host bacterium, the phage adsorption is the essential step. Phage adsorption requires both a particular level of specificity (host cell recognition) and efficiency (rate of adsorption). Both of these requirements are influenced by the properties of the bacterial cell surface as well as by phage adsorption protein(s). The two-stage capture model, also known as RREDR (Reaction Rate Enhancement by Dimensional Reduction) and originally introduced by Adam and Delbrück (1968), is a widely accepted kinetic model which describes the phage adsorption process in more details (Fig. 9.2a). On the bacterial cell surface, a three-dimensional (3D) diffusion step of the phage particle is followed by a two-dimensional (2D) "random walk" of the phage particle in the search of an appropriate receptor, until the phage particle through its adsorption protein (antireceptor) is captured (1D) by the specific phage

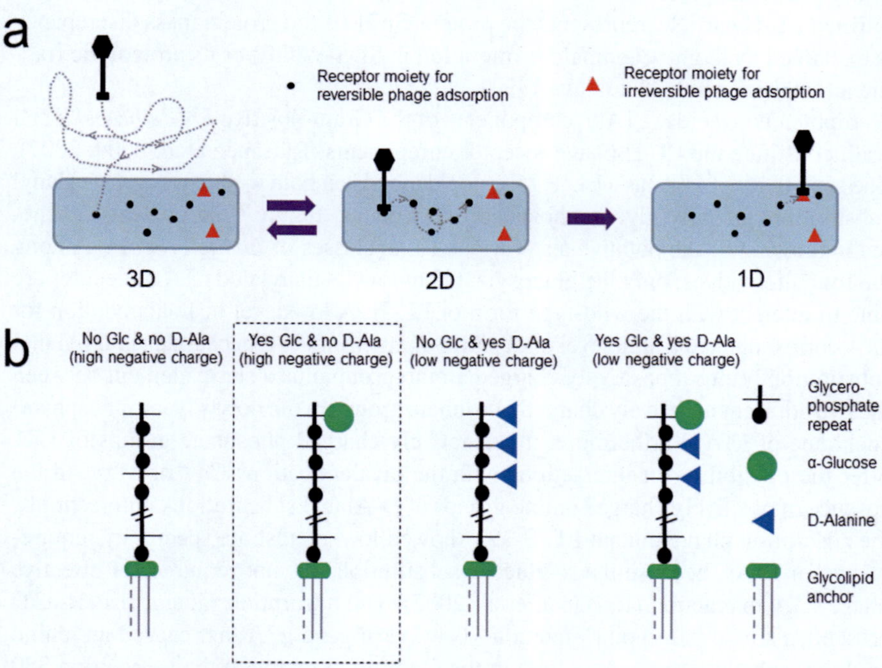

Fig. 9.2 Phage LL-H adsorption. (**a**) The two-stage capture model for LL-H adsorption is adapted and modified from the RREDR model described by Adam and Delbrück (1968): here, a separate specific receptor moiety is responsible for the reversible (sphere) or for the irreversible (triangle) adsorption stages. (**b**) *Lactobacillus delbrueckii* lipoteichoic acids (LTAs) classes differing by their levels of glucosylation and D-alanylation substitutions. Only the LTA class (dotted box) that presents an α-glucose substitution (sphere) together with the absence or a low level of D-Alanine substitutions (triangle) is believed to be the most efficient receptor for the successful adsorption of phage LL-H

receptor moiety. The "walk" phase corresponds, in fact, to a quasi-two dimensional diffusion process step including several local reversible attachments (3D ↔ 2D) on the cell surface, allowing a more effective search for the specific phage receptor structure. Once the adsorption protein of the phage particle is bound to the specific receptor structure, the process is irreversible (2D → 1D) and the phage particle cannot desorb anymore from the receptor, but instead induces structural rearrangements of the phage tail, finally triggering the phage DNA release (Chatterjee and Rothenberg 2012; Storms and Sauvageau 2015). Reversible and irreversible phage-host cell surface interactions can occur at the level of the same receptor or, more frequently, involve different surface components (Moldovan et al. 2007; Plisson et al. 2007; Baptista et al. 2008).

The gene *g71* of phage LL-H encodes the phage minor structural protein Gp71, which is responsible for the specific binding to the *Lb. delbrueckii* host cell wall receptors; consequently, *g71* represents an antireceptor encoding gene. Furthermore based on the LL-H adsorption mutant and gene *g71* homology studies, Gp71 represents the tail fiber protein of phage LL-H (Ravin et al. 2002). Most probably, the minor structural protein of about 58 kDa in molecular mass (SP58), found in the purified LL-H particle, represents the protein Gp71 (if the protein mass discrepancy is explained by slightly anomalous migration in SDS-PAGE) or its proteolytic fragment (Mikkonen and Alatossava 1994).

Lipoteichoic acids (LTAs), components of the Gram-positive *Lb. delbrueckii* cell wall, constitute the LL-H phage receptor components (Räisänen et al. 2004, 2007). The specificity of LTA, as a LL-H receptor, depends on both α-glucose and D-Alanyl substitutions of poly(glycerophosphate) backbones. Figure 9.2b presents a schematic drawing of the putative and various LTAs classes of *Lb. delbrueckii*. Among the four alternatives, only the glucosylated but low D-alanylated LTA molecules are able to interact with the wild-type form of LL-H. A low level of D-alanylation for LTA corresponds to a high level of negative charges on LTA, because each D-Alanyl substitution brings a positively charged amino group plus a steric element between the two adjacent negatively charged phosphate groups in the poly(glycerophosphate) backbone of LTA. Furthermore, the negatively charged phosphate groups of LTA offer the possibility for interactions with the divalent cations Ca^{2+} or Mg^{2+}, in the absence of positively charged amino groups of D-Alanyl substitutions. Interestingly, the adsorption phage mutant LL-H-a21 showed lower substrate specificity requirements for LTAs, because the α–glucose substitution was not required for effective phage-LTA interaction (Räisänen et al. 2007). The adsorption mutant (LL-H-a21) presents a single base point mutation (T → A) in gene *g71*, that caused an amino acid (aa) substitution (Asn → Lys) in the C-terminal region at the aa position 380 of the adsorption protein Gp71 (Ravin et al. 2002). This point mutation could explain the observed wider host-range of phage LL-H-a21 at the level of the adsorption (Räisänen et al. 2007), and the increased number of functional receptor molecules for LL-H-a21 in the host strain cell wall (Räisänen et al. 2004); here two LTA classes, instead of one (among the four classes), are suitable as phage receptors (Fig. 9.2b). However, an increase of the host-range may also cause more frequent abortive infections; consequently, the fitness of this kind of mutated phage may be reduced.

What could be at the basis of the differences between the four considered classes of LTAs regarding their biological properties? The elongation of the poly(glycerophosphate) backbone with the glycerophosphate repeated units happens enzymatically for LTAs on the extracellular side of the cytoplasmic membrane of Gram-positive bacteria (Koch et al. 1984, Schneewind and Missiakas 2017). Hence, further decorations of LTA by D-Alanine and sugars substitutions occur enzymatically on the side, where nondecorated LTA locate (Reichmann et al. 2013; Rismondo et al. 2018). The order of biosynthesis of the four LTA classes could be: nonsubstituted LTA, followed by one type of substituted LTA (D-alanylated or glucosylated), and finally by fully substituted LTA (both D-alanylated and glucosylated). The biosynthesis of LTAs requests energy; the presence of both nonsubstituted and partially substituted LTAs in the cell wall indicates that LTA biosynthesis is in progress in the cell, most probably still during the active growth phase. Interestingly, a direct link between proteins involved in cell division and proteins involved in LTA synthesis has been documented (Reichmann and Gründling 2011). For the success of phage infection, it is important that the targeted host cell is in an active state and finds the necessary energy and nutritional resources in the environment for its growth. On the phage side, this information would be already essential at the adsorption step, in order to avoid the infection of a host cell lacking the necessary resources for the progeny phage production. If the bacterial cell surface could exhibit certain type of indicator structures, typically present in growing cells and absent in resource-poor cells (like cells in a stationary phase), these structures would be greatly beneficial for the success of the infection cycle. Nonsubstituted and partially substituted LTAs could represent such kinds of indicator structures, and similarly fully substituted LTAs could deliver the opposite message.

One α-glucose moiety, located most probably at or near the free end of LTA (Räisänen et al. 2007), is supposed to increase the host specificity regarding the sensitive *Lb. delbrueckii* strains, and accordingly the LTA glucose-LL-H Gp71 protein interaction could correspond to the reversible stage of LL-H adsorption (Fig. 9.2a). For LL-H, the success of an irreversible stage of adsorption is best achieved in the conditions of a low D-alanylation level of LTA (below 30%) combined to external Ca^{2+} and Mg^{2+} concentrations below 5 mM (Fig. 9.2a).

The adsorption of the LL-H-a21 phage mutant does not rely on the LTA glucose-LL-H-a21 Gp71 protein interaction; but instead, this mutant is capable to achieve directly an irreversible adsorption with higher numbers of both glucosylated and non-glucosylated *Lb. delbrueckii* LTAs, provided they have a low-level of D-Alanine substitution (here, the obviously reduced "random walking" phase is compensated by an increased number of receptors). The high level of LTA D-alanylation inhibits the adsorption of LL-H and of LL-H-a21 by at least 100- and ten-fold, respectively (Räisänen et al. 2007).

9.4.4 Phage LL-H DNA Injection and Associated Cation Movements

The successful phage DNA injection, from the virion into the host cell cytoplasm, implies that the phage DNA has to possess a proper gateway through the peptidoglycan and cytoplasmic membrane. In the case of LL-H, a model for the DNA injection and the preceding steps has been presented (Munsch-Alatossava and Alatossava 2013). Up to six negatively charged LTA molecules interact with the same phage tail fiber consisting of six Gp71 (SP58) protein subunits; in following, the negatively charged phosphate groups of six LTAs are cross-linked together with Ca^{2+}/Mg^{2+} ions, which amounts are depending on the environmental Ca^{2+}/Mg^{2+} composition; their concentrations appear to be optimal if they exceed 20 mM. The proposed Ca/Mg-LTA channel formation, triggered by the attachments of up to six LTA molecules to the same phage tail fiber, promotes the re-organisation of the tail fiber itself to the form of a Gp71 protein ring, attached to the base plate, allowing the proposed TMP homolog proteins ORF546 (SP54?) and ORF351 to be released from the core of the phage tail inside the peptidoglycan space, restricted by the Ca/Mg-LTA channel; in following, the peptidoglycan will be degraded inside the channel by ORF546 (with muranolytic activity), to finally provide an open gateway first to ORF351 protein (with fusogenic acitivity) that interacts with the cytoplasmic membrane and makes a pore for the phage DNA transfer. Purified LL-H ghost particles still contain the protein SP58 (the putative tail fiber protein), but not the protein SP54 (the putative TMP homolog protein ORF546) (Trautwetter et al. 1986), which altogether supports the model of phage LL-H antireceptor-LTA interactions.

Optimal environmental concentrations of Ca^{2+} and Mg^{2+} ions, requested for the adsorption of phage LL-H on its host, are five- to ten-fold lower than those necessary for the obtaining of an optimal e.o.p. (Alatossava et al. 1987); this implies that, following the irreversible phage adsorption step, additional Ca^{2+} and/or Mg^{2+} cations at concentrations of at least 20 mM are necessary.

Cation movements across the cytoplasmic membrane accompany infections by viruses including phages. In a bacterial cell, among the intracellular cations, K^+ and magnesium Mg^{2+} are present at highest concentrations and consequently, contents and movements of these cations are of special interest. Also phage LL-H induces transmembrane cation movements during the infection of a *Lb. delbrueckii* host strain (Alatossava et al. 1987). Many of the observed K^+, Mg^{2+} and Ca^{2+} movements are associated with the phage LL-H DNA injection step into the bacterial cytoplasm.

Regarding this subject, the major results, reported by Alatossava et al. (1987), are summarized in Fig. 9.3a. The obtained data are all based on the use of the same purified phage LL-H preparation dialysed against 5 mM Tris-HCl/10 mM $MgCl_2$ buffer (pH 6.8), the use of the same multiplicity of infection (m.o.i. of about 45), the same bacterial cell density, and the consideration of the host strain in exponential growth phase. The cultivation of the host strain was performed at 37 °C in MRS broth, supplemented with 5 mM or 20 mM $CaCl_2$ or with 20 mM $MgCl_2$. Each packaged LL-H DNA is expected to bound only Mg^{2+} ions (about 37,500 per capsid).

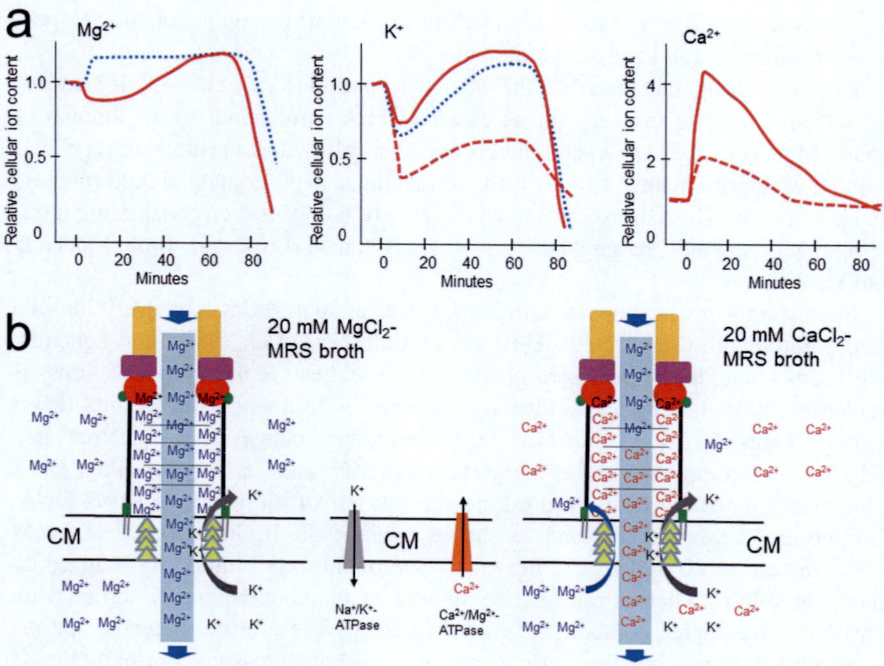

Fig. 9.3 Proposed model of phage LL-H DNA injection into the *Lb. delbrueckii* host cell together with the associated cation movements, in the presence of high external calcium or magnesium concentrations. (**a**) Changes in relative cellular magnesium (Mg^{2+}), potassium (K^+) and calcium (Ca^{2+}) contents when *Lb. delbrueckii* cells, grown in exponential growth phase at 37 °C, were infected with purified phage LL-H (in 5 mM Tris-HCl/10 mM $MgCl_2$ buffer, pH 6.8) at a same m.o.i. (about 45): for the *Lb. delbrueckii* cultivation, the MRS broth was supplemented with 20 mM $MgCl_2$ (dotted line), 20 mM $CaCl_2$ (solid line) or 5 mM $CaCl_2$ (broken line). The results were detailed in the study by Alatossava et al. (1987). (**b**) Proposed model for the LL-H DNA Mg^{2+}/Ca^{2+} counterions movements through the suggested Ca/Mg-LTA channel and further into the bacterial cell through the cytoplasmic membrane (CM) when the MRS broth was supplemented with 20 mM $MgCl_2$ (left) or 20 mM $CaCl_2$ (right). Associated cation leakages and compensating ion pump activities have been indicated, too. Symbols: LL-H TMP homolog ORF351 (triangle), LL-H tail fiber (antireceptor) protein Gp71 (SP58?) subunit (large sphere), α-glucose moiety of LTA (small sphere), no/low-D-alanylated LTA with free, negatively charged phosphate groups (vertical bar) capable of Ca^{2+}/Mg^{2+} binding and further crosslinking with other LTAs (horizontal dotted lines), to form the Ca/Mg-LTA channel. The arrow indicates the direction of the LL-H DNA injection, from the phage head through the tail end into the cytoplasm of the infected cell. *Note*: For simplification purpose, the schematic drawing does not consider other cell wall components like peptidoglycan

After the infection by LL-H, the cellular total Mg^{2+} content increased very rapidly (within 5–10 min) by 15–20% when *Lb. delbrueckii* was grown and further infected in the presence of 20 mM $MgCl_2$ (added to the growth medium); the Mg^{2+} cellular content reduced slightly, by about 10%, when *Lb. delbruckii* was grown and further infected in the presence of 20 mM $CaCl_2$, which altogether suggests that the injected

LL-H DNA carry Mg^{2+} or Ca^{2+} as counterions in amounts depending on their respective concentrations in the external medium.

By considering that each of the 45 LL-H particles (a m.o.i. of 45) carries 37,500 Mg^{2+} ions inside each phage capsid (which corresponds to the amount of bound Mg^{2+} per LL-H DNA) and infects one host cell (with an average intracellular volume of approximately 1.0 μm^3), the intracellular Mg^{2+} content should increase by about 6 mM. This corresponds to an increase of 6–12%, when considering intracellular Mg^{2+} contents ranging between 50 and 100 mM (Lusk et al. 1968, Moncany and Kellenberger 1981).

Interestingly, it was shown that in the presence of 20 mM $CaCl_2$, no Mg^{2+} ions are incorporated with the injected LL-H DNA (Alatossava et al. 1987); consequently, Mg^{2+} ions bound to the packaged phage DNA have been replaced by Ca^{2+} ions, as evidenced by the strong Ca^{2+} influx accompanied with a slight Mg^{2+} efflux (Mg^{2+} leakage?), instead of an influx of Mg^{2+} during the phage DNA injection step (Fig. 9.3a). Consequently, the exchange between Mg^{2+} and Ca^{2+} ions, as phage DNA counterions, needs to take place during the transfer of the uncoiled LL-H DNA, through the Ca/Mg-LTA channel as shown in Fig. 9.3b. In the presence of 5 mM $CaCl_2$ instead of 20 mM $CaCl_2$, the relative amount of Ca^{2+} influx has reduced by about three-fold, whereas the relative amount of the concomitant K^+ efflux with Ca^{2+} influx has increased about 2.5-fold (Fig. 9.3a). These results suggest, that the observed K^+ efflux is not the consequence of an osmotic pressure control by the K^+ pumps, but instead results from K^+ leakage through the LL-H promoted hole(s) of the cytoplasmic membrane during the phage DNA-Ca/Mg injection period. Later, and after the completion of the phage DNA injection step, intracellular K^+ and Ca^{2+} contents will be re-equilibrated obviously by specific ion pumps like cation-specific ATPases of the membrane (Fig. 9.3b). The presence of external cadmium (Cd^{2+}) appeared to be strongly inhibitory for the success of the LL-H infection (Alatossava et al. 1983). During the LL-H DNA injection, external Cd^{2+} cations have a chance to be incorporated into the cell as counterions of phage DNA such as to increase the intracellular Cd^{2+} concentration to levels which are efficiently inhibiting some subsequent step(s) of LL-H infection. In contrast to Cd^{2+}, even high but transient, intracellular Ca^{2+} contents induced by phage LL-H DNA injection (in condition of high m.o.i.s) do not disturb bacterial or phage encoded biosynthesis in the infected cell. This is in agreement with the observation that intracellular Ca^{2+} of up to about 20 mM is partially capable to complement the lack of intracellular Mg^{2+} (Alatossava 1988). Finally, the LL-H DNA injection step does not require any energy from the infected host cell: in an energy depleted host cell (when pretreated with ATP and PMF energy blockers before phage infection), the phage-DNA injection associated Ca/Mg counterion influx is at least as strong as in an untreated and energy rich host cell (Alatossava et al. 1987). In other words, the energy stored in the packaged and coiled state of LL-H DNA, with its high Mg^{2+} counterion concentration (1 M Mg^{2+}, which corresponds approximately to an osmotic pressure of 25 atm, at 37 °C), is sufficient for an effective and complete phage DNA transfer from inside the capsid into the bacterial host cell: this view is supported by the findings from other studied phage systems (Molineux and Panja 2013; Rao and Feiss 2015).

9.5 Conclusion

Even inside a same phage taxonomical group, biological and genetic features of bacteriophages are very diversified. The study of a phage genome sequence constitutes a good start but not more for the characterization of a phage: this is due to the fact that the biological functions of all phage encoded proteins (and especially how each of the phage proteins performs its biological function(s) through its conformational and interactional features) are not directly predictable from the nucleotide sequence data, at least not yet. In addition to genetic studies, the research has to consider biochemical, biophysical, structural biology and high resolution microscopy studies in order to reveal the most significant and specific features for a phage of interest.

To unravel the complexity of phage-host interactions, a multidisciplinary approach is requested irrespectively on whether the research focuses on phage applications (like the use of phages as biocontrol or therapeutic agents), or aims to improve phage resistance and the genetic stability properties of industrial bacterial strains in situations where phages constitute nuisances.

References

Ackermann H-W (2007) 5500 phages examined in the electron microscope. Arch Virol 152:227–243

Ackermann H-W, DuBow MS (1987) Viruses of prokaryotes, General properties of bacteriophages, vol 1. CRC Press, Boca Raton

Ackermann H-W, Cantor ED, Jarvis AW, Lembke J, Mayo J (1984) New species definitions in phages of Gram-positive cocci. Intervirology 22:181–190

Adam G, Delbrück M (1968) Reduction of dimensionality in biological diffusion processes. In: Rich A, Ravidson N (eds) Structural chemistry and molecular biology. WH Freeman & Co, San Francisco, pp 198–215

Alatossava T (1982) Factors affecting *in vitro* DNA ejection of the *Lactobacillus lactis* bacteriophage LL-H. J Gen Virol 59:173–175

Alatossava T (1987) Molecular biology of *Lactobacillus lactis* bacteriophage LL-H (PhD Thesis) Acta Univ Oul A 191. University of Oulu, Oulu

Alatossava T (1988) Effects of cellular calcium and magnesium contents on growth of ionophore A23187-treated *Lactobacillus lactis*. Agric Biol Chem 52:1275–1276

Alatossava T, Pyhtilä M (1980) Characterization of a new *Latobacillus lactis* bacteriophage. IRCS Med Sci 8:297–298

Alatossava T, Juvonen T, Huhtinen R-L (1983) Effect of cadmium on the infection of *Lactobacillus lactis* by bacteriophage LL-H. J Gen Virol 64:1623–1627

Alatossava T, Jütte H, Kuhn A, Kellenberger E (1985) Manipulation of intracellular magnesium content in polymyxin B nonapeptide-sensitized *Escherichia coli* by ionophore A23187. J Bacteriol 162:413–419

Alatossava T, Jütte H, Seiler H (1987) Transmembrane cation movements during infection of *Lactobacillus lactis* by bacteriophage LL-H. J Gen Virol 68:1525–1532

Baptista C, Santos MA, Sao-Jose C (2008) Phage SPP1 reversible adsorption to *Bacillus subtilis* cell wall teichoic acids accelerates virus recognition of membrane receptor YueB. J Bacteriol 190:4989–4996

Bayfield OW, Klimuk E, Winkler DC, Hesketh EL, Chechik M, Cheng N, Dykeman EC, Minakhin L, Ranson NA, Severinov K, Steven AC, Antson AA (2019) Cryo-EM structure and *in vitro* DNA packaging of a thermophilic virus with supersized T7 capsids. Proc Natl Acad Sci U S A 116:3556–3561

Boulanger P, Jacquot P, Plancon L, Chami M, Engel A, Parquet C, Herbeuval C, Letellier L (2008) Phage T5 straight tail fiber is a multifunctional protein acting as a tape measure and carrying fusogenic and muralytic activities. J Biol Chem 283:13556–13564

Chatterjee S, Rothenberg E (2012) Interaction of bacteriophage λ with its *E. coli* receptor, LamB. Viruses 4:3162–3178

Doron S, Melamed S, Ofir G, Leavitt A, Lopatina A, Keren M, Amitai G, Sorek R (2018) Systematic discovery of antiphage defense systems in the microbial pangenome. Science 359:1–11

Forsman P (1994) Genetic variation and evolution of bacteriophages of lactobacilli (PhD Thesis) Acta Univ Oul A 257. University of Oulu, Oulu

Forsman P, Alatossava T (1991) Genetic variation of *Lactobacillus delbrueckii* subsp. *lactis* bacteriophages isolated from cheese processing plants in Finland. Appl Environ Microbiol 57:1805–1812

Koch HU, Haas R, Fischer W (1984) The role of lipoteichoic acid biosynthesis in membrane lipid metabolism of growing *Staphylococcuc aureus*. Eur J Biochem 138:357–363

Korolev N, Lyubartsev AP, Rupprecht A, Nordeshiöld L (1999) Competitive binding of Mg^{2+}, Ca^{2+}, Na^{+}, and K^{+} ions to DNA in oriented DNA fibers: experimental and Monte Carlo simulation results. Biophys J 77:2736–2749

Leforestier A (2013) Polymorphism of DNA conformation inside the bacteriophage capsid. J Biol Phys 39:210–213

Li D, Liu T, Zuo X, Li T, Qiu X, Evilevitch A (2015) Ionic switch controls the DNA state in phage λ. Nucl Acids Res 43:6348–6358

Lusk JE, Williams RJP, Kennedy EP (1968) Magnesium and the growth of *Escherichia coli*. J Biol Chem 243:2618–2624

Mikkonen M (1996) Genes and genome of *Lactobacillus* phage LL-H (PhD Thesis) Acta Univ Oul A 281. University of Oulu, Oulu

Mikkonen M, Alatossava T (1994) Characterization of the genome region encoding structural proteins of *Lactobacillus delbrueckii* subsp. *lactis* bacteriophage LL-H. Gene 151:53–59

Mikkonen M, Räisänen L, Alatossava T (1996) The early gene region completes the nucleotide sequence of *Lactobacillus delbrueckii* subsp. *lactis* bacteriophage LL-H. Gene 175:49–57

Moldovan R, Chapman-McQuiston E, Wu XL (2007) On kinetics of phage adsorption. Biophys J 93:303–315

Molineux IJ, Panja D (2013) Popping the cork: mechanisms of phage genome ejection. Nature Rev Microbiol 11:194–204

Moncany MLJ, Kellenberger E (1981) High magnesium content of *Escherichia coli* B. Experientia 37:846–847

Munsch-Alatossava P, Alatossava T (2013) The extracellular phage-host interactions involved in the bacteriophage LL-H infection of *Lactobacillus delbrueckii* ssp. *lactis* ATCC 15808. Front Microbiol 4:Article 408

Plisson C, White HE, Auzat I, Zafarani A, Sao-José C, Lhuillier S, Tavares P, Orlova EV (2007) Structure of bacteriophage SPP1 tail reveals trigger for DNA ejection. EMBO J 26:3720–4728

Räisänen L (2007) Phage-host interactions in *Lactobacillus delbrueckii*: host recognition and transcription of early phage genes (PhD Thesis) Acta Univ Oul A 484. University of Oulu

Räisänen L, Schubert K, Jaakonsaari T, Alatossava T (2004) Characterization of lipoteichoic acids as *Lactobacillus delbrueckii* phage receptor components. J Bacteriol 186:5529–5532

Räisänen L, Draing C, Pfitzenmaier M, Schubert K, Jaakonsaari T, von Aulock S, Hartung T, Alatossava T (2007) Molecular interaction between lipoteichoic acids and *Lactobacillus delbrueckii* phages depends on D-Alanyl and α-glucose substitution of poly(glycerophosphate) backbones. J Bacteriol 189:4135–4140

Rao V, Feiss M (2015) Mechanisms of DNA packaging by large double-stranded DNA viruses. Ann Rev Virol 2:351–378

Ravin V, Räisänen L, Alatossava T (2002) A conserved C-terminal region in Gp71 of the small isometric-head phage LL-H and ORF474 of the prolate-head phage JCL1032 is implicated in specificity of adsorption of phage to its host, *Lactobacillus delbrueckii*. J Bacteriol 184:2455–2459

Ravin V, Sasaki T, Räisänen L, Riipinen KA, Alatossava T (2006) Effective plasmid pX3 transduction in *Lactobacillus delbrueckii* by bacteriophage LL-H. Plasmid 55:184–193

Reichmann NT, Gründling A (2011) Location, synthesis and function of glycolipids and polyglycerolphosphate lipoteichoic acid in Gram-positive bacteria of the phylum Firmicutes. FEMS Microbiol Lett 319:97–105

Reichmann NT, Picarra Cassona C, Gründling A (2013) Revised mechanisms of D-alanine incorporation into cell wall polymers in Gram-positive bacteria. Microbiology 159:1868–1877

Riipinen K-A (2011) Genetic variation and evolution among industrially important *Lactobacillus* bacteriophages (PhD Thesis) Acta Univ Oul A 589. University of Oulu, Oulu

Rismondo J, Percy MG, Gründling A (2018) Discovery of genes required for lipoteichoic acid glycosylation predicts two distinct mechanisms for wall teichoic acid glycosylation. J Biol Chem 293:3293–3306

Schneewind O, Missiakas D (2017) Lipoteichoic acid synthesis and function in Gram-positive bacteria. In: Geiger O (ed) Biogenesis of fatty acids, lipids and membranes. Springer, Cham, pp 163–180

Storms ZJ, Sauvageau D (2015) Modeling tailed bacteriophage adsorption: insight into mechanisms. Virology 485:355–362

Trautwetter A, Ritzenthaler P, Alatossava T, Mata-Gilsinger M (1986) Physical and genetic characterization of the genome of *Lactobacillus lactis* bacteriophage LL-H. J Virol 59:551–555

Vasala A (1998) Characterization of *Lactobacillus* bacteriophage LL-H genes and proteins having biotechnological interest (PhD Thesis) Acta Univ Oul A 315. University of Oulu, Oulu

Wommack KE, Colwell RR (2000) Virioplankton: viruses in aquatic ecosystems. Microbiol Mol Biol Rev 64:69–114

Chapter 10
Archaeal Viruses and Their Interactions with CRISPR-Cas Systems

Roger A. Garrett, Shiraz A. Shah, Laura Martinez-Alvarez, and Xu Peng

Abstract Our knowledge of archaeal viruses has increased rapidly over the past four decades since the discovery of the archaeal domain. Most surprising has been the morphological diversity of crenarchaeal viruses that generate several forms not previously observed in the viral sphere. Much recent work has focussed on host-virus interactions and especially on the influence on viral replication of the CRISPR-Cas immune systems that are highly complex in many archaea. Here we provide an overview of archaeal virus biology and summarise many of the major findings on archaeal viruses and their cellular interactions, with a special focus on two model viruses, the rudivirus SIRV2 and the bicaudavirus SMV1. Moreover, we explore insights into the interplay between the host CRISPR-Cas systems and viruses infected singly or in pairs. The evolutionary relationships between archaeal viruses and bacteriophages are also considered.

10.1 Introduction

The time-scale for investigating archaeal viruses has been short compared to that spent on studying bacterial phages. It dates back to the time when the seminal work of Woese and Fox, performed on partial 16S ribosomal RNA sequences, led to the discovery of the archaeal domain (Woese and Fox 1977). At that time archaeal viruses received little attention. The main drive of the research groups, centred in Germany, that first focused on archaea, was to characterise the genetic, biochemical, molecular and cellular properties that distinguished archaea from bacteria. Very important at that time was the demonstration by Zillig's group that the archaeal RNA polymerases were more complex than the bacterial RNA polymerases, and that their subunit composition closely resembled that of eukaryotic RNA polymerases, later confirmed by sequencing studies (reviewed in Pühler et al. 1989). Many

R. A. Garrett (✉) · S. A. Shah · L. Martinez-Alvarez · X. Peng
Danish Archaea Centre, Department of Biology, Copenhagen University, Copenhagen, Denmark
e-mail: garrett@bio.ku.dk

G. Witzany (ed.), *Biocommunication of Phages*,
https://doi.org/10.1007/978-3-030-45885-0_10

other differences were established at an early stage including the demonstration of varied but essentially simple cell wall structures of archaea (Kandler 1994). Morever, archaeal membranes were shown to carry ether-linked isopranyl glycerol lipids, mainly in monolayers, in contrast to ester-linked glycerol lipids of bacteria and eukaryotes (Langworthy and Pond 1986). In addition, similarities were found between the archaeal and eukaryal cellular DNA replication mechanisms that differed from those of bacteria (reviewed by Forterre 2013), and most crenarchaea use the eukaryotic-like ESCRT complex for cell division (Samson et al. 2008). Furthermore, methanogenesis is specific to euryarchaeal methanogens, and several archaea-specific enzymes were characterised including an intron splicing and RNA processing enzyme (Lykke-Andersen and Garrett 1997) and an integrase gene that partitions on integration and can lead to genome capture of viruses and plasmids (Muskhelishvili 1994; She et al. 2001a).

Although studying archaeal viruses was not a priority, they were not totally neglected. There were early reports of a few head-tailed phage-like viruses infecting euryarchaea, including the well characterised phage ϕH of *Halobacterium salinarium* and the transducing phage ΨM1 of *Methanobacterium thermoautotrophicum* (Schnabel et al. 1982; Meile et al. 1990). However, the momentum of research changed dramatically when Zillig and colleagues undertook multiple sampling expeditions to hot springs in different countries, and in particular to Iceland, collecting extreme- and hyperthermophilic archaeal species. They observed many virus-like particles in electron micrographs of their cultured samples that exhibited a wide range of exceptional morphologies. The first viruses characterised were three filamentous viruses TTV1 to 3, carrying linear dsDNA genomes, from the hyperthermophile *Thermoproteus tenax* isolated on Iceland (Janekovic et al. 1983). However no plaque assay was developed and they were not studied extensively. More success was achieved with the first fusiform virus SSV1, with a circular dsDNA genome that infected *Sulfolobus shibatae* isolated from a hot spring in Japan (Martin et al. 1984). Plaque assays were developed by Schleper and colleagues who then exploited SSV1 to produce archaeal vector systems (Schleper et al. 1992).

At the end of the twentieth century, the archaea scientific community was still quite small and tightly knit and ideas and emerging data where exchanged relatively freely. Carl Woese, the co-discoverer of Archaea, had an important influence on research directions in the field and strongly encouraged collaborative research efforts. Shortly before he died in 2012 he confided that he had chided Zillig earlier for "wasting his time" looking at viruses when there was much more important and urgent work to be done in characterising archaeal cellular biology (Garrett 2017). He had assumed that the viruses would turn out to be similar to bacteriophages, presumably based on the early reports mentioned above. Graciously, he admitted, regretfully, how wrong he had been, as we will attempt to outline below.

10.2 Early Archaeal Virus Studies

Zillig and colleagues isolated a large variety of different crenarchaeal virus-like particles mainly from Icelandic hotsprings in the period after 1982. They exhibited primarily filamentous or fusiform morphologies, and carried linear or circular dsDNA genomes, respectively (Zillig et al. 1994, 1996, 1998). Most of the viral hosts were from the genera *Sulfolobus* and *Acidianus* with fewer from *Thermoproteus, Aeropyrum, Pyrobaculum* and *Stygiolobus*, and they all grew optimally at about 80 °C or higher. There was a strong interest in characterising the viruses of these crenarchaeal extreme thermophiles. The hosts were generally aerobic and with relatively short doubling times compared to those of many euryarchaea and, moreover, they were especially amenable to the development of archaeal genetic systems (Martin et al. 1984; Schleper et al. 1992).

Many of these viruses could be stably propagated in different hosts and did not induce cell lysis. A consensus developed that the viruses propagated in stable carrier states within the hosts and that they were continually released without inducing cell lysis, thereby minimising exposure of the virions to the extreme temperature and pH conditions of the external environments (Prangishvili and Garrett 2005). This hypothesis required that some degree of regulation of viral replication occurred intracellularly.

SSV1 and other fuselloviruses have been demonstrated to be lysogenic. Their circular genomes integrate into the host chromosome, generally at tRNA genes, via a virus-encoded integrase of the tyrosine recombinase family that, exceptionally, partitions during genome integration (Muskhelishvili 1994; She et al. 2001a). This can lead to capture of the virus in the chromosome and, in *Sulfolobus* genomes in particular, trapped and fragmented fusselloviral genomes are relatively common in the vicinity of tRNA genes (Guo et al. 2011; You et al. 2011). Replication of the integrated viral genomes can be induced through UV irradiation and by growth under suboptimal growth conditions including low temperatures or deficient media or aeration (e.g. Prangishvili et al. 2006c); provirus induction results in growth retardation for fuselloviruses (Zillig et al. 1998).

10.3 Morphological Diversity

Archaeal viruses exhibit a much wider range of morphologies than bacteriophages and the archaea-specific morphotypes include rod-shaped rudiviruses, filamentous lipothrixviruses, fusiform fuselloviruses, tailed fusiform bicaudaviruses and bottle-shaped ampullaviruses. They also include all the morphotypes common to bacteriophages including head-tailed and icosahedral forms, as listed in Table 10.1. Many electron micrographs exemplifying the different virion morphotyes have been depicted in recent reviews (e.g. Prangishvili 2013; Wang et al. 2015; Prangishvili et al. 2017). The archaea-specific viral morphotypes are illustrated schematically in Fig. 10.1.

Table 10.1 Summary of viral families infecting archaea. Selected viruses and their properties are shown for each family

Family	Name	Host	Morphotype	Genome (kb)	Accession number
Archaea-specific					
Rudiviridae	SIRV1,	*Sulfolobus*	Rod-shaped	lin 32.3	NC_004087
	SIRV2			lin 35.5	NC_004086
Lipothrixviridae	SIFV	*Sulfolobus*	Filamentous	lin 41	NC_003214
Clavaviridae	APBV1	*Aeropyrum*	Rod-shaped	cir 5.3	NC_043030
Tristromaviridae	PFV1	*Pyrobaculum*	Rod-shaped	lin 17.7	NC_029548
Ampullaviridae	ABV	*Acidianus*	Bottle	lin 23.9	NC_009452
Spiraviridae	ACV	*Aeropyrum*	Spiral	ss.cir 24.9	HE681887
Bicaudaviridae	ATV	*Acidianus*	Two tailed-fusiform	cir 62.7	NC_007409
	SMV1	*Sulfolobus*		cir 48.5	NC_023585
Monocauda viruses	STSV2	*Sulfolobus*	Single tail-fusiform	cir 76	NC_020077
Fuselloviridae	SSV1	*Sulfolobus*	Fusiform	cir 15.5	NC_001338
Guttaviridae	APOV1	*Aeropyrum*	Ovoid	cir 13.8	NC_028256
Globuloviridae	PSV	*Pyrobaculum*	Spherical	lin 28.3	NC_005872
Pleolipoviridae	HHPV1	*Haloarcula*	Pleomorphic	cir 8.2 nt	NC_013758
Portogloboviridae	SPV1	*Sulfolobus*	Icosahedral	lin 20.2	NC_038017
Turriviridae	STIV1	*Sulfolobus*	Icosahedral	cir 17.7	NC_005892
Cosmopolitan					
Myoviridae	HSTV2	*Halorubrum*	Head-tail	lin 68.5	NC_020159
Siphoviridae	HVTV1	*Haloarcula*	Head-tail	lin 102.3	KC117377
Sphaerolipoviridae	SH1	*Haloarcula*	Icosahedral	lin 30.9	NC_007217
Salterprovirus	His1	*Haloarcula*	Fusiform	lin 14.5	NC_007914
Podoviridae	HSTV1	*Haloarcula*	Head-tail	lin 32.2	KC117378

Most archaea-specific morphotypes are found amongst crenarchaeal viruses. In contrast, euryarchaea carry fusiform viruses, icosahedral viruses with an inner lipid layer (e.g. Jaakkola et al. 2012), pleomorphic viruses and head-tailed viruses classified in the order Caudovirales (e.g. Atanasova et al. 2012). However, studies of natural environments where euryarchaea predominate, including hypersaline lakes and anoxic lake sediments, revealed a wide variety of morphotypes including those of known crenarchaeal viruses as well as some novel morphotypes (Sime-Ngando et al. 2011; Borrel et al. 2012).

Lipothrixviruses, rudiviruses, clavaviruses and tristromaviruses have filamentous structures and of these only the lipothrix- and tristromaviruses are enveloped (Kasson et al. 2017). Virion structures implicated in adsorption to host cells vary considerably between, and within, the different viral families. For example, lipothrixviruses, and rudiviruses, carry thin flexible terminal structures for attachment, either single or multiple at each end and some exhibit more complex terminal structures including claw-like appendages in AFV1 (Bettstetter et al. 2003).

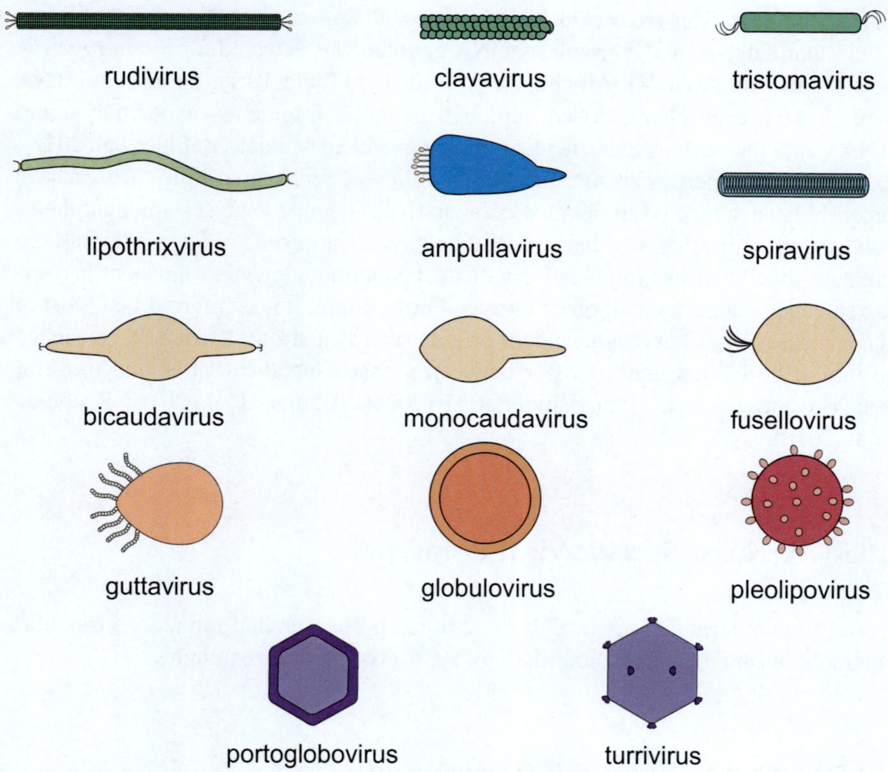

Fig. 10.1 Schematic representation of the diverse morphologies of archaea-specific viruses

10.4 Genomic Diversity

Around 1983 Zillig contacted us in Aarhus, Denmark. We had set up a sequencing lab and he asked us to collaborate on genome sequencing of his new virus isolates and other mobile genetic elements. This led to a long term productive and inspiring collaboration that later continued, in Copenhagen, with his close collaborator David Prangishvili. Progress was not rapid. At that time relatively large amounts of DNA were required for library preparation and the viral yields were often low, and sometimes the virus was lost. Nevertheless several genomes were sequenced and analysed (reviewed in Prangishvili et al. 2006a). All consisted of dsDNA and were circular or linear, and the linear rudiviral genomes exhibited covalently closed ends. The genomes clearly reflected the morphological diversity of the viruses. They carried highly diverse gene contents and relatively few of the gene products could be assigned functions based on sequence comparison studies. To date, most characterised archaeal viruses derive from either crenarchaeal hyperthermophiles or euryarchaeal extreme halophiles. In addition a few viruses have been isolated from euryarchaeal methanogens and, recently, three fusiform viruses were obtained from marine thaumarchaea (Kim et al. 2019). Almost all known genomes consist of linear or circular dsDNA, and a circular ssDNA genome was detected for the *Aeropyrum* coil-shaped virus (ACV) (Mochizuki et al. 2012) (Table 10.1). To date, no RNA viruses have been characterised, although genomic fragments of positive strand RNA viruses were detected in acidic hot springs rich in archaea (Bolduc et al. 2012).

Genomic properties of the different families of archaea-specific viruses have recently been subjected to detailed comparative sequence analyses amongst themselves, and with viruses of bacteria and eukarya. The results indicated that all the archaea-specific virus groups fall in a distinct supermodule which shows little overlap in gene content with all other viruses. Furthermore, it was inferred that most of these viruses must have independent origins such that strong barriers have existed to inhibit horizontal gene transfer between archaea-specific viruses and those of bacteria and eukarya (Prangishvili et al. 2006b; Iranzo et al. 2016; Krupovic et al. 2018).

10.5 Archaea-Specific Viral Properties

Some crenarchaeal viruses exhibit exceptional properties that influence virus-host interactions and they are exemplified by the following three examples.

10.5.1 Extracellular Tail Development

Virions of the bicaudavirus ATV are extruded from *Acidianus convivator* host cells as fusiform particles that then develop tails extracellularly at each end of the particle, and of variable length. This development is triggered at above 75 °C, the temperature of the hot spring of origin (Häring et al. 2005; Prangishvili et al. 2006c). The dynamics of the tail development is dependent on a virus-encoded chaperone system consisting of a MoxR-type AAA-ATPase and a Von Willebrand factor A domain-containing protein (Scheele et al. 2011). Both chaperone proteins are also encoded by another bicaudavirus SMV1 that undergoes similar extracellular tail development illustrated in Fig. 10.2 (Erdmann et al. 2014b; Uldahl et al. 2016). The morphologically related monocaudaviruses STSV1, STSV2 and APSV1 where the single tailed version dominates in infected cultures do not carry these genes, and the tails are formed intracellularly possibly facilitated by host proteins (Xiang et al. 2005; Mochizuki et al. 2011; Erdmann et al. 2014a).

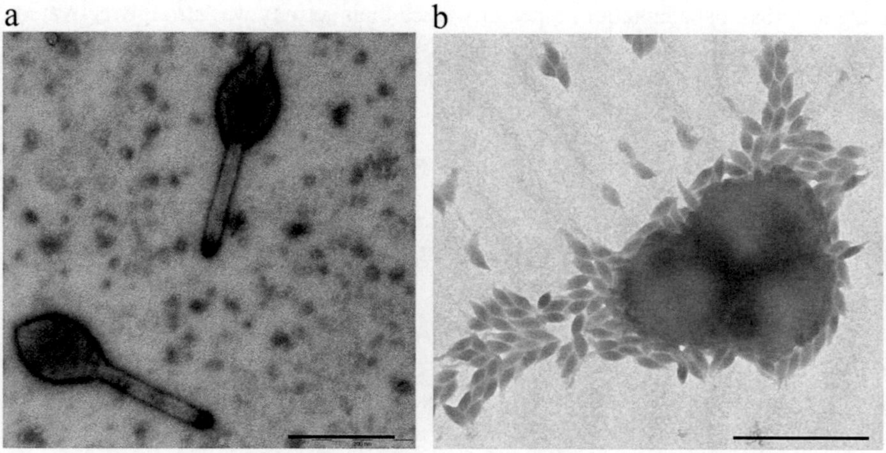

Fig. 10.2 Electron micrographs of: (**a**) SMV1 virions with a single tail (below) and two tails (above): scale bar – 200 nm; (**b**) an *S. islandicus* ΔC1C2 culture (lacking CRISPR loci) infected with SMV1 at high MOI at 22 h post infection: scale bar – 1 µm. Soley Gudbergsdóttir and Kristine Uldahl kindly provided micrographs (**a**) and (**b**), respectively

10.5.2 Virus-Associated Pyramids

Two *Sulfolobus* viruses, the rudivirus SIRV2 and the turrivirus STIV have been shown to generate pyramid-like structures on the surfaces of host cells. These pyramid structures span the surface layer (S-layer) of the host cells and they open at the end of the virus life cycle enabling the release of mature virions. Each of the characterised pyramid stuctures is assembled from a single viral protein. Moreover, the protein can self-assemble into pyramidal structures when expressed heterologously in other organisms but it remains unclear whether other viral, or host, proteins facilitate opening of the pyramid structures prior to virion release (Bize et al. 2009; Quax et al. 2010; Snyder et al. 2011).

10.5.3 12 bp Indels in Viral Genomes

The occurrence of 12-bp indels, or multiples thereof, is common for genomes of both rudiviruses and lipothrixviruses. These were discovered when the rudivirus SIRV1, originally isolated from *S. islandicus* KVEM10H3, was passed through *S. islandicus* strains REN2H1 and LAL14/1. Thereafter, many indels were detected in overlapping clones from shotgun libraries of the viral DNA (Peng et al. 2004). Most were observed in ORFs where they altered the gene size without disrupting the reading frame. Moreover, a comparative genome analysis of several lipothrixviruses revealed multiple examples of 12-bp indels located in otherwise conserved sequence regions (Vestergaard et al. 2008). It is likely that these genome changes relate to virus-host adaptation, especially since their activity was strongly activated by passing SIRV1 though different hosts. Moreover, their formation could be an anti-CRISPR-Cas response given the presence of multiple active CRISPR-Cas systems in these *S. islandicus* hosts. At present the molecular mechanism of indel formation remains unclear (Peng et al. 2004).

10.6 A Model Archaeal Virus – The Rudivirus SIRV2

The archaeal rudivirus, *Sulfolobus islandicus* rod-shaped virus 2 (SIRV2), belongs to the family *Rudiviridae* and is one of the best-studied crenarchaeal viruses (Prangishvili et al. 1999; Peng et al. 2001). The 35.5 kb, linear, dsDNA genome encodes about 54 proteins, many of which show little sequence similarity to proteins in public sequence databases and are of unknown function.

As described above, most newly isolated crenarchaeal viruses, including SIRV2, could be propagated relatively stably in a variety of crenarchaeal hosts with no signs of cell lysis. This picture was later transformed for SIRV2 by the results of a detailed study of its interactions with its natural host *S. islandicus* LAL14/1. Unexpectedly, massive degradation of the host chromosomal DNA was detected from early stages of infection followed by cell lysis (Bize et al. 2009).

10.6.1 SIRV2 Entry

Electron microscopy studies have provided evidence for SIRV2 virions attaching directly to the tips of *S. islandicus* pili structures via the three fibres at each end of the virion; the viruses are then translocated along the pili to the cellular S-layer (Quemin et al. 2013). In order to investigate the cellular involvement in the attachment process, SIRV2-resistant *S. solfataricus* P2 cells, from which active CRISPR-Cas systems had been deleted, were isolated after a prolonged cultivation of an infected culture in which most cells had died. Genome sequencing of colonies plated from the resistant cells revealed mutations in either of two host gene clusters. The first encoded a putative type IV secretion system (*sso2386-sso2387*) and the second encoded proteins with a limited sequence homology to pili proteins (*sso3138-sso3141*). This suggested that the pili-like proteins were secreted by membrane proteins encoded by the former gene operon and they then blocked expression of either one of the two components that are linked genetically to the SIRV2 resistance (Deng et al. 2014).

10.6.2 Transcriptional Regulation

In early studies little evidence was found for temporal control of gene expression during the SIRV2 life cycle as determined mainly by Northern blot analyses of viral genes but this result appeared inconsistent with the multiple putative transcription regulators that are encoded by SIRV2 and other crenarchaeal viral genomes (Kessler et al. 2004). This picture changed significantly when SIRV2 was infected at higher MOIs and more quantitative microarray and RNAseq analyses were employed to assay transcript yields (Okutan et al. 2013; Quax et al. 2013). The results showed that higher transcript levels were produced initially from genes located towards the SIRV2 genomic termini whereas genes lying in the central region, including the structural protein genes, reached highest expression at a later stage of the viral life cycle (Okutan et al. 2013; Quax et al. 2013).

10.6.3 DNA Replication

Exceptional brush-like replication intermediates, composed of multimeric SIRV2 genomes (>30 units), were demonstrated to be produced by employing DNA spreading and fluorescence microscopy techniques (Martinez-Alvarez et al. 2016). Based on these observations and on results from a range of experimental approaches, including 2D neutral-neutral gel electrophoresis, a replication model was developed (Martinez-Alvarez et al. 2016). The model proposed that strand-displacement replication initiates from either genomic terminus and leads to repetitive replication from single genome templates. The resultant ssDNA genome intermediates then attach to one end of a mature SIRV2 genome forming a brush-like structure. The mechanism by which the ssDNA intermediates are subsequently converted to mature dsDNA genomes remains uncertain but *in vitro* studies on three ssDNA processing proteins, SIRV2 gp17, gp18 and gp19, suggested that they participate in SIRV2 replication, repair and/or maturation (Guo et al. 2015). Furthermore, replication of SIRV2 DNA was shown to take place in a localised cellular region near the periphery of the *S. solfataricus* host cells, using the selective staining procedures illustrated in Fig. 10.3. This seminal result localised, for the first time, a viral factory within archaeal cells (Martinez-Alvarez et al. 2017), as has been shown for some bacteriophages and eukaryal viruses.

10.6.4 Virion Release

Over the first 12 h of infection in *S. islandicus* LAL14/1 the host genome is extensively degraded and the virus replicates rapidly. Moreover, the virus overtakes control of the host cell converting it to a complex viral factory. The cell surface is

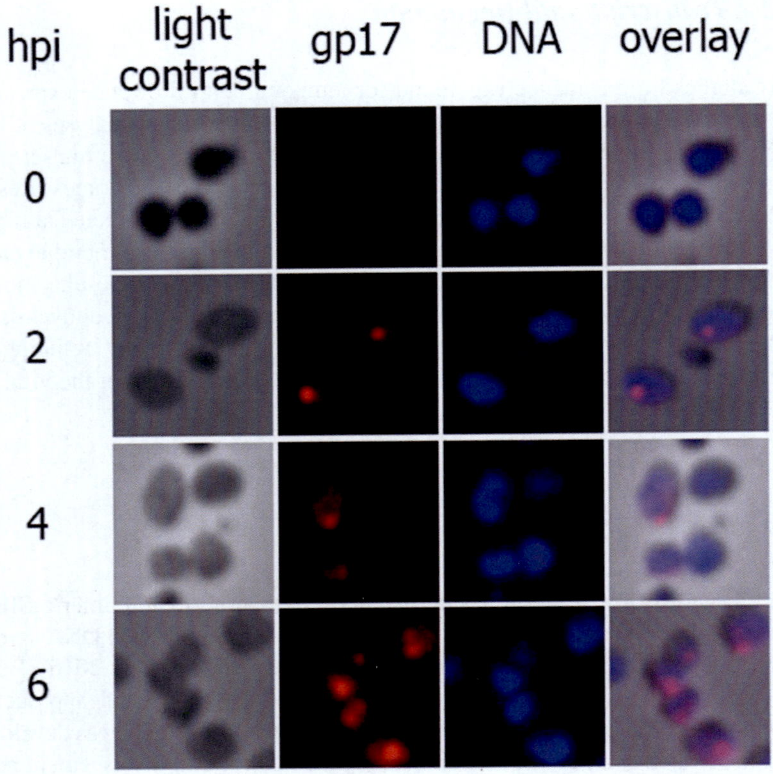

Fig. 10.3 Viral DNA replication focus in *S. islandicus* LAL14/1 cells infected with SIRV2. Cell aliquots were taken at 0, 2, 4 and 6 h. post infection (hpi) and fluorescence microscopy was performed as described (Martinez-Alvarez et al. 2017). Light contrast – differential interference contrast image of the cells; gp17 – viral replication foci located with SIRV2 ssDNA-binding protein gp17; DNA – total DNA staining using DAPI; overlay – merging of the three panels

modified and pyramid structures are formed at the cell surface. A similar development of viral factories has been observed in virus-infected eukaryotic cells (Novoa et al. 2005). At the end of the infection cycle mature virions are released via the pyramid-like structures which open causing localised disruption of the cell envelope.

10.6.5 Anti-CRISPR Proteins

The natural host of SIRV2, *S. islandicus* LAL14/1, carries three different CRISPR-Cas systems subtype I-A, I-D and III-B and five CRISPR arrays. Despite the host CRISPR arrays carrying 13 potentially active spacers matching the SIRV2 genome, SIRV2 was still lytic post-infection and reached a burst size of approximately 30 virions/infected cell (Bize et al. 2009). Hence, SIRV2 was suspected of producing anti-CRISPR proteins and this, in turn, led to the discovery and characterisation of one anti-CRISPR protein, AcrID1, SIRV2 gp03, that specifically inhibited the subtype I-D CRISPR-Cas activity and another protein SIRV2 gp48, AcrIIIB1, that inhibited subtype III-B interference. The former protein AcrID1 binds to the large Cas10d protein of the subtype I-D effector complex, and inhibits either the nuclease activity or the PAM recognition capacity of Cas10d. In contrast, protein AcrIIIB1 binds to, and inhibits, the subtype III-B effector complex by an, as yet, unknown mechanism (He et al. 2018; Bhoobalan-Chitty et al. 2019).

10.7 *Sulfolobus* CRISPR-Cas Immunity and the Bicaudavirus SMV1

10.7.1 The Choice of a Crenarchaeal Virus Host

During an archaea meeting, in Martinsried, Germany, in 1995, a group of senior researchers discussed the possibility of exploiting the recent advances in DNA sequencing techniques to sequence the genome of an archaeal viral host in order to facilitate studies on virus-host interactions. Many of the new viral isolates and plasmids from the Zillig lab were propagated in *Sulfolobus* species and especially in *S. solfataricus* P2 and this was considered a promising host for developing vector systems (e.g. Schleper et al. 1992). A consensus was reached at the meeting to sequence the approximately 3 Mb genome of *S. solfataricus* P2 which, in turn, led to European Union-Canadian NRC grants to finance five laboratories to complete the genome, using the Sanger techniques, over a three year period (She et al. 2001b). (Using modern sequencing technology the closely similar strain P1 was sequenced in 2 weeks (Liu et al. 2016)). An exceptional genomic feature that emerged at an early stage was the occurrence of multiple long interspaced repeat structures (later CRISPR) carrying >100 repeats.

10.7.2 Characterisation of Host CRISPR-Cas Systems

A transcriptome study performed on similar CRISPR arrays in another crenarchaeote *Archaeoglobus fulgidus* demonstrated that they were both transcribed and processed, and the approximate processing sites within the repeats could be localised (Tang et al. 2002). A single protein Cbp1 was also characterised from *S. solfataricus* P2 that selectively bound to the arrays (Peng et al. 2003) and evidence was presented later indicating that a function of the protein was to facilitate uninterrupted transcription of long CRISPR loci (Deng et al. 2012). The biological significance of the repeats and spacer regions was deduced later from studies of Mojica et al. (2005) who examined genomes of numerous archaeal and bacterial species and their viruses and, independently by Pourcel et al. (2005), studying two bacterial species and their phages. The spacer sequences were shown to exhibit matches to sequences in viruses that could potentially invade the corresponding hosts. Thus arose the concept of an anti-viral adaptive immune system, present in most archaea (Lillestøl et al. 2006) and in many bacteria, whereby each new spacer uptake created resistance against an incoming virus while the host retained spacers conferring immunity against previously infecting viruses (reviewed in Mojica and Garrett 2013).

The immune complex consists of multiple Cas proteins that generate two functional modules, an adaptation, spacer-acquisition, module and an interference module (Garrett et al. 2011) where the latter module complexes crRNAs generated from spacers and targets and cleaves the matching viral sequence. Many crenarchaea encode both type I and type III interference modules, and they often share both adaptation modules and CRISPR loci. The former targets viral DNA, and the latter viral DNA or RNA transcripts (Zhang et al. 2012; Peng et al. 2014; Vestergaard et al. 2014). Protospacer recognition and excision requires the presence of a specific protospacer adjacent motif (PAM) of 2–5 bp (reviewed in Shah et al. 2013). Protospacers are selected more or less randomly from either DNA strand (Shah et al. 2009) as illustrated for the viruses SIRV2 and SMV1 infecting *Sulfolobus* hosts in Fig. 10.4a. Spacers in *Sulfolobus* species can still be effective in repression of viral DNA replication even when carrying multiple mismatches (Gudbergsdóttir et al. 2011; Manica et al. 2013; Mousaei et al. 2016).

10.7.3 CRISPR Loci

CRISPR loci are preceded by a leader sequence that is essential for spacer acquisition and most archaea carry multiple large CRISPR loci with leaders. Viral infections in different *Sulfolobus* species can induce uptake of multiple *de novo* spacers, often with duplicated or overlapping sequences, during a single infection (Lillestøl et al. 2006, 2009; Erdmann and Garrett 2012). Nevertheless, the overall CRISPR locus sizes appear to be limited. The size limitation is likely to be controlled by recombination events occurring between repeat sequences. Comparison of the corresponding loci of closely related *Sulfolobus* strains revealed evidence for large deletions and also provided evidence of recombination occurring between different intracellular CRISPR loci as illustrated in Fig. 10.4b (Lillestøl et al. 2006, 2009). Moreover, the spacer acquisition process from *Sulfolobus* viruses can be actively reversed and/or rapidly selected against (Gudbergsdóttir et al. 2011). Furthermore, when cells are no longer disadvantaged by mutated elements there must be a strong selection for retaining a few *de novo* spacers that are unique and that target relatively conserved genomic sites (Erdmann et al. 2014b; Liu et al. 2016).

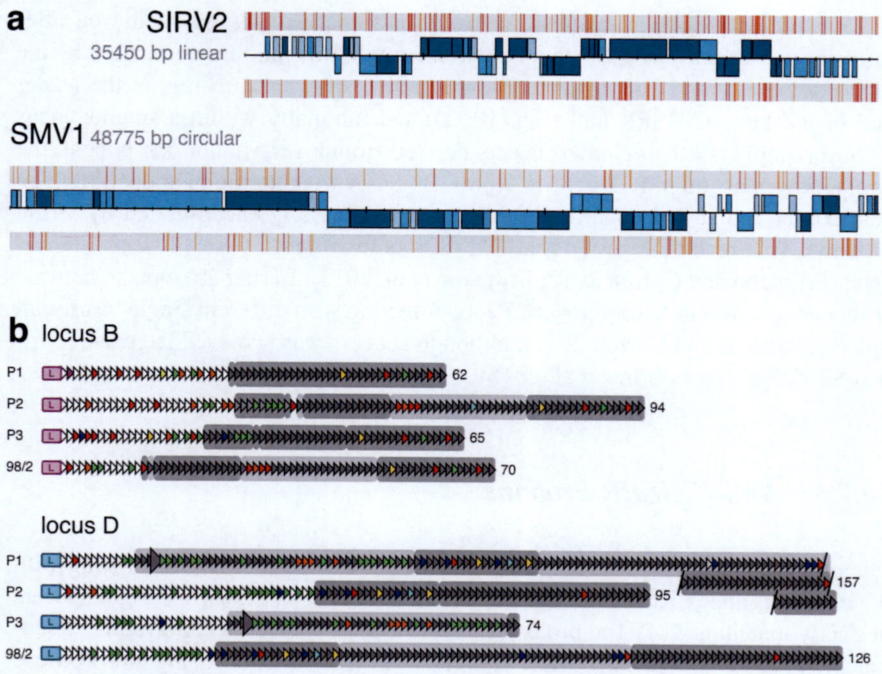

Fig. 10.4 CRISPR spacers of *Sulfolobus* hosts matching infecting viral genomes. (**a**) Scheme of the rudivirus SIRV2 and the bicaudavirus SMV1 genomes with matching CRISPR spacers. Spacers were extracted from 118 complete or draft genomes of Sulfolobales in GenBank 27880 (Nov. 2019), and duplicates removed, yielding 9367 unique spacers. They were aligned with SIRV2 and SMV1 using FASTA and an e-value cutoff of 0.05; a maximum of 10 mismatches were allowed including up to two in the seed region using an earlier approach (Shah et al. 2009; Mousaei et al. 2016). Red vertical lines – matches to spacers of *S. islandicus* LAL14/1 for SIRV2 and *S. solfataricus* P2 for SMV1. Orange lines – matches to spacers of other members of the Sulfolobales. Viral ORFs are shaded and darker colours indicate higher conservation levels within the viral families. (**b**) CRISPR loci B and D in *S. solfataricus* strains P1, P2, P3 and 98/2 are aligned where arrowheads denote spacers. Colour coding for spacer sequence matches to major *Sulfolobus* viruses: red – rudiviruses, orange – lipothrixviruses-, yellow – fuselloviruses, green – bicaudaviruses (using the same criteria as in (**a**)). Matches to conjugative plasmids (blue) are also included. CRISPR sequence regions shared between two or more strains are shaded light or dark grey, respectively

10.7.4 The Bicaudavirus SMV1; A Special CRISPR Scenario

SMV1 is a tailed fusiform virus that generates one or two tails extracellularly (Uldahl et al. 2016). It carries a large circular dsDNA genome and at least seven virion proteins. On infection with SMV1 growth retardation occurred in *Sulfolobus* species that was dependent on the strain and the MOI and it was concurrent with the onset of viral replication but no cell lysis was observed on virion release. In contrast, infection of CRISPR-minus mutants of two different *Sulfolobus* strains produced rapid growth retardation (Erdmann et al. 2013, 2014b; Uldahl et al. 2016).

The first active spacer acquisition from an archaeal virus was achieved after infecting *S. solfataricus* P2 with an environmental virus mixture dominated by the bicaudavirus SMV1. This resulted in hyperactive spacer acquisition at the leader end of the large CRISPR loci (Fig. 10.1a), and internally within a smaller locus E. Surprisingly, all the *de novo* spacers derived from a very minor component, the conjugative plasmid pMGB1 (SMV1:pMGB1 DNA weight ratio of about 3300:1 in cell extracts), that had copurified with the virus, possibly encapsulated by virion proteins. This result suggested that SMV1 was resistant to CRISPR-Cas interference (Erdmann and Garrett 2012; Erdmann et al. 2013). Earlier attempts to activate spacer acquisition in *S. solfataricus* P2, by infecting with different single viruses all failed (Erdmann and Garrett 2012), although success was achieved later with purified STSV2 (León-Sobrino et al. 2016).

10.7.5 Anti-CRISPR Proteins

The observed lack of spacer acquisition from SMV1 became more intriguing with the discovery that CRISPR loci of the host *S. solfataricus* P2 carried eight spacers perfectly matching SMV1 at protospacer sites carrying specific PAM motifs essential for spacer uptake (Erdmann and Garrett 2012). This was surprising, first because the host was isolated from Naples, Italy, while the virus was sampled in Yellowstone National Park, USA, suggesting that the viruses and/or hosts are more mobile geographically than previously imagined. More importantly, it indicated that SMV1 must carry an anti-CRISPR system against the host subtype I-A and/or subtype III-B immune systems.

This led to experiments designed to test the extent to which the different CRISPR-Cas immune systems were effective in inhibiting SMV1 DNA replication. *S. islandicus* strains were constructed that constitutively expressed antiviral immunity from either subtype I-A, or subtype III-B, or from both subtype I-A and III-B systems. Initially, there was strong inhibition of SMV1 replication, but later, whereas inhibition by the I-A system was gradually lost, the III-B system strongly limited both viral DNA replication and virion formation (Guo et al. 2019). The absence of escape mutants (Gudbergsdóttir et al. 2011) in the former construct reinforced that SMV1 carried anti-subtype I-A activity.

Although the anti-CRISPR agent has not yet been identified, a viral mutant SMV1v was isolated from *S. islandicus* REY15A-infected cells earlier that, unlike wild-type SMV1, was susceptible to spacer acquisition and it lacked a specific virion protein ORF114. This implicated the protein, at least indirectly, in anti-CRISPR activity (Erdmann et al. 2014b).

10.7.6 Induction of Cell Death

Infection of *Sulfolobus* species with the mutant SMV1v resulted in DNA loss and cell death in an increasing fraction of the cell population (Erdmann et al. 2014b). Subsequently, it was proposed that the conflict between the rapidly replicating SMV1 and subtype I-A CRISPR-Cas interference generates a signal mimicking the DNA damage stress response that induces DNA loss and cell death via a programmed series of cellular events (Han et al. 2017). According to this hypothesis, high intracellular levels of SMV1 virus activate programmed cell death in the host cells, thereby reducing virion production in the cellular population (Guo et al. 2019). A similar scenario is possible for the type III CRISPR-Cas system that can potentially cause degradation of both DNA and RNA at high virus levels. This hypothesis receives support from the observation that, at high SMV1v copy numbers, enhanced self-targeting of *S. islandicus* chromosomal DNA by the CRISPR-Cas systems occurred (Erdmann et al. 2014b) and, moreover, a similar loss of DNA and cell death was also induced in *S. islandicus* by high copy numbers of a conjugative plasmid pKEP9 (Liu et al. 2016).

10.7.7 Inter-viral Conflicts Exploiting Host CRISPR-Cas Systems

Experimental evidence indicated that while SMV1 is resistant to CRISPR-Cas adaptation in different *Sulfolobus* strains, it can induce a CRISPR-Cas response, both adaptation and interference, against coinfecting genetic elements, including the monocaudavirus STSV2, the rudivirus SIRV3 and the conjugative plasmid pMGB1, all three of which are then lost from the cellular population (Erdmann et al. 2013, 2014b; Papathanasiou et al. 2019). To explain this phenomenon, it was proposed first that CRISPR-Cas adaptation is triggered by viral DNA replication and, second, that SMV1 replicates very slowly after infection while the coinfecting elements replicate strongly such that the host cell exclusively targets the latter. The demonstration that SMV1 actively inhibits subtype I-A (but not subtype III-B) interference (Guo et al. 2019) would enable the coinfecting element to replicate partially uninhibited until the cellular stress levels became too high when either the CRISPR-Cas response dominates or programmed cell death is initiated (Han et al. 2017). It was concluded that the infecting SMV1 facilitates host cells in eliminating coinfecting viruses other than SMV1 (Erdmann et al. 2014b).

Recently, evidence for another type of interviral warfare was presented in which two *Sulfolobus* viruses were shown to carry mini-CRISPR loci carrying single virus matching spacers that could potentially exploit the host *cas* genes to target coinfecting viruses (Medvedeva et al. 2019). Such mechanisms for resolving conflicts between coinfecting genetic elements are likely to be relatively widespread. This was exemplified earlier for the *Sulfolobus* conjugative plasmid pKEF9 that carries a small CRISPR locus with spacers matching both the rudivirus SIRV1/2 and a fusellovirus SSV5, all of which originated from the same Icelandic hot spring environment (Lillestøl et al. 2006; Liu et al. 2016).

10.7.8 Transposable Element Conflicts

The ongoing conflict between SMV1 and its hosts is also likely to involve transposable elements. *Sulfolobus* hosts, bicaudaviruses, and conjugative plasmids, tend to be rich in active transposable elements that have been shown to be activated by viral infections (e.g. León-Sobrino et al. 2016) and during CRISPR-Cas adaptation (Erdmann et al. 2013). SMV1 carries one insertion sequence element (IS200/IS650) and two SMN2 MITEs (miniature inverted repeat transposable element) that can all potentially target and inactivate host *cas* genes (Erdmann et al. 2014b). Conversely, transposable elements of the host can target and inactivate viral and plasmid genes (Liu et al. 2016) and, moreover, they commonly disrupt and inactivate proviruses integrated at tRNA genes (Guo et al. 2011; You et al. 2011).

In conclusion, it was always puzzling that *S. solfataricus* P2 and *S. islandicus* species had been traditionally used as hosts for propagating newly isolated crenarchaeal viruses despite their both carrying type I and type III CRISPR-Cas immune systems (Zillig et al. 1998; Garrett et al. 2015). However, more recent insights show that viral DNA replication can, for some viruses at least, continue at a reduced level over long periods despite continuing activity of the immune system (e.g. Papathanasiou et al. 2019). As for SMV1, one explanation is that viruses produce anti CRISPR-Cas agents but it also appears that the immune systems often just maintain mutually beneficial reduced virus levels without elminating the virus, as exemplified by the partial effect of the active type III systems of *S. islandicus* on SMV1 levels and the lack of detectable viral escape mutants (Guo et al. 2019).

10.8 Evolutionary Perspectives

Morphologically, archaeal viruses are much more diverse than bacteriophages (Table 10.1). Several different archaea-specific viral morphotypes are not found amongst bacteriophages whereas all known phage morphotypes appear to be repesented amongst archaeal viruses (cosmopolitan viruses) (reviewed in Krupovic et al. 2018). These findings are especially remarkable given that relatively few archaeal viral species have currently been characterised, in contrast to the several thousand bacteriophage species. Moreover, this morphological diversity of the archaeal viruses is likely to increase with time as indicated by metagenomic sequencing analyses performed on samples extracted from extreme environments in different geographical locations (e.g. Sime-Ngando et al. 2011; Borrel et al. 2012).

It has been proposed that the exceptional morphological diversity of archaeal viruses may reflect that many of them derive from an ancient virosphere that occurred at the time of the last universal common ancestor (LUCA) (Prangishvili et al. 2006a; Pina et al. 2011). The inference being that whereas many ancient viral morphotypes infected and propagated in primitive archaea a limited fraction of these viral morphotypes, and in particular those of the shared viral morphotypes (cosmopolitan viruses), infected the evolving bacteria. This general hypothesis for the origin of the archaea-specific viruses does not preclude that some of these virus types may have evolved later from plasmids or other non-viral mobile genetic elements (e.g. Arnold et al. 1999; Erdmann et al. 2017).

This scenario also provides a possible explanation for the few homologous virus-specific genes/proteins that archaeal viruses share with bacteriophages or eukaryal viruses (Prangishvili et al. 2006b; Iranzo et al. 2016); they are also likely to have derived from a viral gene pool that existed in the ancient virosphere (Prangishvili 2013). In an attempt to rationalise the above findings it was proposed that bacteria may have evolved to counter infection by some virus types by developing complex cellular membranes carrying, for example, peptidoglycans and outer membrane capsids (Krupovic et al. 2018). In contrast, archaea, and especially crenarchaea, carry relatively simple membrane structures; many only exhibit a highly structured protein lattice (S-layer) with surface proteins that are modified and commonly N-glycosylated (Albers and Meyer 2011).

References

Albers SV, Meyer BH (2011) The archaeal cell envelope. Nat Rev Microbiol 9:414–426

Arnold HP, She Q, Phan H, Stedman K, Prangishvili D, Holz I, Kristjansson JK, Garrett R, Zillig W (1999) The genetic element pSSVx of the extremely thermophilic crenarchaeon *Sulfolobus* is a hybrid between a plasmid and a virus. Mol Microbiol 34:217–226

Atanasova NS, Roine E, Oren A, Bamford DH, Oksanen HM (2012) Global network of specific virus-host interactions in hypersaline environments. Environ Microbiol 14:426–440

Bettstetter M, Peng X, Garrett RA, Prangishvili D (2003) AFV1, a novel virus infecting hyperthermophilic archaea of the genus *Acidianus*. Virology 315:68–79

Bhoobalan-Chitty Y, Johansen TB, Di Cianni N, Peng X (2019) Inhibition of type III CRISPR-Cas immunity by an archaeal virus-encoded anti-CRISPR protein. Cell 179:448–458

Bize A, Karlsson EA, Ekefjard K, Quax TE, Pina M, Prevost MC, Forterre P, Tenaillon O, Bernander R, Prangishvili D (2009) A unique virus release mechanism in the Archaea. Proc Natl Acad Sci U S A 106:11306–11311

Bolduc B, Shaughnessy DP, Wolf YI, Koonin EV, Roberto FF, Young M (2012) Identification of novel positive-strand RNA viruses by metagenomic analysis of archaea-dominated Yellowstone hot springs. J Virol 86:5562–5573

Borrel G, Colombet J, Robin A, Lehours AC, Prangishvili D, Sime-Ngando T (2012) Unexpected and novel putative viruses in the sediments of a deep-dark permanently anoxic freshwater habitat. ISME J 6:2119–6127

Deng L, Kenchappa CS, Peng X, She Q, Garrett RA (2012) Modulation of CRISPR locus transcription by the repeat-binding protein Cbp1 in *Sulfolobus*. Nucl Acids Res 40:2470–2480

Deng L, He F, Bhoobalan-Chitty Y, Martinez-Alvarez L, Guo Y, Peng X (2014) Unveiling cell surface and type IV secretion proteins responsible for archaeal rudivirus entry. J Virol 88:10264–10268

Erdmann S, Garrett RA (2012) Selective and hyperactive uptake of foreign DNA by adaptive immune systems of an archaeon via two distinct mechanisms. Mol Microbiol 85:1044–1056. Corrigendum. Mol Microbiol 86:757

Erdmann S, Shah SA, Garrett RA (2013) SMV1 virus-induced CRISPR spacer acquisition from the conjugative plasmid pMGB1 in *Sulfolobus solfataricus* P2. Biochem Soc Trans 41:1449–1458

Erdmann S, Chen B, Huang X, Deng L, Liu C, Shah SA, Le Moine BS, León Sobrino C, Wang H, Wei Y, She Q, Garrett RA, Huang L, Lin L (2014a) A novel single-tailed fusiform *Sulfolobus* virus STSV2 infecting model *Sulfolobus* species. Extremophiles 18:51–60

Erdmann S, Le Moine BS, Garrett RA (2014b) Inter-viral conflicts that exploit host CRISPR immune systems of *Sulfolobus*. Mol Microbiol 91:900–917

Erdmann S, Tschitschko B, Zhong L, Raftery MJ, Cavicchioli R (2017) A plasmid from an Antarctic haloarchaeon uses specialized membrane vesicles to disseminate and infect plasmid-free cells. Nat Microbiol 2:1446–1455

Forterre P (2013) Why are there so many diverse replication machineries? J Mol Biol 425:4714–4726

Garrett RA (2017) In: Archaea and the tree of life. Nature Interviews. https://www.nature.com/collections/cvqkpjwsyd/interviews

Garrett RA, Vestergaard G, Shah SA (2011) Archaeal CRISPR-based immune systems: exchangeable functional modules. Trends Microbiol 19:549–556

Garrett RA, Shah SA, Erdmann S, Liu G, Mousaei M, León-Sobrino C, Peng W, Gudbergsdóttir S, Deng L, Vestergaard G, Peng X, She Q (2015) CRISPR-Cas adaptive immune systems of the Sulfolobales: unravelling their complexity and diversity. Life (Basel) 5:783–817

Gudbergsdóttir S, Deng L, Chen Z, Jensen JVK, Jensen LR, She Q, Garrett RA (2011) Dynamic properties of the *Sulfolobus* CRISPR/Cas and CRISPR/Cmr systems when challenged with vector-borne viral and plasmid genes and protospacers. Mol Microbiol 79:35–49

Guo L, Brügger K, Liu C, Shah SA, Zheng H, Zhu Y, Wang S, Lillestøl RK, Chen L, Frank J, Prangishvili D, Paulin L, She Q, Huang L, Garrett RA (2011) Genome analyses of Icelandic strains of *Sulfolobus islandicus*: model organisms for genetic and virus-host interaction studies. J Bacteriol 193:1672–1680

Guo Y, Kragelund BB, White MF, Peng X (2015) Functional characterization of a conserved archaeal viral operon revealing single-stranded DNA binding, annealing and nuclease activities. J Mol Biol 427:2179–2191

Guo T, Han W, She Q (2019) Tolerance of *Sulfolobus* SMV1 virus to the immunity of I-A and III-B CRISPR-Cas systems in *Sulfolobus islandicus*. RNA Biol 16:549–556

Han W, Xu Y, Feng X, Liang YX, Huang L, Shen Y, She Q (2017) NQO-induced DNA-less cell formation is associated with chromatin protein degradation and dependent on A0A1-ATPase in *Sulfolobus*. Front Microbiol 8:01480

Häring M, Vestergaard G, Rachel R, Chen L, Garrett RA, Prangishvili D (2005) Virology: independent virus development outside a host. Nature 436:1101–1102

He F, Bhoobalan-Chitty Y, Van LB, Kjeldsen AL, Dedola M, Makarova KS, Koonin EV, Brodersen DE, Peng X (2018) Anti-CRISPR proteins encoded by archaeal lytic viruses inhibit subtype I-D immunity. Nat Microbiol 3(4):461–469

Iranzo J, Krupovic M, Koonin EV (2016) The double-stranded DNA virosphere as a modular hierarchical network of gene sharing. MBio 7:e00978-16

Jaakkola ST, Penttinen RK, Vilén ST, Jalasvuori M, Rönnholm G, Bamford JK, Oksanen HM (2012) Closely related archaeal *Haloarcula hispanica* icosahedral viruses HHIV-2 and SH1 have nonhomologous genes encoding host recognition functions. J Virol 86:4734–4742

Janekovic D, Wunderl S, Holz I, Zillig W, Gierl A, Neumann H (1983) TTV1, TTV2 and TTV3, a family of viruses of the extremely thermophilic, anaerobic, sulfur reducing archaebacterium Thermoproteus tenax. Molec Gen Genet 192:39–45

Kandler O (1994) Cell wall biochemistry and three-domain concept of life. Syst Appl Microbiol 16:501–509

Kasson P, DiMaio F, Yu X, Lucas-Staat S, Krupovic M, Schouten S, Prangishvili D, Egelman EH (2017) Model for a novel membrane envelope in a filamentous hyperthermophilic virus. elife 6:e26268

Kessler A, Brinkman AB, van der Oost J, Prangishvili D (2004) Transcription of the rod-shaped viruses SIRV1 and SIRV2 of the hyperthermophilic archaeon *Sulfolobus*. J Bacteriol 186:7745–7753

Kim J-G, Kim S-J, Cvirkaite-Krupovic V, Yu W-J, Gwak J-H, López-Pérez M, Rodriguez-Valera F, Krupovic M, Cho J-C, Rhee S-K (2019) Spindle-shaped viruses infect marine ammonia-oxidizing thaumarchaea. Proc Natl Acad Sci U S A 116:15645–15650

Krupovic M, Cvirkaite V, Iranzo J, Prangishvili D, Koonin EV (2018) Viruses of archaea: structural, functional, environmental and evolutionary genomics. Virus Res 244:181–193

Langworthy TA, Pond JL (1986) Archaebacterial ether lipids and chemotaxonomy. Syst Appl Microbiol 7:253–257

Lillestøl RK, Redder P, Garrett RA, Brügger K (2006) A putative viral defence mechanism in archaeal cells. Archaea 2:59–72

Lillestøl RK, Shah SA, Brügger K, Redder P, Phan H, Christiansen J, Garrett RA (2009) CRISPR families of the crenarchaeal genus *Sulfolobus*: bidirectional transcription and dynamic properties. Mol Microbiol 72:259–272

León-Sobrino C, Kot WP, Garrett RA (2016) Transcriptome changes in STSV2-infected REY15A undergoing continuous CRISPR spacer acquisition. Mol Microbiol 99:719–728

Liu G, She Q, Garrett RA (2016) Diverse CRISPR-Cas responses and dramatic cellular DNA changes and cell death in pKEF9-conjugated *Sulfolobus* species. Nucl Acids Res 44:4233–4242

Lykke-Andersen J, Garrett RA (1997) RNA-protein interactions of an archaeal homotetrameric splicing endonuclease with an exceptional evolutionary history. EMBO J 16:6290–6300

Manica A, Zebec Z, Steinkellner J, Schleper C (2013) Unexpectedly broad target recognition of the CRISPR-mediated virus defence system in the archaeon *Sulfolobus solfataricus*. Nucl Acids Res 41:10509–10517

Martin A, Yeats S, Janekovic D, Reiter W-D, Aicher W, Zillig W (1984) SAV1, a temperate u.v.-inducible DNA virus-like particle from the archaebacterium *Sulfolobus acidocaldarius* isolate B12. EMBO J 3:2165–2168

Martinez-Alvarez L, Bell SD, Peng X (2016) Multiple consecutive initiation of replication producing novel brush-like intermediates at the termini of linear viral dsDNA genomes with hairpin ends. Nucl Acids Res 44:8799–8809

Martinez-Alvarez L, Deng L, Peng X (2017) Formation of a viral replication focus in *Sulfolobus* cells infected by the rudivirus *Sulfolobus islandicus* rod-shaped virus 2. J Virol 91:e00486–e00417

Medvedeva S, Liu Y, Koonin EV, Severinov K, Prangishvili D, Krupovic M (2019) Virus-borne mini-CRISPR arrays are involved in interviral conflicts. Nature Comm 10:5204

Meile L, Abendschein P, Leisinger T (1990) Transduction in the archaebacterium *Methanobacterium thermoautotrophicum* Marburg. J Bacteriol 172:3507–3508

Mochizuki T, Sako Y, Prangishvili D (2011) Provirus induction in hyperthermophilic archaea: characterization of *Aeropyrum pernix* spindle-shaped virus 1 and *Aeropyrum pernix* ovoid virus 1. J Bacteriol 193:5412–5419

Mochizuki T, Krupovic M, Pehau-Arnaudet G, Sako Y, Forterre P, Prangishvili D (2012) Archaeal virus with exceptional virion architecture and the largest single-stranded DNA genome. Proc Natl Acad Sci U S A 109:13386–13391

Mojica FJM, Garrett RA (2013) Discovery and seminal developments in the CRISPR field. In: Barrangou R, Van der Oost J (eds) CRISPR-Cas systems. Springer Press, Berlin Heidelberg, pp 1–31

Mojica FJM, Díez-Villaseñor C, García-Martínez J, Soria E (2005) Intervening sequences of regularly spaced prokaryotic repeats derive from foreign genetic elements. J Mol Evol 60:174–182

Mousaei M, Deng L, She Q, Garrett RA (2016) Major and minor crRNA annealing sites facilitate low stringency DNA protospacer binding prior to type I-A CRISPR-Cas interference in *Sulfolobus*. RNA Biol 13:1166–1173

Muskhelishvili G (1994) The archaeal SSV integrase promotes intermolecular excisive recombination in vitro. Syst Appl Microbiol 16:605–608

Novoa RR, Calderita G, Arranz R, Fontana J, Granzow H, Risco C (2005) Virus factories: associations of cell organelles for viral replication and morphogenesis. Biol Cell 97:147–172

Okutan E, Deng L, Mirlashari S, Uldahl K, Halim M, Liu C, Garrett RA, She Q, Peng X (2013) Novel insights into gene regulation of the rudivirus SIRV2 infecting *Sulfolobus* cells. RNA Biol 10:1–11

Papathanasiou P, Erdmann S, Leon-Sobrino C, Sharma K, Urlaub H, Garrett RA, Peng X (2019) Stable maintenance of the rudivirus SIRV3 in a carrier state in *Sulfolobus islandicus* despite activation of the CRISPR-Cas immune response by a second virus SMV1. RNA Biol 16:557–565

Peng X, Blum H, She Q, Mallok S, Brügger K, Garrett RA, Zillig W, Prangishvili D (2001) Sequences and replication of genomes of the archaeal rudiviruses SIRV1 and SIRV2: relationships to the archaeal lipothrixvirus SIFV and some eukaryal viruses. Virology 291:226–234

Peng X, Brügger K, Shen B, Chen L, She Q, Garrett RA (2003) Genus-specific protein binding to the large clusters of DNA repeats (short regularly spaced repeats) present in *Sulfolobus* genomes. J Bacteriol 185:2410–2417

Peng X, Kessler K, Phan H, Garrett RA, Prangishvili D (2004) Multiple variants of the archaeal DNA rudivirus SIRV1 in a single host and a novel mechanism of genomic variation. Mol Microbiol 54:366–375

Peng W, Feng M, Feng X, Liang YX, She Q (2014) An archaeal CRISPR type III-B system exhibiting distinctive RNA targeting features and mediating dual RNA and DNA interference. Nucleic Acids Res 43:406–417

Pina M, Bize A, Forterre P, Prangishvili D (2011) The archeoviruses. FEMS Microbiol Rev 35:1035–1054

Pourcel C, Salvignol G, Vergnaud G (2005) CRISPR elements in *Yersinia pestis* acquire new repeats by preferential uptake of bacteriophage DNA, and provide additional tools for evolutionary studies. Microbiology 15:653–663

Prangishvili D (2013) The wonderful world of archaeal viruses. Ann Rev Microbiol 67:565–585

Prangishvili D, Garrett RA (2005) Viruses of hyperthermophilic crenarchaea. Trends Microbiol 13:535–542

Prangishvili D, Arnold HP, Gotz D, Ziese U, Holz I, Kristjansson JK, Zillig W (1999) A novel virus family, the Rudiviridae: structure, virus-host interactions and genome variability of the *Sulfolobus* viruses SIRV1 and SIRV2. Genetics 152:1387–1396

Prangishvili D, Forterre P, Garrett RA (2006a) Viruses of the Archaea: a unifying view. Nat Rev Microbiol 4:837–848

Prangishvili D, Garrett RA, Koonin EV (2006b) Evolutionary genomics of archaeal viruses: unique viral genomes in the third domain of life. Virus Res 117:52–67

Prangishvili D, Vestergaard G, Häring M, Aramayo R, Basta T, Rachel R, Garrett RA (2006c) Structural and genomic properties of the hyperthermophilic archaeal virus ATV with an extracellular stage of the reproductive cycle. J Mol Biol 359:1203–1216

Prangishvili D, Bamford DH, Forterre P, Iranzo J, Koonin EV, Krupovic M (2017) The enigmatic archaeal virosphere. Nat Rev Microbiol 15:724–739

Pühler G, Leffers H, Gropp F, Palm P, Klenk HP, Lottspeich F, Garrett RA, Zillig W (1989) Archaebacterial DNA-dependent RNA polymerases testify to the evolution of the eukaryotic nuclear genome. Proc Natl Acad Sci 86:4569–4573

Quax TE, Krupovic M, Lucas S, Forterre P, Prangishvili D (2010) The *Sulfolobus* rod-shaped virus 2 encodes a prominent structural component of the unique virion release system in Archaea. Virology 404:1–4

Quax TE, Voet M, Sismeiro O, Dillies M-A, Jagla B, Coppée JY, Sezonov G, Forterre P, van der Oost J, Lavigne R, Prangishvili D (2013) Massive activation of archaeal defense genes during viral infection. J Virol 87:8419–8428

Quemin ER, Lucas S, Daum B, Quax TE, Kühlbrandt W, Forterre P, Albers SV, Prangishvili D, Krupovic M (2013) First insights into the entry process of hyperthermophilic archaeal viruses. J Virol 87:13379–13385

Samson RY, Obita T, Hodgson B, Shaw MK, Chong PL, Williams RL, Bell SD (2008) Molecular and structural basis of ESCRT-III recruitment to membrnes during archaeal cell division. Mol Cell 41:186–196

Scheele U, Erdmann E, Ungewickell EJ, Felisberto-Rodrigues C, Ortiz-Lombardía M, Garrett RA (2011) Chaperone role for proteins p618 and p892 in the extra-cellular tail development of *Acidianus* two-tailed virus ATV. J Virol 85:4812–4821

Schleper C, Kubo K, Zillig W (1992) The particle SSV1 from the extremely thermophilic archaeon *Sulfolobus* is a virus: demonstration of infectivity and of transfection with viral DNA. Proc Natl Acad Sci 89:7645–7649

Schnabel H, Zillig W, Pfäffle M, Schnabel R, Michel H, Delius H (1982) *Halobacterium halobium* phage øH. EMBO J 1:87–92

Shah SA, Hansen NR, Garrett RA (2009) Distributions of CRISPR spacer matches in viruses and plasmids of crenarchaeal acidothermophiles and implications for their inhibitory mechanism. Biochem Soc Trans 37:23–28

Shah SA, Erdmann S, Mojica FJM, Garrett RA (2013) Protospacer recognition motifs mixed identities and functional diversity. RNA Biol 10:1–9

She Q, Peng X, Zillig W, Garrett RA (2001a) Genome evolution: gene capture events in archaeal chromosomes. Nature 409:478

She Q, Singh RK, Confalonieri F, Zivanovic Y, Gordon P, Allard G, Awayez MJ, Chan-Weiher C-Y, Clausen IG, Curtis B, De Moors A, Erauso G, Fletcher C, Gordon PMK, Heikamp de Jong I, Jeffries A, Kozera CJ, Medina N, Peng X, Phan Thi-Ngoc H, Redder P, Schenk ME, Theriault C, Tolstrup N, Charlebois RLM, Doolittle WF, Duguet M, Gaasterland T, Garrett RA, Ragan M, Sensen CW, Van der Oost J (2001b) The complete genome of the crenarchaeon *Sulfolobus solfataricus* P2. Proc Natl Acad Sci U S A 98:7835–7840

Sime-Ngando T, Lucas S, Robin A, Tucker KP, Colombet J, Bettarel Y, Desmond E, Gribaldo S, Forterre P, Breitbart M, Prangishvili D (2011) Diversity of virus-host systems in hypersaline Lake Retba, Senegal. Environ Microbiol 13:1956–1972

Snyder JC, Brumfield SK, Peng N, She Q, Young M (2011) *Sulfolobus* turreted icosahedral virus c92 protein responsible for the formation of pyramid-like cellular lysis structures. J Virol 85:6287–6292

Tang TH, Bachellerie JP, Rozhdestvensky T, Bortolin ML, Huber H, Drungowski M, Elge T, Brosius J, Hüttenhofer A (2002) Identification of 86 candidates for small non-messenger RNAs from the archaeon *Archaeoglobus fulgidus*. Proc Natl Acad Sci U S A 99:7536–7541

Uldahl KB, Jensen SB, Bhoobalan-Chitty Y, Martínez-Álvarez L, Papathanasiou P, Peng X (2016) Life cycle characterization of *Sulfolobus* monocaudavirus 1, an extremophilic spindle-shaped virus with extracellular tail development. J Virol 90:5693–5699

Vestergaard G, Garrett RA, Shah SA (2014) CRISPR adaptive immune systems of Archaea. RNA Biology 11(2):156–167

Vestergaard G, Aramayo R, Basta T, Häring M, Peng X, Brügger K, Chen L, Rachel R, Boisset N, Garrett RA, Prangishvili D (2008) Structure of the *Acidianus* filamentous virus 3 and comparative genomics of related archaeal lipothrixviruses. J Virol 82:371–381

Wang H, Peng N, Shah SA, Huang L, She Q (2015) Archaeal extrachromosomal genetic elements. Microbiol Mol Biol Rev 79:117–152

Woese CR, Fox GE (1977) Phylogenetic structure of the prokaryotic domain: the primary kingdoms. Proc Natl Acad Sci U S A 74:5088–5090

Xiang X, Chen L, Huang X, Luo Y, She Q, Huang L (2005) Sulfolobus tengchongensis spindle-shaped virus STSV1: Virus-host interactions and genomic features. J Virol 79:8677–8686

You X-Y, Liu C, Wang S-Y, Jiang C-Y, Shah SA, Prangishvili D, She Q, Liu S-J, Garrett RA (2011) Genomic studies of *Acidianus hospitalis* W1 a host for studying crenarchaeal virus and plasmid life cycles. Extremophiles 15:487–497

Zhang J, Rouillon C, Kerou M, Reeks J, Brügger K, Graham S, Reimann J, Cannone G, Liu H, Albers SV, Naismith JH, Spagnolo L, White MF (2012) Structure and mechanism of the CMR complex for CRISPR-mediated antiviral immunity. Mol Cell 45:303–313

Zillig W, Kletzin A, Schleper C, Holz I, Janekovic D, Hain J, Lanzendörfer M, Kristjansson JK (1994) Screening for Sulfolobales, their plasmids and their viruses in Icelandic solfataras. Syst Appl Microbiol 16:609–628

Zillig W, Prangishvilli D, Schleper C, Elferink M, Holz I, Albers S, Janekovic D, Götz D (1996) Viruses, plasmids and other genetic elements of thermophilic and hyperthermophilic Archaea. FEMS Microbiol Rev 18:225–236

Zillig W, Arnold HP, Holz I, Prangishvili D, Schweier A, Stedman K, She Q, Phan H, Garrett R, Kristjansson JK (1998) Genetic elements in the extremely thermophilic archaeon *Sulfolobus*. Extremophiles 2:131–140

Chapter 11
Filamentous Phages Affect Virulence of the Phytopathogen *Ralstonia solanacearum*

Yuichi Tasaka, Takeru Kawasaki, and Takashi Yamada

Abstract ϕRSS-type filamentous phages are frequently found integrated in *Ralstonia solanacearum* genomes and affect host virulence after infection. ϕRSS1, a known virulence-enhancing phage, was found to be a truncated form of a larger phage (designated as ϕRSS0; 7288 nt) integrated in strain C319. A 626-nt ϕRSS0 sequence missing in ϕRSS1 DNA contains a nucleotide sequence element *att*P, corresponding to *dif* of *R. solanacearum*. Thus, ϕRSS0 was integrated at a *dif* site, similarly to CTXϕ of *Vibrio cholerae*, which uses the host XerC/D recombination system. ϕRSS0 could integrate into both the chromosome and megaplasmid of the host genome. The extra region of ϕRSS0 also contained an open reading frame (ORF13) of 156 amino acids with sequence similarity to DNA-binding phage regulators. The ϕRSS0-*att*P is located within the ORF13-coding region; therefore, integration results in a truncation of the C-terminus of ORF13. ORF13 may function as a phage repressor for immunity, because strain C319 (a ϕRSS0 lysogen) is resistant to second infection by ϕRSS0. C319 is susceptible to ϕRSS1, thus ϕRSS1 (without ORF13) seems to be an escaped superinfective phage derived from ϕRSS0. The diversity and dynamic rearrangements of ϕRSS-type phages/prophages in *R. solanacearum* and their effects on host virulence are discussed.

11.1 Introduction

Ralstonia solanacearum is a gram-negative β-proteobacterium that causes bacterial wilt disease in many important crops. Its wide geographic distribution and unusually broad host range (more than 50 plant families) mean it is responsible for significant crop losses worldwide (Hayward 2000). Once the bacteria enter a susceptible host, they colonize the intercellular spaces of the root cortex and vascular

Y. Tasaka · T. Kawasaki · T. Yamada (✉)
Department of Molecular Biotechnology, Graduate School of Advanced Sciences of Matter, Hiroshima University, Higashi-Hiroshima, Japan
e-mail: tayamad@hiroshima-u.ac.jp

G. Witzany (ed.), *Biocommunication of Phages*,
https://doi.org/10.1007/978-3-030-45885-0_11

parenchyma. The bacteria eventually enter the xylem and spread into the upper parts of the plant, causing wilt (Kang et al. 2002; Vasse et al. 2000; Yao and Allen 2007). The development of bacterial wilt disease depends on the pathogenicity and virulence of the pathogen (Carney and Denny 1990; Denny 2006).

Recently, Yamada et al. (2007) isolated and characterized several different kinds of phage that specifically infect *R. solanacearum* strains belonging to different races and/or biovars. One of them, ϕRSS1, was characterized as an Ff-like phage (Inovirus), based on its particle morphology, genomic ssDNA and infection cycle. It infected 10 of 18 strains tested and gave relatively small and turbid plaques on assay plates. ϕRSS1 particles have a flexible filamentous shape of 1100 ± 100 nm in length and 10 ± 0.5 nm in width, giving a morphology resembling coliphage fd (Buchen-Osmond 2003; ICTVdB). Infection with ϕRSS1 phage does not cause host cell lysis, but establishes a persistent association between the host and the phage, releasing phage particles from growing and dividing host cells (Yamada 2012). Kawasaki et al. (2007) characterized the genome organization of ϕRSS1, which has 6662 bases of ssDNA. There are 11 open reading frames (ORFs) located on the same strand. A database survey revealed ϕRSS1 ORFs with significant sequence similarity to Ff-like phage proteins, such as ORF2 (pII), ORF4 (pVIII), ORF7 (pIII), ORF8 (pVI) and ORF9 (pI). Interestingly, most *R. solanacearum* strains tested contained ϕRSS1-related sequences in their genomes, suggesting a temperate phage nature of ϕRSS1. However, the mechanism of ϕRSS1's integration into the host genome was unclear. Comparison of the nucleotide sequence between ϕRSS1 DNA and its related prophage sequence in strain C319 revealed only a one base (A) overlap between the two phage-chromosome junctions (Kawasaki et al. 2007). This structure cannot be explained by any known mechanisms for temperate filamentous phages (Askora et al. 2011; Askora and Yamada 2015). There is no ORF showing homology with known recombinases or integrases on the ϕRSS1 genome. This raises the question of how ϕRSS1-like phages integrate into the host genome. It is also noteworthy that ϕRSS1 infection enhanced the virulence of strain C319 on tobacco (Yamada et al. 2007), and strains MAFF 106603 and MAFF 106611 on tomato (Addy et al. 2012b). Integration of ϕRSS1-related sequences was detected frequently in many field-isolated strains in Japan (Yamada et al. 2007). Therefore, we aimed to determine the effects of ϕRSS1-like phages on the host pathogenicity or virulence after infection and integration into the host genome.

Other filamentous Ff-like phages that have a lysogenic cycle include *Xanthomonas campestris* phage Cf1c (Kuo et al. 1991), Cf1t (Kuo et al. 1987a, b), Cf16v1 (Dai et al. 1980), and ϕLf (Lin et al. 2000), *Xanthomonas citri* phage XacF1 (Ahmad et al. 2014, 2017), *Xylella fastidiosa* phage Xfϕ f1 (Simpson et al. 2000), *Yersinia pestis* phage CUSϕ-2 (Gonzalez et al. 2002), and *Vibrio cholerae* phage VGJϕ and CTXϕ (Campos et al. 2003; Huber and Waldor 2002). The host bacteria of these phages are pathogenic to plants or animals, and are frequently involved in pathogenesis. The pathogenicity of *V. cholerae* depends on two key virulence factors: the toxin co-regulated pilus (TCP) and cholera toxin. The toxin genes are encoded on filamentous phage CTXϕ and are introduced into the bacterial cells by phage infection/integration mediated by the host *dif*/XerCD recombinase system (Huber and Waldor 2002; Davis and Waldor 2003). However, genomic integration of filamentous phages in other bacteria is mediated by diverse mechanisms, including via phage-encoded transposases and phage-encoded integrases (both tyrosine-recombinase and serine recombinase) (Askora et al. 2011; Askora and Yamada 2015)). The integration mechanisms of filamentous phages in phytopathogenic bacteria are largely unknown.

It was revealed that the ϕRSS1 structure was a truncated form derived from its parent phage (ϕRSS0) integrated in strain C319. ϕRSS0 was integrated into the host genome by the *dif*/XerCD recombination system. ϕRSS-type phages integrated at *dif* sites of both the host chromosome and megaplasmid. Diversity and dynamic rearrangements observed in ϕRSS-type phage/prophage sequences in *R. solanacearum* showed different effects on host virulence.

11.2 Integration Mechanism of ϕRSS-Like Phages

11.2.1 Identification of the Integration Site for a ϕRSS1-Like Phage in Strain C319

In our previous work, a ϕRSS1-like prophage sequence was detected in strain C319 and both right and left phage-chromosome junction sequences were determined. The ϕRSS1 sequence at the integration site corresponded to position 6629 in the IG region (large intergenic region), 34 nt upstream from ORF1, and there was only one overlapping base (A) between the two junctions (Kawasaki et al. 2007). To determine the biological effects of the ϕRSS1-like integrated sequence on the host physiology, we tried to remove it from the lysogenic strain C319 by double crossing-over using the junction sequences. A plasmid (pRSJ1) containing an approx. 500-bp left junction (LJ) and an approx. 500-bp right junction (RJ) (Fig. 11.1A) was introduced into C319 cells and Km-resistant recombinants were analyzed. pRSJ1 cannot replicate in *R. solanacearum* cells; therefore, only integrants can grow on selection plates. All 20 transformants analyzed were not ϕRSS1-sequence disruptants but had the plasmid integrated at the same single site of the host chromosome, as revealed by Southern blot hybridization with ϕRSS1 and pBluescriptII SK+ DNA as probes. Nucleotide sequences determined around the plasmid integration site were amplified by inverse PCR using a primer set of 5′- AAT CTC GTG ATG GCA GGT TGG GCG TCG CTT (forward) and 5′-TAT GAA AGG TTG GGC TTC GGA ATC GTT TTC (reverse), corresponding to Km^R coding regions of the vector, revealed duplication of an approx. 60 bp sequence at both plasmid-chromosome junctions, containing two sequence elements of 19 bp, α and β (Fig. 11.1B). These results indicated that the plasmid pRSJ1 integrated into the C319 chromosome at a site adjacent to the ϕRSS1-like prophage that contained sequence elements α and β. There were a few base changes in both α and β between LJ and RJ (Fig. 11.1B).

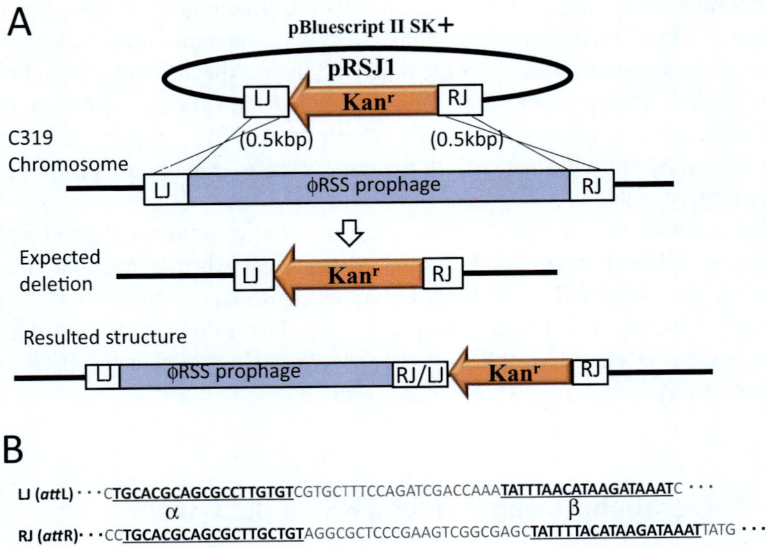

Fig. 11.1 Integration of pRSJ1 into the C319 chromosome and detection of repeated sequences at the integration junctions. (**A**) pRSJ1 containing a 500-bp sequence of the left junction (LJ) and a 500-bp sequence of the right junction (RJ) of ϕRSS1-like prophage in C319 (Kawasaki et al. 2007) was introduced into strain C319 to delete ϕRSS-prophage by double crossing over. However, pRSJ1 was integrated into the C319 chromosome adjacent to the ϕRSS-prophage. (**B**) By inverse PCR, repeated sequences were found on both plasmid-chromosome junctions containing two 19bp sequence elements (α and β)

11.2.2 Detection of ϕRSS0 Phage, the Prototype Phage of ϕRSS1, Excised from Strain C319

Meanwhile, we occasionally detected small turbid plaques on phage-screening plates where ϕRSL1-infected C319 cells were spread. ϕRSL1 is a jumbo phage with a large dsDNA genome of 240 kb and usually generates small clear plaques (Yamada et al. 2010). A phage was propagated from such a single turbid plaque with strain MAFF 106603 as a host, and its genomic DNA was sequenced. The resulting sequence revealed that this phage (designated ϕRSS0) was very similar to ϕRSS1, but contained an extra sequence of 626 nt at a ϕRSS1 position between 6628 and 6629 next to the IG region, giving an entire genomic size of 7288 nt (Yamada 2012, 2013). Within the ϕRSS0 extra region, we found sequence elements α and β (corresponding to the C319-RJ) and an ORF (ORF13) of 468 nt corresponding to 156 amino acid residues in a reversed orientation compared with the other ORFs (Fig. 11.2A, B). The amino acid sequence of ORF13 showed similarity to DNA-binding phage regulators (accession no. B5SCX5, E-value = $1e^{-29}$). The sequence elements α and β were located within the ORF13 coding region (Yamada 2013).

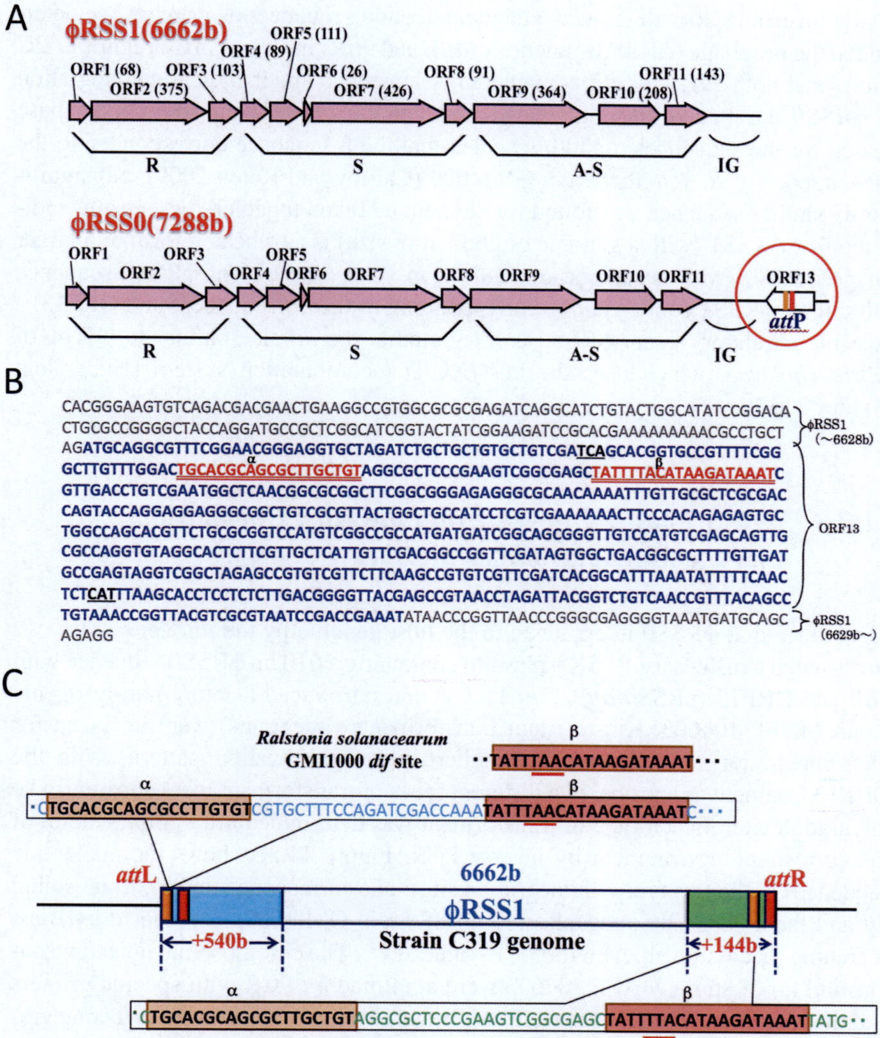

Fig. 11.2 Characterization of ϕRSS0 induced from lysogenic strain C319. (**A**) Comparison of gene arrangements between ϕRSS1 and ϕRSS0. Arrows oriented in the direction of transcription represent ORFs or genes. The functional modules for replication (R), structure (S), and assembly-secretion (A–S) are indicated. ORF sizes (in amino acids) are in parentheses. An extra region found in ϕRSS0 is circled. The α and β elements (putative *att*P) are shown as vertical bars within ORF13 of ϕRSS0. (**B**) Nucleotide sequence of a 626-bp extra region in ϕRSS0 containing ORF13 (reversed orientation) with both 5′- and 3′- flanking ϕRSS1 regions. Both α and β elements of possible *att*P and the initiation and termination codons of ORF13 (reversed orientation) are underlined. (**C**) Prophage junctions (*att*L and *att*R) in strain C319. There are two elements, α and β, in both left and right junctions. The β element corresponds to the core *dif* sequence of *R. solanacearum* GMI1000 chromosome. (Carnoy and Roten 2009)

Using inverse PCR with the new phage nucleotide sequences as primers, we determined the prophage (ϕRSS0)-junctions (*att*L and *att*R) in strain C319. Figure 11.2C shows that both *att*L and *att*R contained the elements α and β, indicating integration of ϕRSS0 at a specific site containing α and β on the C319 chromosome. A database search for the sequences α and β revealed that the β sequence corresponded to the *dif* sequence of *R. solanacearum* GMI1000 (Carnoy and Roten 2009). No significantly similar sequence was found for element α. Taken together, these results indicated that ϕRSS1 (with a genome of 6662 nt in size) is a truncated form of a larger phage ϕRSS0 (with a genome of 7288 nt in size). The 626 nt ϕRSS0 sequence missing in ϕRSS1 contains *att*P (corresponding to the *dif* sequence) and ORF13, a possible regulatory gene. ϕRSS0 is integrated at the *dif* site, similar to CTXϕ of *vibrio cholerae*, which uses the host XC/D recombination system (Huber and Waldor 2002).

11.2.3 *ϕRSS Phages Integrate into the Host Chromosome and the Megaplasmid at* **dif** *Sites*

To confirm that ϕRSS0 integrates into the host genome by the *dif*/Xer system, we constructed a pBluescriptII SK+ plasmid containing a 610 bp ϕRSS0 sequence with *att*P and ORF13 (pRSS0-*att*P, Fig. 11.3A) and introduced it into a non-lysogenic strain MAFF 106603. Km-resistant transformants (integrants) were analyzed for their integration sites. Based on Southern blot hybridization patterns with the ORF13 fragment as a probe, two different types of transformant were apparent. The integration site of each type of transformant was determined after amplification of the corresponding fragment by inverse PCR. Figure 11.3B shows the nucleotide sequences of the two types of integration sites: one corresponds to the chromosomal *dif* and the other to the megaplasmid *dif* of strain GMI1000, as revealed by Blast searching (accession no. AL646053, E-value $2e^{-24}$). The chromosomal *dif* and megaplasmid *dif* of strain MAFF 106603 were amplified by PCR with specific primers and their nucleotide sequences were confirmed to exactly correspond to the integration sites for pRSS0-*att*P. It is notable that the megaplasmid *dif* site lacks the α element of the ϕRSS0 att*P* region. ϕRSS0 integration at the chromosome DNA occurred at a higher frequency (two-fold) than at the megaplasmid DNA. These results strongly suggested that ϕRSS0 integrates into the host chromosome and megaplasmid by the *dif*/Xer system.

In the case of CTXϕ integration, mediated by XerC/D in *V. cholerae*, the direct integration of the single-stranded phage DNA into the double-stranded bacterial genome occurs (Das et al. 2010). There are two *att*P sites (*att*P1 and *att*P2) on the CTXϕ genomic DNA forming a secondary fork and stem structure to serve as a XerC/D substrate. This mechanism renders the CTXϕ integration process irreversible (Das et al. 2010). However, as described in this study, the ϕRSS0 phage was excised from lysogenic strain C319 after infection with ϕRSL1. ϕRSL1 has a

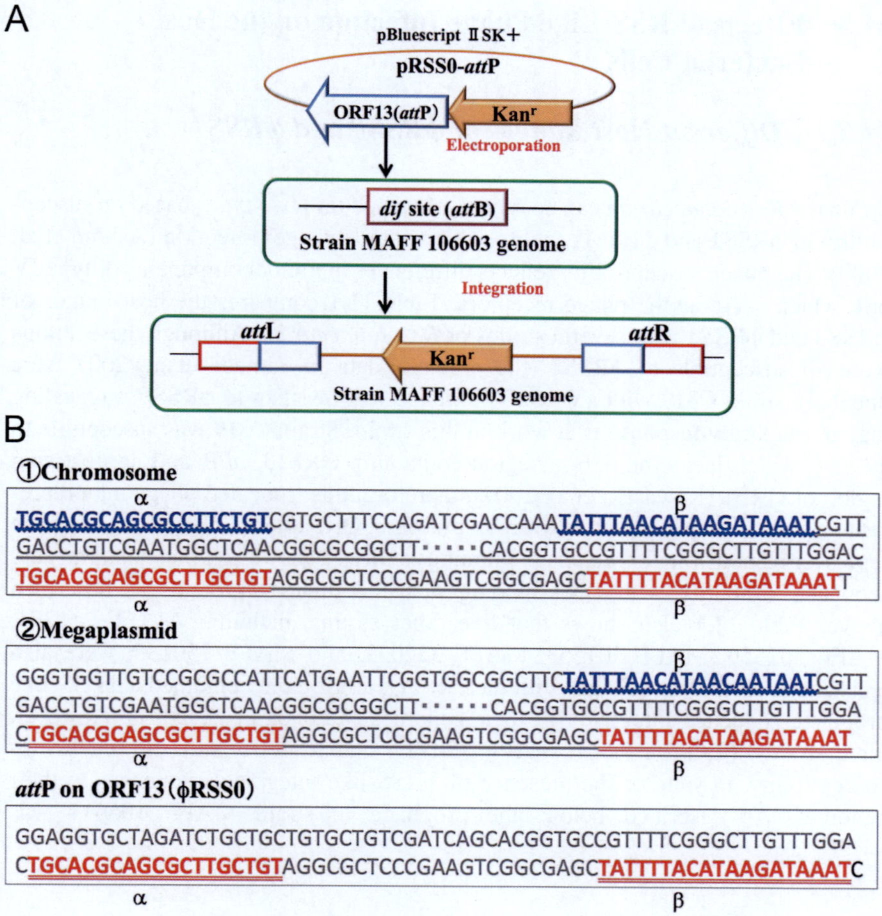

Fig. 11.3 Integration of an *att*P plasmid (pRSS0-*att*P) into host genomes. (**A**) After electroporation, Km^R transformants of MAFF 106603 were analyzed for inegration sites. (**B**) Two types of integration site in C319 for pRSS0-*att*P. One site corresponds to the chromosomal *dif*; bold letters underlined show the α and β elements on both junctions. Integrated pRSS0-*att*P sequence is underlined. The other site corresponds to the megaplasmid *dif*; bold letters underlined shows the α and β elements on both junctions. The α and β elements on the *att*P in ϕRSS0 are also shown (underlines)

relatively long infection cycle, with an eclipse phase of 90 min and a latent period of ~150 min (Yamada et al. 2010). During ϕRSL1 infection, activated XerC/D might have recombined LJ and RJ of the ϕRSS0 prophage and released a ds circular form of ϕRSS0 DNA in a reverse reaction. By contrast to CTXϕ, ϕRSS0 genome contains a single copy of the *att*P sequence; thus, recombination by XerC/D in integration may take place between *att*P on the dsDNA replicative form of the phage and *att*B (*dif*) on the host genome. As demonstrated for the integration/excision of VGJϕ, another filamentous phage of *v. cholerae* (Das et al. 2011), this process is reversible and explains the observations described in this work.

11.3 Effects of RSS-Like Phage Infection on the Host Bacterial Cells

11.3.1 Different Host Ranges of ϕRSS0 and ϕRSS1

Strains of *R. solanacearum* can be roughly grouped into two types based on susceptibility to ϕRSS1 and ϕRSM1 (another filamentous phage) infection (Askora et al. 2009). The phage susceptibility reflects differences in minor components of type IV pili, which serve as the phage receptors. Table 11.1 compares the host ranges of ϕRSS0 and ϕRSS1 with several strains of *R. solanacearum*. Although these strains were all susceptible to ϕRSS1 (the previous data in Yamada et al. 2007 were updated), strain C319 with a ϕRSS0 prophage was resistant to ϕRSS0, suggesting that an immunity response is at work in this strain. Strain C319 was susceptible to ϕRSS1, which lacks the 626-nt region containing ORF13, *att*P, and an upstream region of ORF1 (Kawasaki et al. 2007); therefore, this truncated phage may represent an escaped super-infective form. In this case, ORF13 possibly encodes a phage repressor responsible for the phage immunity: ORF13 shows significant amino acid sequence similarity with DNA-binding phage regulatory proteins as described above. Table 11.1 also shows that five other strains, including MAFF 211271, MAFF 301556, MAFF 301558, MAFF 327032, and MAFF 730135, were also resistant to ϕRSS0, consistent with the fact that all of them contained a ϕRSS-like prophage sequence integrated in their genomes (Yamada et al. 2007). However, three strains, MAFF 106611, MAFF 730139, and RS1002, still showed RSS0-susceptibility, in spite of the presence of ϕRSS-like integrated sequences in their genomes. As described below, the prophage in strain MAFF 106611 was

Table 11.1 *Ralstonia solanacearum* strains used in this study

Strain[a]	Race	Biovar	Phylotype	Sensitivity to phages[b]		Source
				ϕ RSS1	ϕ RSS0	
C319	1	-	I	S	R	Furuya et al.
MAFF 106603	1	3	I	S	S	NIAS[c]
MAFF 106611	1	4	I	S	S	NIAS
MAFF 211271	3	N2	IV	S	R	NIAS
MAFF 301556	1	4	I	S	R	NIAS
MAFF 301558	3	N2	IV	S	R	NIAS
MAFF 327032	1	4	I	S	R	NIAS
MAFF 730135	1	4	I	S	R	NIAS
MAFF 730139	1	4	I	S	S	NIAS
RS1002	1	4	I	S	S	Mukaihara et al.

[a]All strains originated in Japan
[b]Sensitivity to phages: *S* sensitive, *R* resistant
[c]National Institute of Agrobiological Sciences, Japan

considerably different from ϕRSS0. It is not known whether ϕRSS-related sequences in strains MAFF 730139 and RS1002 also contain many mutations, or have even lost ORF13.

11.3.2 Host Virulence Is Affected Differently by ϕRSS0 or ϕRSS1 Infection

Upon infection by the ϕRSS1 phage, the host *R. solanacearum* cells showed several abnormal behaviors, including less turbidity and frequent aggregation in the liquid culture, less coloration of colonies on plates, and a decreased growth rate (approx. 60% of the normal rate). More interestingly, ϕRSS1-infected cells showed enhanced virulence on tobacco plants (Yamada et al. 2007) and tomato plants (Addy et al. 2012b). Surface-associated ϕRSS1 particles (or phage proteins) may change the surface nature (hydrophobicity) of host cells to generate a high local cell density, resulting in early activation of *phcA*, the global virulence regulator (Addy et al. 2012b). These possible changes were also examined for ϕRSS0-infected MAFF106603 cells. The physiological features of ϕRSS0-infected cells were almost the same as ϕRSS1-infected MAFF 106603 cells, except that the ϕRSS0-infected cells formed colonies of more mucoid appearance on CPG plates (data not shown).

When 1 μl of cell suspension containing 10^5 colony forming units (CFU) of MAFF 106603 (wild-type) was inoculated into the major stem of tomato plants, all 12 plants showed wilting symptoms as early as 3 days post-inoculation (p.i.) and died 5–7 days p.i. ϕRSS1-infected cells of MAFF 106603 inoculated into tomato in the same way caused wilting earlier, 2 days p.i. and all 12 plants died by 5 days p.i. In contrast, tomato plants inoculated with ϕRSS0-infected cells showed wilting symptoms much later: most plants (10 of 12) survived after 7 days and a few plants did not show any symptoms until 23 days p.i. Figure 11.4a shows several examples from each experiment. The reduced virulence observed for ϕRSS0-infected cells may be caused by the function(s) of ORF13 encoded by ϕRSS0.

11.3.3 States of ϕRSS Phage in the Host Cell and Their Different Effects on Host Virulence

The protein encoded by ORF13 of ϕRSS0 showed homology to phage repressor proteins and may have functions in the immunity of lysogenic strains, as well as in repression of virulence genes of the host. Loss of virulence in *R. solanacearum* strains after infection with filamentous phages was also reported previously for another kind of phage, ϕRSM3 (Addy et al. 2012a), where a phage-encoded ORF (ORF15) with homology to transcriptional repressors was suggested to

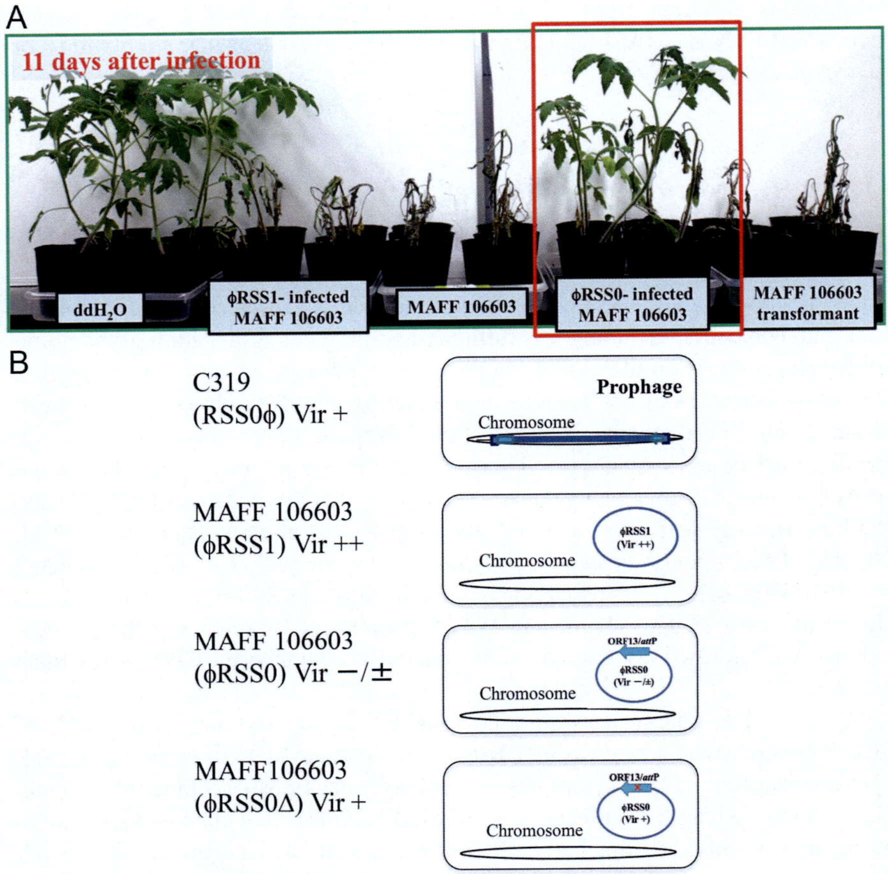

Fig. 11.4 Infection of ϕRSS0 and ϕRSS1 affected host virulence differently. (**A**) The stem of tomato plants were inoculated with bacterial cells. Tomato plants inoculated with wild-type (strain MAFF 106603) or ϕRSS1-infected bacterial cells died by 4 days post inoculation. In contrast, most tomato plants inoculated with ϕRSS0-infected cells survived after 11 days. However, MAFF 106603 cells with a ϕRSS0 mutant where ORF13 was disrupted, killed tomato plants similarly to the wild type cells. (**B**) The different effects on the host virulence by the three states of the ϕRSS phages are compared including strain C314 that contains a ϕRSS0 prophage

downregulate virulence genes via *phc*A repression in the host. A mutant of ϕRSM3 lacking ORF15 (ϕRSM3-ΔORF15) retained virulence in host cells after infection. In that case, ϕRSM3 and ϕRSM3-ΔORF15 correspond to ϕRSS0 and ϕRSS1 of this work, respectively. As shown in Fig. 11.4A, upon infection by ϕRSS0, host cells showed reduced virulence, but lysogenic strain C319 still showed stable virulence. This observation suggests changes in ϕRSS0 gene expression after integration into the host genome. There is one base change between the host *dif* sequence (either chromosomal or megaplasmid) and *att*P on ϕRSS0 (Fig. 11.3B). This difference results in truncating ORF13 after recombination between *dif* and *att*P; by creating a stop codon UAA in the reading frame (within the *dif* sequence in a reverse orientation). The protein encoded by truncated ORF13 reduces from 156 aa to 130 aa, with a 26-aa truncation at the C-terminus (see below). ORF13 encodes a putative regulatory protein; therefore, this change may contribute to the differential regulation of host virulence. It is notable that ϕRSS1 derived from ϕRSS0 but lacking ORF13 (with *att*P) showed enhanced virulence in infected host cells (Addy et al. 2012b). The different effects of the three states of the ϕRSS phages are compared in Fig. 11.4B.

11.4 Diversity of ϕRSS-Like Phages

11.4.1 Other ϕRSS-Phages Also Integrate in the Host Genome by the dif/XerCD System

Another ϕRSS-type phage (designated ϕRSS2), whose replicative form (RF) DNA was initially identified as a circular dsDNA plasmid pJTPS1 in strain M4S (Negishi et al. 1993), showed 95% nucleotide sequence identity with ϕRSS1 DNA. Local nucleotide changes concentrated in ORF7 encoding pIII protein for host recognition and in the IG region (Fig. 11.2A) cause the host ranges of ϕRSS1 and ϕRSS2 to be completely different (Kawasaki et al. 2007). pJTPS1 (ϕRSS2) was spontaneously produced in the lysogenic strain M4S (Negishi et al. 1993); therefore, we considered a similar way to produce sub-phages from a lysogenic ϕRSS0-like phage in this strain. Using the nucleotide sequences around *att*L and *att*R of ϕRSS0 in C319 as primers for inverse PCR, both junction regions of the prophage (designated ϕRSS20) in strain M4S were obtained and sequenced. The ϕRSS20-M4S junction sequences were almost the same as those of ϕRSS0-C319 *att*L and *att*R shown in Fig. 11.2A; even with 30 base changes within the phage sequences. Therefore, ϕRSS20 was also integrated into the M4S chromosome at the *dif* site by XerC/D recombination system. There was also an ORF (156 aa residues) between *att*L and ORF1 of ϕRSS20 in strain M4S, corresponding to ϕRSS0-ORF13; the two ORFS showed 89.7 % amino acid sequence identity (Fig. 11.5A).

As mentioned above, strain MAFF106611 contains a ϕRSS-type prophage (designated ϕRSS30) and is susceptible to ϕRSS0 infection (Table 11.1). Using the ϕRSS0 sequence to design primers for PCR, the MAFF 106611 prophage DNA was amplified and sequenced. The ϕRSS30 nucleotide sequence showed that the entire phage region (8193 nt) was 905 nt larger than ϕRSS0 and contained a conserved ϕRSS-gene cluster ORF1-ORF11 (Fig. 11.2A), except for ORF10; ORF10 was replaced with another ORF with no amino acid sequence similarity to ϕRSS1-ORF10 (DDBJ accession no. AB259124). There was also ORF13 between *att*L and ORF1 in the reverse orientation, but it differed from ϕRSS0-ORF13 in size (157 amino acids residues) and in amino acid sequence (87% identity) (Fig. 11.5A). In addition, there was another ORF in the reverse orientation on the opposite side near

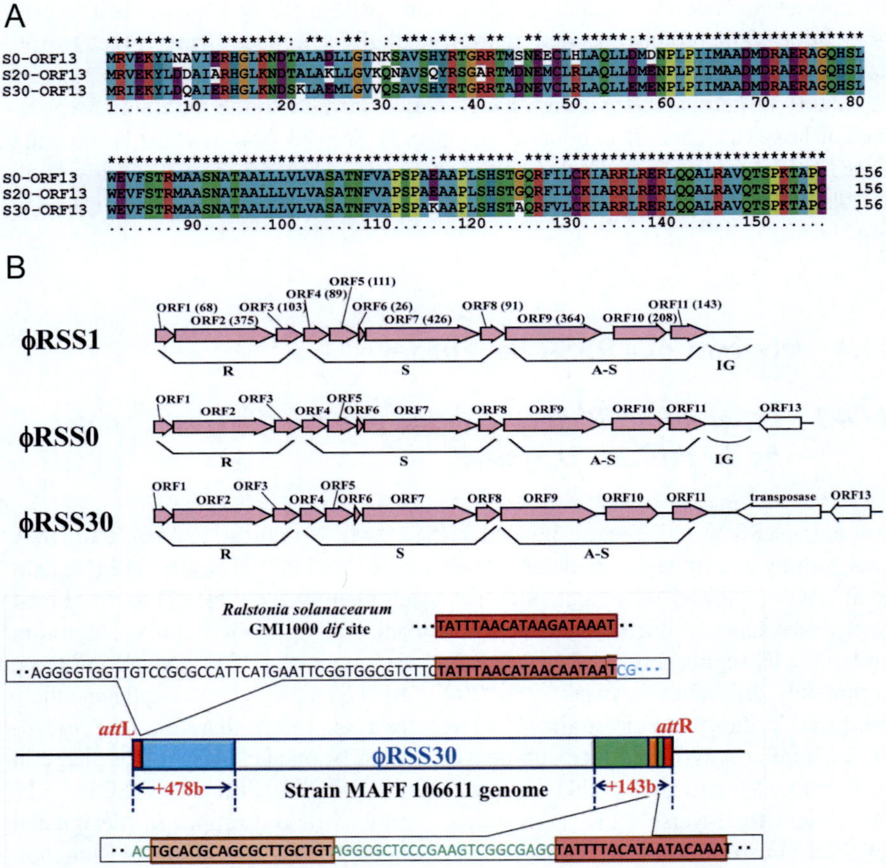

Fig. 11.5 Comparison of ϕRSS01-like filamentous phages. (**A**) Multiple comparisons of amino acid sequences of ORF13 encoded by ϕRSS1-like filamentous phages. The amino acid sequences of ϕRSS0 (JQ408219), ϕRSS20 (AB830321), and ϕRSS30 (AB828698) were aligned with each other using ClustalX. The ClustalX coloring scheme depends on both residue type and the pattern of conservation with column (http://www.cgl.ucsf.edu/chimera/Docs/ContributedSoftware/multalignviwer/cxcolor.html). (**B**) Genetic map of ϕRSS30 (8193 nt) integrated in strain MAFF 106611 compared with those of ϕRSS1 and ϕRSS0. In addition to ORF13, there is an extra ORF encoding a transposase-like sequence (upper panel). The *att*L sequence of ϕRSS30 coincided to the *dif* sequence of megaplasmid shown in Fig. 11.3, indicating that ϕRSS30 was integrated in the megaplasmid of strain MAFF 106611

*att*R, comprising 1173 nt and encoding a protein of 391 aa. The amino acid sequence of this ORF showed high similarity to an transposase IS4 family protein (*R. solanacearum* IS1405, accession no. Q9RBZ6_RALSO, E-value: 0.0) (Fig. 11.5B). It is notable that the *att*L sequence of ϕRSS30 coincided with the *dif* sequence of the megaplasmid shown in Fig. 11.3B, indicating that ϕRSS30 was integrated into the megaplasmid of strain MAFF 106611 (Fig. 11.5B). When a sub-phage DNA (ca. 6.5 kbp) of ϕRSS30 was amplified by PCR with the sequences of the 5′-end of *ORF1* and the 3′-end of *IS1405* as primers and self-ligated, the resulting DNA could successfully replicate in transformed cells of MAFF 106603 and formed phage plaques (ϕRSS30) on plates (data not shown). ϕRSS30-infected cells showed similar properties to ϕRSS1-infected cells, as described above. These diverse ϕRSS-type phages of *R. solanacearum* are compared in Fig. 11.5B.

11.4.2 Formation of ϕRSS1 from ϕRSS0

ϕRSS1 was first detected in screening of phages infecting *R. solanacearum* with strain C319 as a host (Yamada et al. 2007). C319 was found to contain a prophage with a ϕRSS1-like sequence; therefore, ϕRSS1 was thought to be eventually induced from C319 (Kawasaki et al. 2007). As shown in this study, ϕRSS0 is the prophage existing in a single copy in C319, and ϕRSS1 should have been derived from the ϕRSS0 prophage. Using the ϕRSS0-prophage as a template, ϕRSS1 may be produced by a replicative process. In the case of CTXϕ in *V. cholerae*, where multiple tandem copies of CTXϕ-related sequences are integrated, the rolling circle machinery of CTXϕ can drive the production of the (+) ssDNA genome of the phage by initiating replication on a first integrated copy and terminating it on a second copy (Moyer et al. 2001). Without a second copy of the phage, ϕRSS1 ssDNA formation will require a specific sequence substituting for it, if a CTXϕ-like mechanism is responsible. Figure 11.6A represents a putative model for ϕRSS1 DNA synthesis. DNA synthesis may initiate near the 5′ end of ORF1, probably at a nick introduced by the ϕRSS0 encoded replication protein (ORF2). DNA synthesis continues until it encounters an appropriate stop site. Comparing the nucleotide sequences of ϕRSS0 and ϕRSS1, this stop site corresponds to a position 50 nt downstream from ORF13 (3′ end) in the ϕRSS0 prophage (Fig. 11.6A, S1R), which coincides with the point previously identified as the integration point of ϕRSS1 (Kawasaki et al. 2007). The nucleotide sequence around this position in the ϕRSS0 DNA can form a stem-loop structure shown in Fig. 11.6B (S1R), where an arrow indicates the putative stop point. This region may be involved in the rolling-circle replication of circular ϕRSS0 DNA, serving as the nicking site for the replication protein (encoded by ORF2) (Zinder and Horiuchi 1985). A similar sequence element forming a stem-loop structure (designated S1L) was found between ORF13 (5′ end) and ORF1 of the ϕRSS0 prophage, as shown in Fig. 11.6B. There is a conserved sequence, TAGA (or CTAGAT), in the loop of S1L and S1R. If this region serves as a nicking site to start rolling-circle DNA replication by ϕRSS0 replicative protein encoded by ORF2, as shown in Fig. 11.6A, the first 59 nt in the synthesized strand would represent an excess sequence added to the net ϕRSS1 sequence. The extra 59 nt may have been removed during or after circularization of the synthesized ssDNA. We propose a structure model for joining cleavage sites to remove the extra sequence and form the exact ϕRSS1 sequence by certain

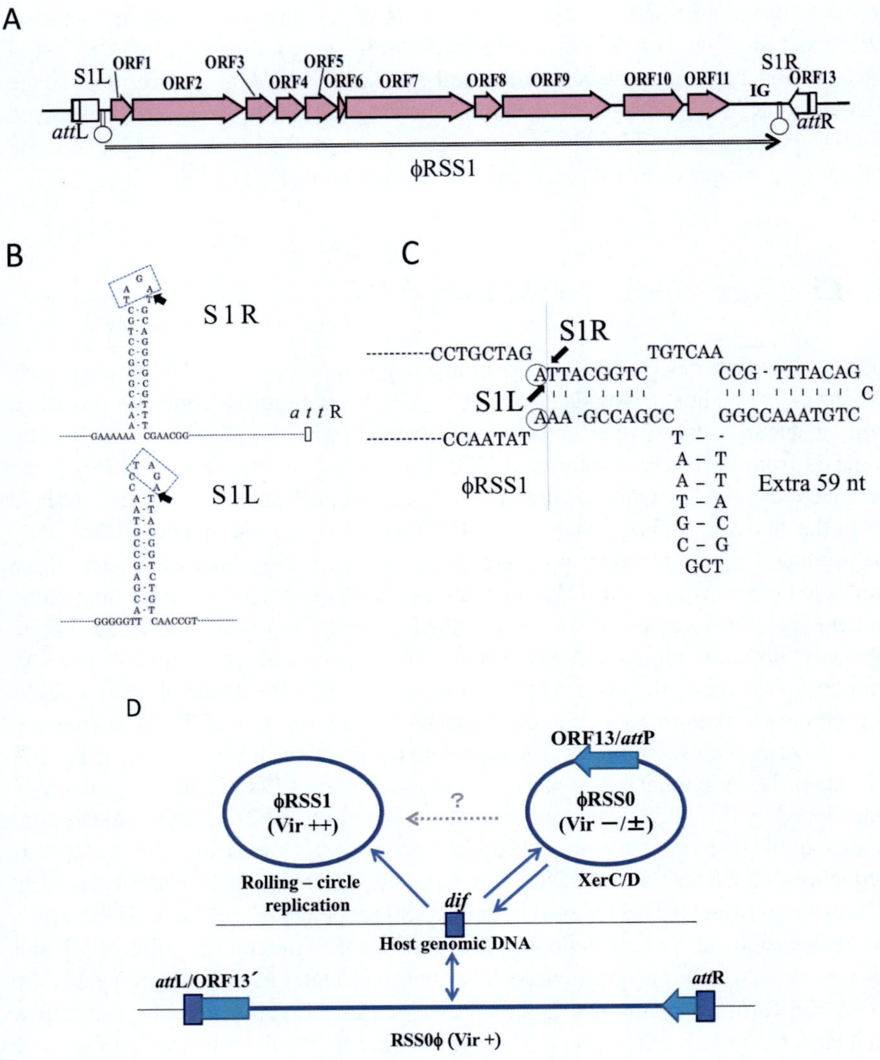

Fig. 11.6 A model for ϕRSS1 formation from ϕRSS0. (**A**) A short-cut DNA synthesis initiated at S1L and terminated at a second terminator S1R by a rolling-circle replication of ϕRSS0 (Moyer et al. 2001) results in a truncated form of ϕRSS1 (shaded). Both regions of S1L and S1R form stem-loop structures, as indicated. An arrow indicates the direction of DNA synthesis from S1L to S1R. (**B**) Stem and loop structures formed by S1L and S1R elements. A conserved sequence TAGA in the loop is boxed. An arrow indicates a putative nicking site (**C**). An extra 59 nt in the primary product of DNA synthesis can form a tight secondary structure, which may be removed during or after circularization of the synthesized ssDNA to produce ϕRSS1 genomic DNA. Arrows indicate putative nicking sites at S1L and S1R. The extra region of 59 nt is on the right of the dotted line. (**D**) ϕRSS1 formation from ϕRSS0 ds replicative forms by rolling-circle mechanism is also possible. As demonstrated in this study, ϕRSS0 can be induced from a lysogenic strain. The three states of ϕRSS0-type phage (ϕRSS0, ϕRSS1, and ϕRSS0 prophage) affect host *R. solanacearum* cells differently after infection, especially in host virulence. Compared wild-type virulence (+), ϕRSS1 enhances (++) and ϕRSS0 reduces (−/±) the host virulence

mechanisms (Fig. 11.6C). The 59-nt sequence can form a tight secondary structure looping out from the ϕRSS1 sequence. This cleavage model to form a truncated sub-phage from a ϕRSS-prophage may also explain the formation of pJTPS1 (ϕRSS2) from lysogenic strain M4S (Kawasaki et al. 2007).

Our observations also suggest another mechanism for producing ϕRSS-like phages from a lysogenic strain of *R. solanacearum* (Fig. 11.6D). The ϕRSS0 phage itself could be excised from the lysogenic strain (C319) after a second infection with lytic phage ϕRSL1, as demonstrated in this study. This excision may be mediated by host XerC/D recombinases, as reported for the VGJϕ prophage in *V. cholerae* (Das et al. 2011). However, the lysogenic strain C319 is very stable under routine culture conditions and produces few phage particles. ϕRSS1 may be produced from replicative dsDNA forms of ϕRSS0 by a rolling circle replication mechanism in the same way as it is produced from lysogenic host DNA (Fig. 11.6D). We do not have any direct evidence for this possibility yet. One interesting question is the genomic change from to ϕRSS0, which provides a simple non-integrative ϕRSS1 phage with many functions including host integration, phage and host regulation, and modifying replication. There is no information about this process. The three states of ϕRSS-type phage affect *R. solanacearum* host cells differently after infection, especially in terms of host virulence, as shown in Fig. 11.6D.

11.5 Conclusion

1. Filamentous ϕRSS1-like phages (inoviruses) infecting *R. solanacearum* have three different states: the free (ϕRSS0), prophage (RSS0ϕ), and superinfective truncated (ϕRSS1) states.
2. ϕRSS0 integrates into the host genome (chromosome and megaplasmid) at the *dif* sequence as *attB* by the host XerCD recombination system to form the RSS0ϕ prophage state.
3. ϕRSS1 was formed from RSS0ϕ through a deletion of ORF13 with *attP* likely through a replication process.
4. ϕRSS0 infection reduces the host virulence by the ORF13 function to repress host virulence genes such as *phcA*.
5. At the state of RSS0ϕ, the ORF13 repressor gene is split by recombination at *attP* within the coding region without affecting the host virulence.
6. ϕRSS1 infection enhances the host virulence through its active replication on the host cell surface, generating a high local cell density in the population, and activating the virulence gene *phcA*.
7. This example of host-phage interaction seen in the *R. solanacearum*- ϕRSS1-like phages system represents a unique mechanism that phages control the host virulence of pathogenic bacteria.

References

Addy HS, Askora A, Kawasaki T, Fujie M, Yamada T (2012a) Loss of virulence of the phytopathogen *Ralstonia solanacearum* through infection by ϕRSM filamentous phages. Phytopathology 102:469–477

Addy HS, Askora A, Kawasaki T, Fujie M, Yamada T (2012b) The filamentous phage ϕRSS1 enhances virulence of phytopathogenic *Ralstonia solanacearum* on tomato. Phytopathology 102:244–251

Ahmad AA, Askora A, Kawaski T, Fujie M, Yamada T (2014) The filamentous phage XacF1 causes loss of virulence in *Xanthomonas axonopodis* pv. *citri*, the causative agent of citrus canker disease. Front Microbiol 5:321

Ahmad AA, Kawabe M, Askora A, Kawasaki T, Fujie M, Yamada T (2017) Dynamic integration and excision of filamentous phage XacF1 in *Xanthomonas citri* pv. *citri*, the causative agent of citrus canker disease. FEBS Open Bio 7:1715–1721

Askora A, Yamada T (2015) Two different evolutionary lines of filamentous phages in *Ralstonia solanacearum*; their effects on bacterial virulence. Front Genet 6:217. https://doi.org/10.3389/fgene.2015.00217

Askora A, Kawasaki T, Usami S, Fujie M, Yamada T (2009) Host recognition and integration of filamentous phage **ϕ**RSM in the phytopathogen, *Ralstonia solanacearum*. Virology 384:69–76

Askora A, Kawasaki T, Fujie M, Yamada T (2011) Resolvase-like serinerecombinase mediates integration/excision in the bacteriophage ϕRSM. J Biosci Bioeng 111:109–116

Buchen-Osmond C (2003) Inoviridae. In: Oracle AZ (ed) ICTVdB-The Universal Virus Database, version 3. ICTVdB Management, The Earth Institute, Biosphere 2 Center, Columbia University

Campos J, Martinez E, Suzarte E, Rodriguez BE, Marrero K, Silva Y, Ledón T, del Sol R, Fando R (2003) A novel filamentous phage of *Vibrio cholerae*, integrates into the same chromosomal site as CTXϕ. J Bacteriol 185:5685–5696

Carney BF, Denny TP (1990) A cloned avirulence gene from *Pseudomonas solanacearum* determines incompatibility on *Nicotiana tabacum* at the host species level. J Bacteriol 172:4836–4843

Carnoy C, Roten C-A (2009) The *dif*/Xer recombination systems in Proteobacteria. PLoS ONE 4(9):e6531

Dai H, Chiang KS, Kuo TT (1980) Characterization of a new filamentous phage Cf from *Xanthomonas citri*. J Gen Virol 46:277–289

Das B, Bischerour J, Val M-E, Barre F-X (2010) Molecular keys of the tropism of integration of the cholera toxin phage. Proc Natl Acad Sci USA 107:4377–4382

Das B, Bischerour J, Barre F-X (2011) VGJϕ integration and excision mechanisms contribute to the genetic diversity of *Vibrio cholerae* epidemic strains. Proc Natl Acad Sci USA 108:2516–2521

Davis BM, Waldor MK (2003) Filamentous phages linked to virulence of *Vibrio cholerae*. Curr Opin Microbiol 6:35–42

Denny TP (2006) Plant pathogenic *Ralstonia* species. In: Gnanamanickam SS (ed) Plant-associated bacteria. Springer, Amsterdam, pp 573–644

Gonzalez MD, Lichtensteiger CA, Caughlan R, Vimr ER (2002) Conserved filamentous prophage in *Escherichia coli* O18:K1:H7 and *Yersinia pestis* biovar. *orientalis*. J Bacteriol 184:6050–6055

Hayward AC (2000) *Ralstonia solanacearum*. In: Lederberg J (ed) Encyclopedia of microbiology, vol 4. Academic, San Diego, pp 32–42

Huber KE, Waldor MK (2002) Filamentous phage integration requires the host recombinases XerC and XerD. Nature 417:656–659

Kang Y, Liu H, Genin S, Schell MA, Denny TP (2002) *Ralstonia solanacearum* requires type 4 pili to adhere to multiple surfaces and for natural transformation and virulence. Mol Microbiol 46:427–437

Kawasaki T, Nagata S, Fujiwara A, Satsuma H, Fujie M, Usami S, Yamada T (2007) Genomic characterization of the filamentous integrative bacteriophage ϕRSS1 and ϕRSM1, which infect *Ralstonia solanacearum*. J Bacteriol 189:5792–5802

Kuo TT, Chao YS, Lin YH, Lin BY, Liu LF, Feng TY (1987a) Integration of the DNA of filamentous bacteriophage Cf1t into the chromosomal DNA of its host. J Virol 61:60–65

Kuo TT, Lin YH, Huang CM, Chang SF, Dai D, Feng TY (1987b) The lysogenic cycle of the filamentous phage Cf1t from *Xanthomonas campestris* pv. *citri*. Virology 156:305–312

Kuo TT, Tan MS, Su MT, Yang MK (1991) Complete nucleotide sequence of filamentous phage Cf1c from *Xanthomonas campestris* pv. *citri*. Nucleic Acids Res 19:2498

Lin NT, Chang RY, Lee SJ, Tseng YH (2000) Plasmids carrying cloned fragments of RF DNA from the filamentous phage ϕLF can be integrated into the host chromosome via site-specific integration and homologous recombination. Mol Genet Genom 266:425–435

Moyer KE, Kimsey HH, Waldor MK (2001) Evidence for a rolling-circle mechanism of phage DNA synthesis from both replicative and integrated forms of CTXϕ. Mol Microbiol 41:311–323

Negishi H, Yamada T, Shiraishi T, Oku H, Tanaka H (1993) *Pseudomonas solanacearum* plasmid pJTPS1 mediates a shift from the pathogenic to nonpathogenic phenotype. Mol Plant-Microbe Interact 6:203–209

Simpson AJ, Reinach FC, Arruda P, Abreu FA, Acencio M, Alvarenga R et al (2000) The genome sequence of the plant pathogen *Xylella fastidiosa*. Nature 406:151–157

Vasse J, Genin S, Frey P, Boucher C, Brito B (2000) The *hrpB* and *hrpG* regulatory genes of *Ralstonia solanacearum* are required for different stages of the tomato root infection process. Mol Plant-Microbe Interact 13:259–267

Yamada T (2012) Bacteriophages of *Ralstonia solanacearum*: their diversity and utilization as biocontrol agents in agriculture. In: Kurtboke I (ed) Bacteriophage. InTech Open Access Publisher, Croatia, pp 113–139

Yamada T (2013) Filamentous phages of *Ralstonia solanacearum*: double-edged swords for pahogenic bacteria. Front Microbiol 4:325. https://doi.org/10.3389/fmicb.2013.00325

Yamada T, Kawasaki T, Nagata S, Fujiwara A, Usami S, Fujie M (2007) New bacteriophages that infect the phytopathogen *Ralstonia solanacearum*. Microbiology 153:2630–2639

Yamada T, Sato S, Ishikawa H, Fujiwara A, Kawasaki T, Fujie M, Ogata, H (2010) A jumbo phage infecting the phytopathogen Ralstonia solanacearumdefines a new lineage of the Myovirus family. Virology 398:135–147

Yao J, Allen C (2007) The plant pathogen *Ralstonia solanacearum* needs aerotaxis for normal biofilm formation and interactions with its tomato host. J Bacteriol 189:6415–6424

Zinder ND, Horiuchi K (1985) Multiregulatory element of filamentous bacteriophages. Microbiol Rev 49:101–106

Chapter 12
Intra-population Interactions and the Evolution of RNA Phages

Ester Lázaro

Abstract As in other RNA viruses, replication of RNA phages takes place with high error rate, which leads to the coexistence within their populations of a wide diversity of interacting mutants. When interactions are mainly cooperative, the replication of the ensemble is favored. Conversely, when interference or competitive interactions are the dominant ones, the viability of the whole population can be compromised. Finally, the fact that high error rates are associated to small and highly compacted genomes also causes that the effect of mutations is not independent. The result is that the evolutionary behavior of RNA phages is modulated by a wide variety of interactions that connect mutants within populations and that add an extra dimension to the process, giving rise to outcomes that differ from what would be expected from the properties of the individual mutants.

12.1 High Error Rates and Small Genomes as a Source of Interactions Within RNA Phage Populations

The last report of the International Committee for the Taxonomy of Viruses included about 5500 virus species that classified into 150 families, of which eight corresponded to single-stranded and double-stranded DNA phages and only two to RNA phages: the single-stranded RNA family *Leviviridae* (Olsthoorn and van Duin 2011), with four recognized species (enterobacteria phages MS2, Qβ, F1, and BZ13), and the segmented, double-stranded RNA family *Cystoviridae*, with seven recognized species infecting *Pseudomonas* (Poranen and Mäntynen 2017). The remarkably low number of RNA phages identified so far contrasts with the fact that viruses with an RNA genome are the most abundant among those characterized in the biosphere, suggesting that there are probably many unrecognized RNA bacteriophages that remain to be discovered (Callanan et al. 2018). Due to the ease of RNA degradation, most approaches for the isolation, selection, and purification of viral

E. Lázaro (✉)
Centro de Astrobiología (CSIC-INTA), Madrid, Spain
e-mail: lazarole@cab.inta-csic.es

G. Witzany (ed.), *Biocommunication of Phages*,
https://doi.org/10.1007/978-3-030-45885-0_12

particles have traditionally focused on DNA phages. The low abundance of RNA phages in databases also hinders the identification of novel gene sequences that could correspond to these viruses (Hatfull 2015). Nevertheless, a recent analysis of a metagenomic dataset led to the identification of 122 partial genomes of novel RNA phages (Krishnamurthy et al. 2016) present in samples collected from a variety of ecological niches, including invertebrates, microbial sediments, and animal-associated microbiomes, illustrating the wide distribution that RNA phages might have in nature. Like DNA phages, RNA phages probably play relevant roles in the environment, contributing to regulate the diversity patterns of microbial populations, influencing the evolution of their hosts, and impacting critical processes in ocean, soil, and animal ecosystems.

The prototype species for the *Cystoviridae* family is phage φ6 (Mindich 2004). The virion consists of an outer lipid envelope surrounding a nucleocapsid shell and an icosahedral procapsid core (Jäälinoja et al. 2007). The genome contains three double-stranded RNA segments (about 13 kb in total) that encode 13 viral proteins (Mindich 1999). The replication cycle of φ6 involves the copy of the RNA segments into complementary RNA molecules that act as templates for protein synthesis and that are also encapsidated into the nascent virus particles (Emori et al. 1983). It is believed that the vast majority of progeny φ6 virions that are produced during the viral life cycle contain a single copy of each of the three genome segments. The best known species of the *Leviviridae* family are phages MS2 and Qβ (Olsthoorn and van Duin 2011). Both possess a non-enveloped icosahedral capsid and infect *Escherichia coli* by adsorbing to the F pilus. Their genome is a positive sense, single-stranded, monopartite RNA of 3569 nucleotides in MS2 and of 4217 nucleotides in Qβ.

Since cells are unable to synthesize complementary strands from RNA templates, all RNA viruses must encode their own enzymes to copy their genomes. In eukaryotic RNA viruses there are two enzymes that can carry out this process, the reverse transcriptase and the RNA-dependent RNA polymerase (RdRp or RNA replicase). Up to now, no reverse transcriptase has been identified in RNA phages. Their genomes are always in the form of RNA, and no DNA stage has been observed. This fact precludes RNA phages from integrating their genomes into the bacterial chromosome to initiate a lysogenic cycle, a fact that clearly distinguishes the biology of RNA and DNA phages, including the interactions that they can establish with their hosts and their capacity to act as vehicles for lateral transfer of genetic material.

A general property of RNA replicases is the lack of a 3′ exonuclease proofreading activity (Steinhauer et al. 1992), which implies that all known RNA viruses replicate their genomes with very high error rates, in the order of 10^{-4} to 10^{-6} errors per nucleotide copied (Drake and Holland 1999; Sanjuan et al. 2010). RNA bacteriophages are not an exception to this rule, as it was soon observed in early studies carried out with MS2 and Qβ that showed the existence of a high variability in some phenotypic traits (temperature sensitivity, plaque size, etc.) (Domingo et al. 1988; Holland et al. 1982). Batschellet et al. determined the mutation rate for a particular substitution in the genome of bacteriophage Qβ, a value that when extrapolated to the rest of genomic positions resulted in about 10^{-4} errors per nucleotide and round of copy (Batschelet et al. 1976). At that time, it became also evident through fingerprint analysis that virus populations contained in lytic plaques, which can be considered viral clones, were genetically heterogeneous, suggesting that multiple mutations can be generated even during the low number of replication rounds necessary to give rise to a lytic plaque (Domingo et al. 1978). A new determination of the error rate of Qβ by means of a fluctuation test yielded a value of 1.4×10^{-4} errors per copied nucleotide (Bradwell et al. 2013), very similar to that found out years before. This high generation of errors implies that most progeny genomes differ from the parent template, and it should take only a few replication rounds to produce a diverse ensemble of mutants. In this scenario, genetic information could be rapidly lost were it not for the action of negative selection, the process that eliminates the worse replicating genomes. The error-prone nature of the replication of RNA genomes has been supported biochemically by fidelity measurements with purified polymerases and through estimates of the mutation frequencies in virus populations (Beaucourt et al. 2011; Cameron et al. 2016; Mansky and Temin 1995).

In addition to the generation of point mutations, other mechanisms that contribute to the generation of genetic variability in RNA phages are genome segment reassortment, in the viruses that have the genome divided into several segments -as it happens in the family *Cystoviridae*-, and recombination, the formation of chimeric genomes from two or more parental genomes. Segment reassortment requires that two or more different viruses co-infect the same host cell and exchange RNA segments, leading to the formation of hybrid viral progeny with genome segments derived from multiple parental strains (Silander et al. 2005). Reassortment provides a mechanism for the exchange of genes or set of genes, in a way analogous to crossing over during sexual reproduction (McDonald et al. 2016). In contrast to observations in some families of animal and plant RNA viruses, recombination is not very frequent in RNA phages. It has been reported that it is almost absent in φ6 (Mindich 2004; Mindich et al. 1976), whereas estimations in members of the *Leviviridae* family give some controversial values, which in any case are several orders of magnitude below the frequency of point mutations (Chetverin 2018; Chetverin et al. 2005; Palasingam and Shaklee 1992).

An evolutionary limitation that occurs in all types of organisms is that the maximum genome size that can be maintained without the loss of genetic information is inversely correlated with the error rate for the replication of that genome (Drake 1991; Eigen 1971). This restriction imposes a limit to the size allowed for the genomes of RNA viruses (Biebricher and Eigen 2005; Bradwell et al. 2013; Gago et al. 2009), which in the case of RNA phages are in the range of 3 to 15 kb. As a consequence, genetic information must be highly compacted, with frequent overlapping reading frames that may also include regulatory regions, multifunctional proteins, and rare non-coding regions (Belshaw et al. 2008; Holmes 2003). In addition, the secondary structure of the genomic RNA is also functional since it participates in its own replication and in the interaction with cellular and viral components. It is easy to understand that highly compacted genomes are more susceptible to the effect of mutations (Domingo-Calap et al. 2009), which generally are more deleterious and more prone to interact than in less compact genomes (Cuevas et al. 2009;

Elena et al. 2010; Sanjuán and Elena 2006). It is a matter of debate whether RNA viruses evolve to select mechanisms of genetic robustness that help preserve the phenotype, despite frequent changes in the genotype (Dolan et al. 2018; Elena et al. 2006; Lauring et al. 2013; Stern et al. 2014). These mechanisms, which probably need long times of evolution in a constant environment to be selected, include the selection of genotypes with few deleterious mutational neighbors, the co-opting of host factors that help to fold protein or RNA correctly, and the generation of a vast progeny in which deleterious mutants can be complemented by functional genomes.

The main consequence of the high error rates of RNA viruses is that within their populations a wide variety of mutants always coexist (Perales et al. 2010). Mutants differ in their genomic sequences, in the nature of the products encoded, and in their fitness values, that is in their ability to produce a viable progeny. All this diversity is a constant source of interactions that add extra information to the system. Its more evident manifestation is the impossibility to predict, in many cases, the behavior of the whole population from the properties of the individual mutants. Interactions can occur among mutants generated within the same cell from the same parental genome or through co-infection by different external mutants. They can be cooperative, when the products of the expression of some virus genomes help other defective viruses to complete their life cycles, or interfering, when defective products impede the functionality of "healthy viruses". Finally, competitive interactions that occur during virus adaptation to new environmental conditions also acquire a new dimension at high error rate. All these interplays can be seen as an unintentional form of communication by which mutants exchange information about their genomes and the products of their expression, influencing the performance of others and even determining whether populations will be able to stand disturbances or, conversely, will be extinguished.

Most of what we know about the role played by intra-population interactions in the behavior of RNA viruses comes from experiments carried out to test evolutionary hypotheses within controlled laboratory conditions. Some of these studies have been performed with plant or animal RNA viruses, and others with RNA phages, especially with Qβ, MS2, and φ6. Since the distinctive features of RNA virus biology are high error rates and high compaction of their genomes, it is possible to extrapolate many of the results obtained with a particular virus to other cases, provided they have comparable error rates and similar genome architectures. Studies carried out with single-stranded DNA phages, which also have small compacted genomes, can give additional information concerning how the density of information contained in a genome influences the frequency and nature of interactions among mutations. In the remaining sections of this chapter we will review the findings of studies carried out with RNA phages where interactions among mutants and mutations were found to be relevant. When necessary, we will also refer to studies performed with other RNA viruses of different specificity, or with DNA phages. Although obtained in much simpler systems than natural environments, it is expected that their conclusions help us understand how RNA phages evolve in complex ecosystems, where multiple selective pressures, including those derived from the biotic environment, act simultaneously.

12.2 The Population Structure of RNA Phages. Sequence Spaces and Fitness Landscapes

RNA virus populations are often described as quasispecies, a term that takes its name from a theory formulated to explain the evolution of self-replicating molecules that might have preceded the origin of life on Earth. According to this theory, error-prone replication combined with natural selection, and acting for a long enough time in infinite populations of replicators differing in their fitness values, would result in the generation of a steady-state mutant distribution in which each mutant has a constant frequency (Eigen 1971; Eigen and Biebricher 1988). The sequence displaying the highest replication rate is known as master sequence and would be the most represented in the ensemble of mutants, also referred to as mutant spectrum. Quasispecies theory is deterministic, and as such, when numerical values are assigned to the parameters defining the system, it is predictable. The first descriptions of RNA virus populations as quasispecies corresponded to the RNA phage Qβ, which as we described in the previous section was the first virus where the high genetic and phenotypic heterogeneity typical of RNA viruses was observed (Domingo et al. 1978). The term quasispecies applied to viruses must be understood as a dynamic distribution of related genomes subjected to a continuous process of genetic variation, competition, and selection that shapes its internal structure (Andino and Domingo 2015; Domingo et al. 2012). There are some features that differentiate virus quasispecies from theoretical molecular quasispecies. Virus populations replicate in complex environments where variables are in continuous change, making it difficult to reach mutation-selection equilibrium (Aguirre and Manrubia 2007; Stich et al. 2007). Although the short replication times of most viruses facilitate the generation of large populations, they are not infinite and frequently experience population bottlenecks that reduce the genetic diversity and introduce stochasticity in the system (Escarmís et al. 2006; Lázaro et al. 2003; Manrubia et al. 2003, 2005). Finally, another element that was not considered in quasispecies theory when it was initially formulated was the existence of a wide variety of interactions both among mutants and among mutations that could be determinant for the evolutionary behavior of the whole population.

Despite the great heterogeneity of RNA virus populations, a consensus sequence corresponding to the most represented nucleotides at each genomic position can be easily determined through standard Sanger sequencing. Consensus sequences are very useful to identify major changes in the genetic composition of populations that may happen as a consequence of population bottlenecks or alterations in the environmental conditions that favor some mutants over others. However, consensus sequences give no idea about the internal composition of the population that is where its evolutionary and adaptive potential resides. The same consensus sequence can be obtained for different mutant distributions, showing the limited applicability of this concept. In an experiment carried out with Qβ, a single clonal mutant, generated through site-directed mutagenesis, was able to displace a clonal virus whose genome had the same sequence as the consensus sequence of a wild-type

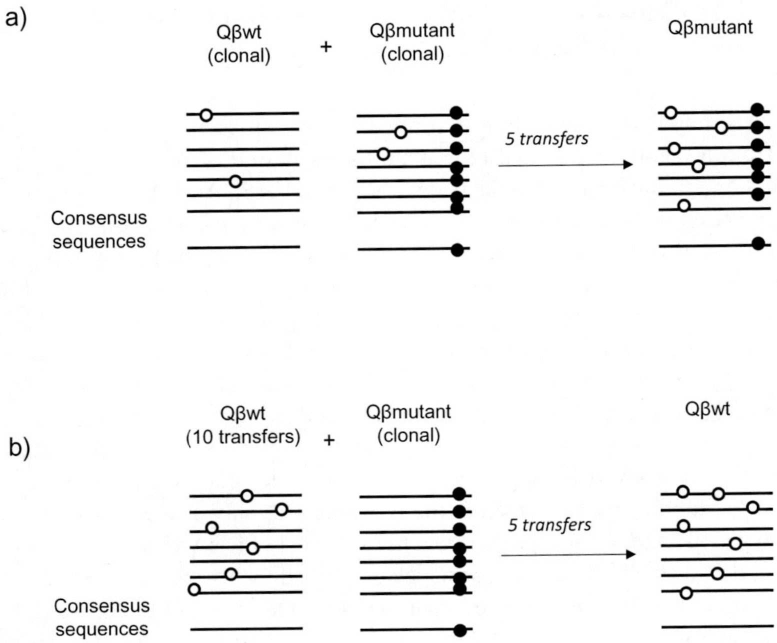

Fig. 12.1 Co-propagation of different Qβ populations. (**a**) Two clonal viruses (Qβmutant, with a mutation fixed in the consensus sequence, and Qβwt, with the wild-type consensus sequence) were propagated together for five serial transfers, after which the virus Qβwt was selected against. (**b**) Previously to being mixed with the mutant, the virus Qβwt was propagated for 10 serial transfers at large population size that allowed it to increase the genetic heterogeneity without introducing changes in the consensus sequence. In this case the outcome of the experiment was the dominance of the virus Qβwt. In both (**a**) and (**b**), populations were composed of genomes (black horizontal lines) containing a variable number of mutations (white circles for low frequency mutations and black circles for fixed mutations). Similar results were obtained for two different Qβ mutants (Arribas et al. 2018)

population. However, when the clonal wild-type virus was allowed to replicate for several generations to generate a diverse population, which happened without introducing changes in the consensus sequence, the result was the opposite, and the mutant was selected against (Fig. 12.1) (Arribas et al. 2018). This simple experiment illustrates how the mutant spectrum and not the consensus sequence is what determines the evolutionary outcome of an RNA virus population.

Selection ranks mutants within the mutant spectrum according to their fitness values. Positive selection is the process by which a mutant increases its frequency in a population due to the fitness advantage that permits it to produce a more numerous progeny than other mutants. In contrast to this, negative selection is the process by which a mutant decreases its frequency as a result of its low fitness value. The high diversity of RNA virus populations means that each mutant has to compete simultaneously with many different types beyond the wild type, which gives a great complexity to the process and can lead to unexpected results.

The fact that virus populations have finite sizes and that RNA genomes are highly susceptible to mutations limits the extension of genetic diversity that they can contain. The amplitude of a mutant spectrum can be viewed as the extent of occupation of the sequence space, which refers to a theoretical representation of all possible variants of a genomic sequence (Eigen 1971; Schuster 2010). The fraction of the sequence space actually occupied by a virus is but a tiny fraction of the whole space. There are many constraints that limit the existence of particular virus variants. In addition to the necessity to synthesize functional proteins that allow the virus to complete its life cycle successfully, there is also the requirement to maintain secondary and tertiary structures in the RNA genome to keep it functional. Following the classical metaphoric description of Sewall Wright (Wright 1931), when a fitness value is assigned to each genome in a sequence space, a fitness landscape is obtained (Acevedo et al. 2014; Schuster 2016). Fitness landscapes have peaks, where the more advantageous genomes are located, and valleys, for the lower fitness sequences. They are dynamics and are continuously modified by the physical-chemical characteristics of the environment, meaning that mutants that occupy the top of the peaks under certain conditions may be at the bottom of the valleys when the environment changes (Catalán et al. 2017; Manrubia and Lázaro 2016).

Populations move through the fitness landscape driven by natural selection (Dolan et al. 2018). The probability that a beneficial mutant increases its frequency and eventually reaches the top of a peak is a function of its fitness value and the effective population size (the number of viruses that will give rise to the next generation). Large population sizes favor competition among mutants and selection of those with highest fitness values (Escarmís et al. 1999; Novella et al. 1995). However, large population sizes also favor the coexistence of different beneficial mutants that can compete among them, delaying adaptation. In contrast to this, the reduction of genetic diversity that is associated to population bottlenecks leads to the frequent fixation of neutral or deleterious mutations, which is associated to fitness losses (Escarmís et al. 1996, 2002, 2006; Manrubia et al. 2005). This increased accumulation of mutations in the virus genomes causes that interactions among mutations acquire more relevance when viruses are propagated at small population sizes. The positive side of population bottlenecks is that they can move populations to distant positions in the fitness landscape, from where they can access to new adaptive pathways. They also eliminate many of the interactions among mutants that occur within the mutant spectrum, something that can be positive when interfering interactions are the predominant (Cases-González et al. 2008).

12.3 Epistatic Interactions in RNA Phages

We have already mentioned that the high degree of compaction of genetic information in RNA genomes gives rise to complex patterns of interactions among different regions, which can also result in that the effect of mutations is not independent (Cuevas et al. 2009; Elena et al. 2010; Sanjuán and Elena 2006). Interactions

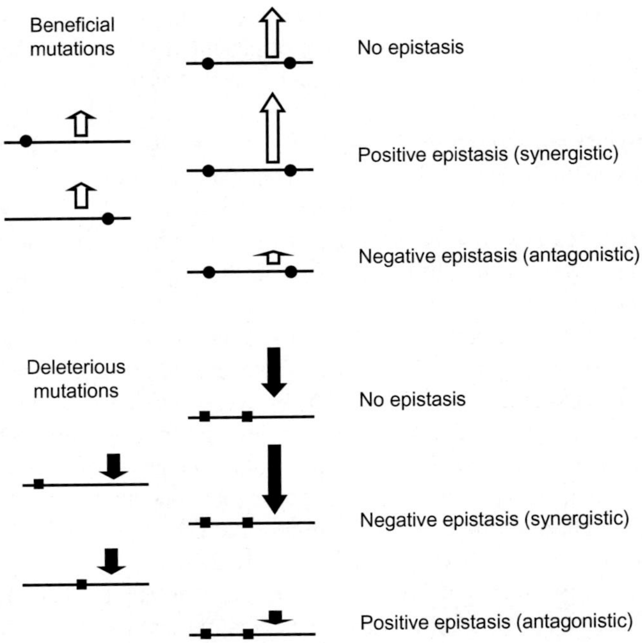

Fig. 12.2 Different types of epistasis. Positive or negative epistasis refer to whether the fitness of genotypes carrying multiple mutations is respectively higher or lower than expected if mutations had independent effects. Black horizontal lines represent genomes and symbols represent beneficial (circles) or deleterious (squares) mutations. The arrows represent the fitness effect in a logarithmic scale conferred by the mutations contained in a particular genome

between mutations take the general name of epistasis, a term that applies when the phenotypic effect of a mutation varies depending on the genetic background, this is the rest of mutations that are present in the same genome. In the absence of epistasis mutations have a multiplicative effect on fitness (additive in a logarithmic scale). Any deviation from this expectation is probably due to some kind of interaction among them.

There are multiple forms of epistasis. A first distinction classifies epistasis in magnitude epistasis, when the fitness value of a mutation changes depending on the genetic background, and sign epistasis, when the sign of the mutation is what changes. Compensatory mutations are a special class of sign epistasis that refers to neutral or deleterious mutations that become beneficial in the presence of others. Magnitude epistasis can be further classified as positive or negative, depending on whether the fitness of a particular combination of mutations is higher or lower than expected from the fitness of each mutation when it is alone in a genome (Fig. 12.2). When talking about deleterious mutations positive epistasis corresponds to antagonistic interactions and negative epistasis corresponds to synergistic interactions. Conversely, positive epistasis of beneficial mutations is synergistic and negative epistasis is antagonistic. All these types of interactions have been observed in RNA viruses, including also RNA phages, sometimes through indirect evidences and also in specific assays (Cuevas et al. 2009; Elena et al. 2010; Michalakis and Roze 2004; Sanjuán et al. 2004). Synergistic epistasis of deleterious mutations increases the strength of negative selection because it would eliminate any combination of deleterious mutations more rapidly than the same mutations when they are alone in a genome. On the contrary, antagonistic epistasis of deleterious mutations can be considered a form of robustness since the effects of mutations will be lower when they are gathered in the same genome. In the case of beneficial mutations, synergistic epistasis would accelerate adaptation, something that disagrees with the hyperbolic trajectory of the fitness dynamics observed in most experiments of RNA virus adaptation, which is better explained by a dominance of antagonistic interactions.

Epistasis determines the form of fitness landscapes (Poelwijk et al. 2011) as well as the accessibility of adaptive pathways (Weinreich et al. 2005). In the absence of epistasis fitness landscapes are smooth and single-peaked. Populations would move always uphill, towards the single global optimum that is reached when all possible beneficial mutations have been fixed. The result of evolution is deterministic and it is only a matter of time that the population achieves the best adaptive solution. In contrast to this, epistasis introduces ruggedness in the fitness landscapes and can make certain evolutionary pathways inaccessible (Kvitek and Sherlock 2011; Whitlock et al. 1995). An evolution experiment carried out with Qβ showed that adaptation of this virus to increased temperature involved the fixation of five silent mutations that provided a suitable genetic background for subsequent mutations that changed their fitness effect from deleterious to beneficial (Kashiwagi et al. 2014). This is a clear example of sign epistasis in which an adaptive pathway only was accessible after fixation of a set of mutations that changed the genetic background where new mutations arose. Sign epistasis implies that the order of mutation fixation is highly relevant for adaptation, which can explain the great genetic convergence detected in many experimental evolution studies (Cuevas et al. 2002; Remold et al. 2008). It is also clear evidence that selection acts more on combinations of mutations than on particular mutations. Other example of how epistasis modifies the availability of adaptive pathways in the fitness landscape comes from a study carried out with phage φ6 in which the impact of preexisting host range mutations on subsequent host range expansion was analyzed (Zhao et al. 2019). The results showed that preexisting mutations in the P3 protein that it is involved in host interaction dramatically narrowed the selection of new host range mutations compared to wild-type φ6. It is easy to imagine that the preexisting mutations destabilize the P3 protein, making it less tolerant to secondary mutations. Probably, if the virus had had enough time to generate and select compensatory mutations, the P3 protein would have recovered its stability, and with that the ability to acquire new mutations. An additional study carried out with the single-stranded DNA phage ID11, which resembles RNA phages in that it has a small, highly compacted genome, also showed the importance of stabilizing mutations to open new pathways for adaptation (Lee et al. 2011).

Mutation accumulation experiments provide a suitable system to test the effect of deleterious mutations that fix progressively in a genome. The experimental procedure consists in subjecting a virus population to repeated population bottlenecks that can be as small as a single individual, as it happens in plaque-to-plaque transfers. In this case, negative selection is not efficient, since it can only operate during the few replication rounds that take place during the development of the lytic plaques. Under this condition, fixation of mutations is enhanced and happens independently of their selective value. Since most mutations are deleterious, serial plaque-to-plaque transfers usually produce significant fitness losses, the dynamics of which can be analyzed to determine whether the effect of the new mutations incorporated is additive or not. The most complete study of accumulation of mutations of this type was carried out with foot-and-mouth disease virus (FMDV) and involved four different lineages that were subjected to hundreds of plaque-to – plaque transfers (Escarmís et al. 2002). The kinetics of fitness decay throughout the transfer series showed a biphasic kinetics, with an initial phase of exponential fitness drop, followed by a stationary fitness state in which the virus experienced large fluctuations around an average constant value (Lázaro et al. 2002, 2003). Fluctuations were attributed to an increased fixation of compensatory mutations in low fitness viruses that rescued them from extinction due to the increased mutational load (Escarmís et al. 2008; Manrubia et al. 2003). Other study carried out with the plant virus tobacco etch virus (TEV) also showed that the pattern of fitness decay as a function of the number of plaque-to-plaque transfers was better explained through antagonistic epistasis than through an additive model for the effect of mutations (Carrasco et al. 2007). In the case of RNA phages, an experiment performed with phage φ6 also showed clear evidences of antagonistic interactions for the new mutations incorporated in clonal lineages that had been previously subjected to 40 plaque-to-plaque transfers (Burch and Chao 2004).

Concerning compensatory mutations, the first evidences of their relevance in RNA genomes come from observations obtained with several animal viruses that adapted to drug treatments through the fixation of a first mutation that conferred the resistance phenotype, followed by the subsequent fixation of other mutations that compensated the side effects of the first one fixed (Nijhuis et al. 2009). When tested alone, the fitness effects of the secondary mutations were usually deleterious. The relevance of compensatory mutations to revert fitness losses in RNA phages was evaluated in an early study that involved several lineages of phage φ6 previously propagated through a series of extreme population bottlenecks that reduced their fitness values through the accumulation of deleterious mutations. Subsequent propagation of the bottlenecked lineages at large population sizes allowed them to recover fitness, which mainly took place through compensatory mutations (Burch and Chao 1999). The high frequency of compensatory mutations in RNA viruses is probably due to the great compaction of their genomes, where if a mutation produces a positive effect in one function it is highly probable that it has a side effect on other function, creating the necessity of a new mutation to restore overall fitness.

Direct approaches to quantify epistasis involve building genotypes containing the same mutations either alone or in combination and measuring the fitness values of the phenotypes generated upon their expression. A strategy of this type was used in an experiment carried out with the animal virus vesicular stomatitis virus (VSV) that showed that epistasis was quite common, with most interactions being antagonistic both for deleterious and for beneficial mutations (Sanjuán et al. 2004). In a few instances, the combination of two deleterious viable mutations created a lethal genotype, an extreme case of synergism known as synthetic lethality. Evolution of Qβ in the presence of a mutagenic nucleoside analogue showed the selection of two different mutations in the replicase gene that were deleterious in the absence of the drug and beneficial in its presence. Intriguingly, both mutations remained as long term polymorphisms in the Qβ population, and several attempts to build the double mutant through site-directed mutagenesis were infructuous, suggesting that this was an example of synthetic lethality (Cabanillas et al. 2013).

A similar strategy to that used for VSV was applied for the characterization of epistasis between beneficial mutations in DNA phages of the family *Microviridae*. This family contains viruses with a single-stranded genome of small size (from 4.5 to 6 kb) in which information is highly compacted and probably also subjected to the same constraints as in RNA genomes, except for the restrictions operating on the RNA secondary structure. In a study carried out with phage ID11, Rokyta et al. analyzed the fitness effects of nine beneficial mutations both alone and combined in pairs, finding that all combinations analyzed displayed antagonistic epistasis (Rokyta et al. 2011). A few years later, Caudle et al. measured epistatic interactions for these same mutations under different temperatures that imposed a selective pressure of different intensity on virus growth. They found clear evidences of antagonistic interactions under all conditions that were more pronounced at the highest temperature assayed, which was also the most stringent selective pressure (Caudle et al. 2014). Finally, a recent study analyzed the nature of epistatic interactions between beneficial mutations in 30 double-mutants of the phage ID8, also from the family *Microviridae*, evolving under different selective regimes that allowed to decompose the fitness measures into two different traits: growth rate and capsid stability. All mutants displayed antagonistic epistasis with regard to growth rate and fitness, but synergistic epistasis and absence of epistasis were common for more stable phenotypes, such as the capsid structure (Sackman and Rokyta 2018). The conclusion that arises from these studies is that although epistasis is common in all types of organisms, its frequency and relevance probably increase in RNA and single-stranded DNA phages, as a result of the pleiotropic interactions that arise between the individual components of complex phenotypes in small size genomes.

12.4 Interactions That Arise as a Consequence of the Intracellular Sharing of Resources

Complementation is a process by which a genome expressing a functional protein can promote replication of another closely related genome whose corresponding protein is suboptimal. When different viruses infect the same cell, the products encoded by their genomes diffuse within the cell and contribute to a common pool, which prevents any individual virus from exclusive access to its own gene products. The result of this intracellular sharing of resources is that low fitness or even defective viruses that could not replicate or did it at low levels can give rise to a regular progeny. Complementation among different components of the mutant spectrum was early suggested in studies that showed that the fitness of individual viruses isolated from a population was lower than the fitness of the whole population (Domingo et al. 1978; Duarte et al. 1994). At low error rates, complementation usually requires co-infection, which is favored at high multiplicity of infection. However, if the error rate is high enough, as it happens in RNA viruses, complementation could also occur among the descendants generated intracellularly from the same parental virus.

Complementation can buffer the harmful fitness effects of deleterious mutations, reducing the efficiency of negative selection. Therefore, one of its evolutionary consequences is an increased presence in populations of low fitness mutants that otherwise would have been eliminated. This reduction in the effect of deleterious mutations can be viewed as a form of robustness that would be only effective when conditions for complementation are adequate. If the conditions change, and complementation is no longer possible, that could result in the opposite: an increased fragility of the population, due to the high frequency of low fitness mutants and the lack of mechanisms of robustness that could have selected if deleterious mutations had shown a larger effect. These ideas were tested in an experiment in which several populations of phage φ6 were propagated at low and high multiplicity of infection to examine whether evolution of mutational robustness occurs differently for viruses evolved at low and high levels of co-infection that allow for different levels of complementation (Montville et al. 2005). Robustness was assayed by subjecting the evolved viruses to serial plaque-to-plaque transfers and measuring the rate of fitness decay. Viruses evolved under high co-infection decreased fitness faster than those evolved under low co-infection, indicating that they were less robust. The conclusion is that co-infection is probably advantageous in the short term, because it allows to mask the cost of deleterious mutations, and detrimental in the long term because it reduces the rate at which deleterious mutants are eliminated.

The question that can be asked is what would happen in a population in which the defective products were more abundant than the functional ones. In that case the whole population could collapse due to an interference effect of defective virus (or the products of their expression) on the replication of the rest of viruses. Given the high error rates of RNA viruses, additional increases in this parameter could render a mutant spectrum enriched in defective viruses that would exhaust the resources produced by the viable ones, leading the complete population to extinction. The existence of interference interactions of this type has been demonstrated in populations of lymphocytic choriomeningitis virus (LCMV) and FMDV propagated at increased error rate through the presence of mutagenic agents (Grande-Pérez et al. 2005; Perales et al. 2007). There are few studies concerning the consequences of the increase of the error rate in RNA phages. Nevertheless, in a study carried out with Qβ propagated through serial transfers in the presence of a mutagenic nucleoside analogue it was observed that individual viruses isolated from the mutagenized population could form lytic plaques and keep their viability under conditions that promoted the extinction of the whole population (Cases-González et al. 2008). This results suggests that the separation of individual viruses from their defective partners helps keep them viable, due to the elimination of interfering interactions.

12.5 Interactions That Arise as a Result of Segment Reassortment

When a single particle of a segmented phage infects a host cell, it reproduces clonally, but when multiple phages co-infect a cell, they can generate hybrid progeny through segment reassortment that allows for a new type of interactions similar to those occurring during sexual exchange in polyploid organisms

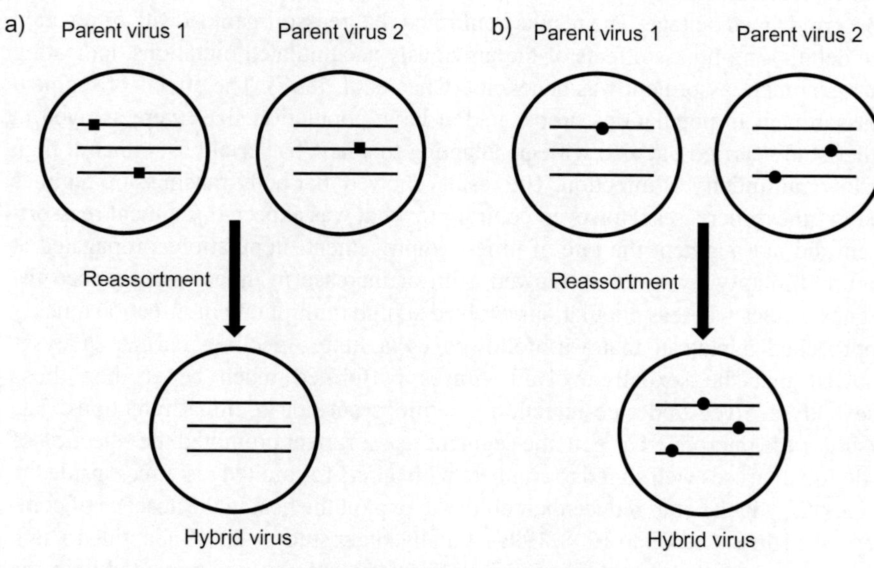

Fig. 12.3 Genome segment reassortment in φ6. (**a**) Co-infection of two viruses carrying deleterious mutations (shown as squares) in different genome segments can lead to the generation of mutation-free viruses through reassortment. (**b**) In the example shown in this part of the figure, co-infection leads to the generation of a hybrid virus containing all the beneficial mutations (shown as circles) present in the parental viruses

(Fig. 12.3). There are many evolution studies devoted to understanding the advantages of genome segmentation in RNA viruses. In them, the probability of reassortment is usually controlled through the manipulation of the multiplicity of infection: high values promote co-infection and reassorment whereas low values favor clonal multiplication. Below we will focus on the studies carried out with φ6, the prototype RNA phage with a segmented genome that can also be considered an appropriate model for the evolution of other segmented viruses, such as influenza.

The possible advantages of dividing the genetic information into several molecules depend on the population size. In small populations, where mutations accumulate independently of their selective value, the frequency of mutation-free individuals is progressively reduced and they can be lost by genetic drift. In asexual populations that loss is irreversible and could lead to extinction, which could be avoided if the wild-type genome were recreated through segment reassorment (Fig. 12.3a). In contrast, in large populations the advantages of genome segment reassorment are mainly due to the possibility of combining in the same genome several beneficial mutations that arose in different ones (Fig. 12.3b), accelerating adaptation in this way (Turner 2003).

To test whether segment reassortment provides advantages in small populations, several lineages of phage φ6 were propagated through plaque-to-plaque transfers with the aim of increasing the accumulation of deleterious mutations in the virus genomes. Then, the mutated viruses were grouped in pairs and used to co-infect the virus host under conditions that promote reassortment. If the reassortant viruses had higher fitness than either of the two parents, then it could be concluded that sex has conferred an advantage. The results confirmed that reassortment slowed or reverted the debilitating fitness effects of the previously accumulated mutations, indicating that genome reassorment was beneficial (Chao et al. 1997). The effects of segment reassortment in populations propagated at large population sizes were assayed in other study carried out also with φ6 adapting to a new bacterial host either at high or low multiplicity of infection. The results showed that both transmission regimes led to fitness increases. However, contrary to what was expected, segment reassortment did not augment the rate of fitness improvement. Populations propagated at low multiplicity of infection showed a linear increase in fitness and reached the highest values whereas populations evolved at high multiplicity of infection quickly approached a plateau that was followed by a fitness decline. Further analyses showed that the sexually-evolved viruses performed much better than those asexually-evolved under co-infection conditions but not in clonal infections. The results were interpreted as genome segment reassortment promoted the selection of individual viruses well suited to compete with others for limited resources inside the same cell, which could reduce their ability to exploit the host in the absence of competitors (Turner and Chao 1998, 1999). Finally, other study with φ6 adapting to high temperature showed that the beneficial effect of reassortment attenuated with the population size (Poon and Chao 2004). The conclusion is that genome segment reassortment seems to be more effective to remove deleterious combinations of mutations than to create new beneficial combinations that could accelerate adaptation.

A confounding factor in these studies is that since genome segment reassortment requires co-infection, it cannot be separated from complementation, meaning that the advantages conferred by a process may be obscured by the disadvantages conferred by the other. Genome segment reassortment promotes the generation of wild-type genomes from viruses containing mutations in different segments. In contrast, complementation favors the permanence of the mutants, at the expense of the resources produced by viable viruses. The relative contributions of complementation and reassortment were tested in a study in which wild-type φ6 was mixed with three single mutants, each carrying a deleterious mutation on a different segment, either in the absence or in the presence of co-infection (Froissart et al. 2004). It was found that the mutational load was eliminated faster in the absence of co-infection, suggesting that the disadvantages of complementation are higher than the benefits of sex.

12.6 Competitive Interactions During Adaptation

Populations adapt to changes in the environmental conditions through the generation of beneficial mutations that, in case they are not lost by genetic drift, subsequently increase their frequencies, until they become fixed. It is generally assumed that in well-adapted populations most mutations have deleterious effects on fitness, and those with beneficial effects represent only a small fraction of the total of possible mutations. Most models of adaptation also consider that the fitness effects of beneficial mutations are monotonically decreasing, which means that there is a low number of large-effect mutations and a majority of small-effect mutations (Eyre-Walker and Keightley 2007). This rarity of beneficial mutations would have as a theoretical consequence that in most biological systems there is enough time for the fixation of the first-appearing mutation before a second-one arises. Thus, at any time of an adaptive process, only a single mutation is in the way to fixation. This simple scenario changes if the frequency of beneficial mutations increases, something that is favored when genome replication takes place with high error rates and/or the population size is large, two conditions that are amply fulfilled in the case of RNA viruses. The existence of a wide mutant spectrum, containing or with the ability to generate in a short time a variety of beneficial mutations, has been usually assumed to be on the basis of the great adaptive ability of RNA viruses, and probably so it is. However, there are also some negative consequences of the simultaneous presence of different beneficial mutants within populations, due the fact that any advantageous mutant not only has to compete with the wild-type genome but also with others that are beneficial as well. This interference among mutants might lead to the loss of a fraction of beneficial mutations, delaying the fixation of the rest and slowing the speed of adaptation. The first evidences showing that RNA viruses adapted following a dynamics of this type were obtained with populations of VSV in which the rate of adaptation increased less than expected from the population size and mutation rate values (Miralles et al. 1999). Recombination and reassortment of genomic segments, as it happens in the *Cystoviridae* family, could gather in the same genome mutations that initially were in separate genomes, ameliorating in this way the consequences of interference among beneficial mutants.

There are several theoretical studies focused on modelling the dynamics of fixation of mutations depending on parameters such as the error rate, the genomes on which beneficial mutations can appear, and the different effects that mutations may have on fitness. Without coming into the details, the clonal interference model assumes that secondary beneficial mutations would always occur in the wild-type background and therefore competition would take place among mutants differing in only one mutation (Gerrish and Lenski 1998). In contrast to this, the multiple mutation model assumes that secondary beneficial mutations, differing only slightly in their fitness effects, may arise in any genetic background, independently of whether it already contains another beneficial mutation or not (Brunet et al. 2008; Desai and Fisher 2007). Finally, the full interference model considers that beneficial mutations that can largely differ in their fitness values may occur on any background, which adds extra complexity to the interactions that establish within the mutant spectrum (Park and Krug 2007). What is more probable to occur in RNA phages? Studies carried out with MS2 (Betancourt 2009), Qβ (Cabanillas et al. 2013), and φ6 (Dennehy et al. 2013) point to a complex scenario in which mutants containing different combinations of mutations, beneficial, neutral, and also deleterious, compete among them. Mutations in the same genome also interact, something that modifies their fitness effects and that makes it difficult to predict whether a particular mutation will be fixed or will be lost.

An interesting result is that the high mutation rate in MS2 seemed to alleviate clonal interference due to the frequent recurrence of beneficial mutations in different backgrounds, allowing the rapid combination in a single genome of beneficial mutations that initially appeared on different ones (Bollback and Huelsenbeck 2007). This pattern differs from the pattern observed in studies with DNA bacteriophages in which less complex interactions among genomes allowed for the sequential fixation of mutations (Holder and Bull 2001; Wichman et al. 1999). In the latter studies there was also competition among different mutants, but they differed in fewer mutations, as it corresponded to their lower mutation rates.

12.7 Future Perspectives: Interactions in the Natural World

We have shown that most of what we know about the relevance of interactions in the evolution of RNA phages comes from studies carried out under controlled conditions in the laboratory. The challenge now is to investigate how interactions among mutants and mutations shape RNA phage evolution in the real world. Natural environments are much more complex than laboratory environments and are influenced by a greater number of interacting and mutually depending variables. Throughout this chapter, we have pointed to the high error rates of RNA phages as the main mechanism responsible for the interactions that occur within their populations.

However, we don't know how error rates are modulated in nature by factors such as sun radiation, temperature or the metabolic state of the host, among others. All these factors might introduce variations in the phage error rate and cause changes in the number and type of interactions that arise within their populations. The population size is another element that is easily controlled in the laboratory and that is probably subjected to frequent fluctuations in natural environments. Adverse conditions can cause strong population bottlenecks, which will in all likelihood have negative consequences for adaptation of phages to new selective pressures. Nevertheless, in adapting populations containing different beneficial mutants that compete among them, bottlenecks could be beneficial, by eliminating interference interactions. Finally, the dependence of epistasis on the environment raises uncertainties concerning how the effect of this type of interactions translates to natural environments that are usually less stable than laboratory environments.

We have also shown that co-infection strongly favors interactions in RNA phages. The fact that in most environments viruses outnumber hots by a great margin suggests that the possibilities for co-infection are high in natural environments. The phenomenon of superinfection immunity conferred by lysogenic phage genomes inserted in the host genome limits co-infection in DNA phages. Since RNA phages do not establish this type of interaction with their hosts, it is to be expected that they have less limitations to penetrate in previously infected cells. There is even the possibility that different phage species infect the same cell, establishing inter-species interactions whose study is beyond the scope of this chapter but that can be highly relevant for virus evolution in nature.

Special mention deserve interactions among phages that arise as a consequence of genomic segments exchange in the family *Cystoviridae*. It has been demonstrated that even different virus species of this family are able to reassort. Although not all segment combinations will be viable, this kind of interactions could be a source for the generation of new phage species.

Summarizing, although interactions seem to be highly relevant for RNA phage evolution, their actual importance in nature remains to be elucidated. How frequent is co-infection in natural environments and how is it modulated by ecological parameters? What is the relative frequency of cooperation and defection and how do they manifest? Do error rates vary as a consequence of environmental factors? These and other questions are relevant research topics that will contribute to a better knowledge about how interactions guide the behavior of RNA phages in the real world.

Acknowledgments The author is grateful to Susanna Manrubia for the critical reading of the manuscript. Work was funded by the Spanish Ministerio de Ciencia, Innovación y Universidades, grant numbers FIS2017-89773-P and MDM-2017-0737 Unidad de Excelencia "María de Maeztu" Centro de Astrobiología (CSIC-INTA).

References

Acevedo A, Brodsky L, Andino R (2014) Mutational and fitness landscapes of an RNA virus revealed through population sequencing. Nature 505:686–690. https://doi.org/10.1038/nature12861

Aguirre J, Manrubia SC (2007) Out-of-equilibrium competitive dynamics of quasispecies. EPL 77:38001. https://doi.org/10.1209/0295-5075/77/38001

Andino R, Domingo E (2015) Viral quasispecies. Virology 479-480:46–51. https://doi.org/10.1016/j.virol.2015.03.022

Arribas M, Aguirre J, Manrubia S, Lázaro E (2018) Differences in adaptive dynamics determine the success of virus variants that propagate together. Virus Evol 4:vex043. https://doi.org/10.1093/ve/vex043

Batschelet E, Domingo E, Weissmann C (1976) The proportion of revertant and mutant phage in a growing population, as a function of mutation and growth rate. Gene 1:27–32

Beaucourt S, Bordería AV, Coffey LL, Gnädig NF, Sanz-Ramos M, Beeharry Y, Vignuzzi M (2011) Isolation of fidelity variants of RNA viruses and characterization of virus mutation frequency. J Vis Exp 52:2953. https://doi.org/10.3791/2953

Belshaw R, Gardner A, Rambaut A, Pybus OG (2008) Pacing a small cage, mutation and RNA viruses. Trends Ecol Evol 23:188–193. https://doi.org/10.1016/j.tree.2007.11.010

Betancourt AJ (2009) Genomewide patterns of substitution in adaptively evolving populations of the RNA bacteriophage MS2. Genetics 181:1535–1544. https://doi.org/10.1534/genetics.107.085837

Biebricher CK, Eigen M (2005) The error threshold. Virus Res 107:117–127. https://doi.org/10.1016/j.virusres.2004.11.002

Bollback JP, Huelsenbeck JP (2007) Clonal interference is alleviated by high mutation rates in large populations. Mol Biol Evol 24:1397–1406. https://doi.org/10.1093/molbev/msm056

Bradwell K, Combe M, Domingo-Calap P, Sanjuán R (2013) Correlation between mutation rate and genome size in riboviruses: mutation rate of bacteriophage Qβ. Genetics 195:243–251. https://doi.org/10.1534/genetics.113.154963

Brunet E, Rouzine IM, Wilke CO (2008) The stochastic edge in adaptive evolution. Genetics 179:603–620. https://doi.org/10.1534/genetics.107.079319

Burch CL, Chao L (1999) Evolution by small steps and rugged landscapes in the RNA virus phi6. Genetics 151:921–927

Burch CL, Chao L (2004) Epistasis and its relationship to canalization in the RNA virus phi6. Genetics 167:559–567. https://doi.org/10.1534/genetics.103.021196

Cabanillas L, Arribas M, Lázaro E (2013) Evolution at increased error rate leads to the coexistence of multiple adaptive pathways in an RNA virus. BMC Evol Biol 13:11. https://doi.org/10.1186/1471-2148-13-11

Callanan J, Stockdale SR, Shkoporov A, Draper LA, Ross RP, Hill C (2018) RNA phage biology in a metagenomic era. Viruses 10:E386. https://doi.org/10.3390/v10070386

Cameron CE, Moustafa IM, Arnold JJ (2016) Fidelity of Nucleotide incorporation by the RNA-dependent RNA Polymerase from Poliovirus. Enzyme 39:293–323. https://doi.org/10.1016/bs.enz.2016.02.002

Carrasco P, de la Iglesia F, Elena SF (2007) Distribution of fitness and virulence effects caused by single-nucleotide substitutions in Tobacco Etch virus. J Virol 81:12979–12984. https://doi.org/10.1128/JVI.00524-07

Cases-González C, Arribas M, Domingo E, Lázaro E (2008) Beneficial effects of population bottlenecks in an RNA virus evolving at increased error rate. J Mol Biol 384:1120–1129. https://doi.org/10.1016/j.jmb.2008.10.014

Catalán P, Arias CF, Cuesta JA, Manrubia S (2017) Adaptive multiscapes: an up-to-date metaphor to visualize molecular adaptation. Biol Direct 12:7. https://doi.org/10.1186/s13062-017-0178-1

Caudle SB, Miller CR, Rokyta DR (2014) Environment determines epistatic patterns for a ssDNA virus. Genetics 196:267–279. https://doi.org/10.1534/genetics.113.158154

Chao L, Tran TT, Tran TT (1997) The advantage of sex in the RNA virus phi6. Genetics 147:953–959
Chetverin AB (2018) Thirty years of studies of Qβ replicase: what have we learned and what is yet to be learned? Biochemistry (Mosc) 83(Suppl 1):S19–S32. https://doi.org/10.1134/S0006297918140031
Chetverin AB, Kopein DS, Chetverina HV, Demidenko AA, Ugarov VI (2005) Viral RNA-directed RNA polymerases use diverse mechanisms to promote recombination between RNA molecules. J Biol Chem 280:8748–8755. https://doi.org/10.1074/jbc.M412684200
Cuevas JM, Elena SF, Moya A (2002) Molecular basis of adaptive convergence in experimental populations of RNA viruses. Genetics 162:533–542
Cuevas JM, Domingo-Calap P, Pereira-Gómez M, Sanjuán R (2009) Experimental evolution and population genetics of RNA viruses. The Open Evolution Journal 3:9–16
Dennehy JJ, Duffy S, O'Keefe KJ, Edwards SV, Turner PE (2013) Frequent coinfection reduces RNA virus population genetic diversity. J Hered 104:704–712. https://doi.org/10.1093/jhered/est038
Desai MM, Fisher DS (2007) Beneficial mutation selection balance and the effect of linkage on positive selection. Genetics 176:1759–1798. https://doi.org/10.1534/genetics.106.067678
Dolan PT, Whitfield ZJ, Andino R (2018) Mechanisms and concepts in RNA virus population dynamics and evolution. Annu Rev Virol 5:69–92. https://doi.org/10.1146/annurev-virology-101416-041718
Domingo E, Sabo D, Taniguchi T, Weissmann C (1978) Nucleotide sequence heterogeneity of an RNA phage population. Cell 13:735–744
Domingo E, Holland JJ, Ahlquist P (1988) RNA genetics, vol I, II, III. CRC Press, Boca Raton
Domingo E, Sheldon J, Perales C (2012) Viral quasispecies evolution. Microbiol Mol Biol Rev 76:159–216. https://doi.org/10.1128/MMBR.05023-11
Domingo-Calap P, Cuevas JM, Sanjuán R (2009) The fitness effects of random mutations in single-stranded DNA and RNA bacteriophages. PLoS Genet 5:e1000742. https://doi.org/10.1371/journal.pgen.1000742
Drake JW (1991) A constant rate of spontaneous mutation in DNA-based microbes. Proc Natl Acad Sci USA 88:7160–7164
Drake JW, Holland JJ (1999) Mutation rates among RNA viruses. Proc Natl Acad Sci USA 96:13910–13913
Duarte EA, Novella IS, Ledesma S, Clarke DK, Moya A, Elena SF, Domingo E, Holland JJ (1994) Subclonal components of consensus fitness in an RNA virus clone. J Virol 68:4295–4301
Eigen M (1971) Self-organization of matter and the evolution of biological macromolecules. Naturwissenschaften 58:465–523
Eigen M, Biebricher CK (1988) Sequence space and quasispecies distribution. In: Domingo E, Ahlquist P, Holland JJ (eds) RNA genetics, vol 3. CRC Press, Boca Raton, pp 211–245
Elena SF, Carrasco P, Daròs JA, Sanjuán R (2006) Mechanisms of genetic robustness in RNA viruses. EMBO Rep 7:168–173. https://doi.org/10.1038/sj.embor.7400636
Elena SF, Solé RV, Sardanyés J (2010) Simple genomes, complex interactions: epistasis in RNA virus. Chaos 20:026106. https://doi.org/10.1063/1.3449300
Emori Y, Iba H, Okada Y (1983) Transcriptional regulation of three double-stranded RNA segments of bacteriophage Φ6 *in vitro*. J Virol 46:196–203
Escarmís C, Dávila M, Charpentier N, Bracho A, Moya A, Domingo E (1996) Genetic lesions associated with Muller's ratchet in an RNA virus. J Mol Biol 264:255–267
Escarmís C, Dávila M, Domingo E (1999) Multiple molecular pathways for fitness recovery of an RNA virus debilitated by operation of Muller's ratchet. J Mol Biol 285:495–505
Escarmís C, Gómez-Mariano G, Dávila M, Lázaro E, Domingo E (2002) Resistance to extinction of low fitness virus subjected to plaque-to-plaque transfers: diversification by mutation clustering. J Mol Biol 315:647–661. https://doi.org/10.1006/jmbi.2001.5259
Escarmís C, Lázaro E, Manrubia SC (2006) Population bottlenecks in quasispecies dynamics. Curr Top Microbiol Immunol 299:141–170. https://doi.org/10.1007/3-540-26397-7_5

Escarmís C, Lázaro E, Arias A, Domingo E (2008) Repeated bottleneck transfers can lead to non-cytocidal forms of a cytopathic virus: implications for viral extinction. J Mol Biol 376:367–379. https://doi.org/10.1016/j.jmb.2007.11.042

Eyre-Walker A, Keightley PD (2007) The distribution of fitness effects of new mutations. Nat Rev Genet 8:610–618. https://doi.org/10.1038/nrg2146

Froissart R, Wilke CO, Montville R, Remold SK, Chao L, Turner PE (2004) Co-infection weakens selection against epistatic mutations in RNA viruses. Genetics 168:9–19. https://doi.org/10.1534/genetics.104.030205

Gago S, Elena SF, Flores R, Sanjuán R (2009) Extremely high mutation rate of a hammerhead viroid. Science 323:1308. https://doi.org/10.1126/science.1169202

Gerrish PJ, Lenski RE (1998) The fate of competing beneficial mutations in an asexual population. Genetica 102-103:127–144

Grande-Pérez A, Lázaro E, Lowenstein P, Domingo E, Manrubia SC (2005) Suppression of viral infectivity through lethal defection. Proc Natl Acad Sci USA 102:4448–4452. https://doi.org/10.1073/pnas.0408871102

Hatfull GF (2015) Dark matter of the biosphere: the amazing world of bacteriophage diversity. J Virol 89:8107–8110. https://doi.org/10.1128/JVI.01340-15

Holder KK, Bull JJ (2001) Profiles of adaptation in two similar viruses. Genetics 159:1393–1404

Holland JJ, Spindler K, Horodyski F, Grabau E, Nichol S, VandePol S (1982) Rapid evolution of RNA genomes. Science 215:1577–1585

Holmes EC (2003) Error thresholds and the constraints to RNA virus evolution. Trends Microbiol 11:543–546. https://doi.org/10.1016/j.tim.2003.10.006

Jäälinoja HT, Huiskonen JT, Butcher SJ (2007) Electron cryomicroscopy comparison of the architectures of the enveloped bacteriophages Φ6 and Φ8. Structure 15:157–167. https://doi.org/10.1016/j.str.2006.12.004

Kashiwagi A, Sugawara R, Sano Tsushima F, Kumagai T, Yomo T (2014) Contribution of silent mutations to thermal adaptation of RNA bacteriophage Qβ. J Virol 88:11459–11468. https://doi.org/10.1128/JVI.01127-14

Krishnamurthy SR, Janowski AB, Zhao G, Barouch D, Wang D (2016) Hyperexpansion of RNA bacteriophage diversity. PLoS Biol 14:e1002409. https://doi.org/10.1371/journal.pbio.1002409

Kvitek DJ, Sherlock G (2011) Reciprocal sign epistasis between frequently experimentally evolved adaptive mutations causes a rugged fitness landscape. PLoS Genet 7:e1002056. https://doi.org/10.1371/journal.pgen.1002056

Lauring AS, Frydman J, Andino R (2013) The role of mutational robustness in RNA virus evolution. Nat Rev Microbiol 11:327–336. https://doi.org/10.1038/nrmicro3003

Lázaro E, Escarmís C, Domingo E, Manrubia SC (2002) Modeling viral genome fitness evolution associated with serial bottleneck events: evidence of stationary states of fitness. J Virol 76:8675–8681. https://doi.org/10.1128/jvi.76.17.8675-8681.2002

Lázaro E, Escarmís C, Pérez-Mercader J, Manrubia SC, Domingo E (2003) Resistance of virus to extinction on bottleneck passages: study of a decaying and fluctuating pattern of fitness loss. Proc Natl Acad Sci USA 100:10830–10835. https://doi.org/10.1073/pnas.1332668100

Lee KH, Miller CR, Nagel AC, Wichman HA, Joyce P, Ytreberg FM (2011) First-step mutations for adaptation at elevated temperature increase capsid stability in a virus. PLoS One 6:e25640. https://doi.org/10.1371/journal.pone.0025640

Manrubia S, Lázaro E (2016) Getting to know viral evolutionary strategies: towards the next generation of Quasispecies models. Curr Top Microbiol Immunol 392:201–217. https://doi.org/10.1007/82_2015_457

Manrubia SC, Lázaro E, Pérez-Mercader J, Escarmís C, Domingo E (2003) Fitness distributions in exponentially growing asexual populations. Phys Rev Lett 90:188102. https://doi.org/10.1103/PhysRevLett.90.188102

Manrubia SC, Escarmís C, Domingo E, Lázaro E (2005) High mutation rates, bottlenecks, and robustness of RNA viral quasispecies. Gene 347:273–282. https://doi.org/10.1016/j.gene.2004.12.033

Mansky LM, Temin HM (1995) Lower in vivo mutation rate of human immunodeficiency virus type 1 than that predicted from the fidelity of purified reverse transcriptase. J Virol 69:5087–5094
McDonald SM, Nelson MI, Turner PE, Patton JT (2016) Reassortment in segmented RNA viruses: mechanisms and outcomes. Nat Rev Microbiol 14:448–460. https://doi.org/10.1038/nrmicro.2016.46
Michalakis Y, Roze D (2004) Evolution. Epistasis in RNA viruses Science 306:1492–1493. https://doi.org/10.1126/science.1106677
Mindich L (1999) Precise packaging of the three genomic segments of the double-stranded-RNA bacteriophage Φ6. Microbiol Mol Biol Rev 63:149–160
Mindich L (2004) Packaging, replication and recombination of the segmented genome of bacteriophage Φ6 and its relatives. Virus Res 101:83–92. https://doi.org/10.1016/j.virusres.2003.12.008
Mindich L, Sinclair JF, Levine D, Cohen J (1976) Genetic studies of temperature-sensitive and nonsense mutants of bacteriophage phi6. Virology 75:218–223
Miralles R, Gerrish PJ, Moya A, Elena SF (1999) Clonal interference and the evolution of RNA viruses. Science 285:1745–1747
Montville R, Froissart R, Remold SK, Tenaillon O, Turner PE (2005) Evolution of mutational robustness in an RNA virus. PLoS Biol 3:e381. https://doi.org/10.1371/journal.pbio.0030381
Nijhuis M, van Maarseveen NM, Boucher CA (2009) Antiviral resistance and impact on viral replication capacity: evolution of viruses under antiviral pressure occurs in three phases. Handb Exp Pharmacol 189:299–320. https://doi.org/10.1007/978-3-540-79086-0_11
Novella IS, Duarte EA, Elena SF, Moya A, Domingo E, Holland JJ (1995) Exponential increases of RNA virus fitness during large population transmissions. Proc Natl Acad Sci USA 92:5841–5844
Olsthoorn RCL, Van Duin J (2011) Leviviridae-positive sense RNA viruses-positive sense RNA viruses. In: international committee on taxonomy of viruses (ICTV). Available online: https://talk.ictvonline.org/ictv-reports/ictv_9th_report/positive-sense-rna-viruses-2011/w/posrna_viruses/263/leviviridae. Cited 20 Nov 2019
Palasingam K, Shaklee PN (1992) Reversion of Qbeta RNA phage mutants by homologous RNA recombination. J Virol 66:2435–2442
Park SC, Krug J (2007) Clonal interference in large populations. Proc Natl Acad Sci USA 104:18135–18140. https://doi.org/10.1073/pnas.0705778104
Perales C, Mateo R, Mateu MG, Domingo E (2007) Insights into RNA virus mutant spectrum and lethal mutagenesis events: replicative interference and complementation by multiple point mutants. J Mol Biol 369:985–1000. https://doi.org/10.1016/j.jmb.2007.03.074
Perales C, Lorenzo-Redondo R, López-Galíndez C, Martínez MA, Domingo E (2010) Mutant spectra in virus behavior. Future Virol 5:679–698. https://doi.org/10.2217/fvl.10.61
Poelwijk FJ, Tănase-Nicola S, Kiviet DJ, Tans SJ (2011) Reciprocal sign epistasis is a necessary condition for multi-peaked fitness landscapes. J Theor Biol 272:141–144. https://doi.org/10.1016/j.jtbi.2010.12.015
Poon A, Chao L (2004) Drift increases the advantage of sex in RNA bacteriophage Phi6. Genetics 166:19–24. https://doi.org/10.1534/genetics.166.1.19
Poranen MM, Mäntynen S (2017) ICTV virus taxonomy profile: Cystoviridae. J Gen Virol 98:2423–2424. https://doi.org/10.1099/jgv.0.000928
Remold SK, Rambaut A, Turner PE (2008) Evolutionary genomics of host adaptation in vesicular stomatitis virus. Mol Biol Evol 25:1138–1147. https://doi.org/10.1093/molbev/msn059
Rokyta DR, Joyce P, Caudle SB, Miller C, Beisel CJ, Wichman HA (2011) Epistasis between beneficial mutations and the phenotype-to-fitness map for a ssDNA virus. PLoS Genet 7:e1002075. https://doi.org/10.1371/journal.pgen.1002075
Sackman AM, Rokyta DR (2018) Additive phenotypes underlie epistasis of fitness effects. Genetics 208:339–348. https://doi.org/10.1534/genetics.117.300451
Sanjuán R, Elena SF (2006) Epistasis correlates to genomic complexity. Proc Natl Acad Sci USA 103:14402–14405. https://doi.org/10.1073/pnas.0604543103

Sanjuán R, Moya A, Elena SF (2004) The contribution of epistasis to the architecture of fitness in an RNA virus. Proc Natl Acad Sci USA 101:15376–15379. https://doi.org/10.1073/pnas.0404125101

Sanjuan R, Nebot MR, Chirico N, Mansky LM, Belshaw R (2010) Viral mutation rates. J Virol 84:9733–9748. https://doi.org/10.1128/JVI.00694-10

Schuster (2010) Genotypes and phenotypes in the evolution of molecules. In: Caetono-Anolles G (ed) Evolutionary genomics and systems biology. Wiley Online Library, Hoboken, pp 123–152

Schuster P (2016) Quasispecies on fitness landscapes. Curr Top Microbiol Immunol 392:61–120. https://doi.org/10.1007/82_2015_469

Silander OK, Weinreich DM, Wright KM, O'Keefe KJ, Rang CU, Turner PE, Chao L (2005) Widespread genetic exchange among terrestrial bacteriophages. Proc Natl Acad Sci USA 102:19009–19014. https://doi.org/10.1073/pnas.0503074102

Steinhauer DA, Domingo E, Holland JJ (1992) Lack of evidence for proofreading mechanisms associated with an RNA virus polymerase. Gene 122:281–288

Stern A, Bianco S, Yeh MT, Wright C, Butcher K, Tang C, Nielsen R, Andino R (2014) Costs and benefits of mutational robustness in RNA viruses. Cell Rep 8:1026–1036. https://doi.org/10.1016/j.celrep.2014.07.011

Stich M, Briones C, Manrubia SC (2007) Collective properties of evolving molecular quasispecies. BMC Evol Biol 7:110. https://doi.org/10.1186/1471-2148-7-110

Turner PE (2003) Searching for the advantages of virus sex. Orig Life Evol Biosph 33:95–108. https://doi.org/10.1023/a:1023973015054

Turner PE, Chao L (1998) Sex and the evolution of intrahost competition in RNA virus phi6. Genetics 150:523–532

Turner PE, Chao L (1999) Prisoner's dilemma in an RNA virus. Nature 398:441–443

Weinreich DM, Watson RA, Chao L (2005) Perspective: sign epistasis and genetic constraint on evolutionary trajectories. Evolution 59:1165–1174

Whitlock MC, Phillips PC, Moore FB-G, Tonsor SJ (1995) Multiple fitness peaks and epistasis. Annu Rev Ecol Syst 26:601–629

Wichman HA, Badgett MR, Scott LA, Boulianne CM, Bull JJ (1999) Different trajectories of parallel evolution during viral adaptation. Science 285:422–424

Wright S (1931) Evolution in Mendelian populations. Genetics 16:97–159

Zhao L, Seth-Pasricha M, Stemate D, Crespo-Bellido A, Gagnon J, Draghi J, Duffy S (2019) Existing host range mutations constrain further emergence of RNA viruses. J Virol 93:e01385–e01318. https://doi.org/10.1128/JVI.01385-18

Chapter 13
ssRNA Phages: Life Cycle, Structure and Applications

Kaspars Tars

Abstract ssRNA phages belonging to the family Leviviridae are among the tiniest viruses, infecting various Gram-negative bacteria by adsorption to their pilus structures. Due to their simplicity, they have been intensively studied as models for understanding various problems in molecular biology and virology. Several of the studied ssRNA characteristics, such as coat protein–RNA interactions and the ability to readily form virus-like particles in recombinant expression systems, have fueled many practical applications such as RNA labeling and tracking systems and vaccine development. In this chapter, we review the life cycle, structure and applications of these small yet fascinating viruses.

13.1 Introduction

Bacteriophages belonging to the family *Leviviridae* are among the simplest known viruses, exhibiting positive-sense single-stranded RNA (ssRNA) genomes of just 3.5-4.5 kilobases, typically encoding only 4 proteins. Morphologically, *Leviviridae* particles present a spherical or isometric shape with a diameter of 30 nm.

Due to their simplicity, ssRNA phages have been used as models to study various processes in molecular biology and virology, including translation repression, RNA-protein interactions and virus evolution. Remarkably, the ssRNA phage MS2 was the first form of life for which a complete sequence of the genome was obtained, in 1976 in Walter Fiers's lab (Fiers et al. 1976). Four decades later, MS2 was the first form of life for which a 3D structure of the genome was established (Koning et al. 2016; Dai et al. 2017). In the meanwhile, ssRNA phages and their components have been found to be useful in many practical applications such as MS2 tagging, armoured RNA technology, drug delivery, nanoreactor construction and vaccine development, as discussed later in this chapter.

K. Tars (✉)
Biomedical Research and Study Center, Riga, Latvia
e-mail: kaspars@biomed.lu.lv

G. Witzany (ed.), *Biocommunication of Phages*,
https://doi.org/10.1007/978-3-030-45885-0_13

13.2 Organization of the Genome, Classification and Relationship to Other Viruses

The genomes of all known ssRNA phages encode a so-called maturation, or "A" protein (AP, also known as A2 in genus Allolevivirus), a coat protein (CP) and a replicase subunit (RP) (Fig. 13.1). AP is a minor structural protein necessary for attachment to bacterial pilus structures, and CP is the major building block of the phage capsid. RP is the catalytic subunit of RNA-dependent RNA polymerase (RdRp). Additionally, many studied ssRNA phages encode a separate lysis protein (LP).

Family *Leviviridae* is further divided into two genera – *Levivirus* and *Allolevivirus*. In *Leviviridae* genomes, AP is encoded at the 5′ end, followed by CP and RP, always in the same order. In contrast, the position of LP in the genome may vary. Typically, the LP gene overlaps partially (Atkins et al. 1979; Olsthoorn et al. 1995) or completely (Kazaks et al. 2011; Rumnieks and Tars 2012) with the CP and/or RP genes,

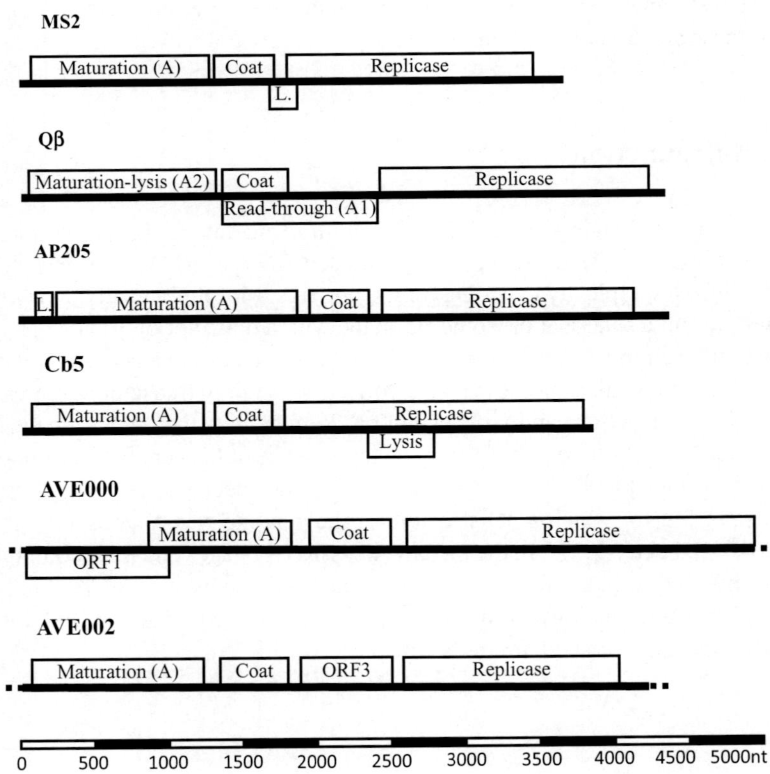

Fig. 13.1 Genomes of ssRNA phages MS2, Qβ, AP205, CB5, AVE000 and AVE002. Genes are shown as boxes. L. stands for lysis protein. In the incomplete sequences of phages AVE001 and AVE002, ORF1 and ORF3 of genes encoding two putative proteins are shown

albeit in different reading frames, except in the phage AP205, in which LP is encoded by a very short ORF at the very 5′ end of the genome(Klovins et al. 2002). Phages belonging to genus *Allolevivirus* lack LP entirely and utilize a bifunctional A2 protein to carry out cell lysis (Karnik and Billeter 1983; Winter and Gold 1983; Bernhardt et al. 2001). Another hallmark of alloleviviruses is a gene encoding the read-through minor coat protein A1, whose function is somewhat unclear, although it has been shown that it is necessary for infection (Hofstetter et al. 1974).

The *Levivirus-Allolevivirus* division was assigned many decades ago when most studies on ssRNA phages were limited to those phages infecting *Escherichia coli* bacteria. Many other ssRNA phages have since been discovered and sequenced, and they do not seem to be particularly similar to any *Levivirus* or *Allolevivirus* representatives. Furthermore, *Levivirus* and *Allolevivirus* members are more closely related to each other than to many of the newly sequenced phages. Therefore, the classical *Levivirus-Allolevivirus* division is clearly outdated, and a new classification based solely on the sequence similarities of proteins should be introduced.

For quite some time, the sequences of relatively few ssRNA phages were known, but this situation changed dramatically in 2016, when two metagenome studies revealed more than 200 novel but somewhat incomplete genomes (Krishnamurthy et al. 2016; Shi et al. 2016). Additionally, 31 partial sequences including full-length CP gene were extracted from the NCBI nucleotide (nt) and environmental nucleotide (env_nt) sequence databases (Lieknina et al. 2019). Although the corresponding phages themselves cannot be reconstructed from these partial sequences and the respective hosts remain unknown, these metagenome sequences revealed new data about the diversity of genomes and their encoded proteins. The sequences were so diverse that in many cases reliable assignment to the *Leviviridae* family was possible due only to the presence of the RP gene, the central part of which is remarkably conserved among all ssRNA phages. In contrast, the sequences of other proteins did not show any similarities to previously characterized members in many cases; for example, the presence of CP ORF could be detected only by assuming its approximate length and placement between the AP and RP genes. Some of the metagenome sequences (AVE002, AVE003, ESO001 and EMM000) seemed to contain two ORFs between the AP and RP genes. The first ORF after AP apparently encodes CP, as the expression of this gene from phages AVE002, ESO001 and EMM000 yielded virus-like particles (VLPs) that were morphologically similar to those of other phages (Lieknina et al. 2019). The expression of the second ORF (designated "ORF3", since it is the third ORF in the genome after AP and CP) of phages found in *E. coli* yielded either an insoluble product (AVE002, ESO001) or no product at all (AVE003, EMM000)(Lieknina et al. 2019). No cell lysis was observed in either case, suggesting that ORF3 is unlikely to be a candidate for encoding LP. It might still be the case that ORF3 is responsible for achieving lysis in host bacteria but is unable to do so in *E. coli*. However, the expression of LPs from other phages from different hosts has so far always led to cell lysis even when expressed in *E. coli*. Therefore, the identity and function of ORF3 remain unknown. Alternatively, ORF3 may be an artifact produced by sequencing errors. Thus, the ORF3 gene could actually be a 5′ portion of the RP gene or a 3′ extension of the CP gene, similar to the

A1 read-through protein in alloleviviruses. ORF3 could also be a separate protein performing the same function as A1 in alloleviviruses. However, this is rather unlikely since the "error model" would require multiple sequencing errors, and ORF3 does not display any homology to known proteins, including RP or A1, of any phage. Additionally, at least in AVE002, there are clearly defined SD sequences prior to the ORF3 and RP genes, further suggesting that ORF3 encodes a separate protein.

One of the metagenome sequences (AVE000) exhibited other unusual traits (Krishnamurthy et al. 2016). First, the incomplete genome was 4950 nucleotides long; therefore, the actual size of the full-length genome could be over 5000 bases. Second, the sequence included the another putative ORF with unknown function at the 5′ end, partially overlapping with AP in a different reading frame. The larger-than-usual size of the AVE000 genome might require larger particles, for which there is some indirect evidence, as discussed further in the VLP section.

Leviviridae members are not evolutionarily closely related to any other bacteriophages except for a very distant relationship to the only known dsRNA phages belonging to family *Cystoviridae*, since it is believed that all RNA viruses share common ancestry. The closest *Leviviridae* relatives appear to occur within genus *Mitovirus*, belonging to the family *Narnaviridae*, which consists of ssRNA viruses that infect the mitochondria of fungi (Hillman and Cai 2013). Mitoviruses do not have a protein capsid, and their genomes encode only one protein, RNA-dependent RNA polymerase (RdRp), which displays some sequence similarities to the RP of *Leviviridae* phages. The RdRp of the other *Narnaviridae* genus *Narnavirus* (infecting the cytoplasm of fungi) is even more distantly related to RP. Ourmia-like viruses, which infect plants, fungi and inverteberates, also harbor an RdRp that is distantly related to RP (Shi et al. 2016). Based on these similarities it has been suggested to create a separate 'Narna-Levi' clade consisting of *Leviviridae*, *Narnaviridae* and Ourmia-like viruses (Shi et al. 2016). However, the coat proteins of Ourmia-like viruses are clearly different from those of *Leviviridae* phages and, judging from similarities to other families, present a jelly roll β-barrel topology typical of many, if not most, viruses (Rastgou et al. 2009). Therefore, it appears that some of the mentioned viruses might carry RdRp genes that have been "borrowed" from each other via horizontal gene transfer. RP is the only protein from the *Leviviridae* family that clearly presents relatives within other viruses, while CP, AP, A1 and LP do not seem to be related to proteins from any other viruses or, indeed, to any other proteins in general.

13.3 Infectious Cycle

13.3.1 Adsorption, Genome Ejection and Penetration

ssRNA phages infect various Gram-negative bacteria by attachment to the sides of their pilus structures. In the case of *Escherichia coli*, most commonly conjugative plasmid-encoded F pili are used (Crawford and Gesteland 1964), but ssRNA phages infecting other bacteria have been reported to attach to completely different genome-encoded pili, such as polar pili in *Pseudomonas* (Bradley 1966) or swarmer cell -specific pili in *Caulobacter* (Schmidt 1966). Some phages display specificity to several hosts carrying certain conjugative pilus-encoding plasmids. For example, the bacteriophage PRR1 is specific to a variety of bacteria, provided that they harbour the incompatibility group IncP plasmid, which produces so-called P pili on the cell surface that serve as receptors for the phage (Olsen and Thomas 1973). Likewise, phage M is able to infect various bacteria carrying the IncM plasmid (Coetzee et al. 1983). However, most of the available knowledge about the initial steps of infection comes from studies of the closely related F-pili-specific phages MS2, R17 and f2. The physiological function of F-pili is to bind and bring donor and recipient bacteria close enough together for the subsequent exchange of genetic material. This is accomplished by retraction, which is essentially the shortening of F-pili, composed of many identical F-pilin protein monomers, via depolymerization. After retraction, the F plasmid enters bacteria through the conjugative pore formed by plasmid-encoded proteins. Apparently, ssRNA phages are able to repurpose this plasmid-transfer machinery to deliver their genomes inside the target cells.

After attachment to pili on bacterial cell surface, AP is cleaved in two parts (Krahn et al. 1972), is ejected from particles together with the genome and enters bacteria via a poorly understood mechanism. A complex of only the AP and the genomic RNA is infectious to the cell (Shiba and Miyake 1975), suggesting that CP only protects viral RNA from environment and plays no other role in infectivity process. Although phage MS2 particles attach to both isolated and cell-bound F-pili (Valentine and Strand 1965), this attachment results in AP cleavage and genome ejection only if pili are attached to viable *E. coli* cells (Danziger and Paranchych 1970). Additionally, there is no evidence that the adsorption of the phage particles to F-pili actively triggers their retraction. However, it is known that F-pili can undergo retraction and elongation spontaneously (Clarke et al. 2008). Therefore, it is reasonable to assume that after adsorption, the occasional retraction of F-pili leads phage particles to be transported close to cell surface, triggering AP cleavage and genome ejection. The exact mechanism of further transport across the cell membrane remains unknown, but it can be speculated that AP, together with RNA, uses a conjugative pore intended for the transport of F-plasmid DNA.

13.3.2 Synthesis of Proteins, Replication and Assembly

Once the genomic RNA enters the cytoplasm of the host cell, it can act directly as mRNA for protein synthesis. The amounts of proteins to be translated are tightly regulated by a variety of mechanisms, including the accessibility of ribosome binding sites, the translational coupling of genes, the formation of RNA secondary structure elements, protein-RNA interactions and read-through of translation stop codons. The known expression regulation mechanisms of the MS2 phage are shown in Fig. 13.2. In the folded genome of the bacteriophage MS2, only the CP gene is initially accessible to the ribosome (Van Duin and Tsareva 2006). During the translation of the CP gene, the RNA secondary structure is partially disrupted, allowing access to the ribosome binding site of RP. Once the RP gene has been translated, the replication of the genome may begin. RP is the catalytic subunit of RNA-dependent RNA polymerase (RdRp), but the fully functional RdRp holoenzyme contains 3 additional proteins hijacked from the cellular translation machinery – the elongation factors EF-Tu and EF-Ts (Blumenthal et al. 1972) and ribosomal protein S1 (Wahba et al. 1974). Structural and functional studies have revealed that cellular components participate in the template binding, recognition and stabilization of the holoenzyme complex. By far the best-studied RdRp of ssRNA phages is that of phage Qβ due to its stability. The function, mechanism and structure of Qβ replicase have been studied in great detail – see (Tomita 2014) for a review. Qβ RdRp is among the fastest known RNA polymerases and can generate up to 10^{10} copies of some artificially selected templates in 10 min (Chetverina and Chetverin 1993). For this reason, it has been proposed that Qβ RdRp could be used for the "room-temperature PCR" amplification of RNA, but template specificity requirements have prevented this type of application so far (Ugarov and Chetverin 2008). The RdRps of *Leviviridae* phages are remarkably specific for both the (+) and (−) strands of their own genomic RNAs and generally do not recognize other arbitrary templates. Additionally, it has been shown that MS2 RdRp does not recognize Qβ RNA or *vice versa* (Haruna and Spiegelman 1965). The specificity factors include sequences at the ends and in the middle of the genome, along with a high degree of secondary structure and long-distance base-pairing interactions (see (Rumnieks and Tars 2018) for a review).

In the folded RNA genome of phage MS2, AP translation is prevented by the formation of an extensive secondary structure around the ribosome binding site (Groeneveld et al. 1995). However, at the beginning of replication, the 5'end of the genome adopts other temporary conformations in which RBS is exposed to ribosomes, allowing the translation of the AP gene (van Meerten et al. 2001). This mechanism ensures that on average only one copy of AP is generated during each replication cycle, which makes sense since only one AP molecule must be present in the mature virion.

Although the initiation codon of the CP gene is accessible in the folded genome, the CP gene lacks a strong RBS sequence. Instead, the so-called S site is located prior to the CP gene. The S site shows affinity for the S1 protein, which is present

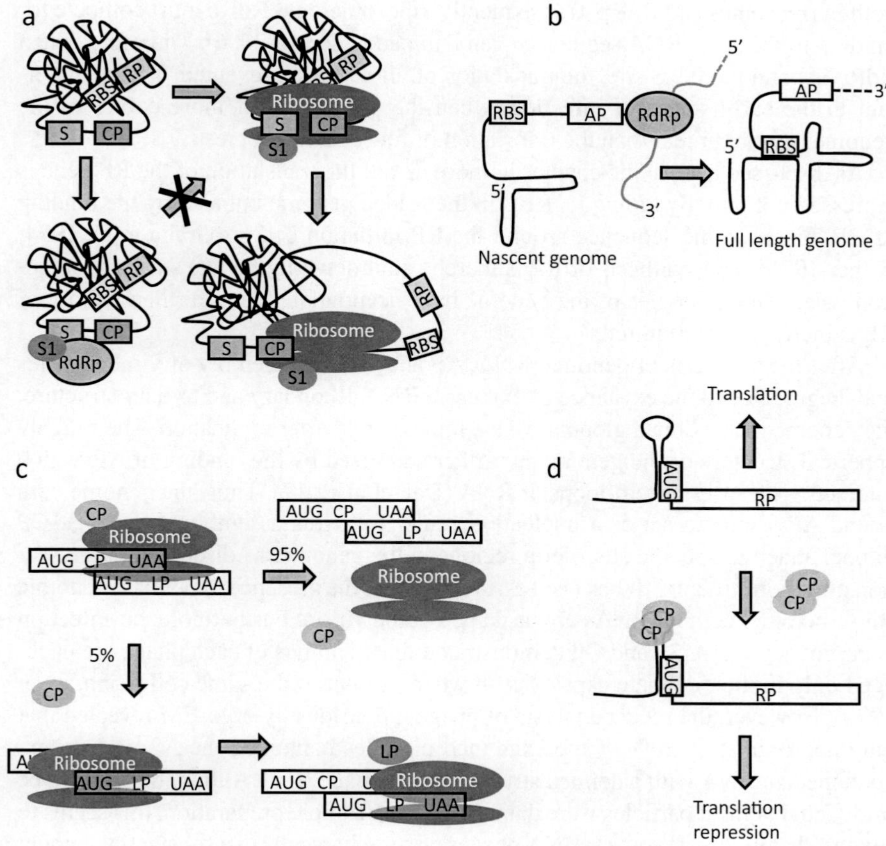

Fig. 13.2 Regulation of gene expression in phage MS2. (**a**) In the folded genome, only the CP gene is available for translation. The CP gene is preceded by the S site, displaying affinity for the S1 protein, which is present in both the ribosome and RdRp. If RdRp is bound to the S site, translation from all genes is disabled. If the ribosome binds to the S site, upon translation, parts of the genome become unwound, exposing the RBS of the RP gene to make it accessible for translation. (**b**) In a full-length genome, AP RBS is normally hidden in the RNA secondary structure and not accessible to the ribosome. However, at the beginning of replication, when 5′ end of the genome has been just synthetized by RdRp an alternative secondary structure is formed in the nascent genome, and AP RBS is temporarily accessible to the ribosome once per replication cycle. (**c**) The LP gene lacks its own RBS and can therefore be accessed only via the backsliding action of the ribosome. In <5% of cases, after the termination of CP synthesis, the ribosome slides backward and initiates LP translation. (**d**) The transcription of the RP gene is repressed by the TR-CP interaction around the initiation codon. The formation of the TR loop itself does not prevent translation, but the binding of the CP dimer to TR abolishes the binding of ribosomes

both in ribosomes and RdRp. Consequently, ribosomes and RdRp must compete for binding to the same RNA sequence (Van Duin and Tsareva 2006). Therefore, when RdRp is bound to the S site, the translation of all genes is prevented. This is important in the early stages of infection, when the production of more copies of the genome is more urgent than the translation of few existing copies.

As discussed later in the chapter in more detail, the translation of the RP gene is repressed not only by hiding its RBS in the folded genome but also by the binding of CP dimers to the sequence around the RP initiation codon (Gralla et al. 1974; Weber 1976). The synthesis of RP is thereby shut down in the late stages of infection when enough copies of the genome have accumulated, and further activity of RP is therefore not required.

After the synthesis of genomic RNA, AP and CP, the assembly of viral particles may begin. Due to the existence of extensive RNA secondary and tertiary structure, the genome must adopt a globular shape immediately after replication. The roughly spherical structure of the genome is further stabilized by the binding of AP, which interacts with several stem-loops of RNA (Dai, et al. 2017). Thus, the genome with bound AP seems to act as a nucleation center for virus assembly. After that, CP dimers attach to specific stem-loop regions in the genome, as discussed in the section on virion structure. It has been assumed that in the presence of cognate genomic RNA, no other cellular RNAs are packaged in the virion. Furthermore, co-infection experiments with MS2 and Qβ demonstrated that genomes of each phage are packaged only in their cognate capsids even when present in the same cell (Ling, et al. 1970). However, the recent analysis of phage Qβ virions by cryo-EM revealed that only approximately 20% of the examined particles contained the A2 protein and contained an RNA with a defined structure (Cui, et al. 2017). Although it can not be excluded that most particles were damaged during sample preparation, this seems to suggest that in at least some ssRNA phages, assembly results in a substantial amount of defective, noninfectious particles, possibly filled with cellular RNAs in a similar fashion to their recombinant VLPs, as discussed in Sect. 13.5.1.

13.3.3 Diverse Lysis Strategies in Leviviridae *Phages*

The last step in the viral life cycle of lytic phages is the lysis of the host cell. dsDNA phages typically make use of several genes to accomplish cell lysis, including genes encoding holins, lysins, spanins and additional proteins involved in the regulation of host lysis. In contrast, *Leviviridae* phages utilize only a single gene for the same purpose (see (Chamakura and Young 2019) for a review). Surprisingly, various clearly related ssRNA phages use at least three very different strategies to accomplish bacterial cell lysis. The LP of MS2 and other representatives of the *Levivirus* genus are inserted into the cellular membrane, creating channels that are permeable to ions (Goessens et al. 1988). This disrupts the electrostatic potential of the cellular membrane and activates cellular autolysins.

The target of the multifunctional A2 protein of phage Qβ and presumably other *Allolevivirus* members is the MurA enzyme, which catalyzes the first cytoplasmic step in proteoglycan wall synthesis, the transfer of enolpyruvate from phosphoenol pyruvate to UDP-N-acetylglucosamine. A2 binds to MurA, blocking the entry of its active site and thereby preventing the synthesis of the cell wall (Karnik and Billeter 1983; Winter and Gold 1983; Bernhardt et al. 2001). Remarkably, the A2 protein is able to block the enzymatic activity of MurA both in its free form and when present within the virion. Therefore, the mature particles of phage Qβ seem to be able to leave the host cell without the aid of any other factor.

Similar to MS2, the LP of phage M contains a single transmembrane helix. However, in the case of phage M, LP is able to interact with the MurJ protein, which is a flippase of lipid-linked precursors necessary for peptidoglycan synthesis. Apparently, the interaction of LP with MurJ locks the flippase in one of two conformations that is necessary for transport (Chamakura and Young 2019).

Since LP is potentially deadly to the host cell, it is important to avoid its overproduction in the early stages of infection. In MS2 and presumably other leviviruses, the translation initiation of the LP gene is coupled to the translation termination of the CP gene. The initiation codon of LP is located approximately 50 nucleotides upstream of the CP termination codon and lacks the preceding RBS sequence; therefore, it cannot be accessed by the ribosome directly. However, in approximately 5% of cases, after the termination of CP translation, the ribosome does not dissociate away from the template but slides backward and reinitiates translation at the LP start codon (Adhin and van Duin 1990). It is thereby ensured that LP accumulates in the cell slowly and only after the synthesis of CP.

Since the LP genes of several other ssRNA phages are found at different locations in the genome, they have probably evolved independently; therefore, it might be possible that some of them use other yet to be discovered mechanisms of cell lysis.

The exact 3D structures of the phage LPs other than the Qβ A2 protein are unknown. LPs from different distantly related phages share little, if any, sequence homology, with the only universally common motif being a putative transmembrane helix, or sometimes two transmembrane helices, as in the case of CB5 (Kazaks et al. 2011). In most cases, the LP gene partially or fully overlaps with the RP gene in a different reading frame. This seems to leave little room for the evolution of LPs since they must coevolve with RP, which is the most conservative of the ssRNA phage proteins. Nevertheless, LPs are actually very efficient and are even able to lyse bacteria other than the hosts of their phages. For example, the LPs of AP205 (host, *Acinetobacter* genospecies 16) and CB5 (host, *Caulobacter crescentus*) are both able to lyse *E. coli* bacteria (Klovins et al. 2002; Kazaks et al. 2011).

13.4 Structure of the Virion

13.4.1 Icosahedral Component – Coat Protein

Mature ssRNA phage virions are composed of icosahedral T = 3 protein shells composed of 178 copies of CP and a single copy of AP (Dent et al. 2013; Koning et al. 2016; Dai et al. 2017). Prior to advances in cryo-EM that made asymmetric reconstructions possible, essentially the only available structural information was for the protein capsid, which was initially thought to consist of 180 chemically identical CP subunits arranged in a perfectly icosahedral shell (Valegård et al. 1990) (Fig. 13.3). Because of T = 3 symmetry, CP exists in three slightly different conformations, A, B and C. There are very tight interactions among the two monomers of CP; therefore, the capsid can be regarded as being composed of 89 dimers. There are two types of dimers in the capsid – one AB dimer (composed of CPs in the A and B conformations) and another CC dimer (composed of two dimers in the C conformation). The structure of CP in the capsid was first solved for MS2 (Valegård et al. 1990; Golmohammadi et al. 1993) and related phages fr (Liljas et al. 1994), Qβ (Golmohammadi et al. 1996), GA (Tars et al. 1997), PP7 (Tars et al. 2000), PRR1 (Persson et al. 2008) and CB5 (Plevka et al. 2009) and seemed to be quite conserved with respect to secondary structure elements. In the CP monomer, the N-terminal hairpin is always followed by a 5-stranded beta-sheet and two C-terminal helices. In dimers, two beta sheets are joined together, forming a single 10-stranded beta-sheet.

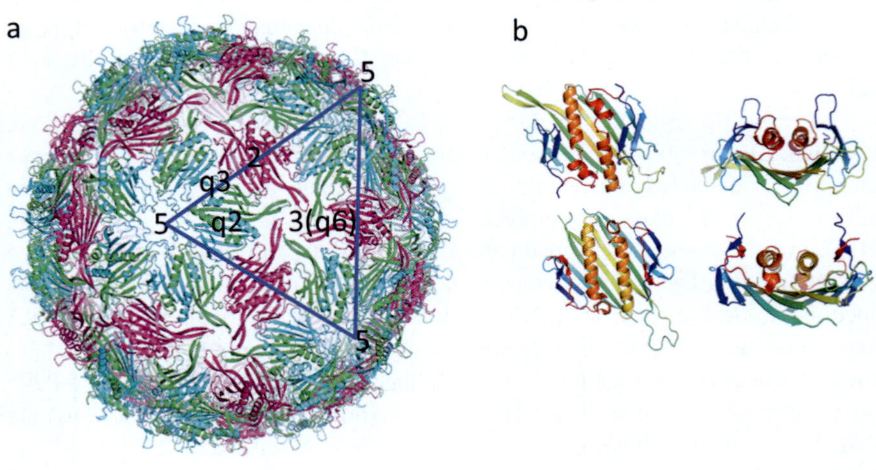

Fig. 13.3 Structure of ssRNA phage capsids. (**a**) Structure of MS2 capsid. One facet of icosahedral particles is shown as a triangle. Approximate positions of icosahedral five-fold, three-fold and two-fold axes and quasi-symmetric q3 (relating three CP dimers), q6 (relating 6 CP dimers, coinciding with the icosahedral three-fold axis) and q2 (relating CP monomers within the AB dimer) axes are shown as well. (**b**) Structure of CP dimers in phages MS2 and AP205. Note the different placement of the terminal strand in the two cases

Deviation from this order is observed in the CP of phage AP205 (Shishovs et al. 2016), in which the N-terminal beta strand has "travelled" to the C-terminus, a circular permutation made possible by the close proximity of the C- and N-termini of the two monomers in the CP dimer. This permutation effectively exposes the C- and N- termini on the surface of AP205, whereas they are not well exposed and cluster together around quasi-threefold axes in other phages.

The exact interactions between CP dimers in capsids are surprisingly variable among even relatively closely related ssRNA phages. In the capsids of the studied *Levivirus* phages, CP dimers are held together merely by noncovalent protein-protein interactions. In some phages, like Qβ (Golmohammadi et al. 1996) and PP7 (Tars et al. 2000) covalent disulphide bonds link CP dimers around fivefold and quasi-sixfold axes, thereby significantly stabilizing the particles. In other cases, metal ions stabilize the structure of capsids in a similar way to that observed for many plant viruses (Tars et al. 2003). The capsid of bacteriophage CB5 exhibits calcium ions located in quasi-three-fold axes, which aid in holding 3 coat protein dimers together (Plevka et al. 2009). In some ssRNA phages, protein-RNA interactions also help to hold the dimers together; for example, CB5, in addition to being stabilized by metal ions, is further stabilized by RNA bases inserted between the CP dimers (Plevka et al. 2009).

13.4.2 Asymmetric Parts – AP, A1 and the Genome

Although it was initially believed that AP must traverse the prominent capsid pores around fivefold or quasi-sixfold axes, later cryo-EM studies revealed that one CC dimer in the capsid is replaced by a single copy of AP (Dent et al. 2013; Koning et al. 2016; Dai et al. 2017). The structure of AP is not similar to that of the CP dimer, yet AP is able to form relatively tight interactions with four different CP dimers. The introduction of AP in the capsid leads to some deviations from the icosahedral architecture – not only is a single CP dimer missing, but the neighboring CP dimers are also displaced from their perfectly icosahedral positions (Gorzelnik et al. 2016). Structurally, AP forms a single domain but can be described as being composed of two regions (Dai et al. 2017; Rumnieks and Tars 2017) – one that is predominantly helical, facing the genomic RNA, and a second containing beta strands (β part), facing the capsid exterior (Fig. 13.4) and interacting with the pilus receptor. While the helical parts are somewhat conserved among the related MS2 and Qβ phages, the β parts are very different. To some extent, this is reflected in their physiological functions – while the β part of the AP of MS2 interacts only with F pili, the β part of the A2 protein of Qβ also interacts with the MurA protein to promote cell lysis. While the structures of MS2 AP and Qβ A2 are quite different, their overall shapes are quite similar, forming a fairly extended, bent structure with their β part, extending tangentially from the virion surface. Recently, a 5 Å-resolution cryo-EM structure was reported for MS2 in complex with the F- pilus receptor (Meng et al. 2019) (Fig. 13.5). The β part of AP was found to be involved in

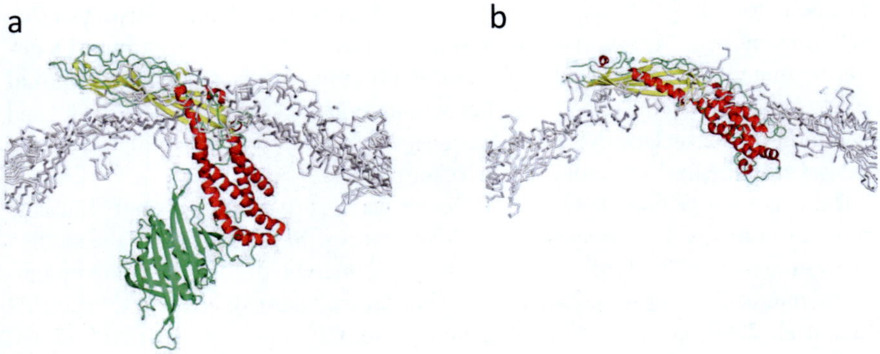

Fig. 13.4 Placement of maturation protein in ssRNA phages Qβ **(a)** and MS2 **(b)**. A section of phage particles around the maturation protein is shown. The particle shell is represented as a Cα trace of CP molecules. Helices of the maturation protein are shown in red, strands are shown in yellow, and loops are shown in green. Note that the helical part faces the interior of the particles, while the β part is exposed on the surface. In Qβ, a single CP dimer is located in proximity to the the maturation protein

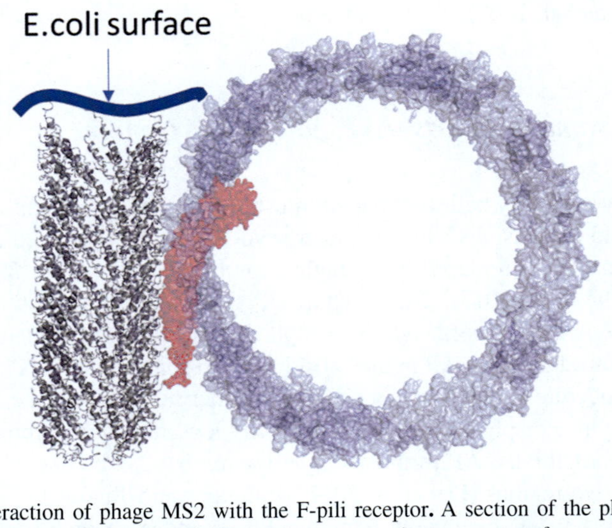

Fig. 13.5 Interaction of phage MS2 with the F-pili receptor**.** A section of the phage is shown, revealing the position of AP. AP and CP are shown as semitransparent surface models in red and blue, respectively. F-pili are shown as a cartoon model. The location of the bacterial surface with respect to the orientation of the F-pili is indicated with a thick line

extensive hydrophobic and charged interactions with at least 6 pilin subunits. No significant changes in AP or genome structure were observed compared to unbound MS2 particles, confirming the previous observation that binding to F-pili *per se* does not induce AP cleavage or genome ejection. The interaction of phages with F-pili seems to be rather flexible, as three classes of MS2 particles with slight differences in their orientations with respect to the F-pili have been observed. The particles bind to the pili so that the tip of AP bends away from the cell surface. Considering the relative orientation towards the cell and the hook-like structure of AP, it has been speculated that upon arrival at the cell surface, the particle could be mechanically opened like a coke can using AP as a pop tab.

In addition to a normal CP, *Allolevivirus* particles exhibit 3-10 copies of the A1 protein, which is an extended version of CP produced by the ribosome through an occasional read-through mechanism involving a leaky termination codon in which translation continues for an additional 600 nucleotides (Weiner and Weber 1971). The exact function of the A1 protein is unknown, although it has been shown that it is required for infection (Hofstetter, et al. 1974). Structurally, the A1 extension presents a roughly globular shape with a mixed α/β architecture that is not observed in any other protein (Rumnieks and Tars 2011) (Fig. 13.6). The N-terminal part of the extension forms an unusually long polyproline type II helix. In cryo-EM reconstructions of Qβ, there are no traces of the A1 extension (Gorzelnik et al. 2016), suggesting that copies of A1 might be flexibly attached and/or randomly distributed in individual phage particles.

Although ssRNA phages nominally present a single-stranded RNA genome, this is correct only in the sense that a single strand of RNA is indeed packaged in each virion. However, more than 70% of the genome is involved in short- and long-distance base-pair interactions forming stem-loops, pseudoknots and other regions that essentially consist of dsRNA (Skripkin et al. 1990). Furthermore, the genome has a well-defined 3D structure that is more or less identical in all virions. The genome structure was first visualized by cryo-EM in phage MS2, initially at medium resolution (Koning et al. 2016) and later at high resolution (Dai et al. 2017), making it possible to observe most of the genome, although only some regions, particularly those forming interactions with the protein shell, were visible at a near-atomic resolution. Later, similar asymmetric cryo-EM structures were reported for phage Qβ

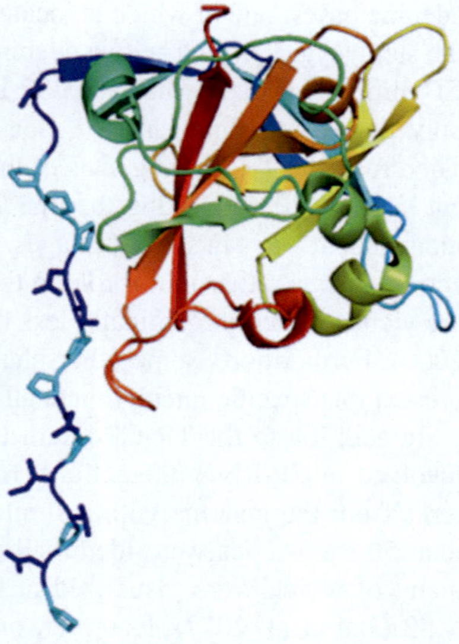

Fig. 13.6 Structure of the Qβ A1 extension**.** The structure is shown as a cartoon model, rainbow colored from the N- to C-termini. In the N-terminal polyproline helix, the side-chains are shown as stick models with prolines colored in cyan and other residues in blue

(Gorzelnik et al. 2016; Cui et al. 2017). Interestingly, unlike MS2, one copy of an isolated CP dimer was found inside the particle (Cui et al. 2017) (Fig. 13.4a), located near the A2 protein and bound to genomic RNA. Therefore, the Qβ virion is actually composed of 180 CP subunits, although a single CP dimer is not a part of the protein shell.

13.4.3 RNA-Coat Protein Interactions

For quite some time, the only available structural information about protein-RNA interactions in ssRNA phage particles came from the studies of CP dimers in complex with a 19 nucleotide-long stem-loop fragment known as TR (translation repression) from the genome region located around the replicase start codon in bacteriophage MS2 (Valegård et al. 1994). The main physiological role of this interaction seems to be the repression of the translation of the replicase gene in late stages of infection, when the presence of the replicase is no longer needed (Gralla et al. 1974; Weber 1976). Additionally, the same interaction provides a nucleation site for capsid assembly and contributes to the packaging of correct RNA inside the virions. The CP-TR interaction in phage MS2 has been studied in great detail both structurally and functionally and remains one of the best understood RNA-protein interactions in general (see (Rumnieks and Tars 2018) for a review), which has fuelled several practical applications, as discussed in later sections. Similar interactions with the same biological purpose exist in some (but not all) other ssRNA phages, including PP7 (Chao et al. 2008), PRR1 (Persson et al. 2008) and Qβ (Rumnieks and Tars 2014). Surprisingly, while the TR-binding sites of the MS2, PRR1 and Qβ CPs are clearly similar, the PP7-binding site is quite different. In both MS2 and PP7, the main specificity determinants of CP-RNA interactions are two adenine bases, one of which is located in the loop, while the other forms a bulge in the stem (Fig. 13.7a). The two adenines occupy symmetrical binding pockets in the CP dimer. However, while in MS2 both pockets is formed by residues belonging only to one monomer, in PP7, pockets are found at the interface of monomers. Therefore, the TR-binding sites of the two phages are located at completely different sites of CP, suggesting that the CP-TR interaction might have evolved in two independent ways in ssRNA phages. However, it is now clear that the CP-TR interaction plays a rather minor role in the life cycle of phage, as TR-deficient mutants are viable and only marginally less fit than wt phages (Peabody 1997a; Licis et al. 2000). Furthermore, some other phages, such as AP205 and CB5, do not seem to present this specific interaction at all.

In addition to the TR-CP interaction, numerous other fragments of ssRNA are involved in CP-RNA interactions. In most cases, these are other stem-loop structures from the genome with a similar appearance. In a cross-linking study, more than 50 stem loops were identified as being bound to CP (Rolfsson et al. 2016), many of which were visualized in later asymmetric cryo-EM reconstructions of MS2 (Dai et al. 2017). However, only 15 stem loops were visible in a cryo-EM

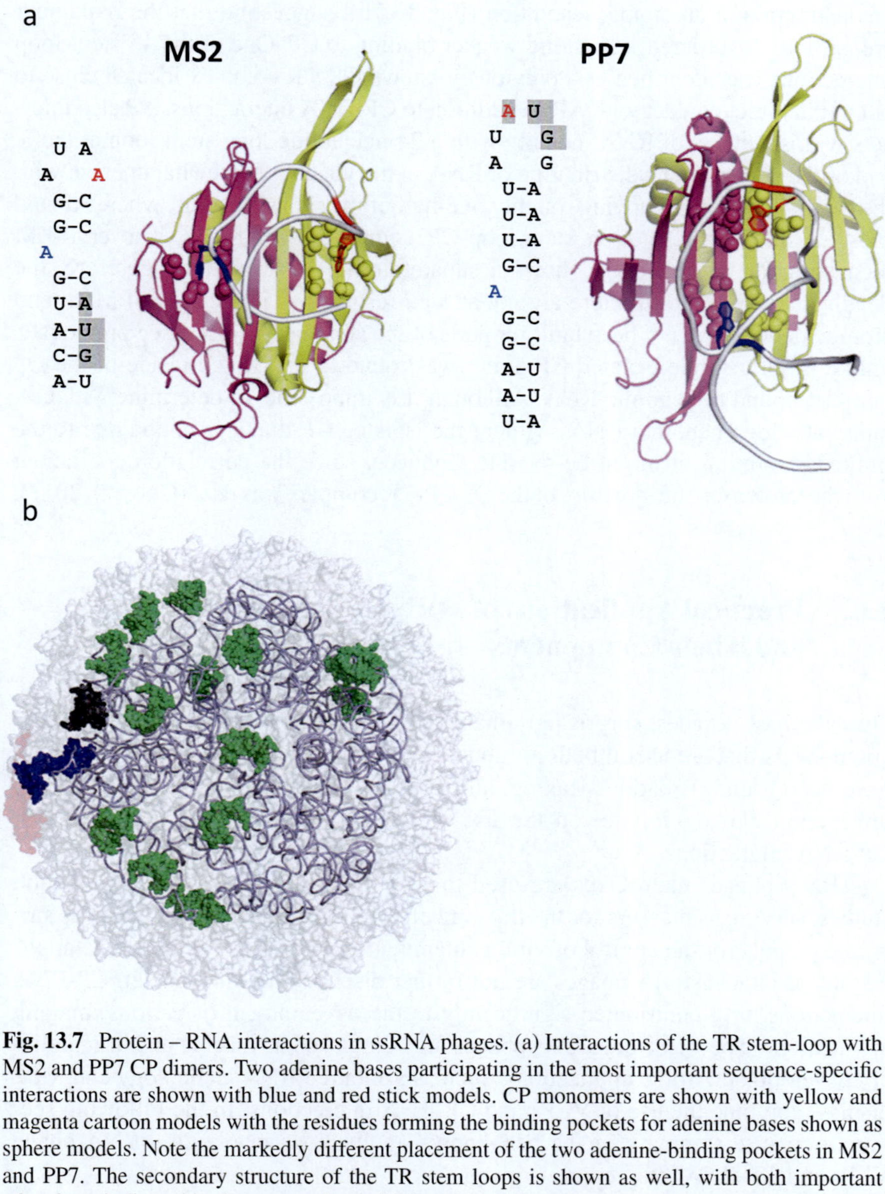

Fig. 13.7 Protein – RNA interactions in ssRNA phages. (a) Interactions of the TR stem-loop with MS2 and PP7 CP dimers. Two adenine bases participating in the most important sequence-specific interactions are shown with blue and red stick models. CP monomers are shown with yellow and magenta cartoon models with the residues forming the binding pockets for adenine bases shown as sphere models. Note the markedly different placement of the two adenine-binding pockets in MS2 and PP7. The secondary structure of the TR stem loops is shown as well, with both important adenylates indicated in the same color as in the stick models. The replicase gene initiation codon is highlighted in gray. (b) Asymmetric cryo-EM reconstruction of phage MS2. CP and AP are shown as semitransparent light gray and red surface models, respectively. High-resolution CP-binding genome stem loops are shown as green sphere models. TR (same as in panel a for MS2) is shown in black and AP interacting stem-loop in blue. A lower-resolution genome model is shown as a coil

reconstruction at an atomic resolution (Fig. 13.7b), suggesting that the remainder are partially disordered, indicating weaker binding to CP. One of the 15 stem-loop interactions was identified as a previously known CP-TR complex located close to but not in direct contact with AP. In addition to CP-RNA interactions, AP also interacts with segments of RNA, notably with a 24 nucleotide-long stem-loop at the 3′ end of the genome. The distribution of RNA in the virion is somewhat uneven, with the densest portion occupying roughly one-half of the capsid volume, where AP and most of the high-resolution stem-loop-CP complexes are found. The cryo-EM reconstruction of phage Qβ shows a subatomic 4.7 Å resolution; therefore, the details of its genome structure are not as well resolved as in the case of MS2, and atomic models have not been built for parts of the genome. As discussed previously, unlike MS2, a single isolated CP dimer was found to be located inside of the Qβ particle, bound to genomic RNA. Although it is impossible to determine with certainty at which genomic RNA segment the isolated CP dimer is bound due to the limited resolution, it might be the TR sequence, since the correlation coefficient with the known crystal structure of the Qβ CP-TR complex was 0.91 (Cui et al. 2017).

13.5 Practical Applications of ssRNA Phages and Their Components

Over the past decades, surprisingly many applications of ssRNA phages and their components that are useful both in fundamental research and product development have been found. Broadly speaking, all of these applications can be grouped into three main classes, relying on the use of intact ssRNA phages, their VLPs or CP-RNA interactions.

ssRNA phages themselves are used mainly in environmental and disinfection studies, serving as markers for the tracking of viral and microbial sources or as surrogate models for the control of viral contamination in various samples. The applications of intact ssRNA phages are not further discussed in this chapter. CP-RNA interactions have contributed significantly to the development of various imaging applications. The VLPs of ssRNA phages are being used in vaccine and drug delivery development. Some applications, such as armoured RNA technology and MS2 display, combine the use of VLPs and CP-RNA interactions. In the following section, a concise summary of the best-known applications involving ssRNA phage VLPs and CP-RNA interactions is given.

13.5.1 *VLPs of ssRNA Phages*

Several applications of ssRNA phages rely on the production of recombinant VLPs from the phages. In most cases, the expression of the CP gene alone in bacteria or yeast is sufficient to produce soluble CP, and CP dimers then spontaneously assemble into VLPs in host cells. Recombinant VLPs of numerous ssRNA phages have been produced in this way (Kastelein et al. 1983; Kozlovskaya et al. 1986; Peabody 1990; Kozlovska et al. 1993). Recently, VLPs of 80 previously unknown ssRNA phages were produced using CP sequences from metagenome sequencing data (Lieknina et al. 2019). In most cases, recombinant VLPs are morphologically indistinguishable from the respective phages when imaged via conventional negative-staining EM. Furthermore, in two known cases involving phages MS2(Golmohammadi et al. 1993; Valegård et al. 1997) and CB5(Plevka et al. 2009), when the icosahedrally averaged crystal structures of both VLPs and the respective phages were determined, the capsid structures were found to be virtually identical. However, in some cases, recombinant VLPs are somewhat heterogeneous as previously observed in a cryo-EM reconstruction of bacteriophage AP205 VLPs and may include a mixture of T = 3, T = 1 and somewhat irregular particles (Shishovs et al. 2016). Mutant VLPs of phages MS2 and PP7 have been shown to exhibit T = 1 (Asensio et al. 2016) and T = 4 (de Martin Garrido, et al. 2019; Zhao, et al. 2019) symmetries. During the characterization of 80 VLPs from metagenome data, several VLPs were found to display considerable deviations from classical T = 3 particles. In two cases, only smaller T = 1 particles could be observed, while in the case of AVE000, somewhat larger heterogeneous particles were observed, some of which may exhibit T = 4 symmetry. In the case of AVE016, elongated T = 3 Q = 4 VLPs were produced. Since VLPs originate from metagenome sequences, the actual phages are not available. Therefore, it is difficult to judge whether the observed shapes and sizes of larger-than T = 3 particles are merely a consequence of the artificial production system or the respective phages themselves indeed exhibit T = 4 or T = 3 Q = 4 symmetry. Intriguingly, as discussed in the previous section, AVE000 presents a very long genome, possibly of more than 5000 nucleotides. Therefore, it might be that larger T = 4 particles are actually present not only in artificial VLPs but also in the phage itself, to enable the packaging of larger genomes. In contrast, the smaller T = 1 particles could not possibly be present in native phages since there would not be enough space for their genome.

Although ssRNA phage VLPs obviously do not contain a genome, they are packaged with various RNAs that are acquired during assembly following expression in host cells (Pickett and Peabody 1993). This is largely due to the nonspecific interactions of CP with any RNA, in which positively charged lysine and arginine residues interact with negatively charged RNA phosphate groups, and some stacking interactions among aromatic residues and RNA bases seem to contribute to unspecific binding as well. Nevertheless, it has been shown that in addition to other cellular RNAs, the recombinant VLPs of ssRNA phages contain substantial amounts of CP mRNA, a property that can be used in several applications, as discussed below.

13.5.2 MS2 Tagging: Identification of Protein-RNA Interactions, RNA Labeling and Tracking

MS2 tagging refers to a variety of techniques relying on specific CP-TR interactions in phage MS2. For this purpose, the formation of VLPs is not desired; therefore, assembly-deficient MS2 CP mutants are used. One MS2-tagging technique referred to as MS2 BioTRAP is a method for the identification of protein-RNA interactions (Bardwell and Wickens 1990; Tsai et al. 2011). In RNA molecules of interest, several copies of TR are introduced. The RNA molecules are coexpressed with the MS2 coat protein modified with an HB tag sequence, which enables biotinylation *in vivo*. As a result, the RNA molecule of interest is decorated with biotinylated copies of MS2 CP. Any proteins bound to the RNA molecule of interest can be extracted together with the RNA using streptavidin beads. The identity of the bound proteins can be revealed by mass spectrometry or other suitable techniques. Using similar approach called MS2-TRAP it is possible to identify also RNA-RNA interactions, such as binding of miRNAs to their target mRNAs (Yoon et al. 2012; Yoon and Gorospe 2016).

Another conceptually similar MS2 tagging technology can be used for the tracking mRNA in living cells (Fig. 13.8). At the 3′ end of an mRNA of interest, multiple copies of TR are inserted. The MS2 coat protein is fluorescently labeled by, for example, fusion to green fluorescence protein (GFP). As a result, upon the binding of labeled MS2 CP, the RNA of interest also becomes fluorescently labeled. Using confocal microscopy, it is further possible to track the path of the tagged RNA of interest in the cell (Bertrand et al. 1998). Similar technology has been developed for phage PP7 (Larson et al. 2011), which also displays a specific interaction of its CP with TR. Several adaptations of the described method exist, some of which make use of both MS2 and PP7 TR sequences located close together in the target RNA. In this case, MS2 and PP7 CPs are labeled with different fragments of an engineered GFP or its variants that are unable to emit fluorescent signals by themselves (Wu et al. 2014; Park et al. 2020). However, upon binding to their respective TRs, the MS2 and PP7 CPs bring the two fragments of GFP together, resulting in fluorescence. Compared to standalone MS2 or PP7 techniques, this significantly reduces the background from unbound CP-GFP molecules. In another adaptation of the PP7 and MS2 RNA labeling method, the two CPs are labeled with different fluorescent proteins (Hocine et al. 2013). Thus, two different RNAs harboring MS2 and PP7 TRs can be simultaneously tracked in a living cell.

13.5.3 Design of Riboswitches

Riboswitches are regulatory components of mRNA that are able to alter gene expression upon binding to small molecule effectors. The activity of a riboswitch may lead to various changes in mRNA, such as accessibility of the ribosome

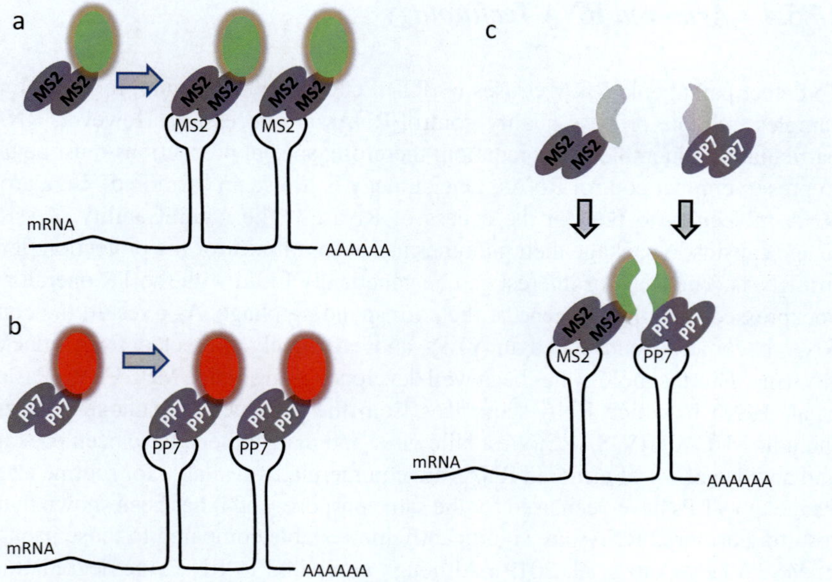

Fig. 13.8 Use of CP-RNA interactions in mRNA visualization. In the 3′-end of target mRNA, several phage-specific TR sequences are inserted. CP dimers of phages MS2 (a) and PP7 (b) are modified by genetic fusion with fluorescent proteins, enabling detection of labeled mRNAs of interest. By combining different specificities of TR interactions in PP7 and MS2 and two different fluorescence proteins, it is possible to track two different mRNAs simultaneously. (c) To reduce background fluorescence resulting from proteins, unbound to target mRNAs, a combined MS2-PP7 approach has been introduced. Both MS2 and PP7 TRs are introduced in the mRNA sequence next to each other. MS2 and PP7 CP dimers are modified by the attachment of GFP segments, which are unable to emit fluorescence signals by themselves. The two halves of GFP are brought together and begin to emit light when the CP dimers bind to their respective TRs

binding site, the formation of transcription terminator hairpins, self-cleavage by induced ribozyme activity or modifications of splice sites. In general, the binding of effector molecules alters the secondary structure of mRNA. This property has been used in selection, where the MS2 TR is inserted in close proximity to the riboswitch sequence (Wu et al. 2019). The experiment can be designed in an "On" or "Off" configuration, so that the RNA structure after the binding of the effector molecule becomes compatible or incompatible, respectively, with the formation of a TR hairpin. The presence or absence of a TR hairpin can in turn be detected by treatment with MS2 CP-GFP.

13.5.4 Armored RNA Technology

To detect pathogenic RNA viruses by RT-qPCR in environmental, food or clinical samples, reliable process quality control RNAs are necessary. However, RNA is particularly vulnerable to degradation; therefore, special precautions must be taken to preserve intact control RNAs. One strategy is to use an "armored" cage around RNA molecules to prevent the access of RNases. The natural ability of ssRNA phage capsids to package their genomes has been utilized for the protection of control RNAs. The RNA of interest can be genetically fused with the TR operator and coexpressed with the CP gene of the corresponding phage. As a result, the control RNA becomes encapsulated in VLPs and effectively protected from attack by RNAses. This technology has been well developed using phage MS2 VLPs (Pasloske et al. 1998) in which RNA sequences from the genomes of pathogenic viruses, including HCV, HIV, SARS, West Nile virus and many others, have been packaged, and a wide variety of armored RNAs are commercially available for routine testing. Phage Qβ VLPs have been used for the same purpose, and it has been shown that the resulting armored RNAs are significantly more stable compared to those packaged in MS2 VLPs (Yao et al. 2019). Although compared to RNA, the degradation of DNA is of significantly less concern, a similar "armored DNA" technique has been developed. In this case, packaging is performed by re-assembling phage MS2 VLPs *in vitro* in the presence of dsDNA fragments (Zhang et al. 2015).

13.5.5 Targeted Delivery

The VLPs of ssRNA phages can be used as nanocontainers for the targeted delivery of various diagnostic or therapeutic agents (see (Pumpens et al. 2016) for a review) (Fig. 13.9). The surface of VLPs can be decorated with an "address", capable of recognizing a particular cell type. Examples of the addresses used for the targeting of ssRNA phage VLPs include cancer cell-targeting proteins(Elsohly et al. 2017) or peptides(Carrico et al. 2008), glycans(Rhee et al. 2012), DNA aptamers(Cohen and Bergkvist 2013), antibodies(ElSohly et al. 2015) and the Z-domain of the *S. aureus* A protein (Zhao et al. 2019). The surface of VLPs can be further modified by the attachment of cell-penetrating peptides(Wei et al. 2009) or PEG chains(Kovacs et al. 2007). Various cargoes can be packaged inside VLPs, including siRNAs(Galaway and Stockley 2013), miRNAs(Wang et al. 2016), dyes(Aanei et al. 2018), small molecule drugs(Finbloom et al. 2018), toxins(Wu et al. 1995), quantum dots(Ashley et al. 2011), nanoparticles(Freivalds et al. 2014), metal ions(Kolesanova et al. 2019), radionuclides (Aanei et al. 2016) and other agents. If an attached address is used in the same VLPs, the cargo can be delivered only to target cells, enabling their selective labeling, activity modulation or destruction. In pioneering studies, it was shown that MS2 VLPs packaged with the ricin A chain and decorated with transferrin on their exterior were able to selectively and

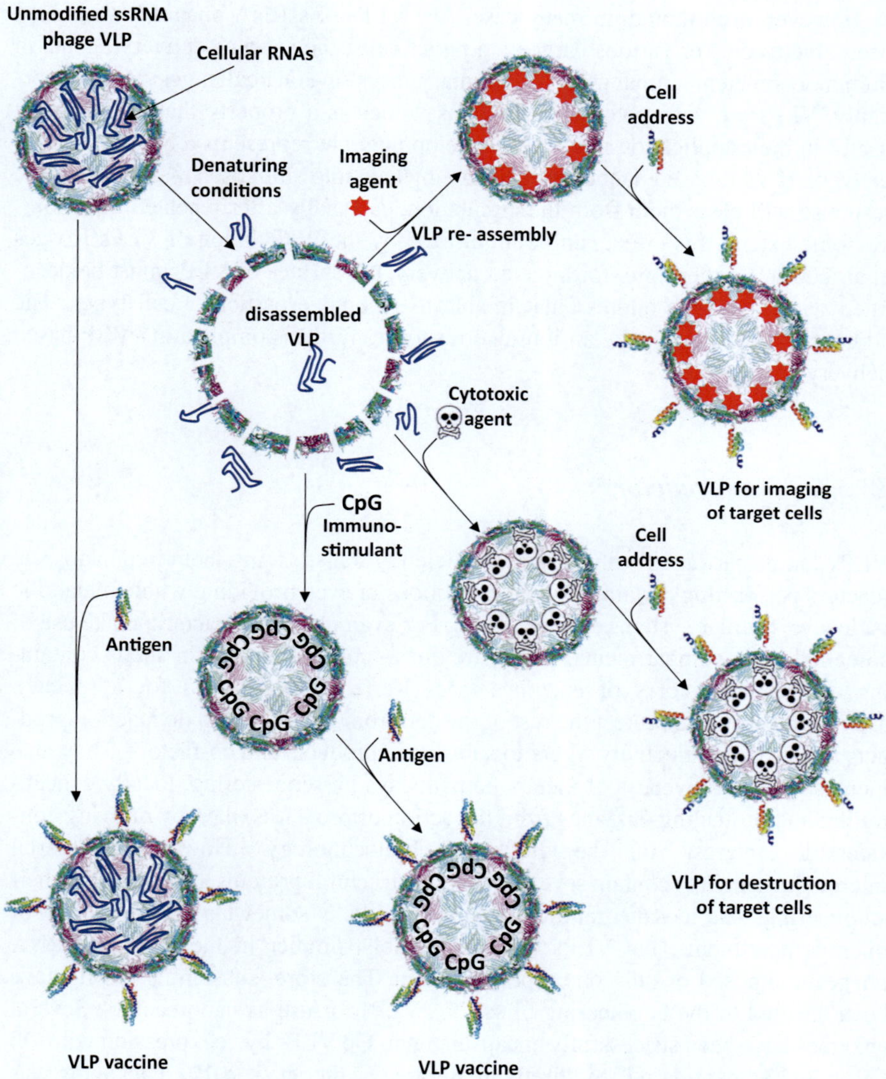

Fig. 13.9 Applications of ssRNA phage VLPs in vaccine development and drug delivery**.** The surface of VLPs can be modified, resulting in antigen presentation for vaccine or attachment of address for targeted nanocontainer delivery. VLPs can be disassembled and reassembled to pack immunostimulatory CpGs for vaccine development or cytotoxic or imaging agents for destruction or visualization of target cells

efficiently kill leukemia cells with exposed transferrin receptors (Wu et al. 1995). Since those studies, many other ssRNA phage VLP-derived targeted nanocontainers with various loads have been developed, most of which are aimed at treating various forms of cancer.

However, even though in many cases, the VLPs of ssRNA phages have indeed been able to deliver various cargoes to target cells, their use as delivery agents in immunocompetent animal models or humans faces several challenges. Most importantly, VLPs are very potent immunogens, which is a property that is extremely useful in their application in vaccine development but represents a hurdle for drug delivery. If VLP-based nanocontainers are injected into animals, a massive immune response will clear them from the circulation, especially after repeated injections. To some extent, this effect can be minimized by the PEGylation of VLPs(Kovacs et al. 2007). Furthermore, for targeted delivery, the surface of VLPs must be decorated with a suitable address that is able to recognize particular cell types. The address itself may provoke an immune response, further complicating VLP-based delivery applications.

13.5.6 Nanoreactors

VLPs can be packaged with enzymes, efficiently transforming them into nanoscale reactors performing certain enzymatic reactions or even providing whole metabolic pathways. Similar systems exist in nature. For example, many bacteria make use of so-called microcompartments, which are quasi-icosahedral protein shells containing encapsulated cores of enzymes (see (Kerfeld et al. 2018) for a review). Microcompartments protect the rest of the cell from potentially toxic reaction products and provide selectivity filters to substrates, products and co-factors. They also increase the effectiveness of metabolic pathways by sequestering, locally concentrating and protecting enzymes from the action of proteases, thereby drawing considerable interest in the field of biotechnology. However, bacterial microcompartments contain several different structural proteins, complicating their engineering. The icosahedral architecture of VLPs is somewhat similar to that of microcompartments, but VLPs are considerably simpler in the case of ssRNA phages composed of only one type of protein. Therefore, substantial efforts have been devoted to the engineering of ssRNA VLPs for use as nanoreactors. Several enzymes have been successfully encapsulated in Qβ VLPs by coexpression with Qβ CP, including peptidase E and firefly luciferase (Fiedler et al. 2010). Packaging was promoted by RNA linker sequence, consisting of Qβ TR and RNA aptamer, able to bind arginine-rich peptide, attached to enzymes. Alternatively, enzymes such as alkaline phosphatase can be modified by attachment to negatively charged oligomers such as the DNA analogue of TR or acidic peptides and packaged *in vitro* in MS2 VLPs (Glasgow et al. 2012). Furthermore, by changing the charges of residues lining the pores around the fivefold and quasi-sixfold axes of VLPs, it is possible to alter the rate of the enzymatic reaction (Glasgow et al. 2015), demonstrating that it is possible to regulate the transport of substrates and products through pores to some extent. In another development, two enzymes from the indigo biosynthesis pathway, pyridoxal phosphate (PLP)-dependent tryptophanase TnaA and nicotinamide adenine dinucleotide phosphate (NADPH)-dependent monooxygenase FMO, were covalently attached to the inner surface of MS2 VLPs using the spy-catcher technique described in Sect. 13.5.7. (Giessen and Silver 2016). The spy tag was genetically inserted in the inner loop of MS2 CP, while the spy-catcher tag was added to two enzymes, TnaA and FMO. The coexpression of all components led to the efficient production of nanoreactors that were 60% more efficient in producing indigo inside the bacterial host compared to naked enzymes. Furthermore, the purified nanoreactors retained 95% of their activity after incubation for 7 days at 25 °C, while isolated enzymes retained only 5% of their activity after similar treatment.

13.5.7 Vaccine Development

Probably the most promising medical application of ssRNA phage VLPs lies in their use in vaccine development (Fig. 13.9). Similar to many other VLPs, the VLPs of ssRNA phages act as very potent immunogens. There are several reasons that VLPs are very immunogenic, the most important of which is their ability to cross-link B-cell receptors, greatly enhancing B-cell proliferation (Bachmann and Zinkernagel 1996). This property can be used to fuse weak antigens to the surface of VLPs to enhance their immune response. Importantly, the coupling of antigens to VLPs may overcome B-cell tolerance to self-antigens (Bachmann et al. 1993) – a property that can be used to create antibodies against certain undesired proteins of organisms; successful examples include but are not limited to the reduction of angiotensin levels in case of high blood pressure (Tissot et al. 2008) and the reduction of proprotein convertase subtilisin/kexin type 9 (PCSK9) levels to reduce low-density lipoprotein cholesterol (Crossey et al. 2015). Although the VLPs of many other viruses can be used in a similar fashion, ssRNA phage VLPs present certain advantages. First, they are exceptionally easy to produce in *Eschericia coli* or yeast systems. Second, they are packaged with unspecific cellular RNAs, which greatly increases their immune response by activating TLR3 and TLR7 toll-like receptors. If necessary, their RNA content can be exchanged for something else – for example, immunostimulatory agents such as CpG oligonucleotides, to achieve an enhanced TLR9 response (Bachmann et al. 2003). Third, antigens can be easily coupled to the VLPs of ssRNA phages by either genetic fusion or chemical coupling. Owing to these advantages, various ssRNA phage VLP scaffold vaccine candidates for a wide variety of diseases have been reported in well over 100 scientific articles, in many cases including proof-of-principle studies in animals or even clinical trials (see (Pumpens, et al. 2016) for a review). In general, all ssRNA phage VLP vaccine candidates can be divided into two broad groups. The first group is represented by candidate prophylactic vaccines against infectious diseases, such as influenza (Tissot et al. 2010; Jegerlehner et al. 2013), West Nile fever (Spohn et al. 2010), Lyme disease(Marcinkiewicz et al. 2018), cervical cancer caused by human papilloma virus (Tumban et al. 2012; Zhai et al. 2019) or malaria(Ord et al. 2014). The second group comprises therapeutic vaccines targeted against self-antigens, cancer antigens or even small molecules, as exemplified by angiotensin (Tissot et al. 2008), tumour-specific carbohydrates(Yin et al. 2015, 2016) and nicotine(Maurer et al. 2005), respectively. In both cases, the corresponding antigens must be attached to VLPs. In the case of protein or peptide antigens, this can be achieved either by genetic fusion or chemical coupling, while for nonprotein antigens, chemical coupling is the only option. Regarding genetic fusion, tolerability to insertions depends heavily on the choice of the particular VLP. In the VLPs of phages MS2, PP7 and Qβ, the N- and C- termini of three adjacent dimers are closely clustered together around quasi-threefold axes(Shishovs et al. 2016); therefore, longer insertions in the N- and C- termini are usually not tolerated due to steric incompatibility. The so-called AB loop located on the VLP surface has also been utilized for short insertions. It is possible to genetically fuse the N- and C- termini of two copies of CP, resulting in a covalent CP dimer in which genetic manipulations such as insertions can be performed in only one monomer, thereby reducing the density of insertions twofold. This way, MS2 VLPs tolerate insertions in AB loop up to 10aa in length (Peabody 1997b; Peabody et al. 2008). Genetic fusions with longer peptides can be achieved much more easily in the case of bacteriophage AP205 VLPs (Tissot et al. 2010) because, compared to MS2 or Qβ, AP205 CP has both its C- and N-termini exposed on the VLP surface (Shishovs et al. 2016), as discussed in the previous section on VLP structure.

Tolerability to longer insertions in AP205 has enabled the development of another coupling technique known as the spy-catcher approach. This technique relies on the use of the engineered collagen adhesion domain CnaB2 from *Streptococcus pyogenes*(Zakeri et al. 2012). In the CnaB2, a covalent isopeptide bond is autocatalytically formed between the Lys31 and Asp117 residues. It is possible to split the CnaB2 protein into two parts, one peptide with only 13 residues containing Lys31 and another containing the remaining 116 residues of the protein, including Asp117. When mixed together, the two parts spontaneously form covalent isopeptide bonds, similar to native proteins(Li et al. 2014). The 13 residue-long spy tag peptide can be added to the antigen of interest, while the longer spy-catcher sequence can be genetically fused to the C-terminus of AP205 CP, which leaves the VLPs relatively intact. Then, the two components can be mixed together, and covalent bonds between them form autocatalytically within a few minutes(Brune et al. 2016). Using this technique, a variety of antigens have been coupled to AP205 VLPs targeting malaria, HPV, cancer antigens and allergy-associated self-antigens (Brune et al. 2016; Thrane et al. 2016; Janitzek et al. 2019).

Although the genetic coupling technique is superior in the sense that only one round of protein expression and purification has to be performed, it fails frequently due to the formation of an insoluble product or failure to assemble into VLPs. For this reason, VLPs and antigens of choice are very often produced and purified separately and then linked together by chemical coupling. In many cases, amine-to-sulfhydryl cross-linkers, such as succinimidyl 6-((beta-maleimidopropionamido) hexanoate are used (Marcinkiewicz et al. 2018). Succinimide moiety reacts with the free amino groups of lysines or N-termini on the VLP surface, while maleimide reacts with the free sulfhydryl groups of the antigen. If a protein antigen does not contain free surface-exposed cysteines, they can be introduced by genetic engineering, or existing surface lysines can be modified with the SATA (N-succinimidyl S-acetylthioacetate) reagent, which attaches free sulfhydryls to amines (Bachmann et al. 2018). Other chemical coupling methods can also be used, such as click chemistry (Polonskaya et al. 2017).

If chemical coupling is used, the antigens do not need to be proteins or peptides; for example, it has been shown that high levels of high-affinity anti-nicotine antibodies can be produced by immunization with nicotine, chemically coupled to the surface of bacteriophage Qβ, potentially acting as a cure for nicotine addiction (Maurer et al. 2005).

13.5.8 MS2 Display

MS2 display is a development that is somewhat similar to classical phage display techniques. As discussed previously, short peptide sequences can be inserted into the AB surface loop of the MS2 coat protein without compromising the stability of VLPs. Therefore, a library of VLPs can be created with randomized 6-10-residue-long sequences in AB loops (Peabody et al. 2008). Furthermore, the obtained VLPs can be "fished" out with a target object displaying affinity to the particular peptide within the AB loop. Crucially, the VLPs of ssRNA phages are known to package their own mRNA. Therefore, after affinity selection, mRNA from the "fished" modified VLPs can be easily extracted, reverse transcribed and sequenced, revealing the sequences of the peptides in AB loops. MS2 display can be used to screen for peptides with high affinity to monoclonal antibodies (Chackerian et al. 2011). In this case, the affinity-selected VLPs can be further directly used as vaccine candidates. Using the monoclonal antibody 5A8, displaying high activity in *Plasmodium falciparum* growth inhibition assays, Ord et al. were able to create a VLP vaccine candidate for malaria (Ord et al. 2014). Using a similar approach invovling monoclonal antibodies against virulence factor AIP4 from *Staphylococcus aureus*, the same research group was able to create a VLP vaccine candidate for *Staphylococcus aureus* infection (O'Rourke, et al. 2015).

References

Aanei IL, ElSohly AM, Farkas ME, Netirojjanakul C, Regan M, Murphy ST, O'Neil JP, Seo Y, Francis MB (2016) Biodistribution of antibody-MS2 viral capsid conjugates in breast cancer models. Mol Pharm 13:3764–3772

Aanei IL, Huynh T, Seo Y, Francis MB (2018) Vascular cell adhesion molecule-targeted MS2 viral capsids for the detection of early-stage atherosclerotic plaques. Bioconjug Chem 29:2526–2530

Adhin MR, van Duin J (1990) Scanning model for translational reinitiation in eubacteria. J Mol Biol 213:811–818

Asensio MA, Morella NM, Jakobson CM, Hartman EC, Glasgow JE, Sankaran B, Zwart PH, Tullman-Ercek D (2016) A selection for assembly reveals that a single amino acid mutant of the bacteriophage MS2 coat protein forms a smaller virus-like particle. Nano Lett 16:5944–5950

Ashley CE, Carnes EC, Phillips GK, Durfee PN, Buley MD, Lino CA, Padilla DP, Phillips B, Carter MB, Willman CL et al (2011) Cell-specific delivery of diverse cargos by bacteriophage MS2 virus-like particles. ACS Nano 5:5729–5745

Atkins JF, Steitz JA, Anderson CW, Model P (1979) Binding of mammalian ribosomes to MS2 phage RNA reveals an overlapping gene encoding a lysis function. Cell 18:247–256

Bachmann MF, Zinkernagel RM (1996) The influence of virus structure on antibody responses and virus serotype formation. Immunol Today 17:553–558

Bachmann MF, Rohrer UH, Kundig TM, Burki K, Hengartner H, Zinkernagel RM (1993) The influence of antigen organization on B cell responsiveness. Science 262:1448–1451

Bachmann MF, Storni T, Maurer P, Tissot A, Schwarz K, Meijerink E, Lipowsky G, Pumpens P, Cielens I, Renhofa R (2003) Packaging of immunostimulatory cpg into virus-like particles: method of preparation and use. In: Office EP (ed) WIPO patent: WO2004084940. https://patentscope.wipo.int/search/en/detail.jsf?docId=WO2004084940

Bachmann MF, Zeltins A, Kalnins G, Balke I, Fischer N, Rostaher A, Tars K, Favrot C (2018) Vaccination against IL-31 for the treatment of atopic dermatitis in dogs. J Allergy Clin Immunol 142:279–281

Bardwell VJ, Wickens M (1990) Purification of RNA and RNA-protein complexes by an R17 coat protein affinity method. Nucleic Acids Res 18:6587–6594

Bernhardt TG, Wang IN, Struck DK, Young R (2001) A protein antibiotic in the phage Qbeta virion: diversity in lysis targets. Science 292:2326–2329

Bertrand E, Chartrand P, Schaefer M, Shenoy SM, Singer RH, Long RM (1998) Localization of ASH1 mRNA particles in living yeast. Mol Cell 2:437–445

Blumenthal T, Landers TA, Weber K (1972) Bacteriophage Qb replicase contains the protein biosynthesis elongation factors EF Tu and EF Ts. Proc Natl Acad Sci USA 69:1313–1317

Bradley DE (1966) The structure and infective process of a Pseudomonas aeruginosa bacteriophage containing ribonucleic acid. J Gen Microbiol 45:83–96

Brune KD, Leneghan DB, Brian IJ, Ishizuka AS, Bachmann MF, Draper SJ, Biswas S, Howarth M (2016) Plug-and-display: decoration of virus-like particles via isopeptide bonds for modular immunization. Sci Rep 6:19234

Carrico ZM, Romanini DW, Mehl RA, Francis MB (2008) Oxidative coupling of peptides to a virus capsid containing unnatural amino acids. Chem Commun:1205–1207

Chackerian B, Caldeira JD, Peabody J, Peabody DS (2011) Peptide epitope identification by affinity selection on bacteriophage MS2 virus-like particles. J Mol Biol 409:225–237

Chamakura K, Young R (2019) Phage single-gene lysis: finding the weak spot in the bacterial cell wall. J Biol Chem 294:3350–3358

Chao JA, Patskovsky Y, Almo SC, Singer RH (2008) Structural basis for the coevolution of a viral RNA-protein complex. Nat Struct Mol Biol 15:103–105

Chetverina HV, Chetverin AB (1993) Cloning of RNA molecules in vitro. Nucleic Acids Res 21:2349–2353

Clarke M, Maddera L, Harris RL, Silverman PM (2008) F-pili dynamics by live-cell imaging. Proc Natl Acad Sci USA 105:17978–17981

Coetzee JN, Bradley DE, Hedges RW, Fleming J, Lecatsas G (1983) Bacteriophage M: an incompatibility group M plasmid-specific phage. J Gen Microbiol 129:2271–2276

Cohen BA, Bergkvist M (2013) Targeted in vitro photodynamic therapy via aptamer-labeled, porphyrin-loaded virus capsids. J Photochem Photobiol B 121:67–74

Crawford EM, Gesteland RF (1964) The adsorption of bacteriophage R17. Virology 22:165–167

Crossey E, Amar MJA, Sampson M, Peabody J, Schiller JT, Chackerian B, Remaley AT (2015) A cholesterol-lowering VLP vaccine that targets PCSK9. Vaccine 33:5747–5755

Cui Z, Gorzelnik KV, Chang JY, Langlais C, Jakana J, Young R, Zhang J (2017) Structures of Qbeta virions, virus-like particles, and the Qbeta-MurA complex reveal internal coat proteins and the mechanism of host lysis. Proc Natl Acad Sci USA 114:11697–11702

Dai X, Li Z, Lai M, Shu S, Du Y, Zhou ZH, Sun R (2017) In situ structures of the genome and genome-delivery apparatus in a single-stranded RNA virus. Nature 541:112–116

Danziger RE, Paranchych W (1970) Stages in phage R17 infection III Energy requirements for the F-pili mediated eclipse of viral infectivity. Virology 40:554–564

de Martin GN, Crone MA, Ramlaul K, Simpson PA, Freemont PS, Aylett CHS (2019) Bacteriophage MS2 displays unreported capsid variability assembling T = 4 and mixed capsids. Mol Microbiol

Dent KC, Thompson R, Barker AM, Hiscox JA, Barr JN, Stockley PG, Ranson NA (2013) The asymmetric structure of an icosahedral virus bound to its receptor suggests a mechanism for genome release. Structure 21:1225–1234

ElSohly AM, Netirojjanakul C, Aanei IL, Jager A, Bendall SC, Farkas ME, Nolan GP, Francis MB (2015) Synthetically modified viral capsids as versatile carriers for use in antibody-based cell targeting. Bioconjug Chem 26:1590–1596

Elsohly AM, MacDonald JI, Hentzen NB, Aanei IL, El Muslemany KM, Francis MB (2017) Ortho-Methoxyphenols as convenient oxidative bioconjugation reagents with application to site-selective Heterobifunctional cross-linkers. J Am Chem Soc 139:3767–3773

Fiedler JD, Brown SD, Lau JL, Finn MG (2010) RNA-directed packaging of enzymes within virus-like particles. Angew Chem Int Ed Engl 49:9648–9651

Fiers W, Contreras R, Duerinck F, Haegeman G, Iserentant D, Merregaert J, Min Jou W, Molemans F, Raeymaekers A, Vandenberghe A et al (1976) Complete nucleotide sequence of bacteriophage MS2 RNA: primary and secondary structure of the replicase gene. Nature 260:500–507

Finbloom JA, Aanei IL, Bernard JM, Klass SH, Elledge SK, Han K, Ozawa T, Nicolaides TP, Berger MS, Francis MB (2018) Evaluation of three morphologically distinct virus-like particles as nanocarriers for convection-enhanced drug delivery to Glioblastoma. Nanomaterials:8

Freivalds J, Kotelovica S, Voronkova T, Ose V, Tars K, Kazaks A (2014) Yeast-expressed bacteriophage-like particles for the packaging of nanomaterials. Mol Biotechnol 56:102–110

Galaway FA, Stockley PG (2013) MS2 Viruslike particles: a robust, semisynthetic targeted drug delivery platform. Mol Pharm 10:59–68

Giessen TW, Silver PA (2016) A catalytic Nanoreactor based on in vivo encapsulation of multiple Enzymes in an engineered protein Nanocompartment. Chembiochem 17:1931–1935

Glasgow JE, Capehart SL, Francis MB, Tullman-Ercek D (2012) Osmolyte-mediated encapsulation of proteins inside MS2 viral capsids. ACS Nano 6:8658–8664

Glasgow JE, Asensio MA, Jakobson CM, Francis MB, Tullman-Ercek D (2015) Influence of electrostatics on small molecule flux through a protein Nanoreactor. ACS Synth Biol 4:1011–1019

Goessens WHF, Driessen AJM, Wilschut J, van Duin J (1988) A synthetic peptide corresponding to the C-terminal 25 residues of phage MS2 coded lysis protein dissipates the protonmotive force in Escherichia coli membrane vesicles by generating hydrophilic pores. EMBO J 7:867–873

Golmohammadi R, Valegård K, Fridborg K, Liljas L (1993) The refined structure of bacteriophage MS2 at 2.8 Å resolution. J Mol Biol 234:620–639

Golmohammadi R, Fridborg K, Bundule M, Valegård K, Liljas L (MS2 co-authors) (1996) The crystal structure of bacteriophage Qb at 3.5 Å resolution. Structure 4:543–554

Gorzelnik KV, Cui Z, Reed CA, Jakana J, Young R, Zhang J (2016) Asymmetric cryo-EM structure of the canonical Allolevivirus Qbeta reveals a single maturation protein and the genomic ssRNA in situ. Proc Natl Acad Sci USA 113:11519–11524

Gralla J, Steitz JA, Crothers DM (1974) Direct physical evidence for secondary structure in an isolated fragment of R17 bacteriophage mRNA. Nature 248:204–208

Groeneveld H, Thimon K, van Duin J (MS2 co-authors) (1995) Translational control of maturation-protein synthesis in phage MS2: a role for the kinetics of RNA folding. RNA 1:79–88

Haruna I, Spiegelman S (1965) Specific template requirments of RNA replicases. Proc Natl Acad Sci USA 54:579–587

Hillman BI, Cai G (2013) The family narnaviridae: simplest of RNA viruses. Adv Virus Res 86:149–176
Hocine S, Raymond P, Zenklusen D, Chao JA, Singer RH (2013) Single-molecule analysis of gene expression using two-color RNA labeling in live yeast. Nat Methods 10:119–121
Hofstetter H, Monstein H, Weissmann C. (Qb co-authors). (1974) The readthrough protein A1 is essential for the formation of viable Qb particles. Biochim Biophys Acta 374:238–251
Janitzek CM, Peabody J, Thrane S, Carlsen PHR, Theander TG, Salanti A, Chackerian B, Nielsen MA, Sander AF (2019) A proof-of-concept study for the design of a VLP-based combinatorial HPV and placental malaria vaccine. Sci Rep:9
Jegerlehner A, Zabel F, Langer A, Dietmeier K, Jennings GT, Saudan P, Bachmann MF (2013) Bacterially produced recombinant influenza vaccines based on virus-like particles. PLoS One 8:e78947
Karnik S, Billeter M. (Qb co-authors). (1983) The lysis function of RNA bacteriophage Qb is mediated by the maturation (A2) protein. EMBO J 2:1521–1526
Kastelein RA, Berkhout B, Overbeek GP, van Duin J (1983) Effect of the sequences upstream from the ribosome-binding site on the yield of protein from the cloned gene for phage MS2 coat protein. Gene 23:245–254
Kazaks A, Voronkova T, Rumnieks J, Dishlers A, Tars K (2011) Genome structure of caulobacter phage phiCb5. J Virol 85:4628–4631
Kerfeld CA, Aussignargues C, Zarzycki J, Cai F, Sutter M (2018) Bacterial microcompartments. Nat Rev Microbiol 16:277–290
Klovins J, Overbeek GP, van den Worm SH, Ackermann HW, van Duin J (2002) Nucleotide sequence of a ssRNA phage from Acinetobacter: kinship to coliphages. J Gen Virol 83:1523–1533
Kolesanova EF, Melnikova MV, Bolshakova TN, Rybalkina EY, Sivov IG (2019) Bacteriophage MS2 as a tool for targeted delivery in solid tumor chemotherapy. Acta Nat 11:98–101
Koning RI, Gomez-Blanco J, Akopjana I, Vargas J, Kazaks A, Tars K, Carazo JM, Koster AJ (2016) Asymmetric cryo-EM reconstruction of phage MS2 reveals genome structure in situ. Nat Commun 7:12524
Kovacs EW, Hooker JM, Romanini DW, Holder PG, Berry KE, Francis MB (2007) Dual-surface-modified bacteriophage MS2 as an ideal scaffold for a viral capsid-based drug delivery system. Bioconjug Chem 18:1140–1147
Kozlovska TM, Cielens I, Dreilinna D, Dislers A, Baumanis V, Ose V, Pumpens P (1993) Recombinant RNA phage Q beta capsid particles synthesized and self-assembled in Escherichia coli. Gene 137:133–137
Kozlovskaya TM, Pumpen PP, Dreilina DE, Tsimanis AJ, Ose VP, Tsibinogin VV, Gren EJ (1986) Formation of capsid-like structures as a result of expression of coat protein gene of RNA phage fr. Dokl. Akad. Nauk SSSR 287:452–455
Krahn PM, O'Callaghan RJ, Paranchych W (1972) Stages in phage R17 infection. VI Injection of A protein and RNA into the host cell. Virology 47:628–637
Krishnamurthy SR, Janowski AB, Zhao G, Barouch D, Wang D (2016) Hyperexpansion of RNA bacteriophage diversity. PLoS Biol 14:e1002409
Larson DR, Zenklusen D, Wu B, Chao JA, Singer RH (2011) Real-time observation of transcription initiation and elongation on an endogenous yeast gene. Science 332:475–478
Li L, Fierer JO, Rapoport TA, Howarth M (2014) Structural analysis and optimization of the covalent association between SpyCatcher and a peptide tag. J Mol Biol 426:309–317
Licis N, Balklava Z, van Duin J (2000) Forced retroevolution of an RNA bacteriophage. Virology 271:298–306
Lieknina I, Kalnins G, Akopjana I, Bogans J, Sisovs M, Jansons J, Rumnieks J, Tars K (2019) Production and characterization of novel ssRNA bacteriophage virus-like particles from metagenomic sequencing data. J Nanobiotechnol 17:61
Liljas L, Fridborg K, Valegård K, Bundule M, Pumpens P (1994) Crystal structure of bacteriophage fr capsids at 3.5 Å resolution. J Mol Biol 244:279–290

Ling CM, Hung PP, Overby LR. (Q-Beta co-authors). (1970) Independent assembly of Qb and MS2 phages in doubly infected Escherichia coli. Virology 40:920–929

Marcinkiewicz AL, Lieknina I, Kotelovica S, Yang X, Kraiczy P, Pal U, Lin YP, Tars K (2018) Eliminating factor H-binding activity of Borrelia burgdorferi CspZ combined with virus-like particle conjugation enhances its efficacy as a Lyme disease vaccine. Front Immunol 9:181

Maurer P, Jennings GT, Willers J, Rohner F, Lindman Y, Roubicek K, Renner WA, Muller P, Bachmann MF (2005) A therapeutic vaccine for nicotine dependence: preclinical efficacy, and phase I safety and immunogenicity. Eur J Immunol 35:2031–2040

Meng R, Jiang M, Cui Z, Chang JY, Yang K, Jakana J, Yu X, Wang Z, Hu B, Zhang J (2019) Structural basis for the adsorption of a single-stranded RNA bacteriophage. Nat Commun 10:3130

Olsen RH, Thomas DD (1973) Characteristics and purification of PRR1, an RNA phage specific for the broad host range Pseudomonas R1822 drug resistance plasmid. J Virol 12:1560

Olsthoorn RCL, Garde G, Dayhuff T, Atkins JF, van Duin J (MS2 co-authors) (1995) Nucleotide sequences of a single-stranded RNA phage from Pseudomonas aeruginosa: kinship to coliphages and conservation of regulatsory RNA structures. Virology 206:611–625

Ord RL, Caldeira JC, Rodriguez M, Noe A, Chackerian B, Peabody DS, Gutierrez G, Lobo CA (2014) A malaria vaccine candidate based on an epitope of the Plasmodium falciparum RH5 protein. Malar J 13:326

O'Rourke JP, Daly SM, Hall PR (2015) Development of a Mimotope vaccine targeting the Staphylococcus aureus quorum sensing pathway (vol 9, e111198, 2014). PLoS One:10

Park SY, Moon HC, Park HY (2020 Jan) Live-cell imaging of single mRNA dynamics using split superfolder green fluorescent proteins with minimal background. RNA 26(1):101–109

Pasloske BL, Walkerpeach CR, Obermoeller RD, Winkler M, DuBois DB (1998) Armored RNA technology for production of ribonuclease-resistant viral RNA controls and standards. J Clin Microbiol 36:3590–3594

Peabody DS (1990) Translational repression by bacteriophage MS2 coat protein expressed from a plasmid. Development of a Mimotope vaccine targeting the Staphylococcus aureus quorum sensing pathway. J Biol Chem 265:5684–5689

Peabody DS (1997a) Role of the coat protein-RNA interaction in the life cycle of bacteriophage MS2. Mol GenGenet 254:358–364

Peabody DS (1997b) Subunit fusion confers tolerance to peptide insertions in a virus coat protein. Arch Biochem Biophys 347:85–92

Peabody DS, Manifold-Wheeler B, Medford A, Jordan SK, do Carmo Caldeira J, Chackerian B (2008) Immunogenic display of diverse peptides on virus-like particles of RNA phage MS2. J Mol Biol 380:252–263

Persson M, Tars K, Liljas L (2008) The capsid of the small RNA phage PRR1 is stabilized by metal ions. J Mol Biol 383:914–922

Pickett GG, Peabody DS (1993) Encapsidation of heterologous RNAs by bacteriophage MS2 coat protein. Nucl Acids Res 21:4621–4626

Plevka P, Kazaks A, Voronkova T, Kotelovica S, Dishlers A, Liljas L, Tars K (2009) The structure of bacteriophage phiCb5 reveals a role of the RNA genome and metal ions in particle stability and assembly. J Mol Biol 391:635–647

Polonskaya Z, Deng SL, Sarkar A, Kain L, Comellas-Aragones M, McKay CS, Kaczanowska K, Holt M, McBride R, Palomo V et al (2017) T cells control the generation of nanomolar-affinity anti-glycan antibodies. J Clin Investig 127:1491–1504

Pumpens P, Renhofa R, Dishlers A, Kozlovska T, Ose V, Pushko P, Tars K, Grens E, Bachmann MF (2016) The true story and advantages of RNA phage capsids as Nanotools. Intervirology 59:74–110

Rastgou M, Habibi MK, Izadpanah K, Masenga V, Milne RG, Wolf YI, Koonin EV, Turina M (2009) Molecular characterization of the plant virus genus Ourmiavirus and evidence of inter-kingdom reassortment of viral genome segments as its possible route of origin. J Gen Virol 90:2525–2535

Rhee JK, Baksh M, Nycholat C, Paulson JC, Kitagishi H, Finn MG (2012) Glycan-targeted virus-like nanoparticles for photodynamic therapy. Biomacromolecules 13:2333–2338

Rolfsson O, Middleton S, Manfield IW, White SJ, Fan B, Vaughan R, Ranson NA, Dykeman E, Twarock R, Ford J et al (2016) Direct evidence for packaging signal-mediated assembly of bacteriophage MS2. J Mol Biol 428:431–448

Rumnieks J, Tars K (2011) Crystal structure of the read-through domain from bacteriophage Qbeta A1 protein. Protein Sci 20:1707–1712

Rumnieks J, Tars K (2012) Diversity of pili-specific bacteriophages: genome sequence of IncM plasmid-dependent RNA phage M. BMC Microbiol 12:277

Rumnieks J, Tars K (2014) Crystal structure of the bacteriophage qbeta coat protein in complex with the RNA operator of the replicase gene. J Mol Biol 426:1039–1049

Rumnieks J, Tars K (2017) Crystal structure of the maturation protein from bacteriophage Qbeta. J Mol Biol 429:688–696

Rumnieks J, Tars K (2018) Protein-RNA interactions in the single-stranded RNA bacteriophages. Subcell Biochem 88:281–303

Schmidt JM (1966) Observations on the adsorption of Caulobacter bacteriophages containing ribonucleic acid. J Gen Microbiol 45:347–353

Shi M, Lin XD, Tian JH, Chen LJ, Chen X, Li CX, Qin XC, Li J, Cao JP, Eden JS et al (2016, Dec 22) Redefining the invertebrate RNA virosphere. Nature 540(7634):539–543

Shiba T, Miyake T (1975) New type of infectious complex of E. coli RNA phage. Nature 254:157–158

Shishovs M, Rumnieks J, Diebolder C, Jaudzems K, Andreas LB, Stanek J, Kazaks A, Kotelovica S, Akopjana I, Pintacuda G et al (2016) Structure of AP205 coat protein reveals circular permutation in ssRNA bacteriophages. J Mol Biol 428:4267–4279

Skripkin EA, Adhin MR, de Smit MH, van Duin J (1990) Secondary structure of bacteriophage MS2. Conservation and biological significance. J Mol Biol 211:447–463

Spohn G, Jennings GT, Martina BE, Keller I, Beck M, Pumpens P, Osterhaus AD, Bachmann MF (2010) A VLP-based vaccine targeting domain III of the West Nile virus E protein protects from lethal infection in mice. Virol J 7:146

Tars K, Bundule M, Fridborg K, Liljas L (1997) The crystal structure of bacteriophage GA and a comparison of bacteriophages belonging to the major groups of Escherichia coli leviviruses. J Mol Biol 271:759–773

Tars K, Fridborg K, Bundule M, Liljas L (2000) Crystal structure of phage PP7 from Pseudomonas aeruginosa at 3.7 Å resolution. Virology 272:331–337

Tars K, Zeltins A, Liljas L (2003) The three-dimensional structure of cocksfoot mottle virus at 2.7 Å resolution. Virology 310:287–297

Thrane S, Janitzek CM, Matondo S, Resende M, Gustavsson T, de Jongh WA, Clemmensen S, Roeffen W, van de Vegte-Bolmer M, van Gemert GJ et al (2016) Bacterial superglue enables easy development of efficient virus-like particle based vaccines. J Nanobiotechnol 14:30

Tissot AC, Maurer P, Nussberger J, Sabat R, Pfister T, Ignatenko S, Volk HD, Stocker H, Muller P, Jennings GT et al (2008) Effect of immunisation against angiotensin II with CYT006-AngQb on ambulatory blood pressure: a double-blind, randomised, placebo-controlled phase IIa study. Lancet 371:821–827

Tissot AC, Renhofa R, Schmitz N, Cielens I, Meijerink E, Ose V, Jennings GT, Saudan P, Pumpens P, Bachmann MF (2010) Versatile virus-like particle carrier for epitope based vaccines. PLoS One 5:e9809

Tomita K (2014) Structures and functions of Qbeta replicase: translation factors beyond protein synthesis. Int J Mol Sci 15:15552–15570

Tsai BP, Wang X, Huang L, Waterman ML (2011) Quantitative profiling of in vivo-assembled RNA-protein complexes using a novel integrated proteomic approach. Mol Cell Proteomics 10(M110):007385

Tumban E, Peabody J, Tyler M, Peabody DS, Chackerian B (2012) VLPs displaying a single L2 epitope induce broadly cross-neutralizing antibodies against human papillomavirus. PLoS One 7:e49751

Ugarov VI, Chetverin AB (2008) Functional circularity of legitimate Qbeta replicase templates. J Mol Biol 379:414–427

Valegård K, Liljas L, Fridborg K, Unge T (1990) The three-dimensional structure of the bacterial virus MS2. Nature 345:36–41

Valegård K, Murray JB, Stockley PG, Stonehouse NJ, Liljas L (1994) Crystal structure of an RNA bacteriophage coat protein-operator complex. Nature 371:623–626

Valegård K, Murray JB, Stonehouse NJ, van den Worm S, Stockley PG, Liljas L (1997) The three-dimensional structures of two complexes between recombinant MS2 capsids and RNA operator fragments reveal sequence-specific protein-RNA interactions. J Mol Biol 270:724–738

Valentine RC, Strand M (1965) Complexes of F-pili and RNA bacteriophage. Science 148:511–513

Van Duin J, Tsareva N (2006) Single-stranded RNA Phages. In: Calendar R (ed) The bacteriophages. Oxford University Press, New York, pp 175–196

van Meerten D, Girard G, van Duin J (2001) Translational control by delayed RNA folding: identification of the kinetic trap. RNA 7:483–494

Wahba AJ, Miller MJ, Niveleau A, Landers TA, Carmichael GG, Weber K, Hawley DA, Slobin LI (1974) Subunit I of Qbeta replicase and 30 S ribosomal protein S1 of Escherichia coli Evidence for the identity of the two proteins. J Biol Chem 249:3314–3316

Wang GJ, Jia TT, Xu XX, Chang L, Zhang R, Fu Y, Li YL, Yang X, Zhang K, Lin GG et al (2016) Novel miR-122 delivery system based on MS2 virus like particle surface displaying cell-penetrating peptide TAT for hepatocellular carcinoma. Oncotarget 7:59402–59416

Weber H (1976) The binding site for coat protein on bacteriophage Qb RNA. Biochim Biophys Acta 418:175–183

Wei B, Wei YX, Zhang K, Wang J, Xu RH, Zhan S, Lin GG, Wang W, Liu M, Wang JN et al (2009) Development of an antisense RNA delivery system using conjugates of the MS2 bacteriophage capsids and HIV-1 TAT cell penetrating peptide. Biomed Pharmacother 63:313–318

Weiner AM, Weber K (1971) Natural read-through at the UGA termination signal of Q-beta coat protein cistron. Nature New Biol 234:206–209

Winter RB, Gold L. (Qb co-authors). (1983) Overproduction of bacteriophage Qb maturation (A2) protein leads to cell lysis. Cell 33:877–885

Wu M, Brown WL, Stockley PG (1995) Cell-specific delivery of bacteriophage-Encapsidated ricin-a chain. Bioconjug Chem 6:587–595

Wu B, Chen J, Singer RH (2014) Background free imaging of single mRNAs in live cells using split fluorescent proteins. Sci Rep 4:3615

Wu MJ, Andreasson JOL, Kladwang W, Greenleaf W, Das R (2019) Automated Design of Diverse Stand-Alone Riboswitches. ACS Synth Biol 8:1838–1846

Yao L, Li F, Qu M, Guo Y, Jiang Y, Wang L, Zhai Y (2019) Development and evaluation of a novel Armored RNA technology using bacteriophage Qbeta. Food Environ Virol 11:383–392

Yin Z, Chowdhury S, McKay C, Baniel C, Wright WS, Bentley P, Kaczanowska K, Gildersleeve JC, Finn MG, BenMohamed L et al (2015) Significant impact of Immunogen design on the diversity of antibodies generated by carbohydrate-based anticancer vaccine. ACS Chem Biol 10:2364–2372

Yin Z, Dulaney S, McKay CS, Baniel C, Kaczanowska K, Ramadan S, Finn MG, Huang X (2016) Chemical synthesis of GM2 Glycans, Bioconjugation with Bacteriophage Qbeta, and the induction of anticancer antibodies. Chembiochem 17:174–180

Yoon JH, Gorospe M (2016) Identification of mRNA-interacting factors by MS2-TRAP (MS2-tagged RNA affinity purification). Methods Mol Biol 1421:15–22

Yoon JH, Srikantan S, Gorospe M (2012) MS2-TRAP (MS2-tagged RNA affinity purification): tagging RNA to identify associated miRNAs. Methods 58:81–87

Zakeri B, Fierer JO, Celik E, Chittock EC, Schwarz-Linek U, Moy VT, Howarth M (2012) Peptide tag forming a rapid covalent bond to a protein, through engineering a bacterial adhesin. Proc Natl Acad Sci USA 109:E690–E697

Zhai L, Yadav R, Kunda NK, Anderson D, Bruckner E, Miller EK, Basu R, Muttil P, Tumban E (2019) Oral immunization with bacteriophage MS2-L2 VLPs protects against oral and genital infection with multiple HPV types associated with head & neck cancers and cervical cancer. Antivir Res 166:56–65

Zhang L, Sun Y, Chang L, Jia T, Wang G, Zhang R, Zhang K, Li J (2015) A novel method to produce armored double-stranded DNA by encapsulation of MS2 viral capsids. Appl Microbiol Biotechnol 99:7047–7057

Zhao L, Kopylov M, Potter CS, Carragher B, Finn MG (2019) Engineering the PP7 virus capsid as a peptide display platform. ACS Nano 13:4443–4454

Chapter 14
Phages as Therapy or "Dietary Supplements" Against Multiresistant Bacteria?

Karin Moelling

Abstract Bacteria and phages form an ecosystem and play a role in obesity, in Intestinal bowel disease, neurological disorders, in the brain-gut axis and more recently in anticancer therapies. We have shown that fecal transfer can cure a patient from a life -threatening infection with Clostridium difficile. The microbiome and virome of the feces of a patient before and after fecal transfer has been analyzed, where phages play a role.

Phages form a quasispecies and are highly specialized to specific bacterial hosts. Further studies are required to develop broad-range phages similar to broad-range antibiotics.

Phages do not fit into the regulatory presently required definition as a medicinal product. They should be defined differently to enable scientists and medical doctors to evaluate them for general phage therapy. They should be defined as food supplements or dietary products, or probiotics similar to probiotic bacteria. Then they could be evaluated for more general applications for people with infections. The rules need to be changed.

14.1 A Brief History of Phage Therapy

For thousands of years a disease-curing activity was detected in the river Ganges, which may be a basis for a religious ritual of the Hindus. They dive into the water, rinse their mouth or drink from the water, hoping for some cure of a an infectious disease. Presumably, the river water contained both, the bacteria as infectious agents as well as their respective phages. Infected people could have profited from the phages. Lytic phages destroy the bacteria and are released into the water. Already before the end of the twentieth century, the British biologist Ernest Hanbury Hankin

K. Moelling (✉)
Max-Planck-Institute for Molecular Genetics, Berlin, Germany

Institute for Medical Microbiology, University Zurich, Zurich, Switzerland
e-mail: moelling@molgen.mpg.de

G. Witzany (ed.), *Biocommunication of Phages*,
https://doi.org/10.1007/978-3-030-45885-0_14

(1865–1939) described the curing effect of water from the Ganges river by testing it against cholera bacterial cultures - which had been known from Robert Koch's studies since some decades. The bacteria were lysed by a then unknown activity in the water, which was thermosensitive. Hankin noticed, that the lysis effect disappeared if the river water was boiled, indicating some labile biological activity - which later turned out to be the phages (Hankin 1896).

This was not known until 1917 when the Canadian microbiologist Felix d'Herelle (1873–1949) at the Pasteur Institute in Paris described the effect of phages as killers of bacteria coining the word phages. They are viruses specialized in infecting only bacteria. He isolated the phages from the stool of a soldier suffering from dysentery. Then he applied the filtrate to bacteria which were plated on a Petri dish, where the phages lysed the lawn and caused translucid halos, the plaques, a test still in use today. D'Herelle even swallowed the phages to prove that they had no adverse effects, before he applied them to infected children against dysentery in Paris. He published his results in his first paper on phages in 1917 (d'Herelle 1917).

Till today it is sewage water, preferentially from hospitals, which serves as major source for fishing new phages against bacteria. Where there are bacteria there will also be the respective phages. Almost all bacteria are infected by phages and they cycle in about a day: a peak of bacterial lysis will lead to a peak of phages and reduction of the bacteria, then the phages are reduced and can replicate again when the bacteria multiply (Moelling et al. 2018). Felix D'Herelle traveled wherever some infectious disease outbreaks occurred and treated patients with phages, South America, Africa (Ruanda, Burundi, Congo), passengers on a French ship in the Suez Canal, Mexico, Africa, India (Assam), and Russia (Haeusler 2006). He was invited to Tbilisi, Georgia, in 1936 by the bacteriologist Georgi Eliava, his former student who had initiated the foundation of what became later the Eliava Institute of Bacteriophages, Microbiology and Virology in Tbilisi in 1923. Phages are being produced there to this very day. This was once a major business and trials were performed with many participants including controls. The enormous production scale, the science behind it, the quality of the trials including controls are to some extent underestimated today. Up to 1200 people were employed there at peak times and produced large amounts of phages. Methods were developed how to store or ship the phages, also band-aids and powder were developed to cover wounds and even pills were produced for easier transportation, up to 1.5 Mio pills per year. The military was a major recipient and driving force for large-scale productions and consumption. Already in 1939 in the Finnish-Russian war 18.000 soldiers received phage mixtures, "cocktais" against anthrax, dripped into their open fractures and 80% of them recovered without amputations (Tsulukidze 1941; Haeusler 2006). In 1963 about 30.000 children were recruited, half of them receiving phages against *Shigella* bacteria and the others placebos. The number of children coming down with dysentery was reduced from 6.7 per 1000 to 1.8, thus 3.8 fold. Even prophylactic phage treatments were tested during an ongoing epidemics by *Shigella*. It was noticed, that passage through the stomach quickly reduced the phage titers. It was noticed, that sufficiently high bacterial numbers were required and frequent application of the phages to allow for their replication: the therapy failed with low bacterial titers, but succeeded during ongoing epidemics. The trials performed in the Soviet Union and Tbilisi throughout many decades will now have to be repeated under more stringent test conditions and scientific reporting. It is not so well-known today, that large cohorts including controls have been analyzed already before, much beyond individual case reports, including controls, recording of conditions, etc. The literature and publications in Georgia are not easily available in the Western World today and may deserve translation and more publicity (Haeusler 2006). Some results are secret military information und unaccessible. In Tbilisi the use of phage therapy was continued because of the lack of antibiotics and is maintained till today. They combine phages with antibiotics that are now available. The numerous studies performed within almost 100 years indicate that phages can be successful and have proven to be safe. Adverse events have never been reported.

14.2 Recent Phage Therapy Trials

There has not yet been a single successful Clinical Trial Phase I, II, or III performed world-wide, according to the standards required by present regulatory authorities for approval of a drug. However, there are individual case reports, "compassionate trials", which can be performed with the consent of patient, doctors and authorities, if there are no other options in life-threatening diseases according to the Helsinki Declaration (2013). They document some surprising effects (Moelling 2019):

The PhagoBurn trial was performed against large burns supported by the European Commission, EU, within the Framework 7. This was a combined European effort including nine Burn Centres in European countries, designed in 2014. This effort deserves respect because it was initiated very early on, but difficulties occurred. Large burns are often infected with numerous bacterial types, so that the rule in their protocol, one kind of phage treatment for one target, could not be met. Instead many bacterial types infect large burns which require phage cocktails. Not enough patients met the inclusion criteria. The concept was modified, and a randomized Phase I/II trial was finally performed with a cocktail of 12 lytic natural anti-*Pseudomonas aeruginosa* phages (PP1131, 10^6PFUs/ml, 10 ml total). Thirteen patients were treated daily for 7 days, and 13 control patients received the standard of care antibiotic treatment. The phage production under the required Good Manufacturing Practice (GMP) conditions took more than 2 years, because no experience existed and procedures had to be developed - which also required most of the grant money. Another unexpected difficulty was, that the titer of the phages went down by 100.000 fold under the applied storage conditions. Consequently, the multiplicity of infection for phages against the bacteria, which was intended 10:1, dropped by five logs, so that the patients were treated with much too low titers. The bacterial burden was reduced by twofold and more slowly than in the controls. The trial was terminated prematurely in 2017 because of insufficient efficacy (Jault et al.

2019). This trial increased the awareness of potential problems with selection of treatable diseases or patients and phage production to be avoided in the future.

Furthermore, the company Nestle performed a clinical trial by using the well-known phage T4 against *Escherichia (E.) coli* bacteria for children suffering from diarrhea in Bangladesh. A clinical trial was performed in Bangladesh which fulfilled all requests by the legal Western authorities. Strong requests had to be met twice, healthy volunteers from three age groups were first treated in Switzerland then again in Bangladesh. A comparison to standards of care and alternative therapies had to be included. A commercial phage cocktail from Russia was included, while the Swiss therapy only involved one single phage type, T4. About 300 children were treated, but it turned out, that the diarrhea was not caused by *E. coli* as expected, and that the number of bacteria in the guts were too small, so that the T4 phages had no effect. Several bacteria not responding to this phage type, dominated, they were probably not actively growing, and were in a hibernation state until more nutrients were added (Brüssow 2019a, b).

Urinary Tract infections (UTI) were treated with phages in about 70 cases in Tblisi in collaboration between the Eliava Institute with a Clinical Swiss investigator, Thomas Kessler, as described (Leitner et al. 2017) with the results still to be reported. Recently a proposal for a Clinical Trial built on this experience was applied for in Switzerland. UTI is not a deadly disease, however, it allows testing different parameters, regimens, development of resistance, determination of half life, titers, etc., which cannot be performed in life -threatening situations such as sepsis and which are urgently needed for further studies.

14.3 Successful Individual Case Reports

A soldier from Ukraine was treated in 2016 after an odyssee through several countries for 2 years. He had skin and bone defects and multi-resistant P*aeruginosa* biofilms. He was treated in the Military Hospital in Berlin by Ch. Willy, who applied several methods, surgery, disinfectants, skin transplants, antibiotics and a phage cocktail from Georgia, Pyo-Phage, directed against *St. aureus, Streptococus, P. aeruginosa, E. coli, and Proteus*. The patient received three ampules per day. Finally his drainage secretion was sterile and he recived a prothesis (Vogt et al. 2017).

Several years ago an eye doctor in Yale, USA, Ali Khodadoust, was infected by multidrug-resistant bacteria after an open heart surgery and after 4 years of suffering he was successfully treated by phages against the bacteria and recovered (Chan et al. 2019; Schmidt 2019).

A similar problem with Heart Surgery was recently solved in Hannover, Germany. A phage therapy was applied to a patient with a history of heart surgery who developed sepsis 2 years later at the Clinic for Heart, Thorax, Transplantation at the Medical University of Hannover (MHH). The phages were prepared in-house and made endotoxin-free by Endo-Trap column chromatography. Phage titres reached 10^{13} plaque-forming units per milliliter, PFUs/ml. Phages were immobilized on clinically approved Fibrin carriers which released about 10^8 phages per day and were considered a success for slow phage release. The patient was treated locally and i.v. with phages by Christian Kühn and Evgenii Rublskii (Kuehn et al. 2019). The phage strain was supplied by the Leibniz-Institute in Braunschweig, Deutsche Sammlung von Mikroorganismen und Zellkulturen, DSMZ, by Christine Rohde, where a bank of about 800 phages exist, not yet for clinical use.

Phages are preferentially applicable for open wounds, open fractures, and deep sores such as gangrenes. Thus, phage therapy has also been recently applied for **diabetic toes** of 11 patients who were not in life-danger but had to face amputations and were saved from surgery (Fish et al. 2016). The advocate and coauthor of this trial was Elizabeth (Betty) Kutter, who has been supporting phage progress and therapy, amongst others by the biannual program "Evergreen International Phage Meetings" near Seattle, USA. The phages were directly applied into the toe wounds once a week, and presumably multiplied there. The actual bacterial titers were not known.

Doses were high- and required sufficiently high bacterial titers to be successful, otherwise the phages cannot multiply and die out. In this case an Eliava cocktail of phages was used, which contained only phages against *S. aureus*. In the manuscript it is pointed out, that depending on the deepness of the sore from surface to the bones a number of different bacteria may be involved and the respective phages should be present in future phage cocktails. Estimated 3 Mio Americans suffer from diabetes and may be facing gangrenes and amputations (Fish et al. 2016; Kutter et al. 2010).

One of the most spectacular cases has been described for phage therapy of a patient with sepsis by a multidrug resistant ***Acinetobacter Baumannii,*** Tom Patterson. He is the patient with one of the most famous success stories recently, with an open-access public youtube story, a whole book written by his wife, and broad media coverage. He got infected as a tourist in Egypt by the bacterium *Acinetobacter baumannii* and had health problems with pancreatitis and diabetes. This bacterium is also designated as "Kiel" germ which killed several patients in the clinics at Kiel. Patterson's courageous wife Steffanie Strathdee, an epidemiologist, activated doctors, agencies and colleagues in the US and finally succeeded in finding help for her husband by phage therapy. Out of 200 "natural" phages tested, finally three were selected. They were pretested in special animal systems, a wax worm model, whereby the wax worms are the caterpillars of the wax moth. One phage originated from the Australian company AmpliPhi Corporation and one from Ry Young at a military-linked Center for Phage Technology (CPT) from Texas A & M University, Texas. A third sample was supplied by the US Naval Medical Research Center (NMRC) which has been performing research on phages for military use. A permission for an emergency inverstigational new drug (eIND) from the Food and Drug Administration, FDA, allowed the application of three selected phage types by three consecutive intravenous injections. Within a few days this terminated a 3 months' coma. Phage therapy was continued for up to 8 weeks through a catheter into the patient's abdomen as well as an intravenous line. Interestingly, the phage effect increased the efficacy of the antibiotics, which the authors described as "counterintuitively"! The resistant bacteria became in part resensitized to antibiotics following phage treatment (see also below). The patient recovered after 3 months in total. He even participated on the March for Science in Washington D.C. (Schooley et al. 2017).

The crew with R. Schooley, R. Young, S. Strathdee and others successfully treated several organ transplant patients with transplants which were infected and were successfully treated with phages. They then started an organization IPATH, Center for Innovative Phage Applications and Therapeutics for Phage Therapy at the University of California in San Diego.

In a recent case a 15 year old girl with Cystic Fibrosis received a lung transplant in England and antibiotics. A phage therapy was initiated against her multi-drug-resistant bacterial infection, ***Mycobacterium abscessus***, which had destroyed her lung and still affected her skin and liver. (It is distinct form ***mycobacterium tuberculosis***, which is unfortunately very difficult to treat with phages due to its encapsidation.) She received **three** phage types specifically selected for her case. Some of the phages were delivered from Graham Hatfull from Pittsburg, who had organized a SEA-PHAGES project (Science Education Alliance - Phage Hunters Advancing Genomics and Evolutionary Science), where classes of students in colleges around the country were trained to collect and characterize phages, targeting particular gram-positive bacteria. They accumulated and stored 15.000 phages, about 1800 of them were characterized. Two of them were bio-engineered, to make them lytic and able to kill the infected bacteria for the case in London. They were modified with the CRISPR/Cas system (Makarova et al. 2011) to eliminate the repressor to make the phages lytic. The repressor kept the phages in a non-lytic stage before removal. This treatment did not add any novel genes into the phages and they were therefore possibly not classified as GMOs and did not cause legal problems.

A local sore on the patient's chest was pretested since no animal studies were performed. She received three billion phages by intravenous injections every 12 h for 32 weeks and phage therapy still continues up to now (Dedrick et al. 2019). It should be noted for how long the phage therapy is continued in the patients.

A recent case was published in a Belgium newspaper, the "Saint-Luc baby", a 13 months old baby with liver and blood infections by multi-drug-resistant bacteria. She received phages for 85 days by a military doctor, Colonel Patrick Soentjens, from the Military Neder-over-Heembeck Hospital near Brussels. The phages were described as "trained" and "tailor-made", selected for. The production of the phages is worth mentioning because it was made by as "pharmaceutical compound" (Hope 2019). This is a routine production for cremes or composite material directly prepared under well-defined conditions in a qualified pharmacy, but is normally not used for biologicals such as phages. It is one of the goals of the front runners of phage therapy to make this production procedure standard and circumvent the costly GMP production (Pirnay et al. 2018; Wilbert and Pirnay 2016).

Also other therapies were reported such as a chronic ear infection tested in London with a preliminary report on 12 patients and 12 controls, which showed some efficacy by a phage against resistant *Pseudomonas auruginosa* (Wright et al. 2009).

14.4 Phages Instead of FMT?

Bacteria are to about 80% infected with phages and the turnover of bacteria in the ocean occurs in about 24–48 h (Sunagawa et al. 2015). The microbiome in the oceans was compared with that of the human gut microbiome, comprises about 10^{12} bacteria of about 1000 types and correspondingly an estimated similar number of phages (Sunagawa et al. 2015; Suttle 2005, 2013; Broecker et al. 2013, 2017). Furthermore, the composition and diversity of bacteria and phages differ in normal and non-healthy human feces. In a patient suffering from severe diarrhea and *Clostridium difficile* (recently reclassified as *Clostridoides difficile*) infection, a Fecal Microbiota Transplantation (FMT) was performed with feces from a healthy donor. This is a well-known procedure, which we recently revitalized for patients suffering from untreatable diarrhea and multi-drug resistant microbiota in the guts (Broecker et al. 2013). This has been performed for decades in Australia and some other countries, also for centuries in animals such as cows, designated as transfaunation, and was reinitiated in a Zurich patient in 2010, who had a dental infection and developed after several antibiotic cycles resistant microbiota in the guts. The *C. difficile* bacteria were resistant to the last resort drug Vancomycin. The patient received a fecal transfer, initially against the opposition of medical doctors. The patient received the a spoon-ful of feces from her sister. The material was suspended in buffer and applied to the recipient by enema intrarectally. The patient quickly felt better even before the recovered microbiota was detectable. The timecourse of the restauration of the microbiota and virome was determined and tested by sequencing (Broecker et al. 2013, 2017). The virome consisted of about 20 types of phages and one chlorella-related giant **virus** (Broecker et al. 2016, 2017, Moelling and Broecker 2016). We expected more than 20 types of phages, but may have detected primarily the most abundant core population and not the minor phage species, which may explain the low number (Broecker et al. 2013, 2016, 2017). This was confirmed for the phages in healthy human guts, where about half of a core phage community is a common set, which is shared among more than half the people and may play a major role in maintaining gut microbiome structure and function and may thereby contribute to human health (Manrique et al. 2016). This approach of FMT is worth mentioning, because the role of phages in a healthy and a non-healthy gut is not yet well-studied. Furthermore, no special purification, no GMP-like material was employed, no surgery room was necessary.

Soon after we applied the procedure in Switzerland, it was forbidden by the FDA. A petition was subsequently submitted by patients suffering from *C. difficile* that this was the only treatment in a life-threatening disease with no other therapeutic options. 15.000 people die of refractory *C. difficile* infections annually in the USA. Only in one case a recipient of FMT died (and another recovered), because the donor feces contained drug-resistant *E. coli.* Which was not tested beforehand, but did not affect 22 other patients receiving feces from the same donor (DeFilipp et al. 2019). Testing of stool samples will reduce this risk in the future. FMT is used exclusively for *C. difficile* patients but not against other intestinal diseases and not against obesity. The method has become routine, saves lives and should be made available for other intestinal diseases, soon (Moelling et al. 2018).

More recently a modified "FMT", was performed with phages only, without the bacteria from the donor stool, for fecal transfer (Ott et al. 2017). The results suggest a therapeutic efficacy of the bacterial-free stool preparation, likely due to phages against *C. dificile* infection, which needs to be verified with a larger cohort. One may expect industrial products to be admitted in the near future.

A novel question referring to the usefulness of phages comes from the knowledge of the role of the microbiome in the guts for the success of the anticancer immunotherapy. It has been shown how important the microbiome is for the therapeutic success – and this means also the phages, which play a role (Zitvogel et al. 2018). Phages and bacteria belong together. Phages are easier to apply than feces – hopefully also for the legal authorities. Mice depleted of their microbiome were used in some of these studies. This raises the question about the need of animal studies for studying phages or microbiomes. What kind of animal systems are available or should be used, gnotobiotic mice or mice with humanized microbiota? Indeed, one of the front runners in respect to animal studies with phages is Laurent Debarbieux at the Institute Pasteur, Paris. Not many groups invest time into this approach with animal models. Some important information can be deduced, however (De Sordi et al. 2019). Mouse studies with FMT have allowed to identify *Akkermansia muciniphila* as basis for successful immunotherapy against cancer, demonstrating the importance of animal studies (Routy et al. 2018).

14.5 Some Risks

Phages are natural products and are the most abundant species on Earth amounting to about 10^{33} (Suttle 2005, 2013). They are present in bacteria or free with a high turnover, about 80% in 24 h in the oceans. Our body is used to phages, except if they transfer toxin-coding plasmids, virulence genes or genes for antibiotic resistance. Phage genomes can swap among phage species as well as with hosts. This can lead to phages carrying plasmids with toxin-coding genes and was the cause of a *Shigella* food poisoning by **Brussel or raw sprouts**. They had been grown in the presence of animal feces which caused a food scandal in Germany in 2011, (EHEC, Enterohaemorrhagic *E. coli*) with 855 sick of HUK (hemolytic-uremic Syndrome) and 53 mortalities (Appelt et al. 2011). Another risk for phage therapy is a contamination of phage preparations with endotoxins, which are breakdown products of bacteria. They are composed of lipids and polysaccharides, LPS, from the outer membranes of some bacteria, disrupted by lytic phages or during preparation. Endotoxins are rather stable and need to be removed from the phage preparations by biochemical methods, to prevent toxic reactions in patients. These can, however be controlled by doctors and are not life-threatening (Schooley et al. 2017). Resistance of phages may not be so dangerous for short-term treatments, or if one can change the regimen and use other phages or find therapeutic regimens to avoid resistance.

14.6 Needed Activities

More phages are needed possibly by banking of a phage library in one location accessible by electronic interaction. Interestingly, groups who engineer phages do not agree with banking of natural isolates and think, that collections of natural phages may no longer be needed, because It is faster to produce them by bioengineering in the laboratory tailor made. The admission for animals, food or human use is, however, not trivial presently, because of legal restrictions for GMOs (Genetically Modified Organisms). Natural phages are needed for the time being and should not be limited by Material Transfer Agreements or commercial limits, thus Governments or the Military should help to set these up. In the UK the National Collection of Type Cultures, NCTC, at Culture Collections in London, is building a phage collection. Collections are made by military institutions, in research institutes, some can be obtained from Tbilisi, Georgia, some from University of Helsiniki, where 400 different types of phages of pathogenic bacteria were collected and tested by sequencing by Dr. Mikael Skurnik, some are available in Poland by Dr. A. Gorski, Ludwik-Hirszfeld Institute in Wroclaw, Poland. A phage collection in Germany was mentioned above, Dr. C. Rohde in Braunschweig, furthermore Dr. G. Hatful in Pittsburgh collected 15.000 phages, which are being sequenced. Storage conditions and protocols for isolation of phages, production and characterization should be standardized and harmonized, and made publicly available. Many groups start collecting phages for pretesting them as in Hamburg, Zürich, Berlin etc., possibly for future trials. Immediate availability in life-threatening diseases such as sepsis would be necessary. Phages available on demand!

A serious drawback for evaluation of the efficiency of phages against diverse diseases is the strict request for GMP production for human patient use. The present rules by the European committee EMA (European Medicines Agency, former EMEA), or the country-linked authorities are too restrictive, because they were developed for chemical compounds and are not adequate for biologicals. Phages may mutate during the procedure of production and during treatment and may be replaced by other phages or combinations - no legal rules exist for such a medical compound.

Several groups suggested and tested compound or magistral production in selected and authorized pharmacies. This would be a great advantage, because of less expense and local productions. If all of them need GMP material, this would drastically limit progress. Presently the individual compassionate trials are extremely complicated, expensive, and time-consuming, impossible for broader applications. The conditions need to be such that a sufficient number of trials can be performed to evaluate the potential success of phage therapies. The host range of phages for bacteria is limited and cocktails are required to cope with an infection, sometimes the type of phages need to be altered during a trial, which is not normally allowed within clinical protocols.

14.7 Promising Future Directions

Scientific progress in genetically engineering "universal" phages with a broad host range are progressing with enormous speed and look promising. Phages can be modified in the laboratory by recombinant technologies for an extended host-range. These include also CRISPR/Cas9 gene editing technology. This method was used to delete a repressor from a lysogenic phage to make it lytic by two different deletions using CRISPR/Cas technology. It was apparently permitted, because no new genes were inserted into the bacterial host genome (see above, Dedrick et al. 2019). Broad-range phages are not as universal as antibiotics and can be specific enough, not to destroy healthy bacteria- the major disadvantage of antibiotics.

CRIPR/Cas or gentic engineering allows fast diversification of the tail fibers, the "rebooting" of phage fibers (Kilcher et al. 2018; Kilcher and Loessner 2019; Dunne et al. 2019; Hupfeld et al. 2018) – new boots to bind to new hosts! The Company Johnson & Johnson is investing about 200 Mio $ into a start-up company for this purpose. The group of Timothy Lu develops „synthetic phages "with modified fibres for broader host range (Hiroki et al. 2015). A still rather underestimated diversity generating retrotransposon (DGR) mechanism is the cassette involving an error-prone reverse transcriptase which generates diversity, equivalent to the human immunoglobulin system ranging up to 10^{24} species, where a special reverse transcriptase is involved for diversifying the variable region of the fibers of the phages (Medhekar and Miller 2007; Broecker and Moelling 2019).

Another option is the incorporation of genes into the phage genome, for biofilm degrading enzymes or antimicrobial peptides, by replacement of non-essential phage genomic regions. Phages could also be engineered with antimicrobial agents that attack gene networks (e.g. the SOS system) of antibiotic resistant bacteria.

This field of research is very actively ongoing and very promising. The legal part is hampering fast progress. Industry will be interested because of the possibility of proprietary rights.

Mixtures of more than one phage are often needed and tested, whether they are compatible with each other. Also within one trial and protocol, changes in composition of the phage cocktails may be essential and must become applicable. Apparently in some of the above-mentioned successful cases the infection was caused by one bacterium.

Some diseases allowing topical application of the phages are preferred, such as open fractures, open sores, burns, deep sores, gangrenes. Recently in some treatment regimens phages were applied intravenously in humans and were successful (Schooley et al. 2017). Progress in understanding phage therapy in compassionate trials are normally under time pressure and not including controls. They did already increase our knowledge on phage therapy enormously but are limited by the numbers of cases and lack of statistics and controls. They were applied to patients with no other therapeutic options. Less severe cases without sepsis such as in Urinary Tract Infections could be important for long-term analysis, and more systematic testing, larger numbers of patients than emergency trials with life-threatening infections (Leitner et al. 2017).

Also phage-derived enzymes are under development such as lysins, to destroy the bacteria, and some are already available for skin applications and proved to be effective (personal observation).

14.8 Phages to Combine with Antibiotics

Already in Tbilisi phages have been used in parallel with antibiotics as soon as they became available. In almost all recent phage trials antibiotics were part of the therapy.

It was reported by Schooley that the multidrug resistant bacteria became more susceptible against antibiotics after phage treatment (Schooley et al. 2017).

During a recent meeting by the German Government, the Bundesinstitut for Risk Assessment (BfR), in Berlin in November 2019, this issue was discussed and it was open, whether simultaneous or successive therapeutic regimens of phages and antibiotics are more successful, indicating that systematic studies are still missing. Also, it was unclear whether a high titre of phages should be applied according to the HIV treatment "treat hard and early" to avoid resistance. In Tbilis the phages were 1000 fold dilated in some studies to about 10^6pfu/ml whereas undiluted 10^9 are preferred in some present studies. This may depend on the bacterial load in the patient and needs to be studied systematically.

A similar observation about phages plus antibiotics was described for multidrug resistant *Pseudomonas aeruginosa* bacteria resistant against a special phage OMKO1, which was shown to lead to increased antibiotic sensitivity. The increase of sensitivity of resistant bacteria can be 2–45 fold (Chan et al. 2019). Among the mechanism that have been proposed is that the bacteria have efflux pumps that excel antibiotics. Phages can shut down the efflux pump by attaching to them. Then the exposure time of the bacteria to the antibiotics increases and they become more effective. The mechanism is called "mex", standing for multi-drug-efflux system. Mex System is composed of three components, an antiporter that functions as a transporter, an outer membrane protein that forms a surface-exposed channel, and a periplasmic membrane fusion protein. They together were described to counteract the antibiotic efflux of drug-resistant *P aeruginosa* (Chan et al. 2019). Thus, on one side lytic phages will lyse the bacteria, but then evolving resistant bacteria can exhibit higher antibiotic sensitivity, probably because the antibiotics persist inside the bacterium for longer periods and can kill it. Healthy bacteria can still replicate and dominate (Chan et al. 2019). This is important and needs to be analyzed further.

It is known, that big pharmaceutical companies do not invest into the development of new antibiotics. Therefore it must be good news, that phages can counteract resistance of bacteria significantly. Antibiotic are still needed!

14.9 Phages as "Dietary Supplements"

This is a provkative proposal: Can one redefine the role of phages as supplement or additives or enhancers of antibiotic therapies, even as "wellness factors" or "food or dietary supplement". Could this describe their role! If so, could then phages be released for use without all the present legal restriction, which are based on the concept "phage therapy", which may be frightening to all legal authorities, reminescent of "gene therapy", with many risks and regulations. Would it make life easier to find a new definition for the role of phages other than phage therapy? Phages as "additive" or "dietary supplement" for improvement of an antibiotic treatment.

Indeed, four strains of phages against *E. coli* bacteria have been applied to 28 healthy patients with 10^9 phages per day for 28 days as dietary supplement together with short fatty acids. The result was the reduction of the bacteria target organisms and some changes in the cytokine levels. This demonstrates the modulating effect on human gut microbiota without global disruption of the bacterial composition by dysbiosis (Febvre et al. 2019). The effects are subtle and one may ask how many phages do we need for a therapy especially in diseased microbiota.

Furthermore the Company Intralytix produces phages against "problematic" bacteria of the gastrointestinal tract, oral cavity, vagina, and skin.

Furthermore, Probiophage with 15 mg phages as "dietary supplement" can be purchased from Amazon, together with bifidobcteria and lactobacteria five billions each with guarantee that they eill replicate It is explicitly sated that the ingredients are not produced under GMP conditions. The product is available only in listed countries, not in Germany. Furthermore, food, such as sliced salmon, is reported to be sprayed with phages for preservation- which does not need to be declared. Thus, there is the beginning to supply phages for people as dietary supplements without designating them "phage therapy"!

There is a solid basis for the demand to make phages easily available as we ask for here: the pre- or probiotics! They are commercially existing products, which patients get during antibiotic therapies to maintain some of the necessary microbiota in the guts. The number of pre- and probiotics in the case of diarrhea including antibiotic treatments are numerous (Brüssow 2019a, b). Dozens of companies produce these bacteria as supplements for protection of the gut microbiota during antibiotic treatments. In some countries antibiotics are sold always together with probiotics for protection of the gut microbiome. On the product labels the companies advertise that ten billion of bacteria of each type are present and often 10 different bacteria are listed. They even stress on the label that the bacteria are able to replicate in the guts! And they designate them as "dietary supplements"!

What distinguishes a bacterial treatment from a phage treatment? Bacteria normally contain phages, they both coevolve in every habitat of our planet. The phages can be relieved depending on environmental conditions and the types of bacteria. Healthy people share a pool of conserved intestinal phages - which, however differ from phages found in patients with diseases (de Sordi et al. 2019). Humans habot in their guts about 10^{14} bacteria and estimated 10^3 fold less viruses. Bacteria and phage coevolution is the driver of ecological and evolutionary processes in microbiobial communities (de Sordi et al. 2019). Fifty percent of healthy people share 23 phage groups, only a small set of other phages are either rarely shared among individuals or unique to a person. Shared core and common phage communities are even globally distributed among people and play an important role in maintaining a healthy gut microbiome and thereby contributing significantly to human health (Manrique et al. 2016). A zip of sea water in any ocean can easily comprise 10^6 phages - with no consequences. In the oceans 80% of bacteria are infected with bacteria and have a turnover within about 24 h (Suttle 2005, 2013). Similarities between microbiome of oceans and guts have been described, both harboring dominating core microbiomes and stable core viromes (Broecker et al. 2017).

One may specify the recent call: "We need phage therapy now" (Moelling et al. 2018) and ask to allow phages as "dietary supplement" or call them "phage supplement", just probiotics, which contain phages anyway!

Why is it allowed to target the gut microbiotica with replication competent pre- or probiotics? and: Why is it not allowed to use phages as food additives to improve antibiotic treatments, once we know which ones to use? d'Herelle swallowed a phage cocktail to convince his colleagues of the safety before they treated children in Paris. Not a single adverse event has ever been described as caused by phages. We only need to continue some research on best cocktails for specific indications.

We could start with phages against the most relevant bacteria for human use with the acronyme: ESKAPE published by WHO, indicating*: Enterococcus faecium, Staphylococcus aureus, Klebsiella pneumoniae, Acinetobacter baumannii, Pseudomonas aeruginosa, and Enterobacter.* These germs are considered to be the main causes of hospital infections.

We need to study them better and use them for the benefit of patients without the present restrictions. Phages are friends not foes.

My neighbours diabetic toes were cut off recently and he was recommended by a Hospital in Berlin to fly to Georgia. This needs to end.

Acknowledgement I want to thank Dr. Felix Broecker for discussions and critical reading of the mansucript. I am grateful to "Betty" Kutter for critical reading of the manuscript.

Declaration There are no obligations.

References

Appelt B, Böl GF, Greiner M, Lahrssen-Wiederholt M, Hensel A, EHEC Outbreak (2011) Investigation of the outbreak along the foodchain. Bundesinstitut für Risikobewertung, BfR

Broecker F, Moelling K (2019) Evolution of immune systems from viruses and transposable elements. Front Microbiol. https://doi.org/10.3389/fmicb.2019.00051

Broecker F, Kube M, Klumpp J, Schuppler M, Biedermann L, Hecht J, Hombach M, Keller PM, Rogler G, Moelling K (2013) Analysis of the intestinal microbiome of a recovered Clostridium difficile patient after fecal transplantation. Digestion 68:243–251

Broecker F, Klumpp J, Schuppler M, Russo G, Biedermann L et al (2016) Long-term changes of bacterial and viral compositions in the instestine of a recovered *Clostridium difficile* patient after fecal microbiota transplantation. Cold Spring Harb Mol Case Stud 2:a000448. https://doi.org/10.1101/mcs.a000448

Broecker F, Russo G, Klumpp J, Moelling K (2017) Stable core virome despite variable microbiome afer fecal transfer. Gut Microbes 8:214–220

Brüssow H (2019a) Hurdles for Phage therapy to become a reality-an editorial comment. Viruses 11:557. https://doi.org/10.3390/v11060557

Brüssow H (2019b) Probiotics and prebiotics in clinical tests: an update. eCollection 2019:F1000Res

Chan BK, Sistrom M, Wertz JE, Kaitlyn KE, Narayan D, Turner PE (2019) Phage seletion restores antibiotic sensitivity in *psuedomonas aeruginosa*. Sci Rep 6:26717

d'Hèrelle F (1917) Sur un microbe unvisible antagoniste des bacteries dysentériques, Comptes Rendues Acad Sci Paris 165:373–375, or d'Hèrelle F (1917) On an unvisible microbe antagonist of dysenteric bacteria Comptes Rendues Acad Sci Paris 165:373–375

De Sordi L, Lourenco M, Debarbieux L (2019) I will survive: a tale of bacteriophage-bacteria coevolution in the gut. Gut Microbes 10:92–99

Declaration H (2013) World medical association declaration of Helsinki: ethical principles for medical research involving human subjects. JAMA 310:2191–2194

Dedrick RM, Guerrero-Bustamante CA, Garlena RA, Russell DA, Ford K, Harris K, Gilmour KC, Soothill J, Jacobs-Sera D, Schooley RT, Hatfull GF, Spencer H (2019) Engineered bacteriophages for treatment of a patient with a disseminated drug-resistant Mycobacterium abscessus. Nat Med 25:730–733

DeFilipp Z, Bloom P, Soto MT et al (2019) Drug-resistant E coli bacteremia transmitted by fecal microbiota transplant. New Engl J Med Oct 30:2019. https://doi.org/10.1056/NEJMoa1910437

Dunne M, Rupf B, Tala M, Plückthun A, Loessner ML, Kilcher S (2019) Reprogramming bacteriophage host range through structure-guided design of chimeric receptor binding proteins. Cell Rep 29:1336–1350

Febvre HP et al (2019) PHAGE study: effects of supplemental bacteriophage intake on inflammation and gut microbiota in health adults. Nutrients 11:666

Fish R, Kutter E, Wheat G, Blasdel B, Kutateladze M et al (2016) Bacteriophage treatment of intransigent diabetic toe ulcers: a case series. J Wound Care 25:273. https://doi.org/10.12968/jowc.2016.25.sup7.s27

Haeusler T (2006) Viruses vs. superbugs: a solution to the antibiotics crisis? Palgrave Macmillan, Basingstoke, p 298

Hankin EH (1896) L'action bactericide des eaux de la Jumna et du Gange sur le vibrion du cholera (in French). Ann Inst Pasteur Bacteriophage 10:511–523

Hiroki A, Lemire S, Pires PD, Lu TK (2015) Engineering Moduar viral scaffolds for targeted bacterial population editing. Cell Syst 1:187–195

Hope A (2019) Liver transplant baby saved by "trained" virus at Saint-Luc hospital The Brussels Time, 22.5.2019 Belgium

Hupfeld M, Trasanidou D, Ramazzini L, Klumpp J, Loessner MJ, Kilcher S (2018) A functional type II-A CRISPR-Cas system from Listeria enables efficient genome editing of large non-integrating bacteriophage. Nucleic Acids Res 46:6920–6933

Jault P, Leclerc T, Jennes S, Pirnay JP, Que YA et al (2019) Efficacy and tolerability of a cocktail of bacteriophages to treat burn wounds infected by *Pseudomonas aeruginosa* (PhagoBurn): a randomised, controlled, double-blind phase 1/2 trial. Lancet Infect Dis 19:35–45

Kilcher S, Loessner MJ (2019) Engineering bacteriophages as versatile biologics. Trends Microbiol 27:355–367

Kilcher S, Studer P, Muessner C, Klumpp J, Loessner MJ (2018) Cross-genus rebooting of custom-made, synthetic bacteriophage genomes in L-form bacteria. Proc Natl Acad Sci U S A 115:567–572

Kuehn C, Rubalskii E, Rohde C reported on television, "Visite" 11.06.2019, Norddeutscher Rundfunk, NDR

Kutter E, de Vos D, Gvasalia G, Alavidze Z, Gogokhia L, Kuhl S, Abedon ST (2010) Phage therapy in clinical practice: treatment of human infections. Curr Pharm Biotechnol 11:69–86

Leitner L, Sybesma W, Chanishvili N, Goderdzishvili M, Chkhotua A, Ujmajuridze A, Schneider MP, Sartori A, Mehnert U, Bachmann LM, Kessler (2017) TM bacteriophages for treating urinary tract infections in patients undergoing transurethral ressection of the prostate: a eandomized, placebo-controlled, double-blind clincal trial. BMC Urol 17:90–95

Makarova KS, Haft DH, Barrangou R, Brouns SJ, Charpentier E, Horvath P, Moineau S, Mojica FJ, Wolf YI, Yakunin AF, van der Oost J, Koonin EV (2011) Evolution and classification of the CRISPR/Cas sytems. NatRev Mikrobiol 9:467–477

Manrique P, Bolduc B, Walk ST, van der Oost J, de Vos WM, Young M (2016) Healthy human gut phageome. Proc Natl Acad Sci USA 13:1000–10405

Medhekar B, Miller JF (2007) Diversity-generating retroelements. Curr Opin Microbiol 10:388–395

Moelling K (2019) New case reports with phage therapy- what is needed for more? Nurs Health Care 4:30–32

Moelling K, Broecker F (2016) Fecal microbiota transplantation to fight *Clostridium difficile* infections and other intestinal diseases. Bacteriophage 6:e1251380

Moelling K, Broecker F, Willy C (2018) A wake-up call: we need phage therapy now. Viruses 10:688–692

Ott SJ, Waetzig GH, Rehman A, Moltzau-Anderson J, Bharti R, Grasis JA, Cassidy L, Tholey A, Fickenscher H, Seegert D, Rosenstiel P, Schreiber S (2017) Efficacy of sterile fecal filtrate transfer for treating patients with Clostridium difficile infection. Gastroenterology 152:799–811. e7

Pirnay JP, Verbeken G, Ceyssens PJ, Huys I, de Vos D, Ameloot C, Fauconnier A (2018) The Magistral phage. Viruses 10:64–69

Routy B et al (2018) Gut microbiome influences efficacy of PD-1-based immunotherapy against epithelial tumors. Science 359:91–97

Schmidt CG (2019) Phage therapy's latest makeover. Nat Biotechnol 37:581–586

Schooley RT, Biswas B, Gill JJ, Hernandez-Morales A, Lancaster J, Lessor L, Barr JJ, Reed SL, Rohwer F, Benler S, Segall AM, Taplitz R, Smith DM, Kerr K, Kumaraswamy M, Nizet V, Lin L, McCauley MD, Strathdee SA, Benson CA, Pope RK, Leroux BM, Picel AC, Mateczun AJ, Cilwa KE, Regeimbal JM, Estrella LA, Wolfe DM, Henry MS, Quinones J, Salka S, Bishop-Lilly KA, Young R, Hamilton T (2017) Development and use of personalized bacteriophage-based therapeutic cocktails to treat a patient with a disseminated resistant acinetobacter baumannii infection. Antimicrob Agents Chemother 61:e00954–e00917

Sunagawa S, Coelho LP, Caffron S, Kultimaj R, Labadie K, Salazar G et al (2015) Ocean plankton. Structure and function of the global ocean microbiome. Science 348:1261359

Suttle CA (2005) Viruses in the sea. Nature 437:356–361

Suttle CA (2013) Viruses: unlocking the greatest biodiversity on Earth. Genome 56:542–544

Tsulukidze AP (1941) Experience of use of bacteriophages in the conditions of war traumatism (in Russian). Tbilisi, Gruzmedgiz

Vogt D, Sperling S, Tkhilaishvili T, Trampuz A, Prinay JP, Willy C (2017) Beyond antibiotic therypy – future antiinfective strategies – an update 2017. Unfallchirurg 120:573–584

Wilbert S, Pirnay JP (2016) Silk route to the acceptance and re-implantation of bacteriophage therapy. Biotechnol J 11:595–600

Wright A, Hawkins CH, Anggard EE, Harper DR (2009) A controlled clinical trial of a therapeutic abcteriophage preparation in chronic otitis due to antibiotic-resistnt Pseudomonas aeruginosa; a preliminary report of effiacy. Clin Otolarynngol 34:349–357

Zitvogel L, Ma Y, Raoult D, Kroemer G, Gajewski TF (2018) The microbiome in cancer immunotherapy: diagnostic tools and therapeutic strategies. Science 359:1366–1370

Chapter 15
Bacteriophage Application and Biological Safety (or How Should I Train My Dog Not to Bite Me)

Robert Armon

> *"inoculated agar tubes often showed watery-looking areas, and in cultures that grew micrococci it was found that some of these colonies could not be subcultured, but if kept they became glassy and transparent."*
>
> Twort (1915)

Abstract Following extended antibiotic resistance as a result of bacterial evolution under superfluous antibiotic stress (e.g. hospital, sewage system, agriculture), bacteriophages seem to be the next defense generation against bacterial pathogens in human health, veterinary, agriculture and environmental issues and many other areas. Bacteriophages have several obvious advantages over antibiotics: they multiply, are highly specific and highly diverse, have a continuous genetic evolution, do not interfere with eukaryotic cells (at least up to our present knowledge) and are highly efficient were antibiotics failed, e.g. diabetic foot ulcer or acute infection. The intensive research on bacteriophages revealed new molecular biology processes (e.g. CRISPR-Cas) that are presently implemented in many areas. The application of bacteriophages on foodborne pathogens was shown to be very effective in infection prevention. In our laboratory, we showed that bacteriophages can prevent biofilm development on filtration membranes (UF) and well screens in combination with minimal chemical intervention. Those scientists who have uncertainty on bacteriophage unrestricted application are also right from the point of view of the safety facet of biological treatment. However, it should be stated that processes like resistance and reversion occur continuously in nature in the vast biosphere without our intervention, we only speed up the process under a concentrated milieu. Further studies oriented to safety should be carried out continuously under the metagenomic umbrella to prevent antibiotics outcome.

In the memory of my good friend late Prof. Hans-Wolfgang Ackermann, a world expert on bacteriophages microscopy and systematics.

R. Armon (✉)
Faculty of Civil & Environmental Engineering, Technion, Haifa, Israel
e-mail: cvrrobi@technion.ac.il

G. Witzany (ed.), *Biocommunication of Phages*,
https://doi.org/10.1007/978-3-030-45885-0_15

15.1 Introduction

Ten years ago, at a bi-national conference held at Technion (Israel Institute of Technology) our laboratory presentation on bacteriophage application was well-received (biofilm prevention *via* bacteriophage application on wells and UF membranes). The presentation was primarily on lytic bacteriophage selection from environmental sources to lyse specific bacterial hosts found to clog expensive equipment such as groundwater filters and ultrafilters (UF). Nevertheless, one of the participants commented that in his country, according to regulations, a person or a governmental body is not allowed to add bacteriophages to water and wastewater systems. At first I was surprised but after a second thought I understood that most scientist dealing with bacteriophages are not completely aware of the abundance of bacteriophages in our environment (Table 15.1); the biosphere is believed to contain 10^{31}–10^{32} virions with the following specific distribution: in soil 2.6×10^{30} and rhizosphere 1.5×10^{8} (Ashelford et al. 2003; Weinbauer 2004), 1.2×10^{30} in oceans and 3.5×10^{30} in subsurface of oceans and 0.25–2.5×10^{31} in terrestrial subsurface (Parikka et al. 2016; Day 2004). Bacteriophages can also be found in highly extreme environments like sea ice (9.0×10^{6}–1.3×10^{8}) and algal flocks (5.6–8.7×10^{10}) (Weinbauer 2004), but also in high temperature niches like: geothermal springs, volcanoes, hydrothermal vents, etc.) and cold and hypersaline environments (Clokie et al. 2011). The ubiquity of phages is mostly based on host abundance and variability in various environments, with bacteriophages tittering bacterial population pending environmental conditions (Day 2004). There are several bacteria against which specific phages have not been detected so far, primarily because of their life cycle (e.g. *Legionella* sp. that mostly have an intracellular life, inside protozoa (amoebae) or polymorphonuclear cells (PMN) in our blood, therefore less susceptible to environmental phages, secondly because they can hide as prophages and thirdly because they may be inactivated by CRISPR system (to be discussed later) (Fig. 15.1). Here it is important to emphasize that infections can fail (abortion) due to host lack of specific receptors, own a restriction system (in abortive infection cases) or CRISPR (clustered regularly interspaced short palindromic repeats) defense system (Vale and Little 2010). Most of the bacteriophages known today are lytic phages that multiply and lyse their host cells yielding new offspring that in turn infect adjacent cells. According to Ackermann (2006b), there are 5568 described bacteriophages infecting eubacteria and archaebacteria, mostly tailed (96%) and a minority polyhedral, filamentous, or pleomorphic (3.7%) (Table 15.2). Certain bacteriophage groups reveal a **convergent evolution** due to life in similar habitats resulting in similar morphology. Finally, according to Srinivasiah et al. (2008) despite aquatic environments global predominance, the microbial abundance and diversity within soil environments exceed that of the aquatic sphere therefore it is expected that bacteriophages will be also diverse and abundant pending application of metagenomics beside that yielded methods for the direct extraction and enumeration of viruses within soils.

Table 15.1 Bacteriophage distribution in different environments (Srinivasiah et al. 2008; Parikka et al. 2016; Wommack and Colwell 2000)

Environment	Type	Estimated number (ml^{-1} or g^{-1})	Estimated number in total
Ocean			1.2×10^{30}
	Oligotrophic	$0.03–9.5 \times 10^{7}$	
	Mesotrophic & Eutrophic	$0.28–4.0 \times 10^{7}$	
Ocean subsurface			3.5×10^{30}
Soil			2.6×10^{30}
	Agricultural	$87–110 \times 10^{7}$	
	Forested	$131–417 \times 10^{7}$	
	Antarctica	$23–64 \times 10^{7}$	
Terrestrial subsurface			$0.25–2.5 \times 10^{31}$
Sea Ice			9.0–1.3 × 108/ml
Algal Flocks			$5.6 \times 8.7 \times 10^{10}/cm^{3}$
Soil and rhizosphere			$1.5 \times 10^{8}/g$
Fresh water	Oligotrophic	$0.42–120 \times 10^{7}$	
	Mesotrophic	$0.53–14.2 \times 10^{7}$	
	Eutrophic	$0.56–96.1 \times 10^{7}$	
Extreme	Alkaline & Hypersaline	$10–1000 \times 10^{7}$	
	Hot springs	$0.007–0.7 \times 10^{7}$	
	Deep sea vents	$0.0313–0.148 \times 10^{7}$	
Total in biosphere			$10^{31}–10^{32}$

15.2 Lytic, Lysogenic (Temperate) and Pseudo-Lysogenic Bacteriophage States and Reciprocal Relations

The two main paths of bacterial infection by bacteriophages are: lysogenic and lytic. Both are dependent on environmental conditions existent where bacteriophages reside. The more aggressive path, resulting in host cell destruction (lysis) is the lytic one. Under this path, a bacteriophage attaches to a bacterial host surface envelope (specifically to a certain receptor) and injects its DNA or RNA into the cell, which becomes its host. The injected circular genetic material can cleave the host's DNA *via* bacteriophage self-enzymes, then utilizing host resources to build self-defense of capsid/shell proteins (the primary bacteriophage structural envelope) intrinsically. The host continues to replicate the bacteriophage nucleic acid in several copies (varying with bacteriophage type) that further are packaged into capsid proteins to defend the genetic material from the host attack. Production of new progeny bacteriophages will occur at a constant rate when the host cell (bacterium) grows at a constant rate (exponential phase). A complete phage owns lysins (lytic enzymes) that will blast host cells dispersing newly formed bacteriophages for a

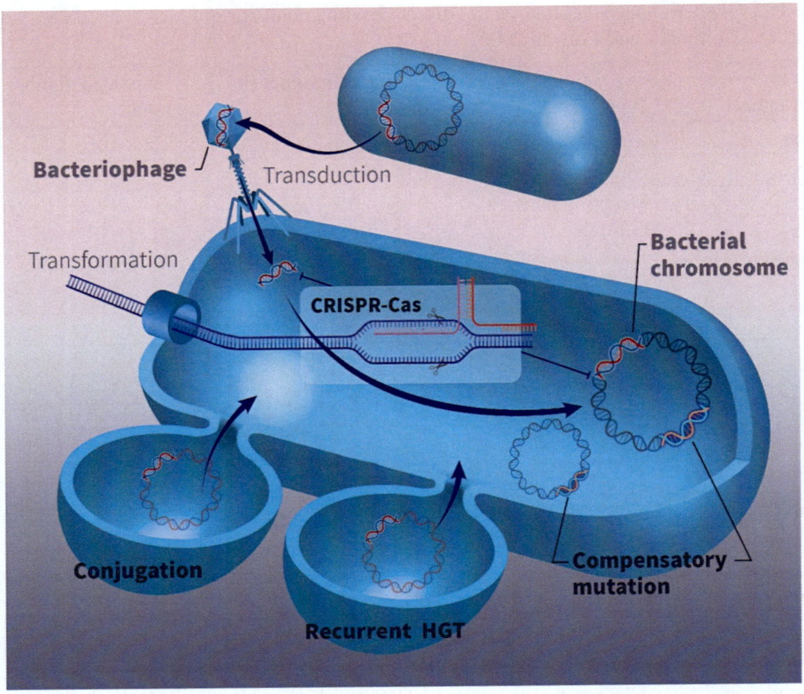

Fig. 15.1 Various mechanisms of acquiring foreign genes, e.g. antibiotic resistance, etc. (Adapted from MacLean and San Millan 2019, by Eitan Ben-Haim)

further attack on the surrounding host cells. Under these lytic circumstances, there are two ways of progeny bacteriophages release: lysis (complete host cell destruction *via* shattered membrane) and shading (while cells are still intact allowing offspring bacteriophage shading *via* membrane holes (e.g. RNA phages: MS2, f2, Qβ, etc). The main visual difference between the two excretion processes is clear plaques for the lytic process [as already reported by d'Hérelle (1917) and Twort (1915)] and fuzzy plaques for shading bacteriophages (Armon and Cabelli 1988).

From the safety point of view of phage application, lytic bacteriophages are much less promiscuous as they lyse the host cell and terminate its multiplication process and life, but they also cause a natural selection in the favor of resistant cells (commonly few colonies left after the infection). From this standpoint, there are three prospects that may impact genes expression related to lytic bacteriophages: host multiple range, therefore non-specific lysis (1), genes transfer between hosts while multiplying and uptake of certain genes into the newly formed phage progenies that will be further transmitted to newly infected host cells (2) and increased selection of phage resistant bacterial cells (3).

Infection and multiplication of bacteriophages in the ecosystems are dependent on available energy found in the system. Under oligotrophic conditions, phage replication is significantly altered, with longer latency periods, reduced burst sizes and diminished overall virulence (Kokjohn et al. 1991). However, the reality shows that

Table 15.2 ICTV taxonomic classification of bacteriophage infecting bacteria and archaea (Ackermann 2006a; Chibani et al. 2019)

Order	Family	Morphology	Nucleic acid	Examples	Subfamilies	Genera
Caudovirales	Ackermannviridae	Non-enveloped, head-tail	**dsDNA**	*Shigella* virus AG3 *Dickeya* virus limestone**,** *Escherichia* virus CBA120, *Salmonella* virus ViI**,** Unassigned (*Erwinia* virus Ea2809, *Serratia* viruses MAM1, IME250, *Klebsiella* virus 0507KN21)	**3** (Aglimvirinae, Cvivirinae, unassigned)	**4**
	Myoviridae	Nonenveloped, contractile tail	**Linear dsDNA**	T4 phage, Mu, PBSX, P1Puna-like, P2, I3, Bcep 1, Bcep 43, Bcep 78	**6**	**41**
	Siphoviridae	Nonenveloped, noncontractile tail (long)	**Linear dsDNA**	λ phage, T5 phage, phi, C2, L5, HK97, N15	**11**	**100**
	Podoviridae	Nonenveloped, noncontractile tail (short)	**Linear dsDNA**	T7, T3, Φ29, P22, P37	**3**	**23**
Ligamenvirales	Lipothrixviridae	Enveloped, rod-shaped	**Linear dsDNA**	*Acidianus* filamentous virus 1		**3**
	Rudiviridae	Nonenveloped, rod-shaped	**Linear dsDNA**	*Sulfolobus islandicus* rod-shaped virus 1		**1**
Unassigned	Ampullaviridae	Enveloped, bottle-shaped	**Linear dsDNA**	*Acidianus bottle-shaped virus*		**1**
	Bicaudaviridae	Nonenveloped, lemon-shaped	**Circular dsDNA**	*Acidianus two-tailed virus***,** possible: *Sulfolobus tengchongensis* spindle-shaped viruses 1 and 2 (STSV1 and STSV2)		**1**
	Clavaviridae	Nonenveloped, rod-shaped	**Circular dsDNA**	*Aeropyrum pernix* bacilliform virus 1		**1**
	Corticoviridae	Nonenveloped, isometric	**Circular dsDNA**	*Pseudoalteromonas* virus PM2		**1**

(continued)

Table 15.2 (continued)

Order	Family	Morphology	Nucleic acid	Examples	Subfamilies	Genera
	Cystoviridae	Enveloped, spherical	**Segmented dsDNA**	*Pseudomonas phage phi6*		**1**
	Fuselloviridae	Nonenveloped, lemon-shaped	**Circular ds DNA**	**Infecting:** *Sulfolobus shibatae, Sulfolobus solfataricus*, and *Sulfolobus islandicus* [environment: extreme temperature ≥ 70 °C and acidic pH ≤4 found in hot springs		**2**
	Globuloviridae	Enveloped, isometric	**Linear dsDNA**	*Pyrobaculum* spherical virus, *Thermoproteus tenax* spherical virus 1		**1**
	Guttaviridae	Nonenveloped, ovoid	**Circular ds DNA**	**Host** *Sulfolobus newzealandicus*		**2**
	Inoviridae	Nonenveloped, filamentous	**Circular ssDNA**	**M13, fd, f1, Pf3**		**7**
	Leviviridae	Nonenveloped, isometric	**Linear ssRNA**	MS2, Qβ**,** fr, f2, GA, JP34, PRR1, and PP7		**2**
	Microviridae	Nonenveloped, isometric	**Circular ssDNA**	ΦX174**,** G4 and α3		**6**
	Plasmaviridae	Enveloped, pleomorphic	**Circular ds DNA**	*Acholeplasma virus L2*		**1**
	Tectiviridae	Nonenveloped, isometric	**Linear dsDNA**	Enterobacteria phage PRD1		**2**

nature takes care of such circumstances by preserving the bacteriophage genome in a state of pseudo-lysogeny (firstly reported by Twort (1915) and substantiated later by Romig and Brodetsky (1961). The pseudo-lysogeny interaction can explain the large numbers of bacteriophages presence also in oligotrophic environments (Ripp and Miller 1997). Briefly, upon infection phage nucleic acid remains in a non-active state inside as long as host cell starves. Nevertheless, when the trophic state alters, and enough energy is available, the residing genome can establish the two known paths: lytic and lysogenic (Lieb 1953). Lysogenic phages are also called temperate (Racine 2014). Temperate bacteriophages play an important role, in most bacterial host, by controlling their gene expression via several mechanisms such as transcription factors, sRNAs, DNA reordering and lysis control! These interrelations are a kind of host-phage adaptation to various environments(Argov et al. 2017). Here is the place to remind that the newly discovered CRISPR system is a genetic configuration aimed to combat bacteriophage invasion by fragmentation of the invader DNA so the host bacteria can defend itself from future attacks by similar invaders (bacteriophages). The CRISPR was discovered in 1987 (Ishino et al. 1987) in conjugation with "*iap*" gene (isozyme conversion of alkaline phosphatase) but without understanding the function of the interrupted clustered repeats, found by these authors. Following additional reports from other parts of the world working on different bacteria, e.g. *M. tuberculosis, Haloferax volcanii*, *Streptococcus thermophilus*, *S. epidermidis*, *Pyrococcus furiosus* and *Archaeoglobus fulgidus,* it became clear that the acronym CRISPR is an important bacterial genetic defense against foreign DNA and RNA invasion of bacterial cell, i.e. bacteriophages and plasmids (*van Soolingen* et al. 1993*;* Mojica and Rodriguez-Valera 2016; Mojica et al. 2000; Barrangou and van der Ost 2013; Charpentier et al. 2015). Recently, hundreds of huge bacteriophages were discovered (also in human guts) with a massive genome (540 kb) that codes for many proteins and CRISPR genes (not present in smaller phages) able to attack their bacterial hosts and competing phages (Ceyssens et al. 2014; Drulis-Kawa et al. 2014; Chang et al. 2005. These phages have an important role in our gut microbiome therefore on human health. This is the place to mention that gut microbiome diversity and abundance are linked to the human brain via vagus nerve, global immune system, short-chain fatty acids (a brain food), serotonins (e.g. tryptophan precursor mostly from the gut). Therefore, any phage activity in this milieu can affect our behavior. Their capsid size and the large genome explain the development of small tailed phages nature. The capsid volume limits genome size, therefore evolutionary development of these phages by acquiring new capsid genes expressed by proteins (size and number) can explain their present huge size (Hendrix 2009). It is very much possible that further research in this direction can explain the evolution of bacteriophages towards a more developed independent microorganism such as bacteria?!

It should be mentioned that CRISPR-Cas system is used presently for genomes editing for different potential medical applications as well in adaptive immunity, but this is not in the scope of the present chapter (Barrangou 2015). A recent report by Zohar et al. (2017) revealed an interesting communication system called "arbitrium" using the host *Bacillus subtilis* infected with different specific bacteriophages (phi3T, spBeta, phi105, rho14 and phi29-spBeta group). These authors showed an existing path whereupon infection a bacteriophage decides what direction to choose: lytic or lysogenic! Briefly, bacteriophages of the spBeta group coordinate the lysis-lysogenic decision based on a small-molecule (6aa-six amino acids) released in the medium to be used in the subsequent infections. These authors revealed that the 6aa molecule concentration is measured by the progeny phages and if quantitatively high enough they will lysogenize. Another interesting finding was that different phages have different communication short molecules all involved in the lysogenic final decision! The so-called "arbritium" system is an interesting communication process among offspring phage with their predecessors to follow the lytic or lysogenic cycle.

15.3 Bacterial Acquisition of Foreign DNA

There are three main routes that a bacterium can acquire foreign DNA: transformation, conjugation, and transduction (Fig. 15.2). The less controlled, from our point of view on safety, is transformation. Following bacterial cell death and its disintegration, DNA is released into the environment. Pieces of DNA can be inserted into another live cell (*via* a special system) and integrated into its own DNA. Thus, antibiotics resistance genes can be acquired by a new bacterium. With the enhanced release of more resistant pathogens into the environment the further transformation capability to get more bacterial antibiotics resistance *via* this process (Domingo-Calap and Delgado-Martínez 2018). The second route is conjugation or interaction between two bacterial cells *via* a "sex pilus" (Zinder and Lederberg 1952). The pilus structure is dictated by a certain genetic feature called episome that can exist independently as a plasmid or integrated into the bacterial chromosome and is related as a fertility factor (Lederberg et al. 1952). When F episome is integrated into a bacterial chromosome it is called Hfr (high-frequency recombinant) and when independent as a plasmid containing just the fertility factor, it is named F^+. This process also takes place naturally but also man-made in the laboratory in order to transfer different marker genes to certain microorganisms. The third route is called transduction and it is bacteriophage dependent. Transduction is a phenomenon that occurs during bacteriophage infection and multiplication that has three directions: lytic, pseudo-lysogenic and lysogenic (see Sect. 15.3 for details). Along its infection route, a bacteriophage can transfer small pieces of the bacterial genome from one bacterium to another and also spread the newly insert to a new progeny and further to other bacteria. This process occurs continuously around the biosphere impacting bacterial evolution. A recent article revealed a new transduction path of a temperate bacteriophage of *Staphylococcus aureus* (a human commensal occasionally turns to be pathogen with multiple antibiotic resistance-MRSA) (Chen et al. 2018; Viertel et al. 2014). These authors termed the transduction of this specific bacteriophage "lateral

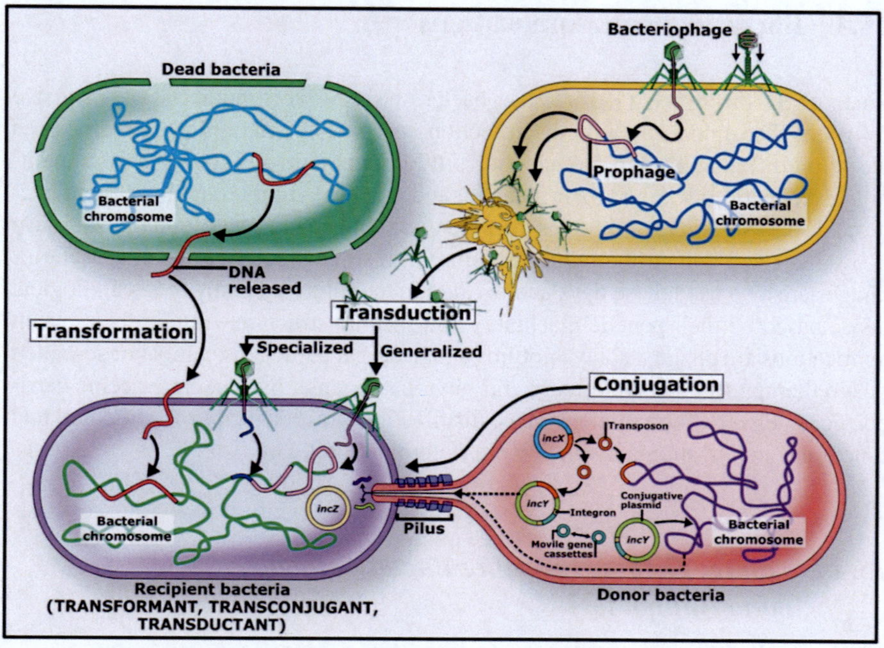

Fig. 15.2 Different genetic exchange interactions between bacteria-bacteria, phage-bacteria, and e-DNA-bacteria that impart antibiotic resistance. (From Bello-López et al. 2019)

transduction" based on the prophage distinct behavior. A prophage is a phase when phage DNA is integrated into the bacterium chromosome. The well-known path of a prophage follows the excision-replication-packaging processes before bacterial lysis and release of new progenies. In their report, the prophage excise late in its lytic program while large metameric spans including large bacterial DNA pieces (several hundred kilobases) are packaged into phage head at very high frequency. Thus, via replication, multiple prophage genomes are produced laterally during phage maturation containing S. aureus chromosome, consequently, parts of *S. aureus* DNA turn into hypermobile regions of gene transfer! Recently, Keen et al. (2017) reported on two novel *Escherichia* coli phage SUSP1 and SUSP, a subset of the natural lytic phage population. Upon infection, these bacteriophages release substantial amounts of intact, transformable plasmid DNA upon lysis, thereby promoting horizontal gene transfer through transformation. The authors called them superspreaders and express their concern on horizontal antibiotics resistance transfer via these kinds of phages if applied for medical treatment. Moreover, they found that addition of phage SUSP2 to a coculture of E. coli^{Kan+} (kanamycin-resistant) and kanamycin-sensitive *Bacillus* sp. bacteria resulted in roughly 1000-fold more kanamycin-resistant *Bacillus* sp. cells than those spontaneously rising in phage-free control.

15.4 Bacteriophages Applications

Since their discovery (1915/1917) bacteriophages were subjects of intensive research. The understanding the infection mechanism, multiplication processes based on different growth methods were the primary steps. From the very beginning, scientists understood their potential and practical ability to treat different bacterial diseases. Akin to antibiotics the potential is huge, and bacteriophages can save many lives and defend the public health. Not restricted solely to health, bacteriophages are also used these days and for sure in the future on many biotechnological tasks, based on their genetic machinery and specific structure. Among most recent applications are phage display, biofilms and bacterial pathogens combat or so-called phage therapy (human, veterinary and phytopathogens), biosensing, vaccine carriers, gene delivery, food safety, surface disinfection, corrosion control, structural and functional stabilization and much more (Ofir and Sorek 2018).

15.4.1 Phage Display (Turning Bacteriophages into Antibodies)

Phage display is a logic development along with the tremendous advance in molecular biology since the 80s (Smith 1985; Smith and Petrenko 1997). Basically, this molecular manipulation is based on phage genetic material and replication machinery to produce foreign proteins that are displayed on its coat. With the introduction of specific genes into phage DNA (expressing a large variety of foreign proteins), these genes are expressed as capsule proteins (the commonly used bacteriophages for display method are *E. coli* filamentous bacteriophages such as f1, fd, M13). These phages were selected because of their structure, infection of well-known (G-) bacteria (e.g. *E. coli* strains that carry an F-conjugative plasmid) and based on preceding information and easiness of genetic manipulation. Currently, the technology is applied for transfusion medicine, neurological disorders, mapping vascular addresses, tissue homing of peptides and cancer gene therapy (Azzazy and Highsmith 2002; Bazan et al. 2012; Bakhshinejad et al. 2014). Bakhshinejad et al. (2014) expressed their enthusiasm that "cancer-specific peptides isolated from phage libraries show huge promise to be utilized for targeting of various gene therapy vectors towards malignant cells. Beyond doubt, bacteriophages will play a more impressive role in the future of medical oncology".

15.4.2 Biofilms and Clogging (Deep Water Wells and Membranes)

Biofilms are common phenomena across the existing biosphere. This form of bacterial coexistence is the optimal organization of different bacteria to survive extreme conditions and resist disinfection processes. Besides survival, biofilms allow bacteria to utilize efficiently organic and inorganic matter and conduct a perpetual dialogue among its components. Biofilms can be looked at as beneficial (e.g. MBR, food fermentation products, removal, and biodegradation of hazardous materials) but also as highly negative (e.g. teeth decay, biocorrosion and biodegradation of construction materials, urinary tract infections (UTI) and shelter to different pathogens). A recent review on phage application to combat biofilms was more restrictive on phages application based on efficacy of the process against multiple bacterial species, cultivation of phages with broad host range, selection of phages with EPS (extracellular polysaccharides)-degrading enzymes, possible genetic modifications and combination of all the above (Kifelew et al. 2019). However, based on personal experience, the author and his laboratory revealed the excellent application of specific bacteriophages and a combination of to combat biofilms in experimental systems such as: well and membrane filtration (UF) clogging by biofilms (Gino et al. 2010; Goldman et al. 2009). Gino et al. (2010) showed that experimental biofilm that produces EPS and clogged well screen can be reduced with excellent results by a combined biological (bacteriophages) and chemical (weak organic acid) process. In the above particular case, it should be remembered that deep wells can attain a depth of 250–300 m (in arid zones) that is not feasible to be cleaned by mechanical methods and the predominant bacterial species was an iron bacteria (*Sphaerotilus natans*) that was highly susceptible to an environmental phage infection resulting in lysis and biofilm disintegration. In this study, no bacteriophage resistance was observed along the process.

A more advanced study, using molecular methods to engineer a certain bacteriophage in order to be able to penetrate biofilms' extracellular polymers (EPS) (Lu and Collins 2007). The newly engineered phage reduced bacterial biofilm cell count ~99.997% (equal to 4,5 orders of magnitude. Lu and Collins (2007) report revealed the capability to manipulate synthetic biology in many areas that encounter bacterial biofilm formation (Fig. 15.3).

15.4.3 Clinical Applications

15.4.3.1 Oncology

Gene therapy in cancer is one of the main targets pursued by medicine. Bacteriophages were shown that can be modified via phage display methodology. The coat modifications of those phages allow the target delivery of therapeutic genes with an excellent safety profile (Garg 2019). Among the large bacteriophages' variety, filamentous bacteriophages (e.g. M13) are leading in phage display and therapy. Genetic manipulation can modify these phages chemical structure that facilitates their application as genes delivery vehicle, drug carriers and as an immunomodulatory agent (Yacoby et al. 2007). Importantly, the peptides or proteins emerging from the surface of a nanocarrier will make the expense of such peptides economically more effective as compared to other immunological tools, and this seems to be a potential approach for developing a new nano drug carrier platform (Bar et al. 2008). Tumors treatment

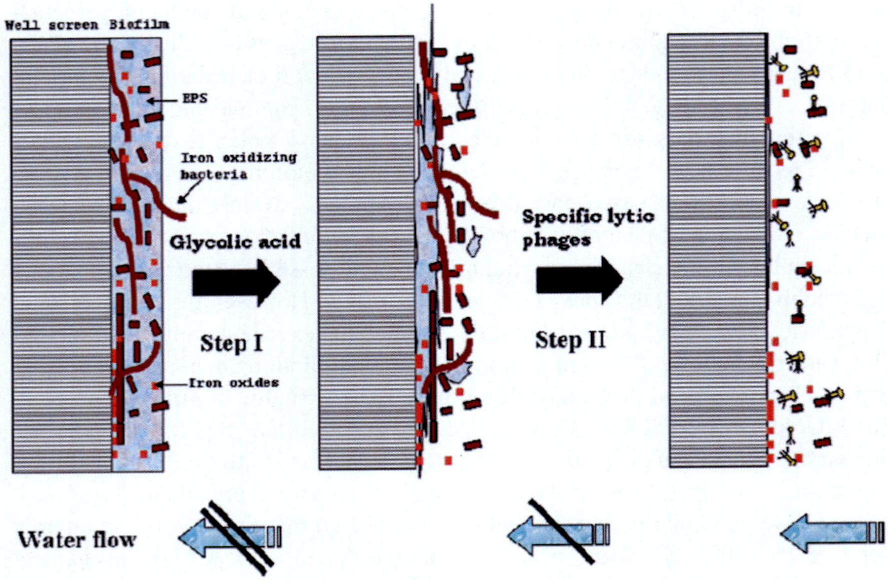

Fig. 15.3 Schematic representation of biotic and abiotic well screen clogging and the application effect of combined glycolate and phages on water flow through biofilm free screen. (Gino et al. 2010)

with chemotherapeutic agents beside its antitumor benefit has tough side effects such as indiscriminate drug distribution with severe toxicity. Bar et al. (2008) presented a new approach based on genetic manipulation of filamentous phage (fUSE5-ZZ) (Yakoby et al. 2007) that displays a host-specificity-conferring ligand. These phages are loaded with high doses of a cytotoxic drug *via* chemical conjugation and targeted to cancer cells membrane markers that by endocytosis are inserted into tumor cells. Following intracellular degradation and drug release target cells growth in vitro inhibited by a potential factor of >1000 in comparison with corresponding free drug.

15.4.3.2 Diabetes

One of the diabetics' serious complication is foot ulcers that often end up in amputation. Antibiotic treatment is challenging by the fact that a heterogeneous biofilm with antibiotic resistance develops at the wound site (Krylov et al. 2013; Morozova et al. 2018). Indeed, bacteriophage's internal application, via the bloodstream is not feasible for obvious reasons, and only the external one is conceivable. Rhoads et al. (2009) performed a one phase study on 39 patients applying bacteriophages infecting *P. aeruginosa*, *S. aureus* and *E. coli* on wounds versus saline as control. Their study didn't reveal "safety concerns with theses bacteriophages along with the treatment" during the 12 weeks direct application and 42 weeks follow up. However, the authors stated that a second phase is necessary to evaluate its efficacy.

An experimental answer came from a group of researchers that use bacteriophages to treat foot ulcers (of diabetic origin) (Fish et al. 2018). These authors presented successful treatment of a series of toe ulcers in diabetic patients accompanied by evidential photos of wounds before and after bacteriophage treatment. The anti-staphylococcal preparation named Eliava BioPreparations, a commercial product containing bacteriophage Sb-1 (a well-studied bacteriophage and safety confirmed) (Rhodes et al. 2009).

Sohrab et al. (2014) raised the potential of bacteriophage application in two important diseases: Alzheimer's disease (a neurodegenerative disease) and type 2 diabetes mellitus. For the first one, application based on phage delivery vectors as well as immunomodulation of anti-amyloid beta pepetide and direct application to heal open ulcers for the second one!

15.4.3.3 Alcoholic Hepatitis

Additional proof of concept on bacteriophage therapy emanated from a study performed on humanized mice (Duan et al. 2019). A humanized mouse is a mouse that carries human genes, cells, tissues and whole organs [e.g. nude or SCID (severe combined immunodeficiency) mice]. Those laboratory animals use as a model for human therapeutics since they are immunodeficient, therefore lack of host immunity and do not reject foreign tissues. Newly developed mice such as NCG, NOG and NSG mice revealed higher efficiency to engraft human cells/tissues. Among these, in two humanized mice strains MITRG and MISTRG four genes encoding cytokines (involved in innate immune cell development) were immobilized. Thus, humanized mice can be the closest simulation to the human immune system for both healthy and diseased humans in order to evaluate different therapeutics. Duan et al. (2019) identified two-subunit exotoxin (cytolysin) secreted by *Enterococcus faecalis* that cause hepatocyte death and consequently liver injury. Patients with alcoholic hepatitis were found to harbor high numbers of the above bacterium that correlated with hepatitis severity and mortality as a result. Using the humanized mice, they targeted cytolytic *E. faecalis* infection in those experimental animals with bacteriophages (*E. faecalis* experimental pathogen isolated from feces of alcoholic hepatitis patients). Their results showed that bacteriophages decreased cytolysin in mice's liver and "abolish ethanol-induced liver disease"! Further clinical studies are requested to validate these results for patients with alcoholic hepatitis, but the path to intestinal microbiota control had been proved!

15.4.3.4 Diarrhea

With the discovery of bacteriophages by Twort and d'Hérelle one of the first assumed applications was to combat gastrointestinal diseases caused by enteric bacteria (e.g. *Salmonella*, *Shigella*, etc.) (d'Hérelle 1917). Application of bacteriophages against diarrhea was reported by several authors (Sarker and Brüssow 2016;

Bruttin and Brüssow 2005; Sarker et al. 2017). Among these studies, the most important one was performed by Sarker et al. (2017) that used a T4-like coliphage cocktail given orally to healthy and diarrheal children from Bangladesh. Briefly, the experimental oral phage cocktail passed mostly passively with no substantial amplification in the intestine. The authors did not find effective therapy and justified it by low numbers of pathogenic E. coli (ETEC and EPEC) and polymicrobial infections in pediatric diarrhea patients in developing countries. Another point raised by these authors was the phenomenon of dysbiosis (alteration of gut microflora) of *E.coli* in the intestine that decreases direct contact between the applied phage and its host. This fact is not new, as d'Hérelle faced the same problem when he tried to cure diarrhea with his newly isolated phages. It should be remembered that the direct application of phages via gastrointestinal tract is a challenging task and needs much more studies oriented to their survival, efficacy and host encounter. However, from the safety point of view, this study showed that serologically (LPS enterotoxin transfer to the bloodstream), vital parameters (e.g. pulse, blood pressure, temperature, and respiration rate), biochemical tests of liver and kidney function and hematology tests did not reveal abnormal reactions to oral phage application. The only parameter that had a lower value (following T4 treatment) was blood urea nitrogen (occasionally found in malnourished children) and neutrophil percentages (not associated with fever or other symptoms) both without clinical relevance substantiated by other studies (Vandenheuvel et al. 2015).

15.4.3.5 Dentistry

An additional application of phage therapy via external usage is tooth decay control. *Streptococcus mutans* an oral cavity bacterium that utilizes diet sucrose by fermentation that results in lactate formation (an organic acid product) that is one of the main causes of enamel demineralization and consequently carries enhancement (Dalmasso et al. 2015). The isolated phage ϕAPCM01 applied against *S. mutans* revealed a decrease of at least 5.6 log CFU/mL between the experimental culture versus control at an MOI (Multiplicity of Infection) lower than 2.5×10^{-3} (Dalmasso et al. 2015).

15.4.3.6 Veterinary

From the empirical point of view, bacteriophage application to fight bacterial infections in animals is in a sense an experimental tool to understand human safety issues and to ameliorate and cure diseased animals (Smith et al. 1987a, b). Smith and Huggins (1983) showed that oral infection of calves with a mixture of 6 pathogenic *E. coli* strains can be controlled by the administration of bacteriophages specific against those pathogens, even at low titer (10^2). Soothill (1992) showed the same trend with mice infected with three pathogens: *Acinetobacter baumanii, P. aeruginosa*, and *S. aureus* and successful control of the pathogens following bacteriophages administration. *P. aeruginosa* a known pathogen infecting skin drafts, that causing their destruction, was successfully inhibited *via* bacteriophage application (Soothill 1994). Finally, 10 dogs with chronic, refractory *P. aeruginosa* otitis media were treated topically (into auditory canal of one ear) with 1×10^5 plaque-forming units (PFU) of each of the 6 bacteriophage types isolated in their study (Hawkins et al. 2010). The authors reported that the treatment was highly effective and safe. An increase in administered bacteriophages was monitored (significant phages multiplication) and no local inflammation or systemic adverse events were observed.

Principi et al. (2019) in his review on bacteriophage application to treat animals and human bacterial diseases summarized that results presented by many authors indicates that bacteriophage treatment seems to be the rational evolution following antibiotics decline, with reservations on safety concern. Besides resistance development, similar to antibiotics, another important point is the formulation standardization for clinical studies not yet established and additional controlled studies are very much needed (Labrie et al. 2010).

15.5 Phytopathogens

There are many bacterial phytopathogens that cause large agricultural losses. Global food demand is increasing with human population growth, therefore new methods are required to combat phytopathogens of microbial origin (Buttimer et al. 2017; Jones et al. 2007). There is a large variety of bacterial pathogens that attack agricultural products such as: *Pectobacterium carotovorum* ssp.*carotovorum*, *Pectobacterium wasabiae*, *Dickeya solani* and *Streptomyces scabies* on potato; *Ralstonia solanacearum* and *Xanthomonas campestris pv.vesicatoria* on tomato; *Xylella fastidiosa* on grapes; *Xanthomonas axonopodis* pv. *Allii* on onion; *Pectobacterium carotovorum* ssp.*carotovorum* on lettuce, radish *Streptomyces scabies* on radish; *Xanthomonas axonopodis* pv.*citri* on grapefruit; *Xanthomonas axonopodis* pv.*citrumelo* on orange; *Pseudomonas syringae* pv.*porri* on leek; *Pseudomonas tolaasi* on mushrooms and *Erwinia amylovora* on pear and apple (a disease called fire blight). The common treatment against these phytopathogens includes the use of antibiotics and traditional chemicals (dispersed by means of aerosol). Next to the development and selection of antibiotic-resistant bacteria that affect the whole plant biosphere, the process is also expensive (large areas to be covered). In a particular case, *Erwinia amylovora* that infected an orchard (in north Israel) was exposed to specific bacteriophages against *E. amylovora* isolated from sewage. (Armon 2015). The Isolated phages were concentrated and sprayed over certain blossom areas. Results showed prevention of the further spread of the disease, with some exceptions, branches that were not reached by our aerosol. Consequently, the idea of using bees as vectors was suggested but not tested in this study. Boulé et al. (2011) applied eight bacteriophage types accompanied by *Pantoea agglomerans* (as a pathogen competitor) to combat fire blight disease with successful results. *E. amylovora* infected detached pear tree blossoms treated with phages cocktail reduced infection by 84–96% with *P. agglomerans* as a carrier. Besides cocktails, there is an additional practical option to apply polyvalent bacteriophages, similar to the one used in clinical practice against antibiotic-resistant Staphylococci in hospitals (O'Flaherty et al. 2005).

15.6 Food

Practically, everybody experienced along his lifetime a foodborne infection. Besides the clinical distress, the cost of treatment is high and is an enormous economic burden (hospitalization, treatment and working days lost). The main bacterial foodborne pathogens are: *Salmonella enterica, Vibrio cholerae, Vibrio parahaemolyticus, Vibrio vulnificas, Campylobacter jejuni, Clostridium botulinum, Escherichia coli,* and *Listeria monocytogenes.* All are evidently causing serious symptoms and can be fatal to humans. Prevention of their proliferation in food products should be strictly prevented. Food production is a complex process starting from animal growth and plants harvest through different processes up to the customer table. Along this path, there are many possibilities of bacterial contamination, especially if the food products are kept under poor conditions (Tan et al. 2014). At present, there are several studies of post-harvest phage biocontrol interventions (direct food applications) for *Escherichia coli* O157:H7, *Listeria monocytogenes*, *Salmonella* spp., *Campylobacter jejuni*, *Cronobacter sakazakii*, *Shigella* spp. and *Staphylococcus aureus* (Sulakvelidze 2013, Bolocan et al. 2018; Gutierrez et al. 2017; Hagens and Loessner 2007).

Progressively, more studies demonstrated that phages therapy/application can be the main approach to reduce foodborne diseases (Goodridge and Bisha 2011). The main questions still to be answered regarding the above approach are (1). better comprehension of phage-host interactions; (2). the efficiency of the treatment and pharmacokinetics and (3). similar to antibiotics development of resistance against bacteriophages! On the third point, additional *in vivo* studies are required, including genomic analyses, in order to assess safety use. In general, in food products biosanitation (e.g. *Listeria* a nasty food pathogen), bacteriophages are highly effective but care should be taken in order to prevent resistance development (Bolocan et al. 2018). Gómez-Gómez et al. (2019) already reported on infectious phage particles packaging antibiotic resistance genes following meat products and chicken feces survey! These authors called not to underestimate phages as a risk factor in global antibiotics resistance transmission. It is interesting to raise a more philosophical question: are antibiotic resistance genes also involved in other bacterial functions, independent of antibiotics presence?

Finally, Hussain et al. (2017) described an encapsulation method to protect bacteriophages against adverse conditions present on the way to their target, with the aim of enhancing their efficacy in food protection.

15.7 Bio-corrosion Control

Like biofouling control, biocorrosion can be also controlled by specific bacteriophages isolated from the environment (Zarasvand and Rai 2014; Baldwin and Summer 2012). From an industrial point of view, biocorrosion has a huge impact on industrial equipment with major loses (e.g. oil and gas wells). Generally, there are two main groups of organisms that especially corrode metal equipment used in different production industries: aerobic and anaerobic, both conditions often encountered by industry. Both conditions result in the production of high corrosive compounds such as sulfuric acid, hydrogen sulfide, and other sulfur compounds. Among the aerobic ones are SOB (sulfur-oxidizing bacteria) like *Thiobacillus thiooxidans* and *Thiobacillus denitrificans* and among the anaerobic SRB (sulfur reduction bacteria) are: *Desulfovibrio*, *Desulfotomaculum*, *Desulfobacter*, and *Desulfuromonas* (Summer and Summer 2012). There are not enough field studies to substantiate this approach, however, it has real potential on direct application following real information on the bacterial components and biofilm structure (Gino et al. 2010; Goldman et al. 2009).

15.8 Nanotechnology

The most advanced application of bacteriophages has been presented by Oh et al. (2014) based on the special structure of certain bacteriophages like phage M13. Phage M13 (a nonenveloped, filamentous, circular ssDNA) related to its structure was used as a template in the synthesis of cobalt manganese oxide nanowires (NWs) to fabricate high capacity $Li - O_2$ battery electrodes. To improve the roundtrip efficiency and cycle life of this kind of battery the researcher hybridized Ni nanoparticles on bio Co_3O_4 NWs. Cobalt manganese spinel oxides (MCO) owe good catalytic activity towards oxygen reduction and oxygen evolution reaction and are also cost-effective, the however conventional method to create electrodes resulted in random sizes and variable geometry. Application of M13 virus with its specific protein structures allowed the formation of filamentous electrodes (NW) at the nanoscale.

Phage M13 was also used in phage display applications (see phage display application) has unique properties ("high aspect ratio of geometry, ~880 nm long and ~6 nm in diameter, genetic tunability of surface protein, and easy replicability"). M13 viral capsid is composed of 2700 copies of helically arranged major coat proteins, and 5–7 copies of each (p3, p6, p9, and p7) located at either ends of M13 virus (filamentous structure). These coat proteins can be genetically modified to display peptide sequences at their respective positions along M13 phage.

15.9 Conclusions, Remarks, and Thoughts About the Future Application of Bacteriophages

Since the discovery of miracle molecules called "antibiotics" countless human lives were spared. There is no doubt, that antibiotics are still the first antibacterial weapon available to physicians! However, antibiotic resistance emerged through extended use in hospitals and other facilities, not always advocated by the prevailing circumstances (e.g. viral infection) (De Vos and Pirnay 2015). While writing this chapter, I stomped on an exciting story related to bacteriophages application that at the first glance support completely the clinical application of those organisms, particularly in emergency circumstances. There is a famous Talmudic expression: 'Whoever destroys a soul, it is considered as if he destroyed an entire world. And whoever saves a life, it is considered as if he saved an entire world.' (from - *Babylonian Talmud*, Sanhedrin 4:8 (37a)" that depicts the subsequent extreme case.

The case was presented by Steffanie Strathdee on TED (https://www.youtube.com/watch?v=AbAZU8FqzX4) that by chance is an epidemiologist (Chief of the Division of Global Public Health in the Department of Medicine at University of California San Diego). Briefly, while on a cruise in Egypt, Tom, her spouse (by himself an HIV researcher), contracted a multiple resistance bacteria *Acinetobacter baumannii*, a Gram-negative opportunistic pathogen and diagnosed with acute peritonitis (Morris et al. 2019). Following a series of unsuccessful hospital treatment, he entered a serious coma with systemic failure. As the last instance, bacteriophage application was suggested by his wife and other researchers. Bacteriophage application is not a new idea and practically used in the former USSR, precisely in Tbilisi, Georgia, where the leading Eliava Institute is located and active. This institute from the start of bacteriophage discovery initiated the application of bacteriophages for human therapy. Many Western scientists mocked this approach by calling it "Soviet antibiotics". However, with the increasing antibiotics resistance all over the world, a second thought started to rise to look at bacteriophages as a possible remedy against infectious diseases. Back in our particular case, it was the first attempt to apply bacteriophages intrinsically as the last resource available. With the consent of Tom's wife and supported by several established bacteriophage researchers and subsequent FDA specific approval, he was administered intravenously (iv) a cocktail of bacteriophages (originally isolated from different environmental sources, including sewage). Intravenous administration of bacteriophages was never tried on humans because of clear immunological reaction, nonetheless this time the decision was correct in order to save patient lives. Undeniably, Tom survived, and his infection was eradicated. The above case is an encouraging example of a bacteriophage application that saved human life. Indeed, there are more examples in which bacteriophages application were effective against human bacterial diseases, 99% of those cases are topical applications (external) (Matsuzaki et al. 2014).

Here is the place to emphasize that despite the chemical differences between an antibiotic and a bacteriophage, the last have an overwhelming advantage over the first one owing to its capability to multiply. Consistent with our roughly 100 years of experience with antibiotics and the following efficiency decline because of excessive use and bacterial resistance emergence, careful thinking should be contemplated/considered, especially in clinical bacteriophage application. The hitherto described case is an excellent example of an emergency application and not a routine one, applications that can get us again to the same antibiotic resistance scenario!

This is the place to suggests several postulates/principles in order to keep a safety border when directing bacteriophages application, mostly in the clinical area.

1. Comprehensive knowledge of bacteriophage genetic material (metagenomic analysis and bioinformation).
2. A worldwide collection of bacteriophages from different sources and biochemical and genetic description including preservation methods (Large database to be available globally).
3. Due to our vast knowledge of most pathogenic bacteria, bacteriophages application should be first tested at laboratory level then followed by laboratory animals e.g. "human mice".
4. More research on bacteriophages genetic material mobility such as transduction and CRISPR-Cas comportment.
5. Follow up of clinical cases treated with bacteriophages for resistance, reversion, toxicity, etc. and impact on the human immune system or pathologies related to the gut microbiome.
6. Development of "artificial" lytic bacteriophages based on nanotechnology with improved efficiency.
7. Application of bacteriophages in clinical cases as the last instance to save a life (in cases of bacteremia or terminal cancer).
8. Prevention of bacteriophage application for mass production and large surface disinfection.
9. Use of cocktails of bacteriophages against a specific pathogenic bacterium to prevent resistance emergence.
10. Based on global research data to answer the following question: how many bacteriophages have one bacterial species. Requires, close collaboration between phage scientists such as workshops and experimental reports under the umbrella of WHO (World Health Organization) or else.

In the last moment, Schmidt (2019) tackled in his review whether phage therapy can be the ultimate antibacterial barrier. According to this author, more and more companies are involved in the development of phage therapy with the earliest one established in 1998 (Intralix), many of these companies involving CRISPR technology. As pointed above, more molecular biology data is needed in order to apply this technology, however, these days due to much more advanced and faster methods in DNA screening, the concept is closer to the real application as a standard.

At this point in scientific history, the above 10 points come to cover the safety application of bacteriophages in many areas. However, a recently published opinion on bacteriophages as an indirect pathogen cannot be overlooked (Tetz and Tetz 2018). These authors raised an interesting issue on bacteriophages as potential human pathogens (mainly indirect) by altering the human microbiome, and possible implication in the progression and maintenance of some pathologies e.g. those associated with proteins misfolding! Figure 15.4 depicts the possible involvement of

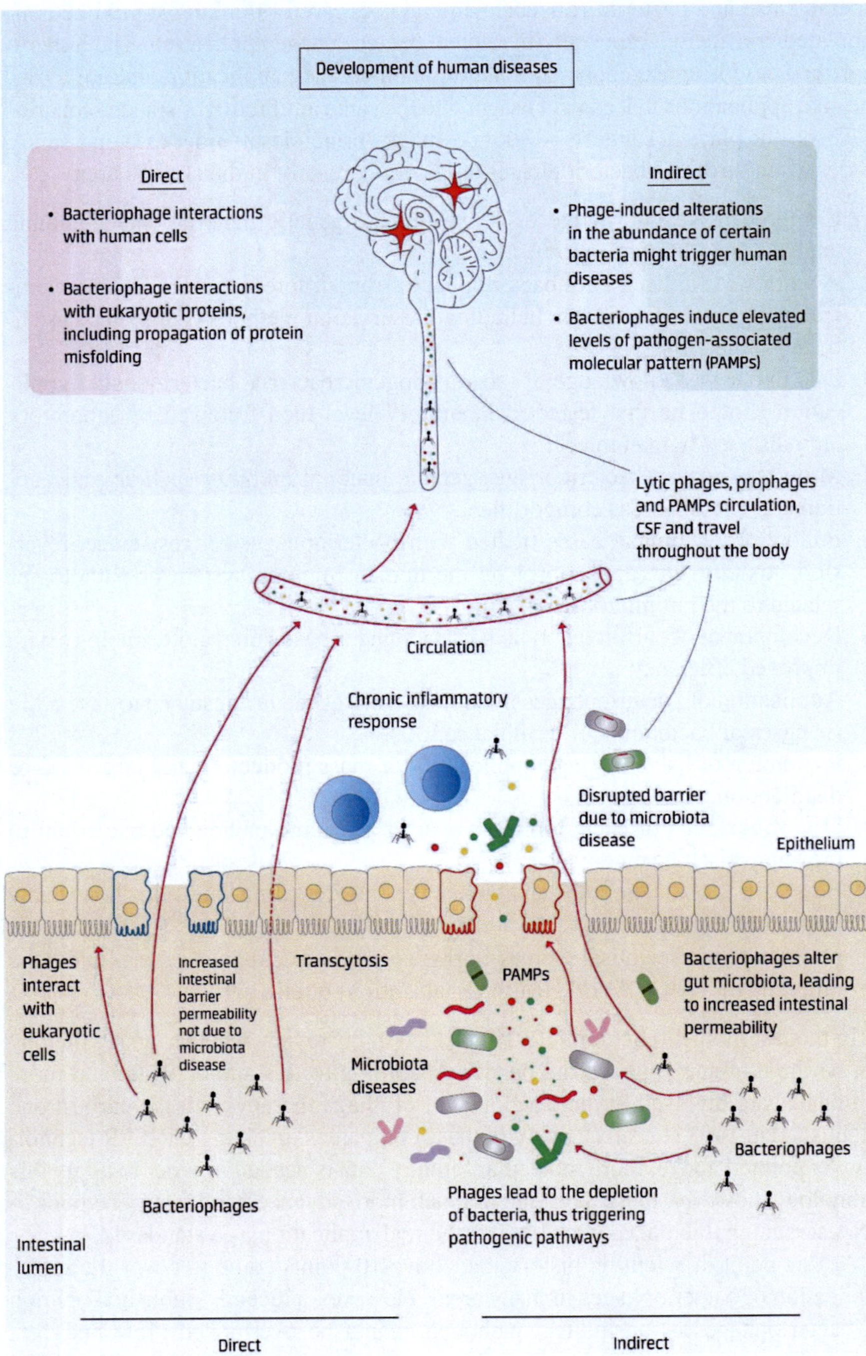

Fig. 15.4 Bacteriophage direct and indirect impact on human diseases concept, as suggested by Tezt and Tetz, (2018). (From Tetz and Tetz 2018)

bacteriophages in human pathologies which have to be explored by further experimental studies.

Finally, it should be pointed out, that while these lines are written, nature doesn't stop its activity consequently bacteriophage-bacteria interactions occur around the globe without our intervention, continuously.

References

Ackermann H-W (2006a) Classification of bacteriophages. In: Calendar R, (ed) The Bacteriophages (2), Oxford University Press, New York, p. 8–16

Ackermann H-W (2006b) 5500 phages examined in the electron microscope. Arch Virol 152:227–243

Argov T, Azulay G, Pasechnek A, Stadnyuk O, Ran-Sapir S, Borovok I, Sigal N, Herskovits AA (2017) Temperate bacteriophages as regulators of host behavior. Curr Opin Microbiol 38:81–87

Armon R (2015) (Personal communication)

Armon R, Cabelli VJ (1988) Phage f2 desorption from clay (bentonite and kaolinite) using non-ionic detergents, beef extract and chaotropic agent. Can J Microbiol 34:1022–1024

Ashelford K, Day M, Fry J (2003) Elevated abundance of bacteriophage infecting bacteria in soil. Appl Environ Microbiol 69(1):285–289

Azzazy HM, Highsmith WE (2002) Phage display technology: clinical applications and recent innovations. Clin Biochem 35(6):425–445

Bakhshinejad B, Karimi M, Sadeghizadeh M (2014) Bacteriophages and medical oncology: targeted gene therapy of cancer. Med Oncol 31(8):110

Baldwin D, Summer NS (2012) Prevention and remediation of petroleum reservoir souring and corrosion by treatment with virulent bacteriophage, patent number: US8168419 B2

Bar H, Yacoby I, Benhar I (2008) Killing cancer cells by targeted drug-carrying phage nanomedicines. BMC Biotechnol 8:37

Barrangou R, van der Oost J (2013) CRISPR-Cas systems : RNA-mediated adaptive immunity in bacteria and archaea. Springer, Heidelberg, p 299

Barrangou R (2015) The roles of CRISPR–Cas systems in adaptive immunity and beyond. Curr Opin Immunol 32:36–41

Bazan J, Całkosiński I, Gamian A (2012) Phage display–a powerful technique for immunotherapy: 1. Introduction and potential of therapeutic applications. Hum Vacc Immunother 8(12):1817–1828

Bello-López JM, Cabrero-Martínez OA, Ibáñez-Cervantes G, Hernández-Cortez C, Pelcastre-Rodríguez LI, Gonzalez-Avila LU, Castro-Escarpulli G (2019) Horizontal gene transfer and its association with antibiotic resistance in the genus *Aeromonas* spp. Microorganisms 7:363

Bolocan AS, Ciolacu L, Oniciuc EA, Draper L, Nicolau AI, Wagner M, Hill C (2018) Chapter 6: Utilization of bacteriophages targeting *Listeria monocytogenes* in the dairy and food industry. In: Verma DK, Patel AR, Srivastav PP (eds) Bioprocessing technology in food and health, potential applications and emerging scope. Apple Academic Press, Oakville

Boulé J, Sholberg PL, Lehman SM, O'gorman DT, Svircev AM (2011) Isolation and characterization of eight bacteriophages infecting Erwinia amylovora and their potential as biological control agents in British Columbia, Canada. Can J Plant Pathol 33:308–317

Bruttin A, Brüssow H (2005) Human volunteers receiving Escherichia coli phage T4 orally: a safety test of phage therapy. Antimicrob Agents Chemother 49:2874–2878

Buttimer C, McAuliffe O, Ross RP, Hill C, O'Mahony J, Coffey A (2017) Bacteriophages and bacterial plant diseases. Front Microbiol 8:34

Ceyssens PJ, Van den Bossche A, Blasdel B, De Smet J, Lavigne R, Minakhin L, Yakunina M, Klimuk E, Noben J-P, Blasi U et al (2014) Development of giant bacteriophage ΦKZ is independent of the host transcription apparatus. J Virol 88(18):10501–10510
Chang HC, Chen CR, Lin JW, Shen GH, Chang KM, Tseng YH, Weng SF (2005) Isolation and characterization of novel giant Stenotrophomonas maltophilia phage φSMA5. Appl Environ Microb 71(3):1387–1393
Charpentier E, Richter H, van der Oost J, White MF (2015) Biogenesis pathways of RNA guides in archaeal and bacterial CRISPR-Cas adaptive immunity. FEMS Microbiol Rev 9(3):428–441
Chen J, Quiles-Puchalt N, Chiang YN, Bacigalupe R, Fillol-Salom A, Su Juan Chee M, Fitzgerald JR, Penadés JR (2018) Genome hypermobility by lateral transduction. Science 372:207–212
Chibani CM, Farr A, Klama S, Dietrich S, Liesegang H (2019) Classifying the unclassified: a phage classification method. Viruses 11(2):195
Clokie M, Millar A, Letarov A, Heaphy S (2011) Phages in nature. Bacteriophage 1:31–45
d'Hérelle F (1917) Sur un microbe invisible antagoniste des bacilles dysentériques. C R Hebd Seances Acad Sci Ser D 165:373–375
Dalmasso M, de Haas E, Neve H, Strain R, Cousin FJ, Stockdale SR, Hill C (2015) Isolation of a novel phage with activity against *Streptococcus mutans* biofilms. PLoS One 10(9):e0138651
Day M (2004) Bacterial sensitivity to bacteriophage in the aquatic environment. Sci Prog 87(3):179–191
De Vos D, Pirnay J (2015) Chapter 18: Phage therapy: could viruses help resolve the worldwide antibiotic crisis? In: Carlet J, Upham G (eds) AMR control 2015: overcoming global antibiotic resistance, 1st edn. Global Health Dynamics Limited, in association with the World Alliance Against Antibiotic Resistance (WAAAR), Woodbridge, pp 110–114
Domingo-Calap P, Delgado-Martínez J (2018) Bacteriophages: protagonists of a post-antibiotic era. Antibiotics 7:66
Drulis-Kawa Z, Tomasz O, Katarzyna D, Grazyna M-S, Ackermann H-W (2014) A giant Pseudomonas phage from Poland. Arch Virol 159(3):567–572
Duan Y, Llorente C, Lang S, Brandl K, Chu H, Jiang L, White RC, Clarke TH, Nguyen K, Torralba M et al (2019) Bacteriophage targeting of gut bacterium attenuates alcoholic liver disease. Nature (London, U K) 575(7783):505–511
Erez Z, Steinberger-Levy I, Shamir M, Doron S, Stokar-Avihail A, Peleg Y, Melamed S, Leavitt A, Savidor A, Albeck S, Amitai G, Sorek R (2017) Communication between viruses guides lysis-lysogeny decisions. Nature 541(7638):488–493
Fish R, Kutter E, Wheat G, Blasdel B, Kutateladze M, Kuhl S (2018) Compassionate use of bacteriophage therapy for foot ulcer treatment as an effective step for moving toward clinical trials. In: Methods in Molecular Biology, Bacteriophage Therapy, vol 1693. Humana Press, New York, pp 159–170
Garg P (2019) Filamentous bacteriophage: a prospective platform for targeting drugs in phage-mediated cancer therapy. J Can Res Ther 15:S1–S10
Gino E, Starosvetsky J, Kurtzbaum E, Armon R (2010) Combined chemical-biological treatment for prevention/rehabilitation of clogged wells by an iron-oxidizing bacterium. Environ Sci Technol 44(8):3123–3129
Goldman G, Starosvetsky J, Armon R (2009) Inhibition of biofilm formation on UF membrane by use of specific bacteriophages. J Membr Sci 342(1–2):145–152
Gómez-Gómez C, Blanco-Picazo P, Brown-Jaque M, Quirós P, Rodríguez-Rubio L, Cerdà-Cuellar M, Muniesa M (2019) Infectious phage particles packaging antibiotic resistance genes found in meat products and chicken feces. Sci Rep 9:13281
Goodridge LD, Bisha B (2011) Phage-based biocontrol strategies to reduce foodborne pathogens in foods. Bacteriophage 1:130–137
Gutierrez D, Rodríguez-Rubio L, Fernandez L, Martínez B, Rodríguez A, García P (2017) Applicability of commercial phage-based products against Listeria monocytogenes for improvement of food safety in Spanish dry-cured ham and food contact surfaces. Food Control 73:1474–1482

Hagens S, Loessner MJ (2007) Application of bacteriophages for detection and control of food-borne pathogens. Appl Microbiol Biotechnol 76(3):513–519

Hawkins C, Harper D, Burch D, Anggård E, Soothill J (2010) Topical treatment of *P. aeruginosa* otitis of dogs with a bacteriophage mixture: a before/after clinical trial. Vet Microbiol 146:309–313

Hendrix RW (2009) Jumbo bacteriophages. Curr Top Microbiol Immunol 328:229–240

Hussain MA, Liu H, Wang Q, Zhong F, Guo Q, Balamurugan S (2017) Use of encapsulated bacteriophages to enhance farm to fork food safety. Crit Rev Food Sci 57(13):2801–2810

Ishino Y, Shinagawa H, Makino K, Amemura M, Nakata A (1987) Nucleotide sequence of the iap gene, responsible for alkaline phosphatase isozyme conversion in Escherichia coli, and identification of the gene product. J Bacteriol 169(12):5429–5433

Jones JB, Jackson LE, Balogh B, Obradovic A, Iriarte FB, Momol MT (2007) Bacteriophages for plant disease control. Annu Rev Phytopathol 45:245–262

Keen EC, Bliskovsky VV, Malagon F, Baker JD, Prince JS, Klaus JS, Adhya SL (2017). Novel "Superspreader" bacteriophages promote horizontal gene transfer by transformation. mBio 8(1):e02115–16

Kifelew LG, Mitchell JG, Speck P (2019) Minireview: efficacy of lytic bacteriophages on multispecies biofilms. Biofouling 35(4):472–481

Kokjohn TA, Sayler GS, Miller RV (1991) Attachment and replication of Pseudomonas aeruginosa bacteriophages under conditions simulating aquatic environments. J Gen Microbiol 137:661–666

Krylov V, Shaburova O, Krylov S, Pleteneva E (2013) A genetic approach to the development of new therapeutic phages to fight *Pseudomonas aeruginosa* in wound infections. Viruses 5(1):15–53

Labrie SJ, Samson JE, Moineau S (2010) Bacteriophage resistance mechanisms. Nat Rev Microbiol 8:317–327

Lederberg J, Cavalli LL, Lederberg EM (1952) Sex compatibility in Escherichia coli. Genetics 37(6):720–730

Lieb M (1953) The establishment of lysogenicity in Escherichia coli. J Bacteriol 65(6):642–651

Lu TK, Collins JJ (2007) Dispersing biofilms with engineered enzymatic bacteriophage. Proc Natl Acad Sci U S A 104(27):11197–11202

MacLean RC, San Millan A (2019) The evolution of antibiotic resistance. Science 365(6458):1082–1083

Matsuzaki S, Uchiyama J, Takemura-Uchiyama I, Daibata M (2014) Perspective: the age of the phage. Nature 509(7498):S9

Mojica FJ, Rodriguez-Valera F (2016) The discovery of CRISPR in archaea and bacteria. FEBS J 283(17):3162–3169

Mojica FJ, Díez-Villaseñor C, Soria E, Juez G (2000) Biological significance of a family of regularly spaced repeats in the genomes of Archaea, Bacteria and mitochondria. Mol Microbiol 36(1):244–246

Morozova VV, Kozlova YN, Ganichev DA, Tikunova NV (2018) Bacteriophage treatment of infected diabetic foot ulcers. Method Mol Biol 1693:151–158

Morris FC, Dexter C, Kostoulias X, Uddin MI, Peleg AY (2019) The mechanisms of disease caused by Acinetobacter baumannii. Front Microbiol 10:1601

O'Flaherty S, Ross RP, Meaney W, Fitzgerald GF, Elbreki MF, Coffey A (2005) Potential of the polyvalent anti-Staphylococcus bacteriophage K for control of antibiotic-resistant Staphylococci from hospitals. Appl Environ Microbiol 71:1836–1842

Ofir G, Sorek R (2018) Contemporary phage biology: from classic models to new insights. Cell 172:1260–1270

Oh D, Qi J, Han B, Zhang G, Carney TJ, Ohmura J, Zhang Y, Shao-Horn IY, Belcher AM (2014) M13 virus-directed synthesis of nanostructured metal oxides for lithium–oxygen batteries. Nano Lett 14:4837–4845

Parikka K, Le Romancer M, Wauters N, Jacquet S (2016) Deciphering the virus-to-prokaryote ratio VPR: insights into virus-host relationships in a variety of ecosystems. Biol Rev 92:1081–1100

Principi N, Silvestri E, Esposito S (2019) Advantages and limitations of bacteriophages for the treatment of bacterial infections. Front Pharmacol 10:513

Racine V (2014) Lysogenic Bacteria as an experimental model at the Pasteur Institute (1915–1965). Embryo Project Encyclopedia (2014-10-10). ISSN:1940-5030 http://embryo.asu.edu/handle/10776/8223

Rhoads DD, Wolcott RD, Kuskowski MA, Wolcott BM, Ward LS, Sulakvelidze A (2009) Bacteriophage therapy of venous leg ulcers in humans: results of a phase I safety trial. J Wound Care 18(237–238):240–243

Ripp S, Miller RV (1997) The role of pseudolysogeny in bacteriophage-host interactions in a natural freshwater environment. Microbiology 143:2065–2070

Romig WR, Brodetsky AM (1961) Isolation and preliminary characterization of bacteriophages for Bacillus subtilis. J Bacteriol 82:135–141

Sarker SA, Brüssow H (2016) From bench to bed and back again: phage therapy of childhood Escherichia coli diarrhea. Ann N Y Acad Sci 1372:42–52

Sarker AS, Berger B, Deng Y, Kieser S, Foata F, Moine D, Descombes P, Sultana S, Huq S, Bardhan PK, Vuillet V, Praplan F, Brüssow H (2017) Oral application of Escherichia coli bacteriophage: safety tests in healthy and diarrheal children from Bangladesh. Environ Microbiol 19(1):237–250

Schmidt C (2019) Phage therapy's latest makeover. Nat Biotechnol 37:581–586

Smith GP (1985) Filamentous fusion phage: novel expression vectors that display cloned antigens on the virion surface. Science 228(4705):1315–1317

Smith HW, Huggins MB (1983) Effectiveness of phages in treating experimental *Escherichia coli* diarrhoea in calves, piglets and lambs. J Gen Microbiol 129:2659–2675

Smith GP, Petrenko VA (1997) Phage display. Chem Rev 97(2):391–410

Smith HW, Huggins MB, Shaw KM (1987a) Factors influencing the survival and multiplication of bacteriophages in calves and in their environment. J Gen Microbiol 133:1127–1135

Smith HW, Huggins MB, Shaw KM (1987b) The control of experimental *Escherichia coli* diarrhoea in calves by means of bacteriophages. J Gen Microbiol 133:1111–1126

Sohrab SS, Karim S, Kamal MA, Abuzenadah AM, Chaudhary AG, Al-Qahtani MH, Mirza Z (2014) Bacteriophage – a common divergent therapeutic approach for Alzheimer's disease and type II diabetes mellitus. CNS Neurol Disord Drug Targets 13(3):491–500

Soothill JS (1992) Treatment of experimental infections of mice with bacteriophages. J Med Microbio 37:258–261

Soothill JS (1994) Bacteriophage prevents destruction of skin grafts by *Pseudomonas aeruginosa*. Burns 20:209–211

Srinivasiah S, Bhavsar J, Thapar K, Liles M, Schoenfeld T, Wommack KE (2008) Phages across the biosphere: contrasts of viruses in soil and aquatic environments. Res Microbiol 159:349–357

Sulakvelidze A (2013) Using lytic bacteriophages to eliminate or significantly reduce contamination of food by foodborne bacterial pathogens. J Sci Food Agric 93:3137–3146

Summer NS, Summer EJ (2012) Process for remediating biofouling in water systems with virulent bacteriophage, patent number: US20120258523 A1

Tan LT-H, Chan K-G, Lee L-H (2014) Application of bacteriophage in biocontrol of major foodborne bacterial pathogens. J Mol Biol Mol Imaging 1(1):9

Tetz G, Tetz V (2018) Bacteriophages as new human viral pathogens. Microorganisms 6(2):54

Twort FW (1915) An investigation on the nature of ultramicroscopic viruses. Lancet 2:1241–1243

Vale PF, Little TJ (2010) CRISPR-mediated phage resistance and the ghost of coevolution past. Proc Biol Sci 277:2097–2103

van Soolingen D, de Haas PE, Hermans PW, Groenen PM, van Embden JD (1993) Comparison of various repetitive DNA elements as genetic markers for strain differentiation and epidemiology of Mycobacterium tuberculosis. J Clin Microbiol 31(8):1987–1995

Vandenheuvel D, Lavigne R, Brüssow H (2015) Bacteriophage therapy: advances in formulation strategies and human clinical trials. Annu Rev Virol 2:599–618

Viertel T, Ritter K, Horz H (2014) Viruses versus bacteria—novel approaches to phage therapy as a tool against multidrug-resistant pathogens. J Antimicrob Chemother 69:2326–2336

Weinbauer M (2004) Ecology of prokaryotic viruses. FEMS Microbiol Rev 28:127–181

Wommack KE, Colwell RR (2000) Virioplankton: viruses in aquatic ecosystems. Microbiol Mol Biol R 64(1):69–114

Yacoby I, Bar H, Benhar I (2007) Targeted drug-carrying bacteriophages as antibacterial Nanomedicines. Antimicrob Agents Chemother 51(6):2156–2163

Zarasvand KA, Rai VR (2014) Microorganisms: induction and inhibition of corrosion in metals. Int Biodeterior Biodegrad 87:66–74

Zinder ND, Lederberg J (1952) Genetic exchange in Salmonella. J Bacteriol 64(5):679–699

Chapter 16
Phage Therapy: An Alternative to Antibiotics

Elena Criscuolo and Sara Spadini

Abstract The emerging of multidrug-resistant bacteria represents nowadays a compelling concern and antibiotics are no longer a satisfactory answer to treat this kind of infections. This menace has made reconsider bacteriophages and their enzymatic products as valid alternatives to traditional drugs. In the present chapter, the efficacy of phage proteins is examined following in vitro and in vivo studies. Novel treatments should take into account the effectiveness of purified and engineered molecules by leveraging their biological features: depolymerases, virion-associated peptidoglycan hydrolases (VAPGHs), holins and endolysins, involved into phage entry and egress, are described given their prospective therapeutic roles.

16.1 Introduction

The appearance of resistant bacteria is occurring worldwide, impairing the efficacy of antibiotics. Many decades after the first patients were treated with these drugs, bacterial infections have become a threat again. The antibiotic resistance crisis has been ascribed to overuse and misuse of these drugs, as well as to the lack of new molecules developed by the pharmaceutical industry. The Centers for Disease Control and Prevention (CDC) classified several bacteria as presenting urgent, serious, and concerning threats. Many of them are already responsible for placing a substantial clinical and financial burden on the U.S. health care system, on patients and their families (Martens and Demain 2017). Coordinated attempts to implement new policies, renew research efforts and pursue steps to manage the crisis are greatly needed. This worldwide crisis has led to a revitalization of bacteriophage (phage) studies in the Western world (Moelling et al. 2018). Phage therapy mainly relies on

E. Criscuolo (✉)
Laboratory of Microbiology and Virology, Vita-Salute San Raffaele University, Milan, Italy
e-mail: criscuolo.elena@hsr.it

S. Spadini
Center for Behavioral Neuroscience and Communication (BNC), Vita-Salute San Raffaele University, Milan, Italy
e-mail: spadini.sara@hsr.it

G. Witzany (ed.), *Biocommunication of Phages*,
https://doi.org/10.1007/978-3-030-45885-0_16

lytic phages to specifically kill their bacterial hosts, leaving human cells unharmed and reducing the impact on commensal bacteria often resulting from broad-spectrum antibiotics use. Phage therapy is rapidly evolving and has resulted in life-saving cases and multiple clinical trials (Dedrick et al. 2019; Schmidt 2019). However, one of its biggest challenges relates to regulations and policy surrounding clinical use and implementation beyond compassionate cases (Cooper et al. 2016). The specificity of phage therapy is its main advantage, but also a limit when translating to clinics (Kim et al. 2019). The few clinical cases described to date prove that phage cocktails are required for the treatment of infections and their success is based on fine patient-specific modulation, making them unfit for commercial development and approval (Schmidt 2019). Moreover, phages are "living" viruses as they replicate within bacteria after administration, arising worries in terms of standardization of production and regulatory policies. Additionally, replicating phages contribute to antibiotic resistance spread due to horizontal gene transfer: phages can accidentally package resistant genes from an infected cell and transduce them into the next host (MacLean and San Millan 2019). In this scenario, recombinant phage-derived enzymes represent a good option. The circulating enzyme concentration in the body declines due to inactivation and clearance. Furthermore, they are more similar to already FDA approved antibiotics and thus more suitable than whole phages for current drug approval processes. To make them more effective or appropriate for treatment, their pharmacokinetics can be better studied and improved than phages, and their spectrum of action may be broadened through protein engineering (Criscuolo et al. 2017).

In this chapter the main types of phage enzymes will be described: depolymerases and virion-associated peptidoglycan hydrolases (VAPGHs), whose action is fundamental for virus to entry into bacteria, and holins and endolysins, involved into virions egress (Fig. 16.1). Their main features will be analyzed in light of their goals achieved as treatments and of their possible clinical use.

16.2 Phage-Derived Proteins

16.2.1 Depolymerases

One of the most important virulence factors of bacteria, especially ESKAPE such as *K. pneumoniae* or *E. coli* pathogens, is the polysaccharide that surrounds the cell. It can be organized as a distinct structure, termed capsule (capsular polysaccharide, CPS), or can be released as exopolysaccharide (EPS). The CPS is composed by repeating subunits from three to six sugars that give rise to different capsule structures. These differences divide them into different serotypes, but genetic variation of polysaccharide synthesis pathways must be considered. To date, 81 capsular types have been identified only for *Klebsiella* spp., characterized by both molecular genotyping and phage typing (Pan et al. 2015). Co-evolution of phages and their

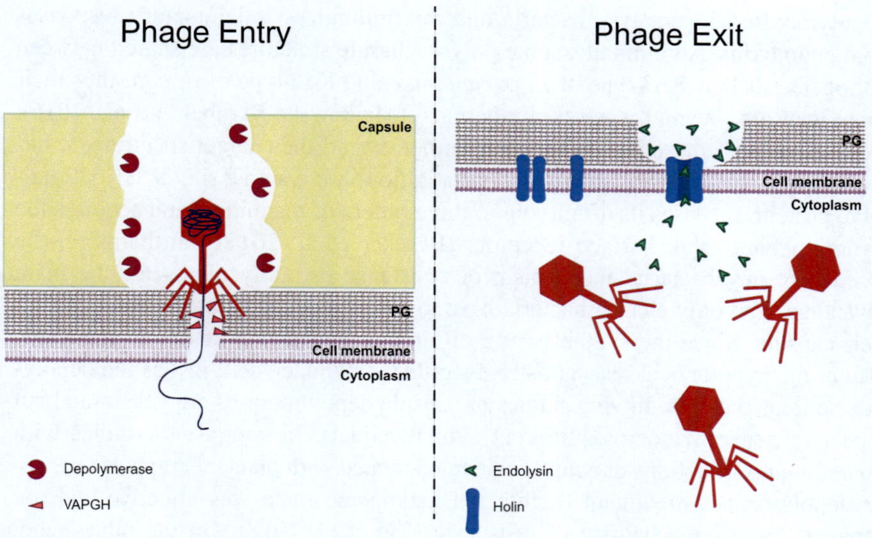

Fig. 16.1 The function of lytic proteins in the phage life cycle. Depolymerases are required to reduce the viscosity of capsular polysaccharide (capsule) and strip it from the bacterial surface. VAPGHs favor the injection of phage genetic material into the cytoplasm by the formation of a hole in the peptidoglycan (PG). Endolysins and holins fulfill their function at the end of the life cycle. Holins form a pore in the bacterial membrane, allowing endolysins to reach and cleave the PG

host bacteria led to the development of specific enzymes that depolymerase this specific capsular polysaccharide structures to make the outer membrane accessible to phage attachment and DNA injection. The capsule is fundamental for pathogenicity because it confers resistance to host immune response mechanisms such as phagocytosis and inflammatory response suppression, allows adherence and biofilm formation. Moreover, it provides a physical barrier to a potentially hostile environment, typically characterized by the presence of antibiotics and phages. Capsular diversity results in high specificity of bacteriophages for their bacterial hosts, meaning that a single phage cannot cover all the pathogens displaying the capsular types mentioned above (Pan et al. 2017).

Some efforts have been made by modifying phage receptors or mixing different viruses into cocktails with positive results against a wider range of clinical infections, but the high specificity is the main burden of phage therapy. Capsular depolymerases represent an interesting antibiotic option: they do not kill per se but reduce CPS viscosity and strip it from the surface of encapsulated bacteria, exposing them to immune components or drugs. Enzyme administration reduced bacteremia in vivo by endorsing neutrophils antibacterial activity and it has been demonstrated that depolymerase treatments can increase bacterial susceptibility to serum neutralization (Wang et al. 2019; Lin et al. 2018). The enzymatic removal of the bacterial capsule leads to the exposure of pathogen-associated molecular patterns on the cell surface, with the consequent enhancing of phagocytosis via activation of the alternative complement pathway (Majkowska-Skrobek et al. 2018; Scorpio et al. 2007).

Depolymerases do not lyse bacteria and this minimizes inflammatory responses from endotoxins. Additionally, some polysaccharide structures are shared between pathogens, such as K63 type of *K. pneumoniae* and K42 type of *E. coli*, thus their range can go beyond species limitations (Majkowska-Skrobek et al. 2016). Modifications of these enzymes could further extend their target spectrum: it has been proved that a single amino acid substitution in the active site of HK620 tail-spike protein increases its affinity up to three orders of magnitude and acquires the ability to bind other surface receptors (Broeker et al. 2013). Furthermore, the absence of any bacterial mechanism of resistance certainly represents the main advantage. The only chance bacteria have to escape depolymerase action is losing their capsule. Since the synthesis of a different capsular type is not a practicable option, the hypothetical resistant strain would be a "nude" cell, just as the depolymerase treated cell is. In vivo studies of capsule depolymerases are yet limited but appear to generalize across different animal models. In contrast to studies with *Pseudomonas* infections of neutropenic mice treated with phage therapy, K1 capsular depolymerases treatment of infected leukopenic mice was effective and not impaired by any mechanism of resistance (Lin et al. 2018). On the other hand, enzyme therapy presents some limitations. Results obtained with therapeutic protocols in immunocompetent mice indicated low efficacy of treatment. Some depolymerases, such as K30, completely lost their antibacterial activity, while others, like K1 enzymes, showed reduced activity compared to prophylactic protocols (Lin et al. 2017). Perhaps significantly, the delayed treatment using cognate K1-specific phages achieved higher rescue than enzymes. No negative consequences of delayed treatment were previously observed in leukopenic mice and data obtained with immune-competent mice imply that other factors must be considered to improve enzyme efficacy when therapeutically administered. On the other hand, reduced efficacy with a delayed treatment is not surprising, especially with rapidly lethal infections. Moreover, in vivo studies have proven the importance of the administration route, as each enzyme has its biochemical properties that may affect pharmacokinetics. For example, differences have been observed between intramuscular and intraperitoneal administrations. Recombinant K1E depolymerase tends to form 18-mers instead of trimers differently from other K1 enzymes. This translates into a higher steric hindrance that affects the in vivo distribution (Lin et al. 2017).

Multiple applications of depolymerases have been proposed, especially against *Klebsiella* spp. These enzymes could be used to define the capsular type for clinical strains, to obtain oligosaccharides from polysaccharides, to prevent and eradicate biofilm formation on living and artificial surfaces, or as alternative therapeutic agents to treat infections (Criscuolo et al. 2017; Gutiérrez et al. 2017). The latter application is of particular significance as this opportunistic pathogen is a growing concern for public health. Moreover, depolymerases do not lyse bacteria and do not release endotoxins as phages do, thus there might be circumstances in which these enzymes could be superior to therapy using whole phages.

16.2.2 Virion-Associated Peptidoglycan Hydrolases (VAPGHs)

Bacteriophages are provided with specific enzymes able to mediate the cleavage of peptidoglycan (PG) of bacteria: endolysins, described below, and virion-associated peptidoglycan hydrolases (VAPGHs). VAPGHs are structural components of the virion particle and participate in the initial steps of infection disrupting the cell wall by local hydrolysis of the PG layer when the phage is adsorbed to the cell surface. Their role is fundamental to allow the phage genetic material entering the target cell, but their function during the lytic cycle is still unclear. PRD1 and T7 mutant phages without VAPGHs activity successfully infected host cells, although more slowly (Moak and Molineux 2000; Rydman and Bamford 2000). On the contrary, antibody neutralization assays resulted in a complete inhibition of bacterial lysis and their structural role could be assumed by the instability of mutant phages lacking these proteins (Kenny et al. 2004). VAPGHs are crucial for successful infection and PG degradation cannot be too efficient because host cell viability is essential for virus replication and propagation. Still, these enzymes are also responsible for the "lysis from without" mechanism, a bacterial lysis induced by high-multiplicity virus adsorption occurring without phage particles production (Rodríguez-Rubio et al. 2013a). VAPGHs have a modular structure composed of one or two N-terminal catalytic domains (EAD, Enzymatically Active Domain) and one C-terminal cell wall-binding domain (CBD, Cell wall-Binding Domain). PG hydrolases show three major activities: glycosidase, amidase, endopeptidase. The CBD is not always present probably because these enzymes are "delivered" directly to the PG by the virion architecture. This modular structure confers extreme specificity to VAPGHs, but the molecular basis of the interaction between these enzymes and their substrate has not yet been fully characterized. Therefore, specificities of VAPGHs might result from the combined interactions of catalytic and binding domains with species-specific host receptors within the PG structure (Gutiérrez et al. 2018). On the other hand, a modular organization is ideal for enzyme engineering. Domain swapping enhances its lytic activity and broadens its host spectrum. VAPGHs act against both Gram-positive and Gram-negative bacteria. The outer membrane needs to be overcome to reach the PG substrate, but the antimicrobial lytic activity of these enzymes has been confirmed after destabilization of the outer membrane. Location of VAPGHs varies in phages infecting these hosts: for example, T4 gp5 protein is anchored to the base plate structure, while T7 gp16 protein is an internal head protein. The effectiveness of VAPGHs has been proven at multiple levels. It has been reported that a VAPGH from TM4 mycobacteriophage with a hydrolase motif allows phage infection in *Mycobacterium smegmatis* during its stationary phase (Piuri and Hatfull 2006). Similarly, the transglycosylase activity of bacteriophage T7 gp16 was shown to be effective against *E. coli* infections (Moak and Molineux 2000). Tuc2009 and TP901-1 virions contain both full-length and truncated VAPGHs to infect bacteria with different levels of cell wall cross-linkage (Stockdale et al. 2013). The most promising results, however, have been achieved against methicillin-resistant *S. aureus* (MRSA) infections, responsible for more than 20% of all infections in

WHO regions. VAPGHs specific for this pathogen were modified to kill intracellular bacteria using a protein transduction domain (PTD) composed of a short cationic peptide sequence facilitating its crossing through the eukaryotic membrane (Becker et al. 2016). Finally, VAPGHs modular structure and their specificity are important to decrease the onset of resistant strains. Data obtained so far with wild type VAPGHs indicate that resistance acquisition is quite rare or even nonexistent, as their target in the PG molecule are essential for bacterial viability and their mutation would be too harmful to the host cell. Chimeric enzymes, obtained adding different catalytic domains, can theoretically reduce the onset probability of resistant bacteria with a double modification in the target structures, as observed against *S. aureus* (Becker et al. 2016). Moreover, VAPGHs from Gram-positive infecting phages have shown high specificity to the same species/genus of their targets (Rodríguez-Rubio et al. 2013b). This characteristic could be a drawback when considering mixed-pathogen infections, but it provides a clear advantage regarding resistance development. Broad range antimicrobials lead to selection for resistant strains in the target pathogen as well as among co-resident commensal bacteria exposed to the same drug. Resistance transfer is not species or genus limited, thus commensal bacteria can be carriers of resistance elements propagating them to neighboring strains, possibly pathogens, and this could be overcome exploiting VAPGHs specificity.

To date, no phage lytic proteins have been approved for human therapeutic use in Europe nor the United States. A set of phage lytic proteins active against *S. aureus* is being studied for their application in critical medical situations, such as bacteremia and endocarditis. Preclinical studies and clinical trials are underway, and also synergistic interactions between antimicrobial compounds have been evaluated (Criscuolo et al. 2017; Rodríguez-Rubio et al. 2012). This can lead to improved efficacy against target bacteria by reducing the required dose of each antimicrobial compound while simultaneously decreasing the probability of resistance development. VAPGHs have not been considered only for their efficacy in eradicating infections, but also as a tool for removing bacterial biofilms (Gutiérrez et al. 2017). According to the National Institutes of Health, biofilms are responsible for more than 60% of microbial infections and 80% of chronic infections in humans, without considering the economic losses in the food industry and food-borne disease outbreaks (Bjarnsholt 2013). In detail, *S. aureus* is one of the most important causative agents of nosocomial infections related to implanted medical devices, as its ability to form biofilm allows the colonization of abiotic surfaces as well as human tissues, hindering both infections treatment and surfaces disinfection (Otto 2013). Biofilm characterization studies identified different susceptibilities depending on strain and biofilm maturation stage and its complete removal is not easy to achieve using phage enzymes. However, promising results have been obtained in the inhibition of bacterial surface attachment and destabilization/disruption of mature biofilms (Miquel et al. 2016). As mentioned, the main advantage of lytic phage enzymes over traditional antibiotics is their ability to lyse bacteria even when they are not actively growing.

16.2.3 Endolysins

Endolysins are enzymes produced by lytic phages at the end of their lifecycle and are essential for their release from the host cell. Endolysins translocate through holes formed by holins, described below, in the plasma membrane of the host bacterium after cytoplasmic accumulation and dismantle the cell wall by targeting and degrading the PG. Seconds after contact, endolysins can make lysis of the target cell occur, causing cell bursting and release of progeny phages: working together with holins, they generate holes in the bacterial wall inducing a consequent cell osmotic lysis and death. Their effectiveness as hydrolases is related to the lack of bacterial outer membrane, as in Gram-positive cells. Endolysins display a variable modular domain structure consisting of two conservative and functional domains, as described for VAPGHs: a N-terminal catalytic domain (EAD) flexibly linked to a C-terminal bacterial wall-binding domain (CBD). The CBD specifically binds ligands on the bacterial wall, tethering the lysin to the proteoglycan in a strong antigen-antibody binding affinity manner. This is particularly representative of endolysins derived from phages infecting Gram-positive bacteria; on the contrary, lysins related to Gram-negative infections mainly have a globular structure (Gutiérrez et al. 2018). Based on their structural organization, hundreds of different endolysins have been classified and their isolation potentially overcomes issues related to whole phage utilization, as the occurrence of resistance events due to transduction of antibiotic resistance genes. Their modular structure allows feasible molecular refinements. Engineering of endolysins aims at the improvement of their bacteriolytic potency, by modifying their specificity, increasing their catalytic efficiency or rising their stability to avoid phage resistance, thus making them the most widely studied phage enzymes. As an example, chimeolysins derived from Gram-positive-infecting phages are safe chimeric enzymes obtained by switching combinations of EAD and CBD domains of pre-existing endolysins (Yang et al. 2014, 2015). The fusion of certain lysins to unrelated peptides or protein domains can also be realized to generate effective hybrid molecules (Lukacik et al. 2012); genetically engineering artilysins are examples of hybrid enzymes targeting Gram-negative strains, carrying a peptide able to cross the outer membrane of bacteria (Briers et al. 2014a). This class is potentially effective toward multidrug-resistant strains and has already been applied in human keratinocytes and *C. elegans* models (Briers et al. 2014b). Enzymatic and antibacterial properties of endolysins were first outlined in 1959 against group A of streptococcal infections (Freimer et al. 1959). Since then, several endolysins have been described and characterized over in vivo studies. LysK, derived from bacteriophage K, is one of the best-characterized staphylococcal endolysins and shows activity against multiple *S. aureus* species commonly colonizing the skin and mucosal membranes, oftentimes responsible for nosocomial infections (O'Flaherty et al. 2005). LysK has been successfully tested in a mouse model of systemic MRSA infection (Schmelcher et al. 2015) and SAL-1, a LysK homolog, showed in vitro successful lytic activity and reduced animal mortality after intravenous administration (Jun et al. 2013).

To date, no phage-derived protein is approved for clinical use even though phage therapy safety has been established over several clinical trials (Criscuolo et al. 2017). After bacterial lysis, pro-inflammatory substances could be systemically released, and this represents the main concern of lysins administration in humans or animals. No adverse effects were nonetheless reported during toxicological in vivo preclinical studies and encouraging results have been obtained in humans. SAL200, based on a recombinant form of SAL-1, is the first endolysin administered to humans: in a Phase I study, intravenous administration showed no serious adverse events, pointing it out as a promising candidate drug for staphylococcal infections (Jun et al. 2017). Another example is CF-301, a lysin that has completed Phase II clinical trial demonstrating activity for the treatment of *S. aureus* bacteremia (ClinicalTrials.gov Identifier: NCT03163446). Rapidity of action and specificity are the main advantages of lysins exploited as therapeutics compared to antibiotics. Their use is feasible on mucosae without altering the resident flora and endolysins coevolution to their species hosts implies high conservation of cell wall targets, as too many changes in peptidoglycan would be deleterious to cell survival. Also, peptidoglycan is absent in eukaryotic cells. Thus, no descriptions of resistance development and no cytotoxicity have been reported following in vitro studies (Nelson et al. 2001). Bacteria are also known to form multi-drug-tolerant biofilms in infection: entrapped in a matrix, they are more resistant to host defense mechanisms and can escape bloodstream or water flow washing. Lysins demonstrated to both disrupt biofilm structure and to kill cells in the biofilm matrix, showing powerful antibacterial activity (Sharma et al. 2018).

All these aspects promote the exploration of endolysins as alternative agents toward bacterial infections in different perspectives.

16.2.4 Holins

As previously mentioned, transport of endolysins across the cytoplasmic membrane at the end of the bacterial life cycle is mediated by holins, hydrophobic proteins responsible for the formation of holes. Holins are generally identified as small hydrophobic proteins provided with a hydrophilic C-terminus and at least one transmembrane domain, and are organized in different families (Roach and Donovan 2015). Their action is carried out by two different pathways (Young 2013). In the holin-endolysin system, the extensive permeabilization of the cytoplasmic membrane, caused by holin formation of large and nonspecific channels, permits endolysins exportation to the cell wall and peptidoglycan degradation. This is followed by the collapse of the cytoplasmic membrane proton motive force and a resulting cell rupture. On the contrary, in the pinholin-SAR (Signal Anchor-Release) system, phage holins trigger and depolarize the cytoplasmic membrane after their accumulation as inactive proteins in the periplasm until a critical concentration. In such a case, holins (or pinholins) are responsible for small pores: they generate protein heptamers with a central channel, too small for protein passage, which allows only ions to diffuse. This causes a local membrane depolarization that activates a specific class of muralytic enzymes called SAR endolysins. The consequence is peptidoglycan degradation, a uniform disruption along the entire cell membrane (Drulis-Kawa et al. 2015; Young 2014).

The biology of holins can be exploited for vaccine generation through the creation of empty Gram-positive bodies, lacking cytoplasm and nucleic acid; this is of particular interest for inducible immunoprotection (Drulis-Kawa et al. 2015). Some holin-like proteins also demonstrated an in vitro bacteriostatic activity, inhibiting bacterial growth and generating holin-promoted lysis, emphasizing their potential potency against several strains of multidrug-resistant cells (Rajesh et al. 2011).

16.3 Conclusions

The progress made in molecular biology addresses the emerging concern of antibiotics resistance through the development of new molecules based on bacteriophages. In recent years, synthetic biology and molecular engineering enabled the design of phage-derived proteins to be used as therapeutic agents instead of whole phage particles. These proteins show a wider antibacterial spectrum, better tissue penetration, lower immunogenicity and almost no chance of bacterial resistance. Results obtained from both preclinical and clinical studies depict these novel compounds as safe and effective. The deepening knowledge of phage genetics and biology will hopefully lead to the successful application of phage-derived proteins in therapy, as it is already happening in diagnostics, food industry and agriculture. The ultimate and challenging objective is the avoidance of antibiotic resistance itself, hypothetically using a combination of biological and chemical treatments.

References

Becker SC, Roach DR, Chauhan VS et al (2016) Triple-acting lytic enzyme treatment of drug-resistant and intracellular Staphylococcus aureus. Sci Rep 6:25063–25010. https://doi.org/10.1038/srep25063

Bjarnsholt T (2013) The role of bacterial biofilms in chronic infections. APMIS Suppl 121:1–51. https://doi.org/10.1111/apm.12099

Briers Y, Walmagh M, Grymonprez B et al (2014a) Art-175 is a highly efficient antibacterial against multidrug-resistant strains and persisters of Pseudomonas aeruginosa. Antimicrob Agents Chemother 58:3774–3784. https://doi.org/10.1128/AAC.02668-14

Briers Y, Walmagh M, Van Puyenbroeck V et al (2014b) Engineered endolysin-based "Artilysins" to combat multidrug-resistant gram-negative pathogens. MBio 5:e01379–e01314. https://doi.org/10.1128/mBio.01379-14

Broeker NK, Gohlke U, Müller JJ et al (2013) Single amino acid exchange in bacteriophage HK620 tailspike protein results in thousand-fold increase of its oligosaccharide affinity. Glycobiology 23:59–68. https://doi.org/10.1093/glycob/cws126

Cooper CJ, Khan Mirzaei M, Nilsson AS (2016) Adapting drug approval pathways for bacteriophage-based therapeutics. Front Microbiol 7:1209. https://doi.org/10.3389/fmicb.2016.01209
Criscuolo E, Spadini S, Lamanna J et al (2017) Bacteriophages and their immunological applications against infectious threats. J Immunol Res 2017:1–13. https://doi.org/10.1155/2017/3780697
Dedrick RM, Guerrero-Bustamante CA, Garlena RA et al (2019) Engineered bacteriophages for treatment of a patient with a disseminated drug-resistant Mycobacterium abscessus. Nat Med 25:1–12. https://doi.org/10.1038/s41591-019-0437-z
Drulis-Kawa Z, Majkowska-Skrobek G, Maciejewska B (2015) Bacteriophages and phage-derived proteins–application approaches. Curr Med Chem 22:1757–1773. https://doi.org/10.2174/0929867322666150209152851
Freimer EH, Krause RM, McCarty M (1959) Studies of L forms and protoplasts of group A streptococci. I. Isolation, growth, and bacteriologic characteristics. J Exp Med 110:853–874
Gutiérrez D, Fernández L, Martínez B et al (2017) Real-time assessment of Staphylococcus aureus biofilm disruption by phage-derived proteins. Front Microbiol 8:6213–6210. https://doi.org/10.3389/fmicb.2017.01632
Gutiérrez D, Fernández L, Rodríguez A, García P (2018) Are phage lytic proteins the secret weapon to kill Staphylococcus aureus? MBio 9:63. https://doi.org/10.1128/mBio.01923-17
Jun SY, Jung GM, Yoon SJ et al (2013) Antibacterial properties of a pre-formulated recombinant phage endolysin, SAL-1. Int J Antimicrob Agents 41:156–161. https://doi.org/10.1016/j.ijantimicag.2012.10.011
Jun SY, Jang IJ, Yoon S et al (2017) Pharmacokinetics and tolerance of the phage endolysin-based candidate drug SAL200 after a single intravenous administration among healthy volunteers. Antimicrob Agents Chemother 61:1763. https://doi.org/10.1128/AAC.02629-16
Kenny JG, McGrath S, Fitzgerald GF, van Sinderen D (2004) Bacteriophage Tuc2009 encodes a tail-associated cell wall-degrading activity. J Bacteriol 186:3480–3491. https://doi.org/10.1128/JB.186.11.3480-3491.2004
Kim B-O, Kim E, Yoo Y-J et al (2019) Phage-derived antibacterials: harnessing the simplicity, plasticity, and diversity of phages. Viruses 11:268–220. https://doi.org/10.3390/v11030268
Lin H, Paff ML, Molineux IJ, Bull JJ (2017) Therapeutic application of phage capsule Depolymerases against K1, K5, and K30 capsulated E. coli in mice. Front Microbiol 8:2257. https://doi.org/10.3389/fmicb.2017.02257
Lin H, Paff M, Molineux I, Bull J (2018) Antibiotic therapy using phage depolymerases: robustness across a range of conditions. Viruses 10:622–617. https://doi.org/10.3390/v10110622
Lukacik P, Barnard TJ, Buchanan SK (2012) Using a bacteriocin structure to engineer a phage lysin that targets Yersinia pestis. Biochem Soc Trans 40:1503–1506. https://doi.org/10.1042/BST20120209
MacLean RC, San Millan A (2019) The evolution of antibiotic resistance. Science 365:1082–1083. https://doi.org/10.1126/science.aax3879
Majkowska-Skrobek G, Łątka A, Berisio R et al (2016) Capsule-targeting depolymerase, derived from Klebsiella KP36 phage, as a tool for the development of anti-virulent strategy. Viruses 8:324. https://doi.org/10.3390/v8120324
Majkowska-Skrobek G, Latka A, Berisio R et al (2018) Phage-borne Depolymerases decrease Klebsiella pneumoniae resistance to innate defense mechanisms. Front Microbiol 9:2517. https://doi.org/10.3389/fmicb.2018.02517
Martens E, Demain AL (2017) The antibiotic resistance crisis, with a focus on the United States. J Antibiot 70:520–526. https://doi.org/10.1038/ja.2017.30
Miquel S, Lagrafeuille R, Souweine B, Forestier C (2016) Anti-biofilm activity as a health issue. Front Microbiol 7:592. https://doi.org/10.3389/fmicb.2016.00592
Moak M, Molineux IJ (2000) Role of the Gp16 lytic transglycosylase motif in bacteriophage T7 virions at the initiation of infection. Mol Microbiol 37:345–355. https://doi.org/10.1046/j.1365-2958.2000.01995.x
Moelling K, Broecker F, Willy C (2018) A wake-up call: we need phage therapy now. Viruses 10:688–614. https://doi.org/10.3390/v10120688

Nelson D, Loomis L, Fischetti VA (2001) Prevention and elimination of upper respiratory colonization of mice by group A streptococci by using a bacteriophage lytic enzyme. Proc Natl Acad Sci U S A 98:4107–4112. https://doi.org/10.1073/pnas.061038398

O'Flaherty S, Coffey A, Meaney W et al (2005) The recombinant phage lysin LysK has a broad spectrum of lytic activity against clinically relevant staphylococci, including methicillin-resistant Staphylococcus aureus. J Bacteriol 187:7161–7164. https://doi.org/10.1128/JB.187.20.7161-7164.2005

Otto M (2013) Staphylococcal infections: mechanisms of biofilm maturation and detachment as critical determinants of pathogenicity. Annu Rev Med 64:175–188. https://doi.org/10.1146/annurev-med-042711-140023

Pan Y-J, Lin T-L, Chen C-T et al (2015) Genetic analysis of capsular polysaccharide synthesis gene clusters in 79 capsular types of Klebsiella spp. Sci Rep 5:15573. https://doi.org/10.1038/srep15573

Pan Y-J, Lin T-L, Chen C-C et al (2017) Klebsiella phage ΦK64-1 encodes multiple Depolymerases for multiple host capsular types. J Virol 91:684. https://doi.org/10.1128/JVI.02457-16

Piuri M, Hatfull GF (2006) A peptidoglycan hydrolase motif within the mycobacteriophage TM4 tape measure protein promotes efficient infection of stationary phase cells. Mol Microbiol 62:1569–1585. https://doi.org/10.1111/j.1365-2958.2006.05473.x

Rajesh T, Anthony T, Saranya S et al (2011) Functional characterization of a new holin-like antibacterial protein coding gene tmp1 from goat skin surface metagenome. Appl Microbiol Biotechnol 89:1061–1073. https://doi.org/10.1007/s00253-010-2907-6

Roach DR, Donovan DM (2015) Antimicrobial bacteriophage-derived proteins and therapeutic applications. Bacteriophage 5:e1062590. https://doi.org/10.1080/21597081.2015.1062590

Rodríguez-Rubio L, Martínez B, Rodríguez A et al (2012) Enhanced staphylolytic activity of the Staphylococcus aureus bacteriophage vB_SauS-phiIPLA88 HydH5 virion-associated peptidoglycan hydrolase: fusions, deletions, and synergy with LysH5. Appl Environ Microbiol 78:2241–2248. https://doi.org/10.1128/AEM.07621-11

Rodríguez-Rubio L, Quiles-Puchalt N, Martínez B et al (2013a) The peptidoglycan hydrolase of Staphylococcus aureus bacteriophage 11 plays a structural role in the viral particle. Appl Environ Microbiol 79:6187–6190. https://doi.org/10.1128/AEM.01388-13

Rodríguez-Rubio L, Martínez B, Donovan DM et al (2013b) Bacteriophage virion-associated peptidoglycan hydrolases: potential new enzybiotics. Crit Rev Microbiol 39:427–434. https://doi.org/10.3109/1040841X.2012.723675

Rydman PS, Bamford DH (2000) Bacteriophage PRD1 DNA entry uses a viral membrane-associated transglycosylase activity. Mol Microbiol 37:356–363. https://doi.org/10.1046/j.1365-2958.2000.01996.x

Schmelcher M, Shen Y, Nelson DC et al (2015) Evolutionarily distinct bacteriophage endolysins featuring conserved peptidoglycan cleavage sites protect mice from MRSA infection. J Antimicrob Chemother 70:1453–1465. https://doi.org/10.1093/jac/dku552

Schmidt C (2019) Phage therapy's latest makeover. Nat Biotechnol 37:581–586. https://doi.org/10.1038/s41587-019-0133-z

Scorpio A, Chabot DJ, Day WA et al (2007) Poly-gamma-glutamate capsule-degrading enzyme treatment enhances phagocytosis and killing of encapsulated Bacillus anthracis. Antimicrob Agents Chemother 51:215–222. https://doi.org/10.1128/AAC.00706-06

Sharma U, Vipra A, Channabasappa S (2018) Phage-derived lysins as potential agents for eradicating biofilms and persisters. Drug Discov Today 23:848–856. https://doi.org/10.1016/j.drudis.2018.01.026

Stockdale SR, Mahony J, Courtin P et al (2013) The lactococcal phages Tuc2009 and TP901-1 incorporate two alternate forms of their tail fiber into their virions for infection specialization. J Biol Chem 288:5581–5590. https://doi.org/10.1074/jbc.M112.444901

Wang C, Li P, Niu W et al (2019) Protective and therapeutic application of the depolymerase derived from a novel KN1 genotype of Klebsiella pneumoniae bacteriophage in mice. Res Microbiol 170:156–164. https://doi.org/10.1016/j.resmic.2019.01.003

Yang H, Zhang Y, Yu J et al (2014) Novel chimeric lysin with high-level antimicrobial activity against methicillin-resistant Staphylococcus aureus in vitro and in vivo. Antimicrob Agents Chemother 58:536–542. https://doi.org/10.1128/AAC.01793-13

Yang H, Linden SB, Wang J et al (2015) A chimeolysin with extended-spectrum streptococcal host range found by an induced lysis-based rapid screening method. Sci Rep 5:17257. https://doi.org/10.1038/srep17257

Young R (2013) Phage lysis: do we have the hole story yet? Curr Opin Microbiol 16:790–797. https://doi.org/10.1016/j.mib.2013.08.008

Young R (2014) Phage lysis: three steps, three choices, one outcome. J Microbiol 52:243–258. https://doi.org/10.1007/s12275-014-4087-z

Chapter 17
Bacteriophage as a Therapeutic Agent to Combat Bacterial Infection: A Journey from History to Application

Umesh Panwar, Murali Aarthy, and Sanjeev Kumar Singh

Abstract Bacteriophages are considered to be the small viruses possessing an important role to infect bacteria's tending to replicate therefore maintaining the environment of bacteria existence in equilibrium. Bacteriophages were considered to be potent therapeutics for the treatment against the bacterial infections in humans. Initially, bacteriophages were implemented for the treatment of intestinal diseases and later, its implications were carried further for during the surgical practice in the twentieth century. But the role of antibiotics has pushed down the importance of bacteriophage but the hasty increase of multi-drug resistant bacteria has portrayed the importance of phage therapy which is a potential substitute to antibiotics. Recently, various results have been recorded for the successful rates of the bacteriophages in the treatment of various infections in clinical cases and clinical trials. Due to growing importance in combating bacterial infection to diminish the antibiotic resistance, the bacteriophage are proposed as a powerful therapeutic agent in wide range of applied areas such as medicine, biotechnology, agriculture, biosensor, veterinary and even more. Thus, we introduce this chapter to brief the concept of phage, phage history, and phage biology with highlights of phage therapy, phage engineering, as well as phage applications including the future of phage.

17.1 Introduction

The self-replicating essential particles containing nucleic acids in terms of DNA or RNA are known as Viruses, surviving with torpid form in the extracellular situation. In spite of random natural changes (Linear/Circular or Single/Double) in DNA or RNA of viruses, it possess an importance in infecting various microbes namely Archaea, Bacteria and other eukaryotes. This role of infecting other organisms portrays them to be the dynamic force in environment. Billions of viruses in various

U. Panwar · M. Aarthy · S. K. Singh (✉)
Computer Aided Drug Design and Molecular Modelling Lab, Department of Bioinformatics, Alagappa University, Karaikudi, Tamil Nadu, India

G. Witzany (ed.), *Biocommunication of Phages*,
https://doi.org/10.1007/978-3-030-45885-0_17

types are presented on the sphere but the recent scenarios estimates about the number of viruses are more than components in the universe. In most cases, viruses become the key regulator among the host inhabitants and their relations within the environment (Lodish et al. 2000; Aiewsakun et al. 2018; Prasad and Schmid 2012; Breitbart and Rohwer 2005). Amongst the extensive population, viruses infect most of the microbes, and these viruses which are termed to be Phage (widely known as Bacteriophage) are considered to be the natural predators of the bacteria. Researchers found that the viruses infecting microbes plays a significant role and possess intimate relationship with various microbes leading to different changes in the host population and ecosystem (Debarbieux et al. 2017; Ofir and Sorek 2018; Sharma et al. 2017). The first discovery of phage was identified by Felix d'Herelle and Twort about 100 years ago to treat bacterial infections. Initial period of phage has drawn a significant graph in treating human infections such as surgical practice for purulent wounds, skin infections and acute intestinal infections in 1930–1940s periods, and also US pharmaceutical shown interested to developed commercial phage. Nevertheless, the lack of understanding on the phage biology and strong scientific proof of phage therapy generated the varying results to be considered as effective agents over the antibiotics in 1940s (Sharma et al. 2017; Campbell 2003; Salmond and Fineran 2015). Later, the implementation of antibiotics had remarkable utilization and terminated the phage therapy in various western countries. However, the continuation of phage therapy was taken forward by former Soviet Unions and some European countries. The Russian Federation has provided approval to sell phage in clinical experiments. The implication of antibiotics had successful achievements for countless years, but, the rapid growth of drug resistance bacteria has revealed a sudden fall in the use of antibiotics towards the treatment of bacterial infections (Cisek et al. 2017; Sulakvelidze et al. 2001; Wienhold et al. 2019; Monk et al. 2010; Ofir and Sorek 2018). The increased multi-drug resistance bacteria have presented an opportunity for the use of phage after long time ignorance in history as an alternative therapeutics against antibiotics. During the mid-twentieth century, the wide exposure of phage therapy has been gained in numerous western countries after realizing the importance and impact of the relationship of phage in biology. Phages are considered to be most ample and epitomise biological reservoir on globe with the flexibility to survive in the environment possessing extreme heat (Sahara) to cold (Polar Inland water) environment. It has also been founded in larger number of population in the atmosphere like animals or plants, water, and soil etc., which has considered as a chief controller antibacterial therapeutics source towards the bacterial fruition (Monk et al. 2010; Breitbart et al. 2004; Prigent et al. 2005; Säwström et al. 2008). This phage infects specific species of bacteria and further provokes them to replicate using host cell machinery for their multiple growth for the infection of other bacterial cells. Thus bacteriophage are categorised on the basis of their morphology, to be mentioned in specific about the environment and the lifecycles they are termed into two different natural forms namely lytic and lysogenic lifecycle (Tagliaferri et al. 2019; Drulis-Kawa et al. 2012; Du Toit 2017).

In recent times, the modern era of phage has been facilitated as effective, safe, specific and promising therapeutics agents. It has been revealed that the result of phage cocktails and engineered phage to be effective in phage therapy which leads to exterminate the bacterial infection easily through the administration by means of oral, local, interaperitoneal, intramuscular, intravenous or intranasal and topical. Some of the phage therapy are in developmental stage which can be utilized for multi purposes in the field of medical and agriculture. Also, lately, phage based vaccination is now catching up the market as protective strategy in various fields like Biosensor, drug carrier, biotechnology, gene transfer including gene delivery, biocontrol as well as control of biofilm and bio-preservation. The historical contribution of phages has shown its vital role in the bacterial pathogenicity, bacterial evolution and worldwide ecology to create an impact on general biology and in the future of medicine (Abedon et al. 2017a, b; Matsuzaki and Uchiyama 2019; Brown et al. 2017; Chen et al. 2019; Harada et al. 2018; Principi et al. 2019).

At present, the phage development with the better efficiency, self-replication, reduced virulence and less cost has portrayed the interest to pharmaceutical companies which helps them to invest in making the world better with the phages. Thus, we aimed to sketch the features of phage, its history, biology, replication mechanism, therapy, genetically engineered bacteriophage and applications in a single stage, which helps the researchers and the scientific community to have better insights of the action and mechanism of phage.

17.2 Major Historical Events of Phage Era

More than a century ago in 1896, A famous bacteriologist Ernest Hanbury Hankin reported the unexpected antibacterial activity against Cholera in Indian rivers Ganga and Yamuna, and also suggested that this water passed through a thin porcelain filter having unique features (Abedon et al. 2011a, b). Similarly, one more bacteriologist (Gamaleya in 1898) found a bacteriolytic substance specific for *B. Subtilis* (Bardell 1982). Later, during the initial phase of twentieth century, in 1915, a British doctor and bacteriologist Frederick Twort found the same phenomenon during his work with microbial agents and reported the observation of filterable agent with antibacterial. Similarly, in 1917, a French-Canadian microbiologist Felix d'Herelle confirmed the observation of Twort's discovery and proposed the phenomenon as Bacteriophage. Later, by 2 years in 1919, d'Herelle applied the use of bacteriophage to treat dysentery. Many researchers referred the phenomenon of Twort-d'Herelle as Bacteriophage phenomenon'. But, in the same year, Bordet was not satisfactory with the results of d'Herelle's on phage therapy and challenged his conception of phage as a virus. In 1923, the Eliava Institute, Georgia opened an avenue towards briefing about bacteriophage and use of it as a therapy, and soon after in 1925, Arrowsmith defined the imaginary efforts to develop phage therapy (Twort 1915; d'Hérelle 2007; Wittebole et al. 2014; Summers 2012; Parfitt 2005; Löwy 2010). During 1940's, Group of Delbruck and Luria carried out experiments in order to study the phage mechanism and genetic material which proves that the information for viral replication has been carried through DNA. In the year, 1969 Luria, Delbruck and Hershey were awarded with the Noble Prize in Physiology and Medicine for their research on phage structure and replication. All through the phase of 1942, Luria and Thomas move towards the assignment of the visualization of Phage by Electron Microscope (EM) at Columbia University and got succeeded (Salmond and Fineran 2015; The Nobel Prize 1969; Ackermann 2011; Luria and Anderson 1942). Later, in 1951, Esther Lederberg from University of Wisconsin discovered the Lambda Phage and in 1955, Benzer identified the fine structure of phage T4 rII (Casjens and Hendrix 2015; Jayaraman 2008). During the period of 1970–1980 two phages based genome has been sequenced by Fiers and Sanger namely the MS2 ssRNA and ΦX174ssDNA (Pierrel 2012; Sanger and Coulson 1975). Smith developed the Phage display in 1985 and later during 1996, Phage lysogenic conversion had been demonstrated in Cholera (Bazan et al. 2012; Waldor and Mekalanos 1996). In 2003, the first synthetic genome of phage ΦX174 was constructed followed by the construction of CRISPR – Cas adaptive phage immunity in 2007, and Cas9 RNA directed to the nuclease for genome editing in 2012 (Chen 2019; Doetschman and Georgieva 2017). The clinical trials for the phage therapy I/II named the PhagoBurn is into existence and initiated in 2013 and completed in 2019. Rober T. Schooley developed the personalized therapeutic agents based on Bacteriophage to treat patients with disseminated resistant *Acinetobacter baumannii* infection in 2017. Recently in the year 2019, Underwood and his team has genome sequence of LuckyBarnes, a newly isolated singleton siphovirus that infects Brevibacterium iodinum (Jault et al. 2019; Schooley et al. 2017; Underwood et al. 2019). Based on this research, it was clearly stated that the phages can be utilized as antimicrobial and antibacterial agents to treat infections through the application of phage therapy.

17.3 Family Classification, Structure and Lifecycle of Bacteriophage

In general, the classification of virus is completely depending on the host organism, type of infection, type of replication and nucleic acid. Thus, classification of phages also possessed based on the morphologies, type of nucleic acid and replications. The genome of phage is constructed either with the single or double stranded DNA/RNA with the linear or circular form and also possess gene in various number. Also, numerous proteins are embedded in the phage which contains capsid and they are represented in the filamentous form, icosahedral or head – tail in shape which is unique. The genetic materials enclosed in phage capsid directly connects with the tunnel tails, composed of protein and, tail is having a six fibres holding receptor with ability to bind with specific targeted bacteria (White and Orlova 2019; Eiserling 1979). The structure of the phage has been represented in Fig. 17.1. In few cases, it has been identified that the specific phages possess different mechanism in binding with the host bacteria instead of its dependence on the binding to tail mechanism.

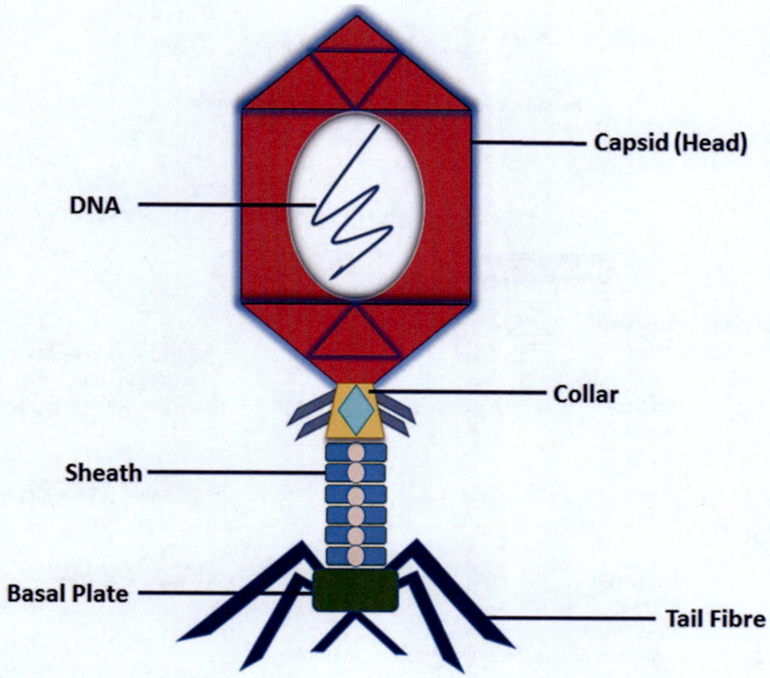

Fig. 17.1 A general structure of bacteriophage

The general classification of bacteriophage has considered in 13 families, as described below (Ackermann 2003; Ackermann 2009; Tolstoy et al. 2018; van Duin 1988; Matsuzaki et al. 2005):

- Double Stranded DNA

  (a) Corticovirdae, an icosahedral with circular dsDNA, capsid with lipid layer.
  (b) Fusellovirdae, an enveloped dsDNA with lipid and without capsid.
  (c) Lipothrixvirdae, an enveloped filamentous with linear dsDNA and lipid.
  (d) Microvirdae, a non-enveloped dsDNA with isometric head and contractile tail.
  (e) Plasmavirdae, an enveloped pleomorphic circular dsDNA with lipid and without capsid.
  (f) Siphovirdae, a non-enveloped dsDNA with isometric head and long tail with noncontractile.
  (g) Tectivirdae, a non-enveloped dsDNA with icosahedral capsid.
  (h) Rudivirdae, an enveloped with linear dsDNA, and helical rods.
  (i) Podovirdae, an enveloped dsDNA, with isometric head and short tail, noncontractile.

- Single Stranded DNA

  (j) Inovirdae, a circular ssDNA with long filamentous and short rods.

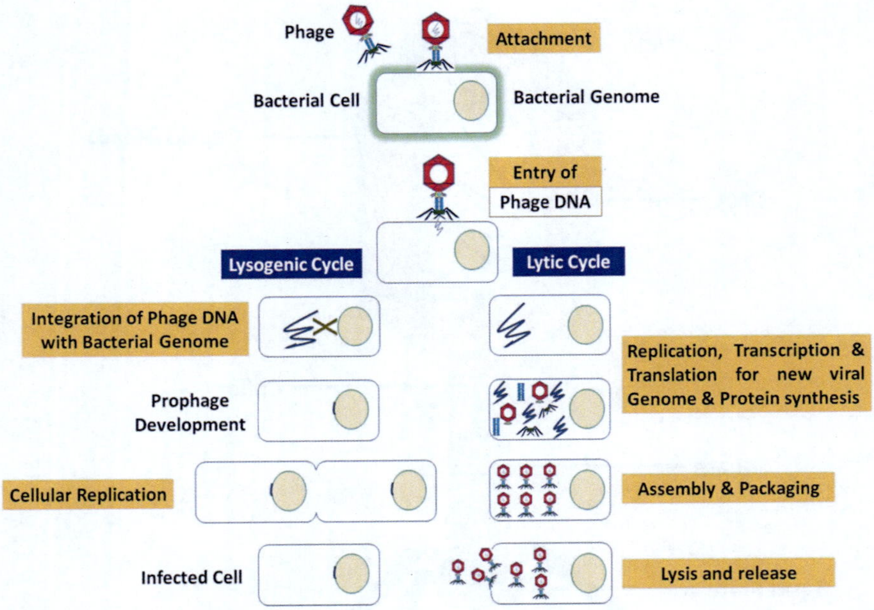

Fig. 17.2 The mechanism of lytic and lysogenic cycle of phage

(k) Microvirdae, a circular ssDNA with icosahedral capsid.

- Double Stranded RNA

(l) Cystovirdae, an enveloped linear dsRNA with icosahedral capsid and lipid.

- Single Stranded RNA

(m) Levivirdae, a ssRNA with qausi icosahedral capsid.

Similar to the viruses, the phages also attacks the surface of the bacterial cell and injects the genetic material into the host cell through the tail and utilize the process of production in progeny which forms the tiny phages that inject the genetic material into the host cell via the tail and make a process of progeny production in the form of tiny phages to further infects the bacteria. This process of infection is generally known as the life cycle of the phage which has been further carried out in two different paths namely the Lytic cycle and the Lysogenic cycle for the reproduction of themselves into the host cell. The Lytic cycle is the representation of the virulent phase whereas the Lysogenic cycle is the temperate form which induces the phenotype (Howard-Varona et al. 2017a, b; Clokie et al. 2011; Young 1992). The schematic representation of Phage life cycle has been represented in Fig. 17.2.

17.3.1 Lifecycle of Bacteriophage

In the lytic cycle, the phage genes are virulent and utilize the host bacterium cell and resources to multiply in number for provoking the bacterial cell for quick lysis and die during the process. This cycle is categorized by the hasty production of fresh viral particles in the host system upon host's lytic infection which is generally found in productive environments, characterized by the abundance host activity. On the other hand, external factors like fluctuations in environment persuade in inducing the events that helps in encountering the viral –host interaction with elevated rates which favours the lytic infections. In general, during the lytic phase of the cell, the proteins present in the tail, develops specific interaction with the surface of the bacterium cell and the genetic material gets injected into the cytoplasm of the host cell. The injected material starts the process of replication, transcription and translation inside the host cell whereas the phage proteins develops into new genome, capsid and tail proteins. Once the phage particles gets developed, the assembling of the new progeny in the later stages of the productive cycle helps utilization of the protein lysin to make lysis of the cell wall to release hundreds of phage progeny that can infects the other cells which continue with similar process. The main focus of the lytic cycle is that it will leads to the instant bursting of the host bacteria which releases the progeny of the bacteriophage (Williamson and Paul 2004; Young 1992; Kara 2018; Karen 2018; Kim and Ryu 2013).

In contrast to the lytic cycle, the cycle of lysogenic is projected to facilitate the survival of the host and prophage under unfavourable conditions, such as low prokaryote abundance and primary production. Constant fluctuations in the host physiology and activity could constitute unfavourable conditions for the phage communities and promote the establishment of lysogeny. In comparison to the Lytic cycle, Lysogenic or temperate phages directly contact with bacterium cell by injecting their viral genetic material with host cell genome to allowing for replication to reproducing the new progeny without killing host cell. The lysogenic process begins with attachment of phage to bacterial cell surface and to inject the genetic material into the cytoplasm. Instead of replicating immediately, the phage DNA integrates itself into the host chromosome by using phage encoded integrase. The integrated phage DNA is well known as prophage which later replicates along with the host DNA as cell divides. At last, the prophage get activated and follows the lytic cycle to reproduce new progeny and release the new phages to infect a new bacteria (Haq et al. 2012; Erez et al. 2017; Clokie et al. 2011; Howard-Varona et al. 2017a, b; Kara 2018; Karen 2018; Chibani-Chennoufi et al. 2004).

17.4 Phage Therapy and Phage Application

The use of antibiotics in the treatment of bacterial infections during the past era, states the importance of antibiotics and numerous uses were recorded. Later, it has directed towards the leading issue of multi-drug resistance in pathogenic bacteria due to chaotic usage. There emerges the need for the alternative solution for the issue of global health to replace antibiotics with high social & economic impacts. Since, viruses are natural predators of bacteria which is very common in nature states that they destroy them as the Bio-control managers. Thus the phages are discovered as a powerful component to safeguard bacterial infection and in recent scenario, it has been understood that the phage therapy has been an effective therapeutic agents to control the bacterial infection on this environment (Domingo-Calap and Delgado-Martínez 2018; Carlton 1999; Doss et al. 2017; Kortright et al. 2019). A phage (bacteriophages) is a unique virus that solely kill specific targeted bacteria and are majorly found in most of the ecosystem. Recently, these are the major biological entities in natural environment which has shown the major therapeutic value to kill the harmful bacteria's without harming the human and other animals. To be more specific, the lytic phages has been imposed in various stages to destroy the cell. An effective treatment on phage therapy might be new after the entry of antibiotics, but it was recognized almost a century ago, which is again brought into use as an alternative to antibiotics. Since the discovery of antibiotics in 1940, the phage therapy was discontinued in numerous western countries but the former soviet nations were on the platform of phage therapy research. Although the discovery of antibiotics holds back the use of phage therapy in medical field, but for almost 100 years, phage therapy was recommended by NIH as one of the advanced strategy for the chief global issue of antimicrobial resistance research and several life-threatening infections. The antibiotic resistance towards the bacterial infections were eradicated with the bacteriophage which possesses the ability to self-replicate and amplifies within the bacteria. This strategy makes the phage to be strong and natural weapon to battle against the infectious diseases even though the colony is meagre. Due to these aspects, the phages will be no longer in survival and removed by the host immune system after the elimination of pathogen (Clokie et al. 2011; Abedon et al. 2017a, b; Doss et al. 2017; Altamirano and Baar 2019; Lin et al. 2017; Kortright et al. 2019; Dixon 2004; Hesse and Adhya 2019). Earlier, the capacity of phage to treat the bacterial infection has not been recognized perfectly whereas when the comparison of antibiotics is executed, it was clearly evident that the phages have unique properties. The high rates of substitution in phage in comparison with bacteria declare that it is not easy for bacteria to resist to the phage therapy. The phage therapy is highly used in the preclinical field and greatly reachable by the blood brain barrier due to greater permissibility. Nowadays, phage are extensively used in cocktails with special interest which is highly active and this makes the host bacteria less resistant to phage therapy in comparison to antibiotics (Keen and Adhya 2015; Morozova et al. 2018; Skurnik and Strauch 2006; Fauconnier 2019; Chan et al. 2013). In Phage therapy, phages are treated as the bio agents that helps

Table 17.1 Properties differences between phage therapy and antibiotics

Property	Antibiotics	Phage therapy
Safety	Not available	Available
Specificity	Broad	Narrow
Immunogenicity	No	Yes
Resistance	High in range	Low in range
Side effects	Present	Not present
Cost & development	Much high and time consuming	Limited and not much
Other infection chance	Yes	No
Patent	Possible	Limited or not much

in eliminating drug-resistance bacteria. In spite of the basic difference in the eradication of bacterial pathogens through the phages and antibiotics, they possess cross-resistance (Zhang and Buckling 2012) Thus, a great combination of phage and antibiotic together fascinates better to treat bacterial infection, while phages also have similar kind of competence to lyse the bacteria with antibiotic resistance (Abedon 2019). Hence, the development of phage therapy is more stable in environment condition, easy to be tailored and cost effective for the promotion of phage evolution against specific disease causing bacteria. Thus, the features of phage and antibiotic have been depicted in Table 17.1. Also, the phages are referred as ecological candidates or tool in molecular biology as well as recycling of cells and organic matter. The ability of bacteriophage as natural therapy to regulate the bacterial populations through the induction of lysis to the bacterial species in the site of infections is clearly understood (Pelfrene et al. 2016; Semler et al. 2012; O'Sullivan et al. 2016; Henry and Debarbieux 2012). Use of broader host range phages would presumably lead to fewer treatment failures due to mismatched host and phage combination in phage therapy. Generally, phage has countering systems so that the host ranges from the dynamic property which can modifies over time. A novel route of linking phage to their specific hosts is the use of genomic sequences to forecast the host ranges. A phage therapy targeting the infectious bacteria can also modulate the immune system, having the resemblance of antibiotics effects, and can also display other regulatory effects on the human body. Also the recent period of advanced strategies known as the Bioengineered phages and purified phage lytic proteins have been suggested for potential phage therapy to treat bacterial infection. Yet, a limited number of phage therapy in clinical trials for eradicating infections in human have been authorized by European Medicine Agency-EMA and Food Drug Administration-FDA, USA (Hyman and Abedon 2010; Ross et al. 2016; Stone et al. 2019; Van Belleghem et al. 2018; Górski et al. 2012; Fauconnier 2019; Debarbieux et al. 2016; Voelker 2019). Thus, over the last decade, a remarkable progress on phage therapy has been taken in fresh look with the reconsideration of old approach in new form to treat bacterial pathogens. Additionally, the future of phage cocktails would be offering into a permissible and profitable territory. Overall, the phage therapy facets have significant and great hope for future as alternative magical medicine bullet (Chan et al. 2013; Keen and Adhya 2015; Kortright et al. 2019).

Bacteriophages or phages which is also termed as viruses are the most copious entities on our planet which is harmless for all organisms including humans except the bacterial hosts that has been targeted. The bacteriophages are highly diverse and infect most of the bacteria present on the earth. Around the world, the enormous development of the bacterial resistance to the antibiotics is becoming renewed for the driving force of bacteriophage therapy which makes use of the lytic phage particles. The emergence of the multidrug resistant bacterial strains to the antibiotics present in the market. The use of antibiotics in various fields has grown drastically and related to numerous people who are in need of health care with various diseases (Harada et al. 2018; Fair and Tor 2014). In twenty-first century, people around the world possess larger use of antibiotics which leads to the selection of bacterial specimens from the normal microbiota and contributing towards the environmental dissemination of resistance genes. Bacteriophages are natural components to treat the bacterial infections which hold the potential to serve as alternatives for the antibiotics (Pirnay et al. 2012). Application of bacteriophages has placed its mark in various fields like arming, agriculture, medical field like gastrointestinal aspects and dental aspects (Fair and Tor 2014). Most people in this world are affected by pulmonary infections which are common due to antibiotic resistant and it has been clearly evident that bacteriophages are capable of combating these infections when they are administered through nasal route instead of antibiotics. Kutateladze and Adamia states that during the year 2008, a 5 year old child was diagnosed with the respiratory infections which is caused due to the association of *P aeruginosa* and *S aureus* but the infection has not responded towards the antibiotic treatment. In the meanwhile, phage therapy has been administered through nasal route which shows significant improvement in the health of the child. Bacteriophages present in the human gut are estimated to be the highest number of biological entities on Earth. The Phage preparations for the intestinal infections by the Salmonella, Shigella, Staphylococcus, Proteus, E. coli and Pseudomonas are termed as the Intesti Phage which is cocktails of bacteriophages (Kutateladze and Adamia 2008; Zschach et al. 2015; Abedon et al. 2011a, b; Deresinski 2009). The main concern of the phage therapy for the gastrointestinal infection is the route of administration, since the gastric acidity can destroy the phages (Stanford et al. 2010). People generally experience breaches in the skin due to the cause of wounds and burns which makes them susceptible to the infection. In order to treat those breaches caused by infections usage of topical solutions is into application which is made up of the lytic bacteriophage particles. These particles could also be used in the cosmetic and pharmaceutical formulations to treat skin infections. The frequent etiological agent in the wound infections is the *Pseudomonas aeruginosa* and the appearance of multiple antibiotic resistant strains has developed noteworthy trouble in the treatment of infected wounds (Krylov et al. 2012). The phage application has been utilized in clinical practice for the treatment of localized infections for example, treating the early stage of acute wound infection, phage specific treatment on infection caused by *Staphylococcus* spp. and *Streptococcus* spp., treatment of infected chronic ulcers and diabetic ulcers, including infected burns (specifically from lymphopenia, sepsis, intoxication, as well as changes in the microbiota) (Morozova et al. 2018). On other hand. Neutralizing mAbs are most important therapeutics for detection of effective epitopes and also can be utilized for the development of vaccine formulations against pathogens. For an example, phage display targeted various proteins from infectious disease for epitope mapping. Additionally, the molecular biology and sequencing approaches shown new opportunities towards the redesign of phage and phage derived proteins, in which, the phage derived proteins has now a great antimicrobial agents and proved as biosensor in pathogen detection, therapy and diagnosis (Criscuolo et al. 2017). Recently, many of foodborne bacterial pathogens are now highly resistant due to unlimited use of antibiotics. Thus, phage based therapy and genetic engineering technology and CRISPR-Cas system have been taken as molecular tool to detect and control the these resistant bacteria and developed organic farming to increasing organic livestock without harming with antibiotics (Easwaran and Ahn 2019; Svircev et al. 2018). Further, the detailed knowledge on phage biology and genetics may lead to a great success of applying phage in many ways for bacterial detection, agriculture, biotechnology and immunology including preservation of food products. Considering the amazing studied done on phage, the upcoming future will be in high in demand of phage as therapeutic agent.

17.5 Phage Engineering

The hasty increase of resistance towards drug for the bacterial infections with the help of the antibiotic pipeline creates an imperative situation to expand alternative therapies. Treatment with antibiotics developed an emergence for the resurging approach that will gain significant interest is the phage therapy where the bacteriophage are used as the antimicrobials (Yehl et al. 2019). The successful use of Phage therapy in recent times depicts the failure of the concerned care cases during the use of antibiotic treatment The selectivity of phage is dependent on the binding to cell surface receptors which recognizes the infection of the host (Bertozzi Silva et al. 2016). Thus, an alternative options phage therapy is under interest to combating antibiotic-resistant pathogens, because, phages are highly anticipated with a specific safety and efficacy in the treatment. Although the importance of phages fall down solely, the development of various hurdles such as developing a phage therapy which are limited to their ability to infect various bacterial strains just because of narrow range targeting to bacterial species, some of phage therapy prohibited due to being lysogenic rather than lytic, and some of disqualified in resulting of toxins factor generation (Chen et al. 2019; Kilcher and Loessner 2019; McGrath et al. 2004; Sousa and Rocha 2019; Karen 2019; Loc-Carrillo and Abedon 2011). So considering this issue, the phage therapy is required for suitable phage in cocktail form to work on different strains without interference. This also enhances the ability of phages in control and detecting the bacterial infection whereas the synthetic biology approach are also applied to design, alter, construct and test the phage to treat infectious disease. Also, the engineered phages are highly standard in safety and efficacy to make a notable clinical impact. When the natural phages are compared, the engineered phages possess more advantages in point of therapeutic suitability due to its extremely patentable to get a profitable value in the world as compare to natural phage (Brown et al. 2017; Pires et al. 2016; Nair and Khairnar 2019; Kilcher and Loessner 2019). Therefore, the phage engineering will be promising to accelerate the phage research in near future. Few of major engineering techniques applied to develop novel synthetic phages to make a better environment with healthy life.

- Homologous Recombination
  Homologous recombination is well known technology in genetic engineering for making exchange in similar bacterial genetic material. It is a classical genetic methodology and natural phenomenon to develop various mutant phenotypes and recombine heterologous DNA into the cells with native genetic material. Homologous recombination mechanism have tendency to incorporate foreign genes into the phage genomes. This approach utilize makes new recombinants of phages with gene deletions, insertions or alternation in plasmid with phage DNA. Generally, the gene will be introduced into the phage DNA flanked by two region of homologous and includes in host to be further phage engineering. Then the recombinant phages will be resulting in range of 10^{-10}–10^{-4} containing with designed mutations. The process of homologous recombination takes time to make new recombinants with low rate of frequency (Pires et al. 2016; Chen et al. 2019; Brown et al. 2017; Pines et al. 2015).
- BRED (Bacteriophage Recombineering of Electroporated DNA)
  The recent era of advance DNA sequencing and phage genetics has shown an important role towards the complete genome determination includes novel construction of bacterial genetics, and mutation for examination of gene function. Thus, an effective technology with the name of BRED (Bacteriophage Recombineering of Electroporated DNA) has highly promoted in efficient construction of direct mutation of bacteriophage genome and vast utilization in gene addition, deletions, insertion, and replacements as well to build base substitutions. In 2012, Marinelli was utilized this methodology to make changes in mycobacteriophage. It is highly recommended technique which can further enhance the frequency of homologous recombination using recombineering substrates with phage encoded recombination system for an example Red system of Phage lambda and RecE/RecT system of Rac prophage. Similarly like homologous recombination, this process applies co-electroporation of the phage genome into the host cell DNA in the presence of plasmids. Recombination occurs exactly after the phage genome replication into the host in containing with designed mutations. Thus, BRED offers various opportunities to understanding the phage biology and genetics with to build efficient bacteriophage recombinants (Pires et al. 2016; Chen et al. 2019; Marinelli et al. 2008, 2019).
- CRISPR-Cas-Mediated Genome Engineering
  Clustered regularly interspaced short palindromic repeats(CRISPR)-Cas system is a genetic tools from class of prokaryotic immune system have widely utilized in various domain of research field to developing world class model organism. CRISPR –Cas genes directly involve in development of immune

system in bacteria and archaea with help of CRISPR-Cas technology, which allows to better examination of prokaryotic viruses. The CRISPR-Cassytems is made up of two major components, Cas proteins and CRISPR RNA, in which, Cas proteins works as catalytic core to cleaves the DNA and the CRISPR RNA functions as mediator to bind with specific target DNA. The CRISPR-Cas system are divided into two broad classes: Class I (with three types – I, III, IV) and Class II (with three types – II, V, VI). In CRISPR-Cas system the Class I contains multiple Cas proteins, and Class II contains single Cas protein. Further, CRISPR-Cas system comprises into three mode of action such as: CRISPR adaption, RNA biogenesis including the last one CRISPR-Cas interference. CRISPR-Cas system technology were applied in many ways to develop phage engineering, for example: a deletion of nonessential gene 1.7 in T7 phage mutant, to make a V. Choleraelytic phage using I – CRISPR-Cas system from Vibriocholerae, and Various recombinants selected after utilizing type II-A CRISPR-Cas in Streptococcus thermophiles. Soon, the future will be with advanced technology CRISPR-Cas to develop novel engineered phage (Pires et al. 2016; Chen et al. 2019; Hatoum-Aslan 2018; McDonald et al. 2019; Makarova et al. 2019; Hao et al. 2018).

- Reconstruction of phages using assembled Phage Genomes

  One more advanced technology in phage engineering that phage genome can be directly engineered before introducing into host. In this process, the assembled phage genome DNA with desired mutation in vivo or in vitro and directly introduced into the host cell genome for further replication, transcription and translation of genomic DNA, will develop the new progeny of phages. The genome assembly can be done based on the genome size, such as (i). In the case of small phage genome, the genomic DNA assembles in vitro using PCA (polymerase cycling assembly) by applying synthetic oligonucleotides with overlap in whole genome, example asphiX174 (ii). In case of large phage genome, the genomic DNA can be assembled via ligation of separate genome fragments with restriction enzymes, example T7 (iii). and, in some of cases, the whole genome DNA can be assembled with overlapping fragments in vivo using TAR (Transformation Associated Recombination) (Pires et al. 2016; Chen et al. 2019; Kilcher et al. 2018).

Additionally, some of engineered bacteriophage has been mentioned in Table 17.2.

It is very difficult to file for patent for the natural phages due to its wide implications in the environment. Hence, the recent technology of genetic engineering paves way for the development of novel synthesis of phages as antimicrobial agents with the isolation from any source. The developed phages generally possess higher efficacy and sustainability with the greater protection in the process of filing the Intellectual properties. Several pharmaceutical companies and business tycoons started showing interest in investing towards the phage engineering for developing novel products (Pires et al. 2016; Nair and Khairnar 2019; Kilcher and Loessner 2019). Thus, the upcoming future of phage will be more beneficial for the

Table 17.2 Some of engineered bacteriophage and their application

Name of Engineered Bacteriophage	Application	References
M13	Phage display, engineered protein purification, nanomaterial's, Vaccinology	Bhattarai et al. (2012), Muzard et al. (2012), and Oh et al. (2014)
M13KE	Pathogen detection	Derda et al. (2013) and Van der Merwe et al. (2014)
T7	Phage display, delivery vector, biofilm control	Xu et al. (2018), Deng et al. (2018), and Sillankorva et al. (2007)
Lambda (λ)	Phage display, vaccinology, biocontrol	Sternberg and Hoess (1995), Gamage et al. (2009), and Yosef et al. (2015)
T4	Phage display, vaccinology	Efimov et al. (1995) and Shivachandra et al. (2006)
ϕA1122	Pathogen detection	Schofield et al. (2009)
A511	Pathogen detection	Loessner et al. (1997)
HK620	Pathogen detection	Cormack et al. (1996)
PBSPCA1	Agriculture	Schofield et al. (2012)
fd	Phage display, nano device fabrication and bottom-up manufacturing	Tornetta et al. (2010) and Zirpel et al. (2015)

eradication of bacterial infections and develops the world with the new paradigms of gaining better profit than other methods of treating bacterial infections.

17.6 Phage Display

In 1985, George Smith developed the first molecular technique termed as phage display, which is for presenting polypeptides on the surface of lysogenic filamentous bacteriophages. It has direct contact with phage phenotype to leading vast technology of molecule screening on phage surface. Phage display technology has a massive performance in the various fields such as cell biology, molecular biology, immunology, and even more. The major applications of phage display is in the identification of receptor-ligand interaction, improving the binding site on proteins to increasing the binding affinity, and vaccine design for infectious diseases. Phage display is also allowing to make changes of phages, developing antibody, and peptides generation (Smith 2019; Pande et al. 2010; Koivunen et al. 1999; Rowley et al. 2004; Molek et al. 2011). The forte of phage display is have ability to interact with specific region of proteins, presentation of numerous peptide on the surface of phage with high attraction and specificity. Rapidly growing phage display technology has major advantage to utilize the phage as designed nanocarries for targeted delivery of chemotherapeutics agents to cancer and can eradicate the other bacterial infections. Phage display has a great advantage in designing of epitope mapping and monoclonal antibodies to utilize in clinical diagnosis. The epitope mapping and antibodies can be widely utilize to develop the vaccine to a particular target, improving the immunological process, and the understanding the relation between antigen-antibody interactions, as well as therapeutic tools for basic and clinical research (Sioud 2019; Rowley et al. 2004; Aghebati-Maleki et al. 2016; Bazan et al. 2012; Hess and Jewell 2019).

17.7 Factor Affecting Phage Therapy

Phage, a potent alternative approach has controlled many of pathogenic bacteria in the growing multi-drug resistant. An effective phage with various advantages over antibiotics has several challenging factors while developing a powerful phage therapy as an antimicrobial therapeutic agents, such as (Ly-Chatain 2014; Dąbrowska 2019):

(a) Specificity: phages should be a specific to target the bacteria.
(b) Isolation: One of major challenging task to isolate a phage against a bacterial host.
(c) Accessibility: The delivery of phages to reach in intracellular pathogenic bacteria is affecting phage therapy.
(d) Toxicity: Most of phages are non-toxic, but, it may have toxic effects in the presence of host. Thus, a careful alternation is necessitate while developing phage agents.
(e) Neutralization: Sometimes phages may become neutralized by immunogenic antibodies or other specific component presence in bacterial host. Thus, developing phage with anti-immunogenic or protect the phage with encapsulation is one of the task to shows difficulties in phage therapy.
(f) Resistance: Bacterial resistance to phage is another hurdle on the way to developing phage therapy. Thus, requirement of successful phage cocktails are in demands to solve this issue.
(g) Stability: One more challenging task that some of physiochemical factors like pH, temperature including moisture effects shows difficulties in phage activity and their survival.
(h) Phage concentration: The phage concentration should require to confirm the MOI (multiplicity of infection) in the ratio of phage number per bacteria.
(i) Dose: A single dose in early stage has shown the issue, thus applying a multi dose in starting itself may reduce the infection rate.
(j) Phage Acceptance: One of tedious issue to provide an acceptance for regular use in medical field.
(k) Awareness: Phage utilization is still in dilemma for people due to lack of awareness.

17.8 Mode of Administration

Using phage therapy is one simplest and easiest way to target the bacterial infections in a particular concentration. The phage therapy as in form of antimicrobial agents can be deliver with various routes of administration to host, which can be specified based on either target infection or organised delivery. In general, the phage can be delivered via oral route, Octic delivery, and topical including subcutaneous, intramuscular, intraperitoneal, intravenous, or intranasal administration. Even though, it is very difficult to present the most effective routes for phage administration among them. Since, all the ways are depending on the site of infection, concentration, dose, preparation of phage and various environmental factors. For example, the oral routes are effective for treating GI tract topical for skin infections, intravenous for local and systematic disease, octic route for chronic otitis infection, and lung infection by inhalation route (Morozova et al. 2018; Speck and Smithyman 2016; Ryan et al. 2011; Matsuzaki and Uchiyama 2019).

17.9 The Future of Phage

Based on over the past century research done on phages, the question arise immediately that what will be the next? Thus, possible prediction of the rapidly growing phage applications will be have a flourishing future in various filed such as basic medicine, agriculture, and even more. Further consideration necessitate to this great natural talent.

- Will be a widely utilization therapy in medicine field to treat multi drug resistant or bacterial infection.
- Will be a greater stage to find out the potent vaccine and drug delivery.
- Will be a powerful gift to maintain the human health.
- Will be more active through genetic engineering to solve the various hurdles in human health.
- Will be a new platform to understand the viral appearance and its defeat.
- Will be an incredible and innovative approaches for the development of nanomedicine and natural products in the field of biotechnology, agriculture and clinical field.

The upcoming century will be having a new generation of phages along with highest achievements in new frames to solve the biological issues (Henein 2013; Chan et al. 2013; Svircev et al. 2018; Roshan 2018; Mark 2019).

17.10 Conclusion

In conclusion, phages are a very diverse set of viruses that can be utilize as an alternative therapeutic agents to treat the specific-disease causing bacteria and preventing from multi-drug resistant pathogen. The success of phage therapy with several advantage are also increasing in various other fields such as clinical, agriculture, biotechnology and many more, which depends on the general public acceptance. It is clear that the further research is of utmost necessary to extend the existing information about the phage and their future utilization should be beneficial to resolving the complex issues of biology to create a healthy world with healthy live.

Note
Despite the major contribution towards the phage and its impact on socio-economic in the world, a great scientific agreement is required to understand in deep about the phage with existing environment, to prove the safety and efficacy in solving the current challenges of phage therapy to treat bacterial infection in broad success.

Acknowledgments SKS, UP, and MA gratefully acknowledge the Alagappa University for providing research facility. SKS thank MHRD RUSA – Phase 2.0 (grant sanctioned vide letter No. F. 24-51 / 2014 – U, Policy (TNMulti-Gen), Dept. of Edn. Govt. of India, Dt. 09.10.2018), DST-PURSE 2nd Phase Programme Order No. SR/PURSE Phase 2/38 (G dated 21.02.2017 and FIST (SR/FST/LSI – 667/2016) for providing the financial assistance. UP thank to Indian Council of Medical Research (ISRM/11/(19)/2017,dated:09.08.2018) for providing ICMR-SRF (Senior Research Fellowship). MA thanks UGC for providing RGNF fellowship award no: F1-17/2016-17/RGNF-2015-17-SC-TAM-18759.

Conflict of Interest The Author(s) declare that there is no conflict of interest.

References

Abedon ST (2019) Phage-antibiotic combination treatments: antagonistic impacts of antibiotics on the pharmacodynamics of phage therapy? Antibiotics (Basel) 8(4):pii: E182. https://doi.org/10.3390/antibiotics8040182

Abedon ST, Thomas-Abedon C, Thomas A, Mazure H (2011a) Bacteriophage prehistory: is or is not Hankin, 1896, a phage reference? Bacteriophage 1(3):174–178. https://doi.org/10.4161/bact.1.3.16591

Abedon ST, Kuhl SJ, Blasdel BG, Kutter EM (2011b) Phage treatment of human infections. Bacteriophage 1(2):66–85. https://doi.org/10.4161/bact.1.2.15845

Abedon ST, Garcia P, Mullany P, Aminov R (2017a) Editorial: phage therapy: past, present and future. Front Microbiol 8:981. https://doi.org/10.3389/fmicb.2017.00981

Abedon ST, García P, Mullany P, Aminov R (2017b) Phage therapy: past, present and future. Front Microbiol 8:981. https://doi.org/10.3389/fmicb.2017.00981

Ackermann HW (2003) Bacteriophage observations and evolution. Res Microbiol 154(4):245–251. https://doi.org/10.1016/S0923-2508(03)00067-6

Ackermann HW (2009) Phage classification and characterization. Methods Mol Biol 501:127–140. https://doi.org/10.1007/978-1-60327-164-6_13

Ackermann HW (2011) The first phage electron micrographs. Bacteriophage 1(4):225–227. https://doi.org/10.4161/bact.1.4.17280

Aghebati-Maleki L, Bakhshinejad B, Baradaran B, Motallebnezhad M, Aghebati-Maleki A, Nickho H, Yousefi M, Majidi J (2016) Phage display as a promising approach for vaccine development. J Biomed Sci 23(1):66. https://doi.org/10.1186/s12929-016-0285-9

Aiewsakun P, Adriaenssens EM, Lavigne R, Kropinski AM, Simmonds P (2018) Evaluation of the genomic diversity of viruses infecting bacteria, archaea and eukaryotes using a common bioinformatic platform: steps towards a unified taxonomy. J Gen Virol 99:1331–1343. https://doi.org/10.1099/jgv.0.001110

Altamirano FL, Barr JJ (2019) Phage therapy in the postantibiotic era. Clin Microbiol Rev 32(2):pii: e00066-18. https://doi.org/10.1128/CMR.00066-18

Bardell D (1982) An 1898 report by Gamaleya for a lytic agent specific for Bacillus anthracis. J Hist Med Allied Sci 37(2):222–225. https://doi.org/10.1093/jhmas/xxxvii.2.222

Bazan J, Całkosiński I, Gamian A (2012) Phage display—A powerful technique for immunotherapy: 1. Introduction and potential of therapeutic applications. Hum Vaccin Immunother 8(12):1817–1828. https://doi.org/10.4161/hv.21703

Bertozzi Silva J, Storms Z, Sauvageau D (2016) Host receptors for bacteriophage adsorption. FEMS Microbiol Lett 363(4):pii: fnw002. https://doi.org/10.1093/femsle/fnw002

Bhattarai SR, Yoo SY, Lee SW, Dean D (2012) Engineered phage-based therapeutic materials inhibit Chlamydia trachomatis intracellular infection. Biomaterials 33(20):5166–5174. https://doi.org/10.1016/j.biomaterials.2012.03.054

Breitbart M, Rohwer F (2005) Here a virus, there a virus, everywhere the same virus? Trends Microbiol 13:278–284. https://doi.org/10.1016/j.tim.2005.04.003

Breitbart M, Wegley L, Leeds S, Schoenfeld T, Rohwer F (2004) Phage community dynamics in hot springs. Appl Environ Microbiol 70(3):1633–1640. https://doi.org/10.1128/aem.70.3.1633-1640.2004

Brown R, Lengeling A, Wang B (2017) Phage engineering: how advances in molecular biology and synthetic biology are being utilized to enhance the therapeutic potential of bacteriophages. Quant Biol 5(1):42–54. https://doi.org/10.1007/s40484-017-0094-5

Campbell A (2003) The future of bacteriophage biology. Nat Rev Genet 4(6):471–477. https://doi.org/10.1038/nrg1089

Carlton RM (1999) Phage therapy: past history and future prospects. Arch Immunol Ther Exp 47(5):267–274

Casjens SR, Hendrix RW (2015) Bacteriophage lambda: early pioneer and still relevant. Virology 479:310–330. https://doi.org/10.1016/j.virol.2015.02.010

Chan BK, Abedon ST, Loc-Carrillo C (2013) Phage cocktails and the future of phage therapy. Future Microbiol 8(6):769–783. https://doi.org/10.2217/fmb.13.47

Chen Y, Batra H, Dong J, Chen C, Rao VB, Tao P (2019) Genetic engineering of bacteriophages against infectious diseases. Front Microbiol 10:954. https://doi.org/10.3389/fmicb.2019.00954

Chibani-Chennoufi S, Bruttin A, Dillmann ML, Brüssow H (2004) Phage-host interaction: an ecological perspective. J Bacteriol 186(12):3677–3686. https://doi.org/10.1128/JB.186.12.3677-3686.2004

Cisek AA, Dąbrowska I, Gregorczyk KP, Wyżewski Z (2017) Phage therapy in bacterial infections treatment: one hundred years after the discovery of bacteriophages. Curr Microbiol 74(2):277–283. https://doi.org/10.1007/s00284-016-1166-x

Clokie MR, Millard AD, Letarov AV, Heaphy S (2011) Phages in nature. Bacteriophage 1:31–45. https://doi.org/10.4161/bact.1.1.14942

Cormack BP, Valdivia RH, Falkow S (1996) FACS-optimized mutants of the green fluorescent protein (GFP). Gene 173:33–38. https://doi.org/10.1016/0378-1119(95)00685-0

Criscuolo E, Spadini S, Lamanna J, Ferro M, Burioni R (2017) Bacteriophages and their immunological applications against infectious threats. J Immunol Res 2017:3780697. https://doi.org/10.1155/2017/3780697

Dąbrowska K (2019) Phage therapy: what factors shape phage pharmacokinetics and bioavailability? Systematic and critical review. Med Res Rev 39(5):2000–2025. https://doi.org/10.1002/med.21572

Debarbieux L, Pirnay JP, Verbeken G, De Vos D, Merabishvili M, Huys I, Patey O, Schoonjans D, Vaneechoutte M, Zizi M, Rohde C (2016) A bacteriophage journey at the European Medicines Agency. FEMS Microbiol Lett 363(2):fnv225. https://doi.org/10.1093/femsle/fnv225

Debarbieux L, Fischer M, Quax TEF (2017) Viruses of microbes. Viruses 9:263. https://doi.org/10.3390/v9090263

Deng X, Wang L, You X, Dai P, Zeng Y (2018) Advances in the T7 phage display system. Mol Med Rep 17(1):714–720. https://doi.org/10.3892/mmr.2017.7994

Derda R, Lockett MR, Tang SK, Fuller RC, Maxwell EJ, Breiten B, Cuddemi CA, Ozdogan A, Whitesides GM (2013) Filter-based assay for Escherichia coli in aqueous samples using bacteriophage-based amplification. Anal Chem 85(15):7213–7220. https://doi.org/10.1021/ac400961b

Deresinski S (2009) Bacteriophage therapy: exploiting smaller fleas. Clin Infect Dis 48(8):1096–1101. https://doi.org/10.1086/597405

d'Hérelle F (2007) On an invisible microbe antagonistic toward dysenteric bacilli: brief note by Mr. F. D'Herelle, presented by Mr. Roux. 1917. Res Microbiol 158(7):553

Dixon B (2004) New dawn for phage therapy. Lancet Infect Dis 4(3):186. https://doi.org/10.1016/S1473-3099(04)00951-X

Doetschman T, Georgieva T (2017) Gene editing with CRISPR/Cas9 RNA-directed nuclease. Circ Res 120(5):876–894. https://doi.org/10.1161/CIRCRESAHA.116.309727

Domingo-Calap P, Delgado-Martínez J (2018) Bacteriophages: protagonists of a post-antibiotic era. Antibiotics (Basel) 7(3):pii: E66. https://doi.org/10.3390/antibiotics7030066

Doss J, Culbertson K, Hahn D, Camacho J, Barekzi N (2017) A review of phage therapy against bacterial pathogens of aquatic and terrestrial organisms. Viruses 9(3):50. https://doi.org/10.3390/v9030050

Drulis-Kawa Z, Majkowska-Skrobek G, Maciejewska B, Delattre AS, Lavigne R (2012) Learning from bacteriophages-advantages and limitations of phage and phage-encoded protein applications. Curr Protein Pept Sci 13(8):699–722. https://doi.org/10.2174/138920312804871193

Du Toit A (2017) The language of phages. Nat Rev Microbiol 15:135. https://doi.org/10.1038/nrmicro.2017.8

Easwaran M, Ahn J (2019) Application of bacteriophages in organic farm animal production. Safety Pract Organic Food 1:365–375. https://doi.org/10.1016/B978-0-12-812060-6.00018-0

Efimov VP, Nepluev IV, Mesyanzhinov VV (1995) Bacteriophage T4 as a surface display vector. Virus Genes 10(2):173–177. https://doi.org/10.1007/bf01702598

Eiserling FA (1979) Bacteriophage structure. In: Fraenkel-Conrat H, Wagner RR (eds) Comprehensive virology volume 13: structure and assembly, Comprehensive virology. Springer, Boston. https://doi.org/10.1007/978-1-4684-3453-8_8

Erez Z, Steinberger-Levy I, Shamir M et al (2017) Communication between viruses guides lysis–lysogeny decisions. Nature 541:488–493. https://doi.org/10.1038/nature21049

Fair RJ, Tor Y (2014) Antibiotics and bacterial resistance in the 21st century. Perspect Med Chem 6:25–64. https://doi.org/10.4137/PMC.S14459

Fauconnier A (2019) Phage therapy regulation: from night to dawn. Viruses 11(4):352. https://doi.org/10.3390/v11040352

Gamage LN, Ellis J, Hayes S (2009) Immunogenicity of bacteriophage lambda particles displaying porcine Circovirus 2 (PCV2) capsid protein epitopes. Vaccine 27(47):6595–6604. https://doi.org/10.1016/j.vaccine.2009.08.019

Górski A, Międzybrodzki R, Borysowski J, Dąbrowska K, Wierzbicki P, Ohams M, Korczak-Kowalska G, Olszowska-Zaremba N, Łusiak-Szelachowska M, Kłak M, Jończyk E (2012) Phage as a modulator of immune responses: practical implications for phage therapy. Adv Virus Res 83:41–71. https://doi.org/10.1016/B978-0-12-394438-2.00002-5

Hao M, Cui Y, Qu X (2018) Analysis of CRISPR-Cas system in Streptococcus thermophilus and its application. Front Microbiol 9:257. https://doi.org/10.3389/fmicb.2018.00257

Haq IU, Chaudhry WN, Akhtar MN, Andleeb S, Qadri I (2012) Bacteriophages and their implications on future biotechnology: a review. Virol J 9:9. https://doi.org/10.1186/1743-422X-9-9

Harada LK, Silva EC, Campos WF, Del Fiol FS, Vila M, Dąbrowska K, Krylov VN, Balcão VM (2018) Biotechnological applications of bacteriophages: state of the art. Microbiol Res 212–213:38–58. https://doi.org/10.1016/j.micres.2018.04.007

Hatoum-Aslan A (2018) Phage genetic engineering using CRISPR-Cas systems. Viruses 10(6):pii: E335. https://doi.org/10.3390/v10060335

Henein A (2013) What are the limitations on the wider therapeutic use of phage? Bacteriophage 3(2):e24872

Henry M, Debarbieux L (2012) Tools from viruses: bacteriophage successes and beyond. Virology 434(2):151–161. https://doi.org/10.1016/j.virol.2012.09.017

Hess KL, Jewell CM (2019) Phage display as a tool for vaccine and immunotherapy development. Bioeng Transl Med 5. https://doi.org/10.1002/btm2.10142

Hesse S, Adhya S (2019) Phage therapy in the twenty-first century: facing the decline of the antibiotic era; is it finally time for the age of the phage? Annu Rev Microbiol 73:155

Howard-Varona C, Hargreaves KR, Abedon ST, Sullivan MB (2017a) Lysogeny in nature: mechanisms, impact and ecology of temperate phages. ISME J 11(7):1511

Howard-Varona C, Hargreaves K, Abedon S et al (2017b) Lysogeny in nature: mechanisms, impact and ecology of temperate phages. ISME J 11:1511–1520. https://doi.org/10.1038/ismej.2017.16

Hyman P, Abedon ST (2010) Bacteriophage host range and bacterial resistance. Adv Appl Microbiol 70:217–248. https://doi.org/10.1016/S0065-2164(10)70007-1

Jault P, Leclerc T, Jennes S, Pirnay JP, Que YA, Resch G, Rousseau AF, Ravat F, Carsin H, Le Floch R, Schaal JV (2019) Efficacy and tolerability of a cocktail of bacteriophages to treat burn wounds infected by Pseudomonas aeruginosa (PhagoBurn): a randomised, controlled, double-blind phase 1/2 trial. Lancet Infect Dis 19(1):35–45. https://doi.org/10.1016/S1473-3099(18)30482-1

Jayaraman R (2008) Seymour benzer and T4 rII. Resonance 13:898–908. https://doi.org/10.1007/s12045-008-0098-6

Kara R (2018) Bacteriophage. https://www.britannica.com/science/bacteriophage. October 12 2018

Karen S (2018) Lytic vs lysogenic – understanding bacteriophage life cycles. https://www.technologynetworks.com/immunology/articles/lytic-vs-lysogenic-understanding-bacteriophage-life-cycles-308094. August 28 2018

Karen DW (2019) Building better bacteriophage with biofoundries to combat antibiotic-resistant bacteria, December 27, 2019

Keen EC, Adhya SL (2015) Phage therapy: current research and applications. Clin Infect Dis 61(1):141–142

Kilcher S, Loessner MJ (2019) Engineering bacteriophages as versatile biologics. Trends Microbiol 27(4):355–367

Kilcher S, Studer P, Muessner C, Klumpp J, Loessner MJ (2018) Cross-genus rebooting of custom-made, synthetic bacteriophage genomes in L-form bacteria. Proc Natl Acad Sci U S A 115(3):567–572. https://doi.org/10.1073/pnas.1714658115

Kim M, Ryu S (2013) Antirepression system associated with the life cycle switch in the temperate podoviridae phage SPC32H. J Virol 87(21):11775–11786. https://doi.org/10.1128/JVI.02173-13

Koivunen E, Arap W, Rajotte D, Lahdenranta J, Pasqualini R (1999) Identification of receptor ligands with phage display peptide libraries. J Nucl Med 40(5):883–888

Kortright KE, Chan BK, Koff JL, Turner PE (2019) Phage therapy: a renewed approach to combat antibiotic-resistant bacteria. Cell Host Microbe 25(2):219–232

Krylov V, Shaburova O, Krylov S, Pleteneva E (2012) A genetic approach to the development of new therapeutic phages to fight pseudomonas aeruginosa in wound infections. Viruses 5(1):15–53. https://doi.org/10.3390/v5010015

Kutateladze M, Adamia R (2008) Phage therapy experience at the Eliava Institute. Med Mal Infect 38(8):426–430. https://doi.org/10.1016/j.medmal.2008.06.023

Lin DM, Koskella B, Lin HC (2017) Phage therapy: an alternative to antibiotics in the age of multi-drug resistance. World J Gastrointest Pharmacol Ther 8(3):162

Loc-Carrillo C, Abedon ST (2011) Pros and cons of phage therapy. Bacteriophage 1(2):111–114. https://doi.org/10.4161/bact.1.2.14590

Lodish H, Berk A, Zipursky SL, Matsudaira P, Baltimore D, Darnell J (2000) Viruses: structure, function, and uses. In: Molecular cell biology, 4th edn. WH Freeman, New York

Loessner MJ, Rudolf M, Scherer S (1997) Evaluation of luciferase reporter bacteriophage A511::luxAB for detection of Listeria monocytogenes in contaminated foods. Appl Environ Microbiol 63(8):2961–2965

Löwy I (2010) Martin Arrowsmith's clinical trial: scientific precision and heroic medicine. J R Soc Med 103(11):461–466

Luria SE, Anderson TF (1942) The identification and characterization of bacteriophages with the electron microscope. Proc Natl Acad Sci U S A 28(4):127–130.1. https://doi.org/10.1073/pnas.28.4.127

Ly-Chatain MH (2014) The factors affecting effectiveness of treatment in phages therapy. Front Microbiol 5:51. https://doi.org/10.3389/fmicb.2014.00051

Makarova KS, Wolf YI, Iranzo J et al (2019) Evolutionary classification of CRISPR–Cas systems: a burst of class 2 and derived variants. Nat Rev Microbiol 18:67–83. https://doi.org/10.1038/s41579-019-0299-x

Marinelli LJ, Piuri M, Swigoňová Z, Balachandran A, Oldfield LM, van Kessel JC, Hatfull GF (2008) BRED: a simple and powerful tool for constructing mutant and recombinant bacteriophage genomes. PLoS One 3(12):e3957. https://doi.org/10.1371/journal.pone.0003957

Marinelli LJ, Piuri M, Hatfull GF (2019) Genetic manipulation of lytic bacteriophages with BRED: bacteriophage recombineering of electroporated DNA. Methods Mol Biol 1898:69–80. https://doi.org/10.1007/978-1-4939-8940-9_6

Mark T (2019) Are phages the wave of the future? Using viruses to treat bacterial diseases. https://www.biospace.com/article/are-phages-the-wave-of-the-future-using-viruses-to-treat-bacterial-diseases/ May 16, 2019

Matsuzaki S, Uchiyama J (2019) Phage pharmacokinetics: relationship with administration route. In: Phage therapy: a practical approach. Springer, Cham, pp 43–57. https://doi.org/10.1007/978-3-030-26736-0_2

Matsuzaki S, Rashel M, Uchiyama J, Sakurai S, Ujihara T, Kuroda M, Ikeuchi M, Tani T, Fujieda M, Wakiguchi H, Imai S (2005) Bacteriophage therapy: a revitalized therapy against bacterial infectious diseases. J Infect Chemother 11(5):211–219. https://doi.org/10.1007/s10156-005-0408-9

McDonald ND, Regmi A, Morreale DP, Borowski JD, Boyd EF (2019) CRISPR-Cas systems are present predominantly on mobile genetic elements in Vibrio species. BMC Genomics 20(1):105. https://doi.org/10.1186/s12864-019-5439-1

McGrath S, Fitzgerald GF, van Sinderen D (2004) The impact of bacteriophage genomics. Curr Opin Biotechnol 15(2):94–99

Molek P, Strukelj B, Bratkovic T (2011) Peptide phage display as a tool for drug discovery: targeting membrane receptors. Molecules 16(1):857–887. https://doi.org/10.3390/molecules16010857

Monk AB, Rees CD, Barrow P, Hagens S, Harper DR (2010) Bacteriophage applications: where are we now? Lett Appl Microbiol 51:363–369

Morozova VV, Vlassov VV, Tikunova NV (2018) Applications of bacteriophages in the treatment of localized infections in humans. Front Microbiol 9:1696. https://doi.org/10.3389/fmicb.2018.01696

Muzard J, Platt M, Lee GU (2012) M13 bacteriophage-activated superparamagnetic beads for affinity separation. Small 8(15):2403–2411. https://doi.org/10.1002/smll.201200099

Nair A, Khairnar K (2019) Genetically engineered phages for therapeutics: proceed with caution. Nat Med 25(7):1028. https://doi.org/10.1038/s41591-019-0506-3

Ofir G, Sorek R (2018) Contemporary phage biology: from classic models to new insights. Cell 172:1260–1270

Oh D, Qi J, Han B, Zhang G, Carney TJ, Ohmura J, Zhang Y, Shao-Horn Y, Belcher AM (2014) M13 virus-directed synthesis of nanostructured metal oxides for lithium-oxygen batteries. Nano Lett 14(8):4837–4845. https://doi.org/10.1021/nl502078m

O'Sullivan L, Buttimer C, McAuliffe O, Bolton D, Coffey A (2016) Bacteriophage-based tools: recent advances and novel applications. F1000Res 5:2782. https://doi.org/10.12688/f1000research.9705.1

Pande J, Szewczyk MM, Grover AK (2010) Phage display: concept, innovations, applications and future. Biotechnol Adv 28(6):849–858. https://doi.org/10.1016/j.biotechadv.2010.07.004

Parfitt T (2005) Georgia an unlikely stronghold for bacteriophage therapy. Lancet 365(9478):2166–2167. https://doi.org/10.1016/S0140-6736(05)66759-1

Pelfrene E, Willebrand E, Cavaleiro Sanches A, Sebris Z, Cavaleri M (2016) Bacteriophage therapy: a regulatory perspective. J Antimicrob Chemother 71(8):2071–2074. https://doi.org/10.1093/jac/dkw083.v

Pierrel J (2012) An RNA phage lab: MS2 in Walter Fiers' laboratory of molecular biology in Ghent, from genetic code to gene and genome, 1963–1976. J Hist Biol 45(1):109–138. https://doi.org/10.1007/s10739-010-9267-z

Pines G, Freed EF, Winkler JD, Gill RT (2015) Bacterial recombineering: genome engineering via phage-based homologous recombination. ACS Synth Biol 4(11):1176–1185. https://doi.org/10.1021/acssynbio.5b00009s

Pires DP, Cleto S, Sillankorva S, Azeredo J, Lu TK (2016) Genetically engineered phages: a review of advances over the last decade. Microbiol Mol Biol Rev 80(3):523–543. https://doi.org/10.1128/MMBR.00069-15

Pirnay JP, Verbeken G, Rose T, Jennes S, Zizi M, Huys I, Lavigne R, Merabishvili M, Vaneechoutte M, Buckling A, De Vos D (2012) Introducing yesterday's phage therapy in today's medicine. Futur Virol 7(4):379–390. https://doi.org/10.2217/fvl.12.24

Prasad BV, Schmid MF (2012) Principles of virus structural organization. Adv Exp Med Biol 726:17–47. https://doi.org/10.1007/978-1-4614-0980-9_3

Prigent M, Leroy M, Confalonieri F, Dutertre M, DuBow MS (2005) A diversity of bacteriophage forms and genomes can be isolated from the surface sands of the Sahara Desert. Extremophiles 9(4):289–296. https://doi.org/10.1007/s00792-005-0444-5

Principi N, Silvestri E, Esposito S (2019) Advantages and limitations of bacteriophages for the treatment of bacterial infections. Front Pharmacol 10:513. https://doi.org/10.3389/fphar.2019.00513

Roshan N (2018) The future is phage? https://sfam.org.uk/resources/the-future-is-phage.html

Ross A, Ward S, Hyman P (2016) More is better: selecting for broad host range bacteriophages. Front Microbiol 7:1352. https://doi.org/10.3389/fmicb.2016.01352

Rowley MJ, O'Connor K, Wijeyewickrema L (2004) Phage display for epitope determination: a paradigm for identifying receptor-ligand interactions. Biotechnol Annu Rev 10:151–188. https://doi.org/10.1016/S1387-2656(04)10006-9

Ryan EM, Gorman SP, Donnelly RF, Gilmore BF (2011) Recent advances in bacteriophage therapy: how delivery routes, formulation, concentration and timing influence the success of phage therapy. J Pharm Pharmacol 63(10):1253–1264. https://doi.org/10.1111/j.2042-7158.2011.01324.x

Salmond GP, Fineran PC (2015) A century of the phage: past, present and future. Nat Rev Microbiol 13:777. https://doi.org/10.1038/nrmicro3564

Sanger F, Coulson AR (1975) A rapid method for determining sequences in DNA by primed synthesis with DNA polymerase. J Mol Biol 94(3):441–448. https://doi.org/10.1016/0022-2836(75)90213-2

Säwström C, Lisle J, Anesio AM, Priscu JC, Laybourn-Parry J (2008) Bacteriophage in polar inland waters. Extremophiles 12(2):167–175. https://doi.org/10.1007/s00792-007-0134-6
Schofield DA, Molineux IJ, Westwater C (2009) Diagnostic bioluminescent phage for detection of Yersinia pestis. J Clin Microbiol 47(12):3887–3894. https://doi.org/10.1128/JCM.01533-09
Schofield DA, Bull CT, Rubio I, Wechter WP, Westwater C, Molineux IJ (2012) Development of an engineered bioluminescent reporter phage for detection of bacterial blight of crucifers. Appl Environ Microbiol 78(10):3592–3598. https://doi.org/10.1128/AEM.00252-12
Schooley RT, Biswas B, Gill JJ, Hernandez-Morales A, Lancaster J, Lessor L, Barr JJ, Reed SL, Rohwer F, Benler S, Segall AM (2017) Development and use of personalized bacteriophage-based therapeutic cocktails to treat a patient with a disseminated resistant Acinetobacter baumannii infection. Antimicrob Agents Chemother 61(10):e00954-17. https://doi.org/10.1128/AAC.00954-17
Semler DD, Lynch KH, Dennis JJ (2012) The promise of bacteriophage therapy for Burkholderia cepacia complex respiratory infections. Front Cell Infect Microbiol 1:27. https://doi.org/10.3389/fcimb.2011.00027
Sharma S, Chatterjee S, Datta S, Prasad R, Dubey D, Prasad RK, Vairale MG (2017) Bacteriophages and its applications: an overview. Folia Microbiol 62:17–55. https://doi.org/10.1007/s12223-016-0471-x
Shivachandra SB, Rao M, Janosi L, Sathaliyawala T, Matyas GR, Alving CR, Leppla SH, Rao VB (2006) In vitro binding of anthrax protective antigen on bacteriophage T4 capsid surface through hoc–capsid interactions: a strategy for efficient display of large full-length proteins. Virology 345(1):190–198. https://doi.org/10.1016/j.virol.2005.10.037
Sillankorva S, Pospiech H, Azeredo J, Neubauer P (2007) Biofilm control with T7 phage. J Biotechnol 131(2):S252. https://doi.org/10.1016/j.jbiotec.2007.07.458
Sioud M (2019) Phage display libraries: from binders to targeted drug delivery and human therapeutics. Mol Biotechnol 61(4):286–303. https://doi.org/10.1007/s12033-019-00156-8
Skurnik M, Strauch E (2006) Phage therapy: facts and fiction. Int J Med Microbiol 296(1):5–14. https://doi.org/10.1016/j.ijmm.2005.09.002
Smith GP (2019) Phage display: simple evolution in a Petri Dish (Nobel lecture). Angew Chem Int Ed 58:14428–14437. https://doi.org/10.1002/anie.201908308
Sousa JAMd, Rocha EPC (2019) Environmental structure drives resistance to phages and antibiotics during phage therapy and to invading lysogens during colonisation. Sci Rep 9:3149. https://doi.org/10.1038/s41598-019-39773-3
Speck P, Smithyman A (2016) Safety and efficacy of phage therapy via the intravenous route. FEMS Microbiol Lett 363(3):pii: fnv242. https://doi.org/10.1093/femsle/fnv242
Stanford K, McAllister TA, Niu YD, Stephens TP, Mazzocco A, Waddell TE, Johnson RP (2010) Oral delivery systems for encapsulated bacteriophages targeted at Escherichia coli O157:H7 in feedlot cattle. J Food Prot 73(7):1304–1312. https://doi.org/10.4315/0362-028x-73.7.1304
Sternberg N, Hoess RH (1995) Display of peptides and proteins on the surface of bacteriophage lambda. Proc Natl Acad Sci U S A 92(5):1609–1613. https://doi.org/10.1073/pnas.92.5.1609
Stone E, Campbell K, Grant I, McAuliffe O (2019) Understanding and exploiting phage-host interactions. Viruses 11(6):pii: E567. https://doi.org/10.3390/v11060567
Sulakvelidze A, Alavidze Z, Morris JG (2001) Bacteriophage therapy. Antimicrob Agents Chemother 45:649–659. https://doi.org/10.1128/AAC.45.3.649-659.2001
Summers WC (2012) The strange history of phage therapy. Bacteriophage 2(2):130–133
Svircev A, Roach D, Castle A (2018) Framing the future with bacteriophages in agriculture. Viruses 10(5):pii: E218. https://doi.org/10.3390/v10050218
Tagliaferri TL, Jansen M, Horz HP (2019) Fighting pathogenic bacteria on two fronts: phages and antibiotics as combined strategy. Front Cell Infect Microbiol 9:22. https://doi.org/10.3389/fcimb.2019.00022
The Nobel Prize in Physiology or Medicine (1969) Nobel Prize.org. Nobel Media AB. https://www.nobelprize.org/prizes/medicine/1969/summary

Tolstoy I, Kropinski AM, Brister JR (2018) Bacteriophage taxonomy: an evolving discipline. Methods Mol Biol 1693:57–71. https://doi.org/10.1007/978-1-4939-7395-8_6

Tornetta M, Baker S, Whitaker B, Lu J, Chen Q, Pisors E, Shi L, Luo J, Sweet R, Tsui P (2010) Antibody fab display and selection through fusion to the pIX coat protein of filamentous phage. J Immunol Methods 360(1–2):39–46. https://doi.org/10.1016/j.jim.2010.06.001

Twort FW (1915) An investigation on the nature of ultra-microscopic viruses. Lancet 2:1241–1243. https://doi.org/10.1016/S0140-6736(01)20383-3

Underwood SL, Foto A, Ray AF, Nelms AE, Kennedy KL, Hartley SG, Ryals LM, Gurung C, D'Angelo WA, Pope WH, Mavrodi DV (2019) Discovery and characterization of bacteriophage LuckyBarnes. Microbiol Resour Announc 8(25):pii: e00330-19. https://doi.org/10.1128/MRA.00330-19

Van Belleghem J, Dąbrowska K, Vaneechoutte M, Barr J, Bollyky P (2018) Interactions between bacteriophage, bacteria, and the mammalian immune system. Viruses 11(1):pii: E10. https://doi.org/10.3390/v11010010

Van der Merwe RG, Van Helden PD, Warren RM, Sampson SL, van Pittius NG (2014) Phage-based detection of bacterial pathogens. Analyst 139(11):2617–2626. https://doi.org/10.1039/c4an00208c

Van Duin J (1988) Single-stranded RNA bacteriophages. In: Calendar R (ed) The bacteriophages, The viruses. Springer, Boston, MA. https://doi.org/10.1007/978-1-4684-5424-6_4

Voelker R (2019) FDA approves bacteriophage trial. JAMA 321(7):638. https://doi.org/10.1001/jama.2019.0510

Waldor MK, Mekalanos JJ (1996) Lysogenic conversion by a filamentous phage encoding cholera toxin. Science 272(5270):1910–1914. https://doi.org/10.1126/science.272.5270.1910

White HE, Orlova EV (2019) Bacteriophages: their structural organisation and function. In: Bacteriophages-biology and applications. IntechOpen. https://doi.org/10.5772/intechopen.85484

Wienhold SM, Lienau J, Witzenrath M (2019) Towards inhaled phage therapy in Western Europe. Viruses 11:295

Williamson SJ, Paul JH (2004) Nutrient stimulation of lytic phage production in bacterial populations of the Gulf of Mexico. Aquat Microb Ecol 36(1):9–17. https://doi.org/10.3354/ame036009

Wittebole X, De Roock S, Opal SM (2014) A historical overview of bacteriophage therapy as an alternative to antibiotics for the treatment of bacterial pathogens. Virulence 5(1):226–235. https://doi.org/10.4161/viru.25991

Xu H, Bao X, Wang Y, Xu Y, Deng B, Lu Y, Hou J (2018) Engineering T7 bacteriophage as a potential DNA vaccine targeting delivery vector. Virol J 15(1):49. https://doi.org/10.1186/s12985-018-0955-1

Yehl K, Lemire S, Yang AC, Ando H, Mimee M, Torres MD, de la Fuente-Nunez C, Lu TK (2019) Engineering phage host-range and suppressing bacterial resistance through phage tail fiber mutagenesis. Cell 179(2):459–469.e9. https://doi.org/10.1016/j.cell.2019.09.015

Yosef I, Manor M, Kiro R, Qimron U (2015) Temperate and lytic bacteriophages programmed to sensitize and kill antibiotic-resistant bacteria. Proc Natl Acad Sci 112(23):7267–7272. https://doi.org/10.1073/pnas.1500107112

Young RY (1992) Bacteriophage lysis: mechanism and regulation. Microbiol Rev 56(3):430–481

Zhang QG, Buckling A (2012) Phages limit the evolution of bacterial antibiotic resistance in experimental microcosms. Evol Appl 5(6):575–582. https://doi.org/10.1111/j.1752-4571.2011.00236.x

Zirpel NK, Arslan T, Lee H (2015) Engineering filamentous bacteriophages for enhanced gold binding and metallization properties. J Colloid Interface Sci 454:80–88. https://doi.org/10.1016/j.jcis.2015.05.006

Zschach H, Joensen K, Lindhard B, Lund O, Goderdzishvili M, Chkonia I, Jgenti G, Kvatadze N, Alavidze Z, Kutter E, Hasman H (2015) What can we learn from a metagenomic analysis of a Georgian bacteriophage cocktail? Viruses 7(12):6570–6589. https://doi.org/10.3390/v7122958

Chapter 18
Phagetherapy: Clinical Applications – Critical Appraisal of Randomized Controlled Trials

Xavier Wittebole and Steve Opal

Abstract Phagotherapy, defined as the use of bacteriophage to treat bacterial infections, was initially proposed by Felix d'Herelle, a French-Canadian who did a lot of research on this topic in Paris, but also travelled the world to treat patients in different settings and conditions. While the literature on this topic is quite extensive, the number of randomized controlled trials is rather limited. Large studies were performed and published in the former USSR republics more than 50 years ago. Since then, there were few trials performed, enrolling a limited number of patients and assessing potential phage efficacy in different settings such as chronic otitis or infected burn wounds. In this chapter, we review and discuss these different randomized trials. While the results might look disappointing at first sight, they all confirm safety of phage used for treatment of difficult clinical situations. Those trials should help define further studies in order to obtain the best possible results and hopefully confirm that phages could be used as an alternative to treat difficult to treat multi-drug-resistant bacterial infections.

18.1 Introduction

Initially described by Ernest Hanbury Hankin, and later on by Frederick Twort, and Félix d'Herelle who proposed to use it to treat bacterial infections, bacteriophages were extensively studied in the first 40 years of the last century (Wittebole et al. 2014). The advent of antibiotics let to a decreased interest in the western countries while an ongoing research was pursued in the USSR and eastern countries. However, due to the emergence of multi-drug resistant bacteria a new appeal to control these

X. Wittebole (✉)
Critical Care Department, Clinique universitaire St Luc, Brussels, Belgium
e-mail: xavier.wittebole@uclouvain.be

S. Opal
Infectious Disease Division, Alpert School of Medicine, Brown University, Providence, RI, USA

G. Witzany (ed.), *Biocommunication of Phages*,
https://doi.org/10.1007/978-3-030-45885-0_18

bacteria was launched and bacteriophage could represent one of the solutions. However, despite a high number of animal models publications, a huge number of cases report and some case series, the number of randomized controlled trials is still limited. In this chapter, we will analyze those randomized trials and discuss how it should be interpreted.

18.2 Results of Randomized Controlled Trials

18.2.1 USSR Trials

Professor Nina Chanishvili, a scientist working at the Laboratory for Genetics of Microorganisms and Bacteriophages, in the Eliava Institute of Bacteriophage, Microbiology and Virology, in Tbilisi, Georgia performed a historic review of the literature (Chanishvili 2012). She also presented the randomized trials performed in the former Soviet Union (USSR) during the European Congress of Clinical Microbiology & Infectious Diseases meeting, hold in Madrid, Spain in 2018 (Chanishvili 2018). Other authors (Morozova et al. 2018) also reviewed this interesting literature, which allows scientists not familiar with Russian, the main language of publication for many early phage studies, to discover some of the very first, large randomized clinical trials assessing phage efficacy.

For instance, the application of phage therapy to surgical and wound treatment, originally described by GA Kokin, was initially studied during the Finnish Campaign in 1939–1940, when Russian soldiers were randomized to receive phage solutions directed against anaerobes, Staphylococcus and Streptococcus (produced at the Eliava Institute for Bacteriology, Microbiology and Virology in Tbilissi, Georgia), or, placebo, to prevent gangrene from infected war wounds. In the 3 mobile sanitary brigades where a prophylactic administration was evaluated, the number of infected soldiers significantly decreased: In the first brigade of 10,418 soldiers, the incidence of gangrene was 4.3% (342/7918) in the placebo group while it was only 1.5% (35/2500) in the group of phage-treated soldiers. Similar results were obtained for the 2 others brigades. Comparison of data described for these three different brigades showed an average 30% decrease in the number of gas gangrene cases with prophylactic phage treatment. When used as a treatment, a decrease of gangrene disease related mortality from 42.2% in the placebo group, down to 18.8% in the phage group was observed in a cohort of 767 infected soldiers. Interestingly, a smaller report demonstrated that early treatment (immediately on the battlefield) benefited better than treatment upon admission in the hospital.

Another very large, randomized and placebo-controlled trial conducted in children from Tbilisi/Republic of Georgia during the 1960s describes prevention of Shigella species dysentery and Escherichia coli diarrhea with orally applied Shigella phages (Babalova et al. 1968; Sulakvelidze et al. 2001). More than 30,000 children, aged between 6 months and 7 years, were randomized during two 5-months periods in 1963 and 1964. Children living on one side of the street received a dose of phage (10–20 ml according to age: 5-years old children and above received 20 ml) every 5–7 days while children living on the other side of the street did not receive the treatment and formed the control group. Doctors provided close follow-up with weekly visits to every child from both experimental and control groups. Information regarding potential side effects or cases of disease were sent to the center. Treatment effect was based on clinical symptoms rather than on bacteriological analysis, as only suspected cases of intestinal disease were studied bacteriologically. The incidence of acute dysentery was 3.8 times higher in the control group, the treatment being more effective in smaller children as the incidence of dysentery dramatically decreased with age. Microbiologically confirmed episodes of diarrhea also occurred with an incidence 2.8 fold higher in the control group. The authors of this study suggested using these phage tablets of dysenteric bacteriophages for mass application.

18.2.2 First Randomized Trials in the Western World

While the first use of phage application was reported in the early 1920's (Bruynhoge and Maisin 1921), there were two small randomized trials reported in the late 60's (Wittig et al. 1966) and very early 70's (Monsur et al. 1970). In the first study, 44 children suffering from what the authors defined as infective asthma (a condition characterized by asthmatic attacks with episodes of infection and by lack of demonstrable allergic etiology) were randomized to receive either placebo or weekly injection of Staphylococcus lysate with ten billion bacteriophages per milliliter (Wittig et al. 1966) While a significant decrease in wheezing days was described, a close analysis of the scattergram did not reveal significant difference. The occurrence of nasopharyngeal staphylococcal positive culture taken at the end of the study was also not different between groups.

In the second trial, (Monsur et al. 1970) very high dose of phage directed against Vibrio cholerae were administered to 8 patients after in vitro confirmation that the phage mixture was effective against all strains of Vibrio cholerae present in the patient's stools. Within 90 min after phage administration, vibrio activity disappeared from the stools. Likewise, stools examination demonstrated that the appearance of large numbers of stools bacteriophage coincided with the disappearance of vibrios. For purposes of comparison, the results from the 8 treated patients were presented together with those from a control group of 50 adult cholera patients treated with intravenous fluid alone, and 18 patients treated with tetracycline. An apparent clinical response, defined by a reduction of the stool output and a shorter duration of diarrhea, was observed in only 4 out of the 8 patients. Even in those patients, the reduction in stool volume and duration of diarrhea was less than that produced by tetracycline.

18.2.3 Treatment of Venous Leg Ulcers

The first large western randomized human clinical trial of a bacteriophage therapeutic against its target bacterial infection was published in 2009 (Rhoads et al. 2009). This Phase I, prospective, double blinded, placebo-controlled study was conducted at the Wound Care Center, Lubbock, Texas, in 42 patients presenting with infected chronic venous leg ulcers. The bacteriophage cocktail solution (WPP-201, developed by Intralytix Inc., USA) applied on the ulcer included 8 lytic bacteriophages directed against Pseudomonas aeruginosa, Staphylococcus aureus and Escherichia coli at a concentration of roughly 1×10^9 PFU/ml of each of the component. It contained monophages isolated from environment and not genetically modified. Sensitivity studies were not performed prior to the treatment application. Patients were treated with an ultrasonic debridement device, followed by wound dressing (which contained the phage solution) and bandage for 12 weeks and followed up to 24 weeks. As a phase I study, the primary objective was to evaluate the safety of the topical use of the phage solution on the healing of the full thickness venous leg ulcers of greater than 30 days duration. There were no reported significant adverse effects. Also, while not powered to evaluate efficacy, there were no difference in cure rate. Interferences between the phage solution and the dressing were proposed as hypothesis to explain this absence of curative effect.

18.2.4 Treatment of Pseudomonas aeruginosa Related Acute Otitis Media

Another study on phage was published in 2009 (Wright et al. 2009). It was a randomized, double-blind, placebo-controlled Phase I/II clinical trial approved by the UK Medicines and Healthcare products Regulatory Agency (MHRA) and the Central Office for Research Ethics Committees (COREC). The study was conducted in patients suffering from chronic otitis for several years at the time of enrolment. Before being randomized to a single dose of Biophage-Pseudomonas aeruginosa (PA) directed against antibiotic-resistant Pseudomonas, or placebo, the ear infection was documented and susceptibility to one or more of the six phages present in the Biophage-PA was assessed and confirmed by plaque assay on isolates from ear swabs. Clinical outcome was the primary objective of the study, as warranted by the UK MHRA, the European Medicines Agency and the American Food and Drugs Administration agency. The measure of bacterial count was a secondary outcome.

Twenty-four adult patients (of an original target of 40) were randomized and either 200 μL dose placebo or 200 μL phage solution was administered locally after the ear was meticulously cleaned. Mean indicators for combined patient-reported Visual Analogy Scale (VAS) scores improved in the group of patients treated with the phage solution. Indeed, patients complained less of discomfort, itchiness, wetness and unpleasant odor. Likewise, doctors in charge of the patients, blinded to the treatment arm, also observed significant improvement in clinical outcomes such as erythema/inflammation, ulceration/granulation/polyps and odor, all assessed by VAS scores. Interestingly, treated patients had significant lower Pseudomonas counts on day 21 and day 42, and, bacteriophage clearance was observed after resolution of infection in all cases where this occurred. There were only some mild or moderate adverse events in both groups and none was considered to be related to the investigational product administration by the trial clinician.

This study has several strengths. First, all of the selected patients were difficult to treat patients with prolonged histories of ear problems that had not responded to multiple treatments including topical and systemic antibiotics, and even surgery. Hence, the outcome was challenging. However, prior to treatment, the authors demonstrated that the phage solution was active against the Pseudomonas strains, which was probably a way to increase the chance of success. Moreover, the treatment was applied locally, directly where the infection was active, as a single-dose administration of a very small dose that represented only 2.4 ng of protein given.

However, these researchers screened the patients before enrollment to assure P. aeruginosa was indeed the infecting agent and that their phage cocktail lysed the infecting pathogen in vitro before actually treating the patients. Furthermore, phages were applied locally into the external ear canal, which prevented a potential loss from another route of administration. Additionally, the authors demonstrated that phages were self-replicating as long as the target bacteria were present. When the target bacteria was reduced or eliminated the phage replication ceased.

One of the strength of this study also points out a potential limitation: the need to analyze the pseudomonas strain and to assess its phage-susceptibility before actual treatment begins shows a huge potential for phage treatment in chronic infections. However, this observation also illustrates some caution with respect to the immediate clinical value of phage therapy in acute settings. Another limit of this study is its early termination as the Chief Investigator requested an interim analysis after enrolling 20 patients since he observed some rapid improvements in 10 out of these patients. This independent control, without any unblinding by investigators nor sponsors, confirmed there was evidence suggesting efficacy of the bacteriolytic phage treatment in the setting of chronic otitis media, and consultations with independent statisticians and regulatory advisors indicated that there was enough data to support the design of a larger phase III clinical trial. It was subsequently decided to stop the trial after completion of the latest enrolled patients, ending the trial at patient 24.

18.2.5 Phage Therapy for Acute Bacterial Diarrhea

For years, scientists from the Nestlé Research Centre and the Nestlé Institute of Health Science in Lausanne, Switzerland, collaborated with doctors and scientists working at the International Center for Diarrhoeal Diseases Research in Bangladesh to assess the potential use of alternative treatments such as phage for bacterial diarrheal diseases.

After the demonstration of safety of 2 coliphages cocktail in healthy adults, (Sarker et al. 2012) as well as in adults and children, (McCallin et al. 2013) they started a large phase II randomized controlled trial in children with the hypothesis that oral administration of those 2 cocktails would be effective in reducing symptoms and severity of diarrhea (Sarker et al. 2016). After obtaining a written informed consent from the parents, 120 boys aged 6–24-months with at least four liquid stools during the previous 24 h were randomized, to receive either a placebo (reduced osmolarity oral rehydration solution + zinc as a local standard of care), a T4 phage cocktail or the Microgen ColiProteus phage cocktail. The various endpoints included the safety of oral coliphage in children, the titration of fecal coliphage and E. coli pathogen to assess in vivo lytic phage activity, and finally, the impact of oral phage on quantitative diarrhea parameters. As in their previous studies, the authors confirmed the safety of phage administration, as there were no description of a Jarisch–Herxheimer reaction and liver, kidney and hematological functions assessed by usual standard tests remained normal. Children treated with one of the phage cocktail had higher fecal phage prevalence and titers than seen in placebo recipients but no patient showed higher fecal output than oral input, excluding significant in vivo replication of oral phage and arguing for a passive transit of phage through the gut. Type of infection did not affect fecal phage titers. Finally, there was no significant difference seen between groups when evaluating quantitative diarrhea criteria such as stool frequency, stool weight, amount of oral rehydration solution required to correct dehydration.

The lack of clinical efficacy might partially be explained by a low E. coli titer as viable E. coli colony counts represented only 5% of all fecal bacteria, which is close to the replication threshold of coliphages. Of note, children with non-E. coli diarrhea and healthy controls showed with few exceptions less than 1% Escherichia abundance in their stools. This is certainly an issue as phage therapy requires phage amplification which seems to occur only when a bacterial strain grows exponentially as demonstrated for Vibrio cholerae phages (Sarker and Brüssow 2016). Also, when E. coli cells excreted with stools were tested for phage susceptibility, only 50% of the colonies were lysed, suggesting another explanation for the observed results. Moreover, bacterial coinfections were frequently detected in stool samples, and many contained more than one pathogen, suggesting polymicrobial infections (Sarker and Brüssow 2016). Finally, microbiota analysis revealed a marked outgrowth of fecal streptococci during the acute phase of diarrhea, which was probably unexpected. It has been attributed to a potential nutrient-driven consequence of glucose provided with oral rehydration solution (Kieser et al. 2018). The specific role of these various Streptoccus species is a matter of debate, some authors arguing it might be the cause of diarrhea (Jin et al. 2013) while others could not demonstrate virulence gene in these strains (Sarker and Brüssow 2016).

18.2.6 Treatment of Healthy Staphylococcus aureus Carriers

Also, from the International Center for Diarrhoeal Diseases Research in Bangladash comes a randomized study on the potential of phage treatment to eradicate Staphylococcus carriage (McCallin et al. 2018). A total of 21 healthy women, aged 20–35, gave informed consent (witnessed by husband) and were randomized to one of the 3 following groups: placebo, the Staphylococcal monophage (10^6 PFU/ml) or the Pyobacteriophage cocktail (10^5 PFU/ml) produced by the Eliava Phage Institute in Tbilisi/Republic of Georgia. Either nasal swabs or vaginal swabs were positive for Staphylococcus aureus. Before randomization, the women underwent clinical examination, and, haematology and clinical chemistry lab testing. At enrolment, the participants received a unique sequential random number, which attributed them to one of the three treatment arms before they received the other two treatments sequentially. The first 10 women received 10 ml of the allotted product three times per day for 2 days, together with 50 ml mineral drinking water. On day 3 blood sampling was drawn and after a wash-out period of 4 days (from day 4 until day 7), they came back on day 8 and 15 to receive the second and third treatments. Eleven other women followed the same exposure scheme except that a nasal instead of oral application was applied.

The study participants did not report any adverse events from the nasal phage exposure and this was not associated with higher rates of abnormal values compared to placebo. Four adverse events were noted during oral treatment product application, none being considered by the clinicians to be related to the oral phage treatment (one adverse event occurred during placebo treatment). Unfortunately, to our knowledge, apart from the safety data, efficacy assessment was not reported.

18.2.7 Infected Burn Wounds and the PHAGOBURN Study

A first randomized phase 1 study on the effect of phage to treat infected burn wounds took place at the Queen Astrid Military Hospital in Brussels, Belgium, in 2007–2008 (Rose et al. 2014). Comparison between standard treatment for Pseudomonas aeruginosa and Staphylococcus aureus burn wound colonization with phage treatment was done on the same colonized burn wound because an objective evaluation and classification of burn wounds is impossible and colonization and infection levels can vary significantly.

Just before the application of the phage solution (BFC-1), the colonized burn wound was divided in two: One-half of the wound would receive the standard treatment while the other half got the phage treatment with BFC-1. Two biopsies were taken by the MD in charge of the patient in the centre of the zone where BFC-1 was to be applied, the other in the centre of the zone where the standard treatment was to be applied. Phage treatment was provided as a single-spray of approximately 1 ml of sterile and endotoxin-purified BFC-1 per 50 cm2 on one-half of the burn wound, using a 5 ml syringe with a spray adapter. The other part of the wound received treatment as local treatment protocols, which include local antimicrobial substances. Patients also received IV antibiotics. Two to five hours after treatment application, the burn wound was uncovered, and, two new biopsies were taken within 2 cm to the first ones. Despite indications of colonization or infection through usual swab culture surveillance, the results were disappointing as bacterial cultures of the homogenized biopsies taken before and after BFC-1 application showed only a very small bacterial load in 8 of the 10 applications. Moreover, phage treatment (and standard of care) did not affect bacterial load. There is no report of a potential side effect on clinical or laboratory point of view. Duration of observation is however not reported. A potential explanation of these "negative results" is that the initial bacterial load was surprisingly low. This is potentially related to the time elapsed between the surveillance swabs culture results showing multi-drug resistant Pseudomonas or Staphylococcus, and the moment where the patients were enrolled in the trial. Indeed, in the meantime, patients received potent topical antimicrobials, dressings and systemic antibiotics. One may wonder why the results were published so late (2014) after the study completion (2008). The authors explain they initially abandoned the idea of a widespread scientific report as the study did not go as expected and would use the experience gained during this small pilot study to set up a larger double blind study. Unfortunately, they had to put their plans for the phase II study on hold because in the meantime phages were classified as medicinal products, which forces scientists to comply with the classical pharmaco-economical framework. Fortunately, colleagues of them convinced them to publish their paper as it may help the scientific community in convincing other competent authorities and ethical committees in approving experimental phage applications and in designing future studies.

Phagoburn is a European Research & Development project funded by the European Commission under the 7th Framework Program for Research and Development. This collaborative project launched in June 2013 gathered 5 partners from 3 European countries (Belgium, France and Switzerland). Twenty-seven patients, aged 18 years or older and suffering a burn wound clinically infected with Pseudomonas aeruginosa, were enrolled in this phase I/II trial (Jault et al. 2019).

In the Phagoburn study, 14 patients received standard of care (1% sulfadiazine silver emulsion cream) and 13 other patients were treated with a cocktail of 12 natural lytic anti-P aeruginosa bacteriophages (PP1131;1×10^6 PFU per mL), both given as a daily topical treatment for 7 days, with 14 days of follow-up. Primary endpoint was median time to sustained reduction in bacterial burden (at least two quadrants via a four-quadrant method), assessed by use of daily wound swabs. Adjunctive antibiotic therapy during the 7-day treatment period was allowed at the discretion of the treating physician.

The phage therapy group reached the primary endpoint later than the standard of care group: 144 h versus 47 h (hazard ratio 0.29, 95% CI 0.10–0.79; p = 0.018). Also, at day 7, the primary endpoint was achieved in a non-statistically significant lower proportion of patients in the phage-treated group (six (50%)) as compared to the standard of care group (11 (85%)) (p = 0.097). By the end of follow-up, a higher proportion of patients in the standard of care group achieved successful treatment (13 (76%) of 17 versus nine (53%) of 17; $p = 0.15$). The only positive observed effects were a lower rate of conversion of superficial wound into deep wound and presence of pus that seemed less frequent in the PP1131 group.

Interestingly, phage treatment did not induce any adverse event, the total number of adverse events being lower in the phage-treated patients. The authors concluded that the anti-pseudomonas phage cocktail PP1131 decreased bacterial burden in burn wounds at a slower pace than usual treatment.

There are at least two reasons that might explain what looks like disappointing results. First, the phage cocktail solution concentration was significantly lower than expected. Previous studies of each component of the final phage solution (i.e., each phage) found that it was stable at concentrations of 1×10^9 PFU/mL or higher for over 24 months after manufacturing. However, once assembled, the global titer of phage in the solution decreased from 1×10^9 PFU/mL down to 1×10^4–1×10^5 PFU/ml. Furthermore, the phage cocktail was diluted to decrease the risk of concentrated endotoxin application to the burned wound. Hence, the applied dose of phage was approximately 10–1 00 PFU/mL while the expected dose was 1×10^6 PFU/ml. Actually, the concerns regarding patient exposure to insufficient phage concentrations led the sponsor to stop recruitment and this decision was later confirmed when the second data and safety monitoring board meeting recommended stopping the trial because of insufficient efficacy. Second, 73 colonies of pseudomonas obtained from patient's wound were tested for phage susceptibility. Three different susceptibility patterns were observed: susceptibility, intermediate susceptibility and resistance. Among 45 colonies tested from five (50%) participants who did not reach the primary endpoint by day 7, 11 (24%) were susceptible, 27 (60%) were intermediately susceptible, and seven (16%) were resistant. The resistant pattern was not observed among 28 colonies isolated from five patients who reached the primary endpoint by day 7. This certainly emphasizes the need for a very specific tailored therapy to obtain a potential curative effect.

18.2.8 Treatment of Chronic Staphylococcus aureus Rhinosinusitis

The most recent published randomized trial was performed at a single tertiary referral center from December, 2015, through September 2016 in Adelaide, Australia (Ooi et al. 2019). This prospective, open-label, phase 1 clinical trial was conducted in 9 patients suffering from recalcitrant chronic rhinosinusitis, a condition challenged by difficult-to-treat infections because of the presence of biofilms and multidrug-resistant bacteria, previously treated by endoscopic sinus surgery. On top of usual care (intranasal twice daily irrigation), the patients were randomized to receive an anti-staphylococcus phage solution (AB-SA01) either at the concentration of 3×10^8 PFU for 7 days, 3×10^8 PFU for 14 days, or 3×10^9 PFU for 14 days.

Patients were not enrolled if the culture of nasal swabs were negative, positive for other bacteria or if the Staphylococcus was not sensitive to the phage solution. Treatment safety was assessed by a panel of clinical biochemistry tests conducted at the screening and exit visits, as well as by a close surveillance of fever. Treatment efficacy was evaluated by the semi-quantitative assessment of pretreatment and posttreatment bacterial culture swabs taken under direct endoscopic guidance. Also, the patients completed the Sino-Nasal Outcome Test, as well as a Visual Analog Scale on 6 items. Finally, an entry and exit video-endoscopy were performed, recorded and later analyzed by an independent blinded surgeon. Treatment was very well tolerated with only 6 adverse events reported by the patients. Those included diarrhea, epistaxis, oropharyngeal pain, cough and rhinalgia. Clinical exams did not change along the study and blood chemistry tests were unaffected by treatment (apart one patient who had a decreased bicarbonate level that normalized 8 days after the end of the treatment). Due to the limited number of patients, there were no statistics performed on the efficacy outcome parameters. Nevertheless, a decrease of bacterial growth was observed and 2 of 9 patients had negative cultures after treatment. Also a global improvement (apart for 1 patient) of the Sino-Nasal Outcome Test, as well as the Visual Analog Scale was described. Finally, a significant trend towards improvement in endoscopic results across all cohorts, with greatest improvement noted in the cohort of patients treated with the highest phage concentration, was noted. Four patients underwent a new evaluation 3 months after the last phage administration. All confirmed a continuing trend toward further improvement in all outcome measures.

18.2.9 Ongoing Projects

Leitner and colleagues recently published the protocol of a randomized controlled trial on the use of phage in the treatment of symptomatic urinary tract infection (Leitner et al. 2017). This protocol is a follow-up to a previous prospective two-phase study. The first phase of this latter study is an adaptation of the Pyo bacteriophage solution commercially available in Georgia (Eliava BioPreparations Ltd., Tbilisi, Georgia) while the second phase corresponds to the treatment of 9 out of the 130 initially screened patients (Ujmajuridze et al. 2018). For the first part, urine cultures from 130 patients, planned for transurethral resection of prostate, were evaluated. Ninety-one percent had positive urine cultures with pathogens such as Staphylococcus aureus, E. coli, Streptococcus spp., Pseudomonas aeruginosa or Proteus mirabilis. Bacteria sensitivity to the Pyo bacteriophage solution was assessed before it was further adapted to enhance efficacy and coverage toward uropathogenic strains that initially scored intermediate or resistant in previous in vitro sensitivity study. In the second part of the study, nine patients received the Pyo bacteriophage instillation by a health care provider two times a day, for 7 days, starting the first day after surgery, through a suprapubic catheter that was kept in place for 7 days to enable instillation. The solution of 20 mL was retained in the bladder for approximately 30–60 min. Prior bacterial culture revealed E. coli in four patients, Streptococcus species in two, Enterococcus species in two and Pseudomonas aeruginosa in one patient. Post treatment analysis demonstrated E. coli and Enterococcus spp. were still isolated from the urine culture of four and one patient, respectively. However, bacterial titers decreased significantly in two thirds of the patients and no bacteriophage-associated adverse events have been detected. A single patient required antibiotics as he developed fever above 38 °C at day 3.

With these interesting data in mind, the authors developed a protocol for a randomized, placebo-controlled, double-blind trial investigating bacteriophages in urinary tract infection treatment, to be conducted at the Tzulukidze National Center of Urology, Tbilisi, Georgia with monitoring and statistical support from the Balgrist University Hospital and Zurich University, Zürich, Switzerland. Patients planned for trans-uretral resection of prostate with positive urine culture showing a microorganism sensitive to the Pyo bacteriophage solution will be randomized to placebo, phage treatment or antibiotics for 7 days, starting on the day of surgery. Eighty-one patients will be enrolled, 27 per arm; each withdrawning patient will be replaced. This would be sufficient to demonstrate phages are more effective than placebo and as effective as antibiotics. While the antibiotic treatment arm representing common clinical practice is open label, the other two arms are blinded to the investigator. Pyo bacteriophage solution could be adapted according to the results of urine cultures performed during the study. Primary outcome is treatment efficacy defined by normalization of urine culture after 7 days of treatment. Secundary outcome include safety and adverse events while others outcome will assess bladder and pain assessment using a visual analog scale and an International Prostate Symptom Score at day 7 versus baseline.

Another interesting project of bacteriophage therapy, administered intravenously, has just received approval from the FDA. Scientists at the University of California San Diego (UCSD) School of Medicine will conduct the trial in collaboration with AmpliPhi Biosciences Corporation, a San Diego–based biotechnology company. In this phase 1 and 2 trial, safety, tolerability, and efficacy of an intravenous bacteriophage therapy will be assessed in 10 patients with resistant Staphylococcus aureus infected ventricular assist devices treated with antibiotics (Voelker 2019). The use of AB-SA01 phage solution (which will be used in the San Diego study) has already been administered in an Australian observational open label study of 13 critically-ill patients admitted to a tertiary-referral hospital because of *Staphylococcus aureus* bacteremia including 6 patients with infective endocarditis (Fabijan et al. 2019). In this study (which has not been published in a peer-review journal), and because of the expected poor outcome, compassionate access to therapy was approved by both U.S. and Australian regulators and by the Westmead Hospital Human Research Ethics Committee. Phage solution was administered twice daily for 14 days. Bacteriophage therapy coincided with a marked reduction in staphylococcal blood bacterial DNA and a decrease in sepsis-associated inflammatory responses in the majority of cases. There were no adverse events attributable to bacteriophage and development of staphylococcus strains resistant to phage treatment was not observed. Also, there was no significant effect of therapy on the gut microflora.

The UCSD recently launched with other research institutions and biotechnology companies the Center for Innovative Phage Applications and Therapeutics (IPATH), the first of its kind in North America, with a primary goal to conduct rigorous clinical trials of phage therapies. It follows the positive outcome after the treatment of some patients with phage at the UCSD, including in March 2016 the treatment of one of their colleague who developed life-threatening multi-drug resistant Acinetobacter baumanii infection after he suffered acute pancreatitis while on vacation in Egypt. The FDA approved the emergent treatment which turned successful. Since then, several other patients also were approved for the therapy. Other projects from IPATH include the treatment of cystic fibrosis patients colonized with Pseudomonas aeruginosa.

18.2.10 In Summary

Despite the increased interest in solutions to fight the development of multi-drug resistant bacteria, well-conducted clinical research on bacteriophage remain very limited by the number of studies and by the number of patients included. Furthermore, the results of the published trials may look disappointing.

One certainty which looks quite apparent in all the trials reported above is the potential safety of phage application, whatever the route of administration. Indeed, reported adverse events remain at a very low level.

Nevertheless, it looks like it is time to analyze the potential reasons for the poor results found in these various trials. It might be related to the source of administration, the dose, the way phages are preserved (with a potential decreased concentration), the interference with other treatments (such as in dressing), etc. This is important, in order to avoid that a very promising solution to treat patients with life-threatening infections would be abandoned. Analyzing the high number of successful case-report would probably help in defining the best way to undertake a new study.

References

Babalova EG, Katsitadze KT, Sakvarelidze LA, Imnaishvili NS, Sharashidze TG, Badashvili VA et al (1968) Preventive value of dried dysentery bacteriophage. Zh Mikrobiol Epidemiol Immunobiol 45(2):143–145

Bruynhoge R, Maisin J (1921) Essais de thérapeutique au moyen du bacteriophage du Staphylocoque. Compt Rend Soc Biol 85:1120–1121

Chanishvili N (2012) Phage therapy--history from Twort and d'Herelle through Soviet experience to current approaches. Adv Virus Res 83:3–40

Chanishvili N (2018) Early experience with phage therapy. Personal communication. ECCMID congress 2018, Madrid, Spain

Fabijan AP, Lin RCY, Ho J, Maddocks S, Iredell JR on behalf of the Westmead Bacteriophage Therapy Team (WBTT) and AmpliPhi Biosciences Corporation (2019) Safety and Tolerability of Bacteriophage Therapy in Severe Staphylococcus aureus Infection. https://www.biorxiv.org/content/10.1101/619999v1 ; Cited April 26, 2019

Jault P, Leclerc T, Jennes S, Pirnay JP, Que YA, Resch G et al (2019) Efficacy and tolerability of a cocktail of bacteriophages to treat burn wounds infected by Pseudomonas aeruginosa (PhagoBurn): a randomised, controlled, double-blind phase 1/2 trial. Lancet Infect Dis 19(1):35–45

Jin D, Chen C, Li L, Lu S, Li Z, Zhou Z et al (2013) Dynamics of fecal microbial communities in children with diarrhea of unknown etiology and genomic analysis of associated Streptococcus lutetiensis. BMC Microbiol 13:141

Kieser S, Sarker SA, Sakwinska O, Foata F, Sultana S, Khan Z et al (2018) Bangladeshi children with acute diarrhoea show faecal microbiomes with increased Streptococcus abundance, irrespective of diarrhoea aetiology. Environ Microbiol 20(6):2256–2269

Leitner L, Sybesma W, Chanishvili N, Goderdzishvili M, Chkhotua A, Ujmajuridze A et al (2017) Bacteriophages for treating urinary tract infections in patients undergoing transurethral resection of the prostate: a randomized, placebo-controlled, double-blind clinical trial. BMC Urol 17(1):90

McCallin S, Alam Sarker S, Barretto C, Sultana S, Berger B, Huq S et al (2013) Safety analysis of a Russian phage cocktail: from metagenomic analysis to oral application in healthy human subjects. Virology 443(2):187–196

McCallin S, Sarker SA, Sultana S, Oechslin F, Brüssow H (2018 Sep) 2018. Metagenome analysis of Russian and Georgian Pyophage cocktails and a placebo-controlled safety trial of single phage versus phage cocktail in healthy Staphylococcus aureus carriers. Environ Microbiol 20(9):3278–3293

Monsur KA, Rahman MA, Huq F, Islam MN, Northrup RS, Hirschhorn N (1970) Effect of massive doses of bacteriophage on excretion of vibrios, duration of diarrhoea and output of stools in acute cases of cholera. Bull World Health Organ 42(5):723–732

Morozova VV, Vlassov VV, Tikunova NV (2018) Applications of bacteriophages in the treatment of localized infections in humans. Front Microbiol 9:1696

Ooi ML, Drilling AJ, Morales S, Fong S, Moraitis S, Macias-Valle L et al (2019) Safety and tolerability of bacteriophage therapy for chronic rhinosinusitis due to Staphylococcus aureus. JAMA Otolaryngol Head Neck Surg. https://doi.org/10.1001/jamaoto.2019.1191

Rhoads DD, Wolcott RD, Kuskowski MA, Wolcott BM, Ward LS, Sulakvelidze A (2009) Bacteriophage therapy of venous leg ulcers in humans: results of a phase I safety trial. J Wound Care 18(6):237–238. 240–3

Rose T, Verbeken G, Vos DD, Merabishvili M, Vaneechoutte M, Lavigne R et al (2014) Experimental phage therapy of burn wound infection: difficult first steps. Int J Burns Trauma 4(2):66–73

Sarker SA, Brüssow H (2016) From bench to bed and back again: phage therapy of childhood Escherichia coli diarrhea. Ann N Y Acad Sci 1372(1):42–52

Sarker SA, McCallin S, Barretto C, Berger B, Pittet AC, Sultana S et al (2012) Oral T4-like phage cocktail application to healthy adult volunteers from Bangladesh. Virology 434(2):222–232

Sarker SA, Sultana S, Reuteler G, Moine D, Descombes P, Charton F et al (2016) Oral phage therapy of acute bacterial diarrhea with two coliphage preparations: a randomized trial in children from Bangladesh. EBio Med 4:124–137

Sulakvelidze AZ, Alavidze Z, Morris JG Jr (2001) Bacteriophage therapy. Antimicrob Agents Chemother 45:649–659

Ujmajuridze A, Chanishvili N, Goderdzishvili M, Leitner L, Mehnert U, Chkhotua A et al (2018) Adapted Bacteriophages for Treating Urinary Tract Infections. Front Microbiol 9:1832

Voelker R (2019) FDA Approves Bacteriophage Trial. JAMA 321(7):638

Wittebole X, De Roock S, Opal SM (2014) A historical overview of bacteriophage therapy as an alternative to antibiotics for the treatment of bacterial pathogens. Virulence 5(1):226–235

Wittig HJ, Raffeto JF, Baon R (1966) Bacteriophage therapy in infective asthma. JAMA 196(5):435

Wright A, Hawkins CH, Anggård EE, Harper DR (2009) A controlled clinical trial of a therapeutic bacteriophage preparation in chronic otitis due to antibiotic-resistant Pseudomonas aeruginosa; a preliminary report of efficacy. Clin Otolaryngol 34(4):349–357

Chapter 19
Bacteriophage Therapies Targets Multiple Diseases Caused by Protein Misfolding

Beka Solomon

Abstract Filamentous bacteriophages (Ff) are a group of related viruses which infect only gram-negative bacteria They are flexible filaments of about 900 nm long and 6–10 nm thick, similar to amyloid fibrils The similarity in characteristics and conformation between amyloids that are composed of different proteins without any sequence homology raised the hypothesis that filamentous phages may affect protein amyloids regardless of the protein from which they are composed. Indeed the filamentous phages may bind to a certain conformation or region which is common for several types of amyloids and effect their aggregation similar to conformational antibodies.

Moreover the filamentous bacteriophage proved to be an efficient and non-toxic viral delivery vector of antibodies to the brain, following the olfactory tract and an efficient immunocarrier for raising antibodies. The therapeutic potential of phages in amyloidogenic diseases, stems from their unprecedented ability to access the CNS, to induce a potent anti-aggregating effect, and from their lack of tropism to mammalian cells.

19.1 Introduction

Filamentous bacteriophages are flexible filaments of about 900 nm long and 6–10 nm thick, similar to amyloid fibrils The similarity in characteristics and conformation between amyloids that are composed of different proteins without any sequence homology raised the hypothesis that filamentous phage may affect protein amyloids regardless of the protein from which they are composed.

B. Solomon (✉)
Department of Molecular Microbiology and Biotechnology, Tel Aviv University, Ramat Aviv, Israel
e-mail: beka@tauex.tau.ac.il

G. Witzany (ed.), *Biocommunication of Phages*,
https://doi.org/10.1007/978-3-030-45885-0_19

19.2 Conformational Diseases

The biological function of cells depends on the correct folding of a network of thousands of proteins. The information required to fold a protein into a functional, specific three-dimensional structure is contained in its amino acid sequence. In general, proteins fold properly into their native conformation and, if they do not, the misfolding is corrected by chaperone proteins. Misfolding of a protein results in its aggregation and accumulation as protein deposits in diverse tissues (reviewed in Soto et al. 2006).

The conformational change may promote the disease by either gain of a toxic activity or by the lack of biological function of the natively folded protein (Soto et al. 2006). One of the hallmarks of conformational diseases is the deposition of protein into highly organized protein fibrils/protofibrils and plaques.

Alzheimer's disease, Parkinson's disease, Huntington's disease, transmissible spongiform encephalopathies (TSEs), serpin-deficiency disorders, hemolytic anemia, cystic fibrosis, diabetes type II, amyotrophic lateral sclerosis, and more than 10 other rare diseases belong to this family of diseases where the proteins involved do not share sequence or structural identity, all of them can adopt at least two different conformations without requiring changes in their amino acid sequence. The misfolded protein is rich in β-sheet conformation, while in α-helices the hydrogen bonds are between groups within the same strand, in β-sheets the bonds are between one strand and another. Since the second β-strand can come from a different region of the same protein or from a different molecule, formation of β-sheets is usually stabilized by protein oligomerization or aggregation. Indeed, the misfolded protein self-associates and becomes deposited in amyloid-like aggregates in diverse organs, inducing tissue damage and organ dysfunction. Understanding the mechanisms of structural conversion of a native, soluble protein, to an amyloidogenic state is crucial to the development of strategies to prevent aggregation. However, this goal has often been hampered by the fact that amyloid formation involves the conversion of a native protein into insoluble aggregates under conditions in which there is little accumulation of soluble intermediates in the aggregation process (Dobson 1999). Furthermore, due to their insoluble nature, the resulting amyloid aggregates are not usually amenable to detailed structural investigation. Previous studies of the aggregation of amyloidogenic peptides and proteins have largely relied on the use of various types of perturbing agents such as high temperature, acidic pH, and/or high concentrations of chemical denaturants (urea, guanidine hydrochloride).

Alzheimer's disease AD, the most studied disease in this family, is characterized by two major pathogenic features involving (a) processing of amyloid precursor protein APP to form neurotoxic amyloid-β peptides AβP and an aggregated insoluble polymer of AβP that forms senile plaques, (b) the formation of intraneuronal tau aggregates called tangles yielding deposits of neurofibrillary tangles NFT.

The current dominant theory of AD etiology and pathogenesis is related to the amyloid cascade hypothesis which states that overproduction of amyloid-beta-peptide AβP, or failure to clear this peptide, leads to Alzheimer's disease primarily through amyloid deposition, presumed to be involved in neurofibrillary tangles formation (Selkoe 1991) Amyloid-β Aβ plaque formation, is a complex kinetic and thermodynamic process The dependence of AβP polymerization on peptide-peptide interactions to form a β- pleated sheet fibril, and the stimulatory influence of other proteins on the reaction suggest that amyloid formation can be modulated. According to this hypothesis, the central event in the disease pathogenesis is an imbalance between Aβ production and clearance, with increased Aβ production in familial disease and decreased Aβ clearance in sporadic disease. Aβ oligomers could directly impair synaptic function, in addition to the inflammatory and oxidative stress caused by aggregated and deposited Aβ. These processes impair neuronal and synaptic function with resulting neurotransmitter deficits and cognitive symptoms. Tau pathology with tangle formation is regarded as a downstream event, but could contribute to neuronal dysfunction and cognitive symptoms.

An immunologic approach to lowering the levels of AβP monomers, oligomers, and higher aggregates was developed by Solomon group (Solomon et al. 1996,1997) and was supported by an increasing number of studies in APP transgenic mice (Schenk et al. 1999; Bard et al. 2000). We previously showed that site-directed monoclonal antibodies towards the N-terminal region of AβP bind preformed Aβ fibrils and lead to amyloid disaggregation into an amorphous state and inhibit the neurotoxic effect *in vitro* (Solomon et al. 1996,1997). Parenteral immunization of mice with synthetic human AβP(1-42) was initially shown to lead to an antibody response associated with striking clearance of Aβ deposits, or else their actual prevention if the mice were immunized very early in life (Schenk et al. 1999). Subsequent studies have confirmed and extended this approach by showing that AβP immunization can lower brain AβP burden and also improve learning deficits in the transgenic mice model of AD. The results of the first immunotherapy clinical trial for AD, AN1792, raise concerns about the safety and the efficacy of active immunization in elderly patients. The low number of vaccine responders, the relative low antibody titers generated even after multiple immunizations, and the seriousness of the adverse events all suggest that alternative approaches to immunotherapy should be pursued (Vasilevko and Cribbs 2006) Additional research is required to establish the most effective anti-AβP antibodies with respect to isotype, affinity, and epitope specificity with respect to limiting or reversing cognitive decline in patients with Alzheimer's disease (Weksler et al. 2005).

19.3 Filamentous Phage

Filamentous phage Ff are a group of related viruses which infect only gram-negative bacteria and specifically adsorb to the tip of pili. They are long and thin particles, threadlike, with no organization into head or spikes. Their genetic material consists of a single-stranded, closed circular deoxyribonucleic acid DNA molecule (Marvin and Hohn 1969; Marvin 1998)

The general appearance of bacteriophage fd fl, M13 is that of a flexible filament, about 900 nm long and 6–10 nm thick, with a molecular weight of 12 x 10^6 (Newman et al. 1977). Its coat consists of 2700–3000 copies of a major protein, the product of gene 8 g8p and four minor components. The coat envelopes one molecule of DNA that extends the total length of the virus and constitutes ca. 12% of its weight. The phage filamentous structure is highly sensitive to mechanical shearing and ultrasonic treatment. Also, in contrast to X-like phages, it is sensitive to detergents and organic solvents, especially chloroform. On the other hand, the virus is rather resistant to proteolytic and nucleolytic digestion, with the sole exception of subtilisin, which partially digests one of the minor coat proteins, the gene 3 product (Marvin and Hohn 1969). By using X-ray diffraction techniques, the molecular architecture of Ff has been extensively investigated (Marvin et al. 1974), the filament is a hollow protein cylinder with an outer diameter of 6 nm and an inner diameter of 2 nm. The DNA molecule is embedded in the cylinder along its longitudinal axis. The major coat protein subunits appear as flexible, helical rods elongated in an axial direction and sloping radially, so as to overlap each other and give an arrangement of molecules reminiscent of scales on a fish (Marvin et al. 1974). According to the results of Opella and co-workers (Opella et al. 1980), the DNA-protein interactions substantially immobilize the DNA packaged in phage fd. The coat protein subunits are held rigidly in the structure and display a limited mobility except for some aliphatic side chains with rapid rotations. The g8p in the phage is almost completely and continuously α-helical, with segments that vary somewhat in their orientation relative to the filament axis The native architecture of the coat seems to be primarily maintained by hydrophobic interactions between the individual major coat protein molecules (Rossomando and Zinder 1968).

Ff phages are continuously extruded through the host cell envelope in a process that couples assembly with export.Filamentous phage became a useful tool in biology due to phage display systems in which peptides or proteins are expressed on the surface of filamentous phage The first phage display library was constructed by Smith (1985), who cloned gene fragments into gIII of phage f1, creating a library of virions displaying peptides between the pIII N and C- terminal domains (Smith 1985; Manoutcharian et al. 2001). This technology has considerably facilitated the design of specific peptide ligands or antibody fragments that bind to receptors and target proteins *in vitro* and *in vivo* (Scott and Smith 1990).

Over the past two decades, many important developments in the field of phage display have occurred. Very large libraries of proteins of assorted sizes, ranging from small peptides and antibody fragments to functional enzymes, can now be displayed on various coat proteins of the phage particle, allowing for the identification of natural or artificial ligands for a wide range of receptors, the selection of antibodies with high specificity and affinity against a given antigen, the targeted delivery of foreign DNA into cells, and the selection of enzymes with improved functionality, just to name a few possibilities of phage display technology (Dickerson and Janda 2005). Their further application was hampered by several factors, of which the inability to remove endotoxins from preparations was predominant. More recently, animal studies have demonstrated that bacteriophage can rescue animals from a variety of fatal infections, (Smith and Huggins 1982) while studies conducted in Eastern Europe have shown that phage can be effective in treating drug-resistant infections in humans (Stone 2002).

The use of filamentous phage *in vivo* has been previously reported by Pasqualini and Ruoslahti (1996). In this work, bacteriophage displaying a random peptide library was intravenously injected into mice and subsequently rescued from the internal organs, showing that the integrity of the phage was not compromised. The ability of bacteriophage can penetrate a wide range of vertebrate tissues without detrimentally affecting the host (Dabrowska et al. 2005).

19.3.1 Filamentous Phage as Brain Delivery Vector

19.3.1.1 Intranasal Administration

The olfactory region of the nasal passages has both extracellular and intracellular pathways into the CNS that bypass the blood-brain-barrier. Olfactory sensory neurons are only first order neurons (neuroreceptors of the mucosa) whose cell bodies are located in a distal epithelium. Their dendritic processes are directly exposed to the external environment in the upper nasal passage while their axons project through perforations in the cribriform plate of the ethmoid bone to synaptic glomeruli in the olfactory bulb (Thorne and Frey 2001) Studies have shown that, fibroblast growth factor-2 (Jin et al. 2003), insulin (Sigurdsson et al. 1997), are able to gain access to or have effects in brain tissue or CSF following intranasal administration.

The extraneuronal pathway probably relies on transport through perineural channels, which deliver substances directly to the brain parenchymal tissue, enabling therapeutic agents to reach the CNS within minutes.

Intranasal delivery provides a practical, non- invasive method for delivering therapeutic by the olfactory and trigeminal nerves. Intranasal administration was chosen as a direct delivery route of vectors to the CNS via the olfactory neuron system or by neuron tissue. Olfactory receptor neurons are bipolar close cells that reside in the epithelial lining of the nose, high in the nasal cavity and seem to form a highway by which viruses or other transported substances may gain access to the CNS.

Intranasal administered drugs reach the parenchymal tissues of the brain and spinal cord and/or cerebrospinal fluid CSF within minutes using an extracellular route through perineural channels [(Frey 2002).

In addition to bypassing the BBB, the advantages of intranasal delivery include rapid delivery to the CNS, avoidance of hepatic first-pass drug metabolism, and elimination of the need for systemic delivery, thereby reducing unwanted systemic side effects. Intranasal delivery also provides painless and convenient self-administration by patients, features that encourage its use for delivering therapeutic agents into the CNS and is associated with several advantages:

(1) This mode of application is not invasive and is easy to perform; therefore patient compliance is higher compared to other routes of administration.

(2) Intranasal administration avoids hepatic/gastrointestinal first-pass elimination and/or degradation in the blood compartment, and therefore may provide extensive relative absorption for substrates that have poor oral bioavailability. (3) By being administered nasally these molecules might be able to circumvent the blood-brain barrier and reach a target in the brain to a higher extent than by other routes of administration. (Frey 2002). For example, it was shown that in a mouse model that [^{3}H]-dopamine reached the olfactory lobe after nasal administration and that 4 h after administration the concentration in this tissue after nasal administration was 27-times higher than after intravenous injection (Illum 2002) (4) Allows most of the drug to act in the brain first, decreases peripheral adverse effects and clearance.

For these reasons, many studies employ this strategy for delivering proteins into the CNS.

The nasal absorption of polar compounds is poor, with bioavailabilities not exceeding 10% for small molecular weight drugs (e.g. morphine, sumatriptan) and less than 1% for peptides such as insulin and calcitonin For higher molecular weight proteins, the nasal absorption has be shown to be even lower, although there is some evidence that even large proteins, such as HRP, can pass the membrane, albeit to a small extent (Illum 2002).

Such molecules normally pass the nasal membrane via the paracellular pathway, through the tight junctions. Therefore, it is exceptional and unprecedented that a large particle, filamentous phage MW~12 MDa can penetrate to brain parenchyma. This ability is probably due to its unique filamentous shape. Our data support the idea that phages are delivered to the brain through the extraneuronal pathway since ^{125}I -labeled phages were detected in mice brains within less than 1 h post-intranasal application. The intraneuronal pathway involves axonal transport and requires hours to days for drugs to reach different brain regions (Frey 2002; Frenkel and Solomon 2002).

19.3.1.2 Phage Distribution

Filamentous bacteriophage offers an obvious advantage over other vectors.

The filamentous phages, M13, f1, and fd, are well studied both structurally and genetically (Rodi and Makowski 1998). They were genetically engineered to display antigen and/or antibody in different biological systems (McCafferty et al. 1990; Scott and Smith 1990). The use of bacteriophage in mammalian systems is safe since its propagation cycle is limited to prokaryotic cells. The bacteriophage can neither replicate in, nor show natural tropism for, mammalian cells. This minimizes the chances of non-specific gene delivery and generation of a replication-competent entity in animals and, also, lacks clinical harmful side effects. The second advantage of bacteriophage is its ease of production. Even on a large scale, phage can be produced at high titer in bacterial culture, making production potentially simpler and more economical than either viral or non-viral systems. Phage particles are also stable under a variety of harsh conditions including extremes in pH, DNase, and proteolytic enzymes. But perhaps the most significant advantage is the genetic flexibility of filamentous phage which allows a wide variety of proteins, antibodies, and peptides to be displayed on the phage coat, thus allowing phage to be targeted genetically to a desired organ. In our study we followed phage distribution and elimination after intranasal application. The distribution of phages in peripheral organs was first characterized using radiolabeled phages. Following intranasal application to BALB/C mice, several peripheral organs were removed and their radioactivity, which was correlated to phage content, was measured by gamma counter. Phages were detected in all the investigated organs- lungs, gut, kidney, spleen and liver. Phages were predominantly found in lung and gut, probably because most of the dose was inhaled or swallowed. Loss of dose is one of the main pitfalls in i.n. administration. In order to overcome this obstacle, the mice were held in a supine position during delivery to maximize retention time of the dose in the nasal cavity.

Phage presence in spleen, liver, kidney and lung was also detected following intravenous administration (Molenaar et al. 2002). Molenaar et al. (2002), reported that after i.v. injection of M13 to C57BL/6 mice, phages were up taken mainly by spleen and liver, pointing at uptake by the reticuloendothelial system RES. This analysis is also echoed by Zou et al., who reported phage uptake predominantly by liver and spleen post i.v. injection (Zou et al. 2004). The RES is mainly active in liver, spleen and lungs and contains lysosomal system that is specifically equipped to take up, kill, and degrade invading microorganisms (Molenaar et al. 2002). Recovery of virulent phage in organs such as lungs, liver, and spleen was considerably lower than that of radioactivity, pointing at substantial disintegration of phages in these organs after uptake and processing by the RES

Filamentous bacteriophage proved to be an efficient and non-toxic viral delivery vector to the brain exhibiting penetration properties to the CNS (Frenkel and Solomon 2002) offering an obvious advantage over other mammalian vectors The bacteriophage lacks the ability to infect mammalian cells, unless designed to do so. Due to its linear structure, the filamentous phage is highly permeable to different kinds of membranes and, following the olfactory tract, it may directly target affected sites in the brain.

19.4 Therapeutic Applications of Filamentous Phage

19.4.1 Phage-Single Chain Antibody Against β-Amyloid Peptide

Filamentous bacteriophages are excellent vehicles for the expression and presentation of foreign peptides in a variety of biological systems. Large repertoires of peptides and antibody fragments are displayed on the phages. The ability of anti-β-amyloid single-chain (ScfV) phage to dissolve Aβ plaques in vivo was demonstrated by repeated intranasal administration to transgenic mice model of AD (Frenkel and Solomon 2002).

Intranasal treatment for 6 months with phage anti- AβP ScFv of transgenic mice over-expressing hAPP resulted in reduction of plaque load and considerable reduction of brain inflammation We showed a direct correlation between the number of applications and the amount of phage detected in the brain. The linear structure of the phage is suggested to confer penetration properties via various membranes. No evidence was shown of spread of filamentous phage to other brain sections, strongly emphasizing the olfactory tract as the most probable path in this case. In a control experiment, we performed intranasal administration of chloroform treated spheroid phages to the mice under the same experimental conditions and no presence of phages was detected (Griffith et al. 1981).

Several hypotheses may be considered regarding the disappearance of the filamentous phage from the brain without inducing any toxic effect, as well as the long life-span of challenged animals].

Many studies are being conducted to prove the safety and efficacy of filamentous phage as well as the pharmacokinetics of phage particles (Krag et al. 2002) evaluated the toxicity of repeated administrations of phage libraries to mice. The half-life of phage in plasma (in terms of recovered infectious particles) was calculated to be 3.6 h. After 72 h, phage was cleared from the blood, mainly through hepatic and renal excretion. This study demonstrated that injection of phages using a variety of regimens resulted in minimal toxicity.

This is the first demonstration that filamentous bacteriophage exhibits penetration properties to the CNS while preserving both the inert properties of the vector and the ability to carry foreign molecules. (Frenkel and Solomon 2002).

19.4.1.1 Phage Displaying –EFRH Epitope

. The recombinant filamentous phage approach for obtaining specific peptide antigens has a major advantage over chemical synthesis, as the products obtained are the result of the biological fidelity of translational machinery and are not subject to the 70–94% purity levels common in the solid-phase synthesis of peptides. In vivo administration of filamentous phages induces a strong immunological response to the phage proteins pIII and pVIII in all animals tested without any evidence of toxic effects (Meola et al. 1995, Delmastro et al. 1997, Frenkel et al. 2003) The display of short immunogenic determinants fused to the phage surface provides the basis for the development of novel peptide We have previously shown that the EFRH residues located at positions 3–6 of the N terminus of amyloid-β peptide AβP represent the main epitope of anti-aggregating mAbs (Frenkel et al. 1998) The interaction of this epitope with specific antibodies interferes with pathological effects in the central nervous system CNS, such as inflammatory events and other pathogenic mechanisms in Alzheimer's disease The immunomodulatory effect of the phage on the immune response against AβP is dependent on the number of epitopes displayed on the phage (Zuercher et al. 2000; Willis et al. 1993). Immunization with the EFRH phage may, in a short period (a few weeks), raise antibodies which recognize whole AβP, either if the epitope is displayed on pIII (10 copies) and/or pVIII on the phage 300 copies. The titer of such antibodies is proportional with the number of copies of EFRH on phage.

We have shown here that immunization with a filamentous phage that carries about 300 copies of the EFRH-epitope elicited high titers of antibodies against the Aβ peptide, operationally similar in their *in vitro* and *in vivo* anti-aggregating properties to monoclonal antibodies against the N–terminal region of AβP, raised by direct injection with fibrillar Aβ in the presence of adjuvant (Solomon et al. 1997; Bard et al. 2000).

Behavior of the animals was evaluated by testing the spatial and temporal navigation of each animal in the Morris water maze (MWM). A considerable improvement in their cognitive functions was obtained, dependent on the treatment. Regardless of the exact mechanism, it is clear that the reversal of memory impairment in these two memory tasks are due to the effect of EFRH immunization. The mice with relatively high level of antibodies to EFRH behaved similarly to non-transgenic mice in the MWM test.

19.5 Filamentous Phage Binds and Disaggregates Amyloid Plaques

19.5.1 In Vitro

Filamentous phage not only stabilizes antibody fragments and enables their penetration into the brain but we found that exhibits anti-aggregating properties inhibiting amyloid formation and dissolving already-formed aggregates of amyloid-β. The interactions between phage and Aβ1-40 were revealed by electron microscopy analysis (Fig. 19.1).

Although Aβ fibrils and filamentous phage are very different in their composition, their structures resembled one another- they both appeared as nanotubes. Incubation of phages with AβP in the prevention assay or with Aβ fibrils in the disaggregation test led to the same results – a marked reduction in size and density of Aβ aggregates and the appearance of amorphous depositions of Aβ, as can be viewed in figure. Additionally, the micrographs showed phage interactions with Aβ fibrils at the interface point.

It seemed that phages were organized in bundles (arrow head) which were dispersed to single phage (arrow) upon contact with the amyloid fibrils. The phage are rearranged parallel to the amyloid axis (open arrow). Amyloid cluster around the phages was smaller and composed of shorter narrow amyloid fibrils (Goren 2008).

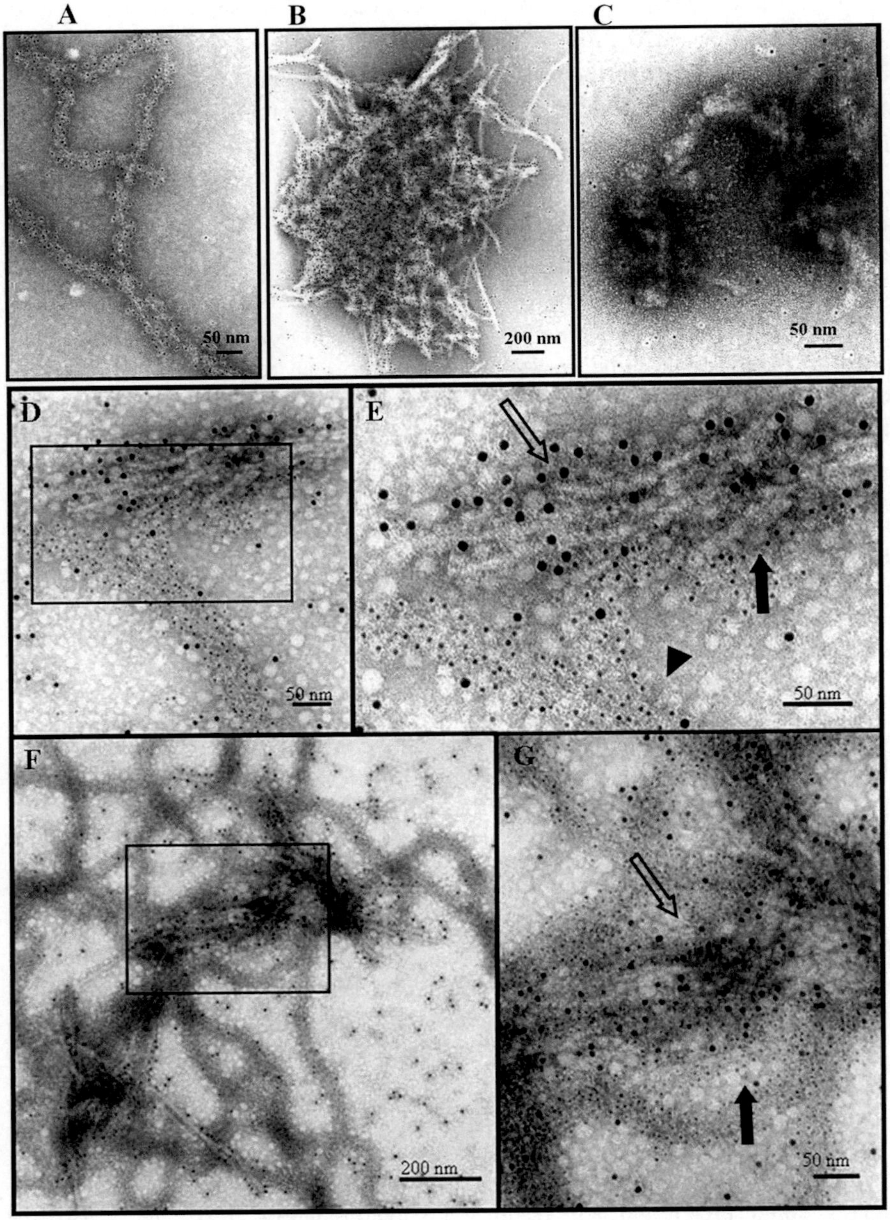

Fig. 19.1 Electron micrographs of Aβ1-40 interactions with M13 filamentous phage Aβ 1-40 was stained with mAb 196 followed by goat anti-mouse 12 nm gold conjugated antibody; M13 particles were labelled with rabbit polyclonal serum followed by goat anti-rabbit polyclonal antibodies (pAb) conjugated to 6 nm gold particles

(**a**) Filamentous phage (10^{12}/ml)

(**b**) Packed aggregates of fibrillar Aβ (97 μM) in PBS after 9 days of incubation at 37 °C

(**c**) Amorphous aggregate of Aβp after incubation with phages

(**d**, **e**) Aβ and filamentous phage interactions. Filamentous phages (1×10^{10} phages) were organized in groups and upon contact with amyloid fibrils dispersed and rearranged along the amyloid fibrilar axis

(**f**, **g**) Magnification of phage and Aβ interface

In vitro anti-aggregating properties of filamentous phage on dissolving and/or preventing Aβ fibril formation were followed by thioflavin ThT experiments. A decline of 26% in Aβ aggregation was observed when the peptide was incubated in the presence of filamentous phage while addition of phage to preaggregated Aβ resulted in 45% reduction in amyloid fibrils. We demonstrate the modulating effect of filamentous phage M13 on amyloid disaggregation of amyloid Aβ. Incubating the phages in chloroform/water solution we modified the linear filamentous structure into a spherical one (Griffith et al. 1981), and abolished its disaggregating as well as penetrating abilities.

19.5.2 *In Vivo*

A large-scale experiment was employed to investigate phage therapeutic effect on PDAPP mice expressing hAPP bearing London mutation that causes an autosomal dominant form of familial of Alzheimer's disease.

Similar to humans with AD, PDAPP mice show an age-related accumulation of diffuse and neuritic plaques beginning at ~6–9 months of age, glial activation, and have learning deficits before plaque deposition (Games et al. 1995; Hartman et al. 2005).

Nine-month old PDAPP transgenic mice were pipette-inhaled with 10^{11} phage in a volume of 20 μl. The mice were treated nasally, every 2 weeks for the first 2 months, and then monthly for an additional 10 months. During this period, cognitive and behaviour trials and odor-treshold examinations were exploited to follow phage therapy efficacy. Olfactory neurons are essential for many functions, as well as orientation and recognition Brain areas involved in olfactory function are situated in medial temporal regions that undergo early neuropathological change in Alzheimer's disease. Correspondingly, patients with Alzheimer's disease develop an olfactory dysfunction early in the disease. Olfactory dysfunction in Alzheimer's disease correlates with disease progression, aids in the differential diagnosis of Alzheimer's disease versus major depression, and may be clinically useful as an early diagnostic marker in predicting the incident of Alzheimer's disease in high-risk individuals (Peters et al. 2003). Based on this data, odor-threshold test was employed, in which the mice were exposed to increasing concentrations of mint oil that normally repel mice and trigger them to turn to the opposite direction.

Once plaques were cleared, neuronal morphology was restored, which may have had a direct impact on cognitive function. The group of mice that were treated with phage, reacted to mint in lower concentrations than the control group. While ED_{50} of the phage-treated mice was 1% mint oil solution, 50% of control mice were repelled only by 10% solution. This experiment demonstrated phage ability to protect neurons in the olfactory tract and connected areas and therefore prevented olfactory dysfunction.

Biochemical and histochemical analysis of the mice revealed additional therapeutic advantages of phage treatment.

A direct correlation was shown between the number of intranasal applications and the amount of phage detected in the hippocampus and olfactory bulb. It is worth mentioning that no pathological signs were observed in those mice brains due to phage administration (Goren 2008).

Here, we demonstrate that filamentous phage not only acts as an inert vector that stabilizes peptide/antibody fragments and enables their penetration into the brain, but also interfered with amyloid formation and dissolved already formed aggregates both *in vitro* and *in vivo*. Phage application to transgenic mice, which are used as a model for the disease, resulted in improvement in "AD-like" symptoms and may thus serve as a potential therapeutic agent.

.Although phages were detected in many peripheral organs, they do not seem to cause any adverse effects. Administration of phages to PDAPP mice, did not lead to any detectable pathological symptoms. Mice weight was not statistically different from control group values, hematoxylin and eosin staining did not reveal abnormal morphology, and neither hemorrhages nor inflammation signs were identified.

Filamentous phage is a very immunogenic particle, however intranasal application prevents its rapid clearance from the body before it reaches its target organ -the brain. The lag period following phage administration in primary and secondary immune response lasts days (4–7 days and 1–3 days respectively), while phage penetrates into brain parenchyma within less than an hour. Therefore, anti-phage antibodies are produced after it leaves the brain.

We trace phage kinetics post intranasal administration to verify that phage penetrates the brain in a reasonable time and in a sufficient amount for therapeutic purposes. To this aim, ^{125}I radiolabelled phage were administered i.n. to BALB/C mice. Surprisingly in spite of its enormous size (thousands of kDa), M13 succeeded to infiltrate the brain in a short period of time and was eliminated almost completely within 24 h. Its clearance rate may have been influenced by the CSF turnover, which is about 2 h in mouse brain (Illum 2000). Since in wild-type mouse brain there are no targets for phage that can lead to its binding and accumulation, it is reasonable that phages were cleared quickly from the brain, and that detected radioactivity was very low 24 h post i.n. application.

The percentage of administered dose taken up per gram of brain (%dose/g) was calculated for each time-point. A maximum of 0.025% dose/g penetrated the brain within an hour. This compares with values of 0.018% for morphine and /or 0.08% for interleukin-1α following i.v. injection. All these substances have profound effects on the CNS (Banks et al. 2002).

The proposed new approach of intranasal application of phage presents several important advantages.

First, it provides a rapid delivery route into the central nervous system due to the unique connection between the nose and brain. A second advantage of using intranasal application is that the olfactory system, which was shown to be one of the first brain parts affected by Alzheimer, s disease, is targeted first. Moreover, phage therapy may overcome some of the drawbacks of β-amyloid immunotherapy, such as haemorrhages and inflammation.

19.5.3 In a Cellular Model

Parkinson's disease PD is the second most common neurodegenerative disease following Alzheimer's disease. It affects approximately 0.3% of the population and 1–2% of those older than 60 years. It is anticipated that the prevalence of PD cases will increase dramatically in the coming decades.

Pathological features of PD include degeneration of dopaminergic neurons in the substantia nigra pars compacta coupled with intracytoplasmic inclusions known as Lewy bodies. (Goedert 2001). Synuclein AS the main component of such aggregates is an abundant neuronal protein predominantly localized in presynaptic terminals in the central nervous system.(El-Agnaf et al. 2006) In order to evaluate the potential of filamentous phages to modulate protein aggregation and clearance we used a cellular model expressing wild-type AS. The exact mechanism by which phages modulate AS aggregation in the cellular model is not fully understood.AS tends to form intracellular aggregates, ranging from small oligomers to larger mature aggregates, in both the soluble and insoluble fractions. Intracellular AS aggregates, unlike extracellular Aβ plaques, require phage internalization into cells. This poses a problem since filamentous phages have no natural tropism to mammalian cells and cannot reach the intracellular AS aggregates. A fd filamentous phage was engineered to present a cyclic RGD sequence on ~150 copies of the phage major coat protein (pVIII) (Hart et al. 1994). This sequence binds integrin receptors on the cell membrane and triggers endosomal internalization into the cells. This route of entry into the cells is utilized by a large number of viruses, leading to the hypothesis that phages that are able to bind integrins will internalize into cells in a similar mechanism. Indeed RGD phage either disaggregate AS oligomers, as demonstrated on AS dimers and trimers, or simply prevent the oligomerization of AS monomers (Dimant et al. 2009; Dimant and Solomon 2010).

The effect of RGD phages on the total amount of AS was examined in Western blot led to the unexpected finding that the phage itself could disaggregate intracellular alpha-synuclein aggregates.

19.6 Mechanism of Action

Whether the clearing of plaques by antibody or phage M13 involves related protein–protein interactions, is largely unknown, but there are clearly striking differences between the two different interactions. Alpha-synuclein like beta-amyloid, transthyretin, and the tau proteins can aggregate to fibrils, which are thought to interfere with their normal neuronal functions. Fibrils can be formed with the central hydrophobic domain of the proteins/peptides and interestingly, fibrils resemble the filamentous structure of the phage. When phage was incubated with a solution of fibrils, a disaggregation of fibrils can be observed The exact mechanism by which phages modulate AS or Abeta aggregation in the cellular model is not fully understood and several issues still need further investigation.

The phage incubated with amyloid fibers showed that its binding was specific to the fibrils and not to the monomer (Krishnan et al. 2014).

By running a series of experiments to test which mutants bound to the amyloid and which ones didn't, Krishnan et al. (2014)were able to figure out that the phage's special abilities involved a set of proteins displayed on the tip of the virus, called GP3. They tested the different variants for examples of phages with or without tip proteins, and found that every time we messed around with the tip proteins, it lowered the phage's ability to attach to amyloids. Because it was already known that the phage minor coat protein, initially called protein A,now gene III product, is critical for the interaction of the phage with the pili of *E. coli*, one could envision that a similar interaction could occur with the fibrils of alpha-synuclein and beta-amyloid.

19.7 Conclusion

The effect of filamentous phage on the protein aggregation was being intensely studied in our laboratory. So far, much study has been invested both *in vitro* and *in vivo*, demonstrating the disaggregation of extracellular and intracellular aggregates, which emphasised the feasibility of filamentous phage as an aggregation modulator of such aggregates, via intranasal administration, opening further possible applications of the phage. Furthermore, according to our data, it appears that phages may also affect aggregated protein present on the cellular membrane and protein misfolding diseases where the protein aggregates within the cell.

To conclude, we present here a novel property of the phages to act as anti-aggregating agents of intra/extracellular aggregates without any adverse effects on treated cells, opening a new research area in the quest to discover new disease modifying agents for the treatment of amyloidogenic diseases in addition of conformational antibodies (Kayed et al. 2010). Astonishingly, the simple M13 virus appeared in principle to possess the properties of a "pan therapy," a universal elixir of the kind the chemist Chris Dobson had imagined.

References

Banks WA, Terrell B, Farr SA, Robinson S, Nonaka N, Morley E (2002) Passage of amyloid beta protein antibody across the blood-brain barrier in a mouse model of Alzheimer's disease. Peptides 23:2223–2226

Bard F, Barbour R, Cannon C, Carretto R, Games D, Guido T, Hoenow K, Hu K, Johnson-Wood K, Khan K, Kholodenko LC, Lee M, Motter R, Nguyen M, Reed A, Schenk D, Tang P, Vasquez N, Seubert P, Yednock T (2000) Epitope and isotype specificities of antibodies to β-amyloid

peptide for protection against Alzheimer's disease-like neuropathology. Proc Natl Acad Sci U S A 100:2023–2028
Dabrowska K, Switala-Jelen K, Opolski A, Weber-Dabrowska B, Gorski A (2005) Bacteriophage penetration in vertebrates. J Appl Microbiol 98:7–13
Delmastro P, Meola A, Monaci P, Cortese R, Galfre G (1997) Immunogenicity of filamentous phage displaying peptide mimotopes after oral administration. Vaccine 15:1276–1285
Dickerson TJ, Janda KD (2005) Recent advances for the treatment of cocaine abuse: central nervous system Immunopharmacotherapy. AAPS J 7:E579–E586
Dimant H, Solomon B (2010) Filamentous phages reduce α-Synuclein Oligomerization in the membrane fraction of SH-SY5Y cells. Neurodegener Dis 7:203–205
Dimant H, Sharon N, Solomon B (2009) Modulation effect of filamentous phage on a-synuclein aggregation Biochem. Biophys Res Comm 383:491–496
Dobson CM (1999) Protein misfolding, evolution and disease. Trends Biochem Sci 24:329–332
El-Agnaf SA, Salem KE, Paleologou MD, Curran MJ, Gibson JA, Court MG, Schlossmacher MG, Allsop D (2006) Detection of oligomeric forms of {alpha}-synuclein protein in human plasma as a potential biomarker for Parkinson's disease. FASEB J 20:419–425
Frenkel D, Solomon B (2002) Filamentous phage as vector-mediated antibody delivery to the brain. Proc Natl Acad Sci U S A 99:5675–5679
Frenkel D, Balass M, Solomon B (1998) N-Terminal EFRH sequence of Alzheimer's β-amyloid peptide represents the epitope of its anti-aggregating antibodies. J Neuroimmunol 88:85–90
Frenkel D, Dewachter I, Van Leuven F, Solomon B (2003) Reduction of beta-amyloid plaques in brain of transgenic mouse model of Alzheimer's disease by EFRH-phage immunization. Vaccine 7:1060–1065
Frey WI (2002) Bypassing the blood-brain barrier to deliver therapeutic agents to the brain and spinal cord. Drug Delivery Technol 2:46–49
Games D, Adams D, Alessandrini R, Barbour R, Berthelette P, Blackwell C, Carr T, Clemens J, Donaldson T et al (1995) Alzheimer-type neuropathology in transgenic mice overexpressing V717F beta-amyloid precursor protein. Nature 373:523–527
Goedert M (2001) Alpha-synuclein and neurodegenerative diseases. Nat Rev Neurosci 2:492–501
Goren O (2008) PhD thesis Submitted to Tel Aviv University
Griffith J, Manning M, Dunn K (1981) Filamentous bacteriophage contract into hollow spherical particles upon exposure to a chloroform-water interface. Cell 23:747–753
Hart AM, Knight RP, Harbottle A, Mistry HD, Hunger DF, Cutler R, Williamson R, Coutelle C (1994) Cell binding and internalization by filamentous phage displaying a cyclic Arg-Gly-Asp-containing peptide. J Biol Chem 269:12468–12474
Hartman RE, Izumi Y, Bales KR, Paul SM, Wozniak DF, Holtzman DM (2005) Treatment with an amyloid-beta antibody ameliorates plaque load, learning deficits, and hippocampal long-term potentiation in a mouse model of Alzheimer's disease. J Neurosci 25:6213–6220
Illum L (2000) Transport of drugs from the nasal cavity to the central nervous system. Eur J Pharm Sci 11:1–18
Illum L (2002) Nasal drug delivery: new developments and strategies. Drug Discov Today 7:1184–1189
Jin K, Xie L, Childs J, Sun Y, Mao XO, Logvinova A, Greenberg DA (2003) Cerebral neurogenesis is induced by intranasal administration of growth factors. Ann Neurol 53:405–409
Kayed R, Canto I, Breydo L, Rasool S, Lukacsovich T, Wu J, Albay R 3rd, Pensalfini A, Yeung S, Head E, March JL, Glabe CG (2010) Conformation dependent monoclonal antibodies distinguish different replicating strains or conformers of prefibrillar Abeta oligomers. Mol Neurodegener 5:57. https://doi.org/10.1186/1750-1326-5-57
Krag DN, Fuller SP, Oligino L, Pero SC, Weaver D, Soden AL, Hebert C, Mills LC, Peterson D (2002) Phage-displayed random peptide libraries in mice: toxicity after serial panning. Cancer Chemother Pharmacol 50:325–332
Krishnan R, Tsubery H, Proschitsky MY, Asp E, Lulu M, Gilead S, Gartner M, Waltho JP, Davis PJ, Hounslow AM, Kirschner DA, Inouye H, Myszka DG, Wright J, Solomon B, Fisher RA

(2014) A bacteriophage capsid protein provides a general amyloid interaction motif (GAIM) that binds and remodels misfolded protein assemblies. J Mol Biol 426:2500–2519
Manoutcharian K, Gevorkian G, Cano A, Almagro JC (2001) Phage displated biomolecules as preventive and therapeutics agents. Curr Pharm Biotechnol 2:217–223
Marvin DA (1998) Filamentous phage structure, infection and assembly. Curr Opin Struct Biol 8:150–158
Marvin D, Hohn B (1969) Filamentous bacterial viruses. Bacteriol Rev 33:172–209
Marvin D, Pigram W, Wiseman R, Wachtel E, Marvin F (1974) Filamentous bacterial viruses. XII. Molecular architecture of the class I (fd, Ifi, Ike) virion. J Mol Biol 88:581–598
McCafferty J, Griffiths AD, Winter G, Chiswell D (1990) Phage antibodies: filamentous phage displaying antibody variable domains. Nature 348:552–554
Meola A, Delmastro P, Monaci P, Luzzago A, Nicosia A, Felici F, Cortese R, Galfre G (1995) Derivation of vaccines from mimotopes. Immunological properties of human hepatitis virus surface antigen mimotopes displayed on filamentous phage. J Immunol 154:3162–3172
Molenaar TJ, Michon T, de Haas SA, van Berkel TJ, Kuiper J, Biessen EA (2002) Uptake and processing of modified bacteriophage M13 in mice: implications for phage display. Virology 293:182–191
Newman J, Swinney H, Day L (1977) Hydrodynamic properties and structure of fd virus. J Mol Biol 116:593–606
Opella SJ, Cross T, Di Verdi J, Sturm C (1980) Nuclear magnetic resonance of the filamentous bacteriophage fd. Biophys J 32:531–548
Pasqualini R, Ruoslahti E (1996) Organ targeting in vivo using phage display peptide libraries. Nature 380:364–366
Peters JM, Hummel T, Kratzsch T, Lotsch J, Skarke C, Frolich L (2003) Olfactory function in mild cognitive impairment and Alzheimer's disease: an investigation using psychophysical and electrophysiological techniques. Am J Psychiatry 160:1995–2002
Rodi DJ, Makowski L (1998) Phage-display technology finding a needle I in a vast molecular haystack. Curr Opin Biotechnol 10:87–93
Rossomando E, Zinder N (1968) Studies on the bacteriophage flI. Alkali-induced disassembly of the phage into DNA and protein. J Mol Biol 36:387–399
Schenk D, Barbour R, Dunn W, Gordon G, Grajeda H, Guido T, Hu K, Huang J, Johnson-Wood K, Khon K, Kholodenko D, Lee M, Liao Z, Lieberburg I, Motter R, Mutter L, Soriano F, Shopp G, Vasquez N, Vandevert C, Walker S, Wogulis M, Yednock T, Games D, Seubert P (1999) Immunization with amyloid-beta attenuates Alzheimer-disease-like pathology in the PDAPP mouse. Nature 400:173–177
Scott JK, Smith GP (1990) Searching for peptide ligands with an epitope library. Science 249:386–390
Selkoe DJ (1991) The molecular pathology of Alzheimer's disease. Neuron 6:487–498
Sigurdsson P, Thorvaldsson T, Gizurarson S, Gunnarsson E (1997) Olfactory absorption of insulin to the brain. Drug Deliv 4:195–200
Smith GP (1985) Filamentous fusion phage; novel expression vectors that display cloned antigens on the virion surface. Science 228:1315–1317
Smith HW, Huggins RB (1982) Successful treatment of experimental E coli infections in mice using phage: its general superiority over antibiotics. J Gen Microbiol 128:307–318
Solomon B, Koppel R, Hanan E, Katzav T (1996) Monoclonal antibodies inhibit in vitro fibrillar aggregation of the Alzheimer beta-amyloid peptide. Proc Natl Acad Sci U S A 93:452–455
Solomon B, Koppel R, Frankel D, Hanan-Aharon E (1997) Disaggregation of Alzheimer beta-amyloid by site-directed mAb. Proc Natl Acad Sci U S A 94:4109–4112
Soto C, Estrada L, Castilla J (2006) Amyloids, prions and the inherent infectious nature of misfolded protein aggregates. Trends Biochem Sci 31:150–155
Stone R (2002) Stalin's forgotten cure. Science 298:728–731
Thorne RG, Frey WH (2001) Delivery of neurotrophic factors to the central nervous system: pharmacokinetic considerations. Clin Pharmacokinet 40:907–946

Vasilevko V, Cribbs DH (2006) Novel approaches for immunotherapeutic intervention in Alzheimer's disease. Neurochem Int 49:113–126

Weksler ME, Gouras G, Relkin NR, Szabo P (2005) The immune system, amyloid-β peptide, and Alzheimer's disease. Immunol Rev 205:244–256

Willis EA, Perham NR, Wraaith D (1993) Immunological properties of foreign peptides in multiple display on a filamentous bacteriophage. Gene 128:79–83

Zou J, Dickerson MT, Owen NK, Landon LA, Deutscher SL (2004) Distribution of filamentous phage peptide libraries in mice Mol. Biol Reprod 31:121–129

Zuercher AW, Miescher SM, Vogel M, Rudolf MR, Stadler MB, Stadler BM (2000) Oral anti-IgE immunization with epitope-displaying phage. Eur J Immunol 30:128–135

Chapter 20
Phage Therapy in Cystic Fibrosis. Challenges and Perspectives

Ersilia V. Fiscarelli, Martina Rossitto, and Paola Rosati

Abstract Multidrug-resistant *Pseudomonas aeruginosa* (MDR PA) produces biofilm thus causing colonization, pulmonary failure and death in patients with cystic fibrosis (CF). Despite prolonging CF survival, antibiotics fail to eradicate MDR PA. A therapeutic alternative is phage therapy (PT). During decades, although testing lytic phages improved their efficacy in vitro to lyse PA biofilm, how they can persistently act in vivo, even in personalized approaches, remain debatable. We report our appraised results from 22 papers on lytic phages tested in vitro and in vivo on pulmonary and non-pulmonary models infected by various PA strains. Challenges and perspectives on PT in CF to advance custom-based approaches to treat or prevent MDR PA, combined with or alternative to antibiotics, are then discussed. To overcome barriers in international regulatory bodies on phages as medicines, we claim a new definition of advanced phage biological therapy aiming to reach a global consensus thus advancing PT in CF.

20.1 Introduction

Cystic fibrosis (CF) is a genetically determined multi-organ disease affecting lungs and the digestive system. CF affects about 30,000 people in the United States, with about 1000 new cases diagnosed every year (NHLBI and CF Foundation 1995). Although CF occurs in persons of any racial and ethnic backgrounds, it is very common in Caucasians and from Northern European ancestry (1/3000 births). In the past decades, generally the diagnosis of CF, happened at an average age of five years, owing to respiratory symptoms, though some people, without a lung disease, could discover to be affected only in their adulthood (NHLBI and CF Foundation 1995). Since the 2000s, several Western countries developed newborn CF screening programs for early diagnoses (Farrell et al. 2001; Merelle et al. 2001). Although screenings and advances in antibiotics and other symptomatic treatments can permit

E. V. Fiscarelli · M. Rossitto · P. Rosati (✉)
Cystic Fibrosis Microbiology, Laboratory Department, Unit of Clinical Epidemiology, Bambino Gesu Children's Hospital, IRCCS, Rome, Italy
e-mail: evita.fiscarelli@opbg.net; martina.rossitto@opbg.net; paola.rosati@opbg.net

G. Witzany (ed.), *Biocommunication of Phages*,
https://doi.org/10.1007/978-3-030-45885-0_20

a prolonged survival in over half of the population affected into at least their fifth decade of life, CF is still a fatal disease (Keogh et al. 2018).

In 1989 the CF gene was identified as the transmembrane conductance regulator (CFTR) gene in chromosome 7 (Kerem et al. 1989; Tsui 1995). The CFTR mutations cause the dysfunction of the chloride channel, preventing the normal chloride ions flow and water into and out of cells, hence causing a thick and sticky mucus which inevitably traps bacteria and leading to progressive respiratory decline and death (Frederiksen et al. 1997; Murray et al. 2007; Moreau-Marquis et al. 2008). Among the bacteria colonizing CF lungs, *Pseudomonas aeruginosa* (PA), an opportunistic environmental pathogen, represents the major pathogen because it typically colonizes 30% of the children affected, and up to 80% of 25-year-old and older adults with CF (Gibson et al. 2003; Stuart et al. 2010). The various PA phenotypes typically produce biofilm, namely a deleterious, complex tower- or mushroom-like matrix enclosing sessile PA aggregates. From these aggregates, some surface motile planktonic PA cells can leave the biofilm and colonize new lung sites, hence new sessile PA micro-colonies can trigger repeated lung infections (Costerton et al. 1999, 2003) (Table 20.1).

The prolonged and repeated PA chronic infections in CF need to be treated with broad-spectrum antibiotic courses to prevent aggressive lung disease progression, thus incurring in antibiotic resistance, and progressively leading to non-mucoid and mucoid multidrug-resistant (MDR), extensively drug resistant (XDR) and the recently isolated pandrug-resistant (PDR) PA strains (Donlan and Costerton 2002; Li et al. 2005; Gómez and Prince 2007; Magiorakos et al. 2012; Rasamiravaka et al. 2015; Djebara et al. 2019). MDR PA is non-susceptible to at least one agent in three or more antimicrobial classes: β-lactam, aminoglycosides, cephalosporins, fluoroquinolones, and carbapenems (Sievert et al. 2013), whereas XDR is non-susceptibility to at least one agent in all but two or fewer antimicrobial categories (bacterial isolates remain susceptible to only one or two categories) (Magiorakos et al. 2012) and PDR is non-susceptible to all agents in all antimicrobial categories (Djebara et al. 2019).

Although international recommendations to avoid repetitive acute and chronic lung infections in CF, including MDR, XDR and PDR organism prevention and control (Saiman et al. 2014), recent findings on PA evolutionary adaptation, diversification and resistance factors in CF lungs, and culture-independent genome sequencing approaches (mainly16S rRNA gene sequencing techniques) have provided new insight into the nature of polymicrobial biofilm-like infections in the CF airway (Smith et al. 2006; Walker et al. 2005; Winstanley et al. 2016; Siegel et al. 2017; O'Toole 2018), research interest is waning on new antibiotics to include in pipeline programs (Theuretzbacher 2009; Bassetti et al. 2011). Hence, to combat CF MDR PA infections, extensive research has reappraised bacteriophage (phage) therapies as a new weapon to treat CF chronic lung infections caused by MDR PA and other multi-resistant superbugs (Sulakvelidze et al. 2001; Thiel 2004; Hurley et al. 2012; Rossitto et al. 2018). Despite suggesting research on phages as promising antibacterial therapy since the first phage discovery about a century ago (Twort 1915; d'Hérelle 1917; Clokie et al. 2011), and the knowledge that phages hosted by

Table 20.1 *Pseudomonas aeruginosa* (PA) phenotypes, infection stages, virulence factors, evolution, biofilm production and composition, metabolic impact, physiopathology, and survival strategies in cystic fibrosis (CF) lungs

PA phenotypes (CF infection stage)[a]	PA virulence factors, and evolution	PA biofilm[b] production, composition, and metabolic impact	Typical CF lung physiopathology, and PA survival strategies	References (first author and publication year)
Environmental non-mucoid PA, wild-type (first acquisition)	Lipopolysaccharide capsule, adhesin, flagella, fimbriae, extracellular enzyme production (i.e. elastase, proteases, exotoxins)	PA biofilm is lacking, and planktonic cell PA are fluctuant (free-swimming)	Thick dehydrated mucus causes muco-ciliary dyskinesis, and progressively favors PA biofilm production	Anderson (2012), Chmiel and Davis (2003), and Gellatly and Hancock (2013)
Non-mucoid and mucoid PA (intermittent and recurrent infection)	Mucoid PA can overproduce alginate and biofilm matrix	Biofilms in tower- and mushroom-like micro-colonies containing embedded self-aggregated cells in a sessile form, and embedded in a matrix including DNA, polymers (i.e. F-actin) from dying neutrophils, extracellular polymeric substance (EPS) (containing three polysaccharides: alginate, Pel and Psl), proteins and extracellular DNA increasing biofilm volume, viscosity and progressive increase in growth	Progressive muco-ciliary dyskinesis and biofilm production and overproduction	Battán et al. (2004), Chmiel and Davis (2003), Cochran et al. (2000), Costerton et al. (1999, 2003), Davies et al. (1989), Donlan and Costerton (2002), Furukawa et al. (2006), Gibson et al. (2003), Høiby et al. (2010), Jesaitis et al. (2003), Parks et al. (2009), Rasamiravaka et al. (2015), Walker et al. (2005), Wei and Ma (2013), and Yoon and Hassett (2004)
		Biofilm preserves PA from leukocyte defensive action, reactive oxygen species, and nitric oxide		
		PA planktonic cells can leave the biofilm, enter the bulk fluid and colonize new lung sites thus initiating new sessile PA micro-colonies		

(continued)

Table 20.1 (continued)

PA phenotypes (CF infection stage)[a]	PA virulence factors, and evolution	PA biofilm[b] production, composition, and metabolic impact	Typical CF lung physiopathology, and PA survival strategies	References (first author and publication year)
Non-mucoid converted into mucoid PA, or small-colony variant PA (chronic colonization)	Small-colony variant PA are slow growing, hyperpiliated, hyperadherent, autoaggregative, and overproduce alginate		Owing to lack of oxygen and nutrients, the CF lung environment affects PA metabolism, and influences oxygen-requiring drug activity and metabolically active cells	Boles et al. (2004), Gooderham and Hancock (2009), Mah and O'Toole (2001), Moskowitz et al. (2004), Smith et al. (2006), Starkey et al. (2009), Walters et al. (2003), Winstanley et al. (2016) and Xu et al. 2000
	PA evolutionary adaptation and diversification by gene mutations involved in lipopolysaccharide and pyoverdine biosynthesis, exotoxin A regulation, flagella-lacking and transcriptional regulation (quorum-sensing) factors. All these changes enable PA to persist, survive and reduce its acute virulence (metabolic fitness)		PA mutations, favor PA persistence in the lungs, immune evasion capacity, support the PA long-term survival "insurance hypothesis", and PA antibiotic resistance in biofilm can become up to 1000-fold higher than the PA planktonic counterpart	

Note: [a]In the various cystic fibrosis infection stages, *Pseudomonas aeruginosa* evolution, biofilm lifestyle, metabolic impact and survival strategies could overlap; [b]Biofilms exhibit a lifestyle that allows *Pseudomonas aeruginosa* (and other Gram-negative and Gram-positive bacteria) to grow and survive, aggregate cells in communities encased within a matrix adherent to the lung surface in cystic fibrosis, and resist immunological and antimicrobial attacks bacteria are deemed unable to determine infections in human organisms (Reardon 2014; Salmond and Fineran 2015), appropriate research on phages to treat CF is actually reappraised (Domingo-Calap et al. 2016), owing to the lack of newly discovered antibiotics (Sulakvelidze et al. 2001; Salmond and Fineran 2015). To advance phage therapy (PT) in CF, we highlight in this chapter the challenging issues on PT, and possibly advocate international regulations to solve ethical concerns on the clinical use of phages in CF, so as to strengthen clinical research on these viruses in patients with MDR, XDR and PDR PA infections (see Sect. 20.5).

20.2 Lytic Phages Used for Therapy in *Pseudomonas aeruginosa* Infections

Phages gained the nickname "intelligent antibiotics" because they can attack the host bacteria, inject their genome but leave the commensal bacteria unchanged (Loc-Carrillo and Abedon 2011; Domingo-Calap et al. 2016). The phage lifecycle inside the bacteria has characteristically two options: the lysogenic cycle (phages using a lysogenic cycle are unsuitable for human therapy), and the lytic cycle (Salmond and Fineran 2015). The phages replicated via the lytic cycle (lytic phages) are the only phages used for decades in clinical research, because they invariably exploit the bacterial molecular system to produce their viral components, and eventually release new virions thus causing bacterial lysis (Bradley and Robertson 1968; Ackermann 2011; Clokie et al. 2011; Salmond and Fineran 2015).

In *Pseudomonas aeruginosa* (PA) infections, lytic phages lyse PA, and through specific depolymerizing enzymes damage bacterial biofilm thus efficiently killing the host bacteria (Criscuolo et al. 2017). Although our comprehensive and argumentative review has synthesized current knowledge on in vitro and in vivo research on lytic phages in CF PA infections (Rossitto et al. 2018), we yet do not know which ready-to-use (*prêt-à-porter*) naturally-occurring (ancestral) or trained (evolved) phages or phages isolated on specific patient PA strains (sur-mesure, personalized approach) alone or in cocktails could increase phage virulence (lytic activity) against CF PA tested in vitro and in vivo (Pirnay et al. 2011; Chan et al. 2013; Hraiech et al. 2015). There is also a need for more information on other open questions including phage gene function and lifecycles, phage-pathogen interactions, genome-sequenced lytic phage characterization, therapeutic commercial phage mixtures stored and certified as safe from central phage banks or libraries that could safely and efficiently combat MDR, XDR and PDR PA strains in patients with CF (Kutateladze and Adamia 2008; Sarhan and Azzazy 2015; Krylov et al. 2016; Górski et al. 2018; Djebara et al. 2019).

20.2.1 *Criteria to Select and Test Lytic Phages in Laboratories for Treating* Pseudomonas aeruginosa *Infections in Cystic Fibrosis (CF)*

Although the Food and Drug Administration recommends lytic phages, before testing in vivo clinical efficacy in CF PA (Brüssow 2012), we lack essential information on the appropriate criteria that should be reported to justify the option for choosing one phage rather than another (Abedon 2017a). These criteria address methods to assess in vitro phage functional characterization (burst size, latency period), activity (host range assay used), bio-informatics genomics (accession numbers from gene sequencing databases, international nucleotide sequence database consortium (INSDC) policy (Brunak et al. 2002; INSDC 2018), how to select and test bacterial hosts to avoid prophage-induced contamination in phage stocks, phage stability, efficacy, and bacterial host resistance in vitro and in vivo (Azeredo and Sutherland 2008; Riou et al. 2010; Ryan et al. 2011; Jassim and Limoges 2014; Hraiech et al. 2015; Weber-Dąbrowska et al. 2016). Criteria for assessing in vivo phage safety include also experiments on purity testing removal of lipopolysaccharides (LPS) derived from PA lysis, and phage administration routes and doses (Abedon 2017a). Only one randomized controlled trial (RCT) tested ex vivo PT in CF sputum and provided evidence that CF PA colonies resistant to a phage cocktail were not resistant to phage *per se* (Saussereau et al. 2014). Although another RCT reported PT in patients with MDR PA-related chronic otitis (Wright et al. 2009), and CF MDR PA is a typical candidate for PT, when we addressed this topic and published are review in 2018 no trials were underway on MDR PA in CF (Rossitto et al. 2018).

Pivotal studies on genomic technologies theorized that genomic engineered phages modified in laboratories to adjust their host range and induce specific depolymerase biofilm-degrading enzymes, interact better than their naturally-occurring species to lyse antibiotic-resistant bacterial hosts (Lu and Collins 2007; Edgar et al. 2012; Pei and Lamas-Samanamud 2014; Reardon 2014; Salmond and Fineran 2015). Equally important, substantial clinical evidence supports that combining phages with antibiotics could restore antibiotic sensitivity in various MDR PA strains (evolutionary synergy) (Chan et al. 2016). Therefore, we need experiments to advance research on selecting phages that use as receptors the bacterial membrane zones responsible for expelling antibiotics (multidrug efflux systems) (Chaturongakul and Ounjai 2014; Chan et al. 2016; Torres-Barcelò and Hochberg 2016; Chaudhry et al. 2017). Advancing research on selecting phages combined with antibiotics is the field of study of our several-year-long multicenter experience on microbiology in patients with CF (Pollini et al. 2011; Bacci et al. 2016; Moretti et al. 2017). Admittedly, our ongoing preliminary observations on phages from a collection stored at Bambino Gesù Children's Hospital tested against mucoid and non-mucoid clinical CF MDR PA strains, combined with antibiotics is yet counteracting, but some evidence on the combination of phages with fluoroquinolones to treat pulmonary CF PA strains holds promises.

20.3 Lytic Phages Tested In Vitro and In Vivo in Various Host Models Infected by Laboratory and Clinical *Pseudomonas aeruginosa* (PA) Strains in Cystic Fibrosis (CF). An Argumentative Review

From our argumentative review, where we synthesized up-to-date reliable information on in vitro and in vivo evidence for and against lytic phage efficacy in various PA strains useful for planning future research into PT for patients with CF (Rossitto et al. 2018), we report in this chapter the reviewed, appraised and synthesized information from the 22 publications collected that described lytic phages tested in vitro in PA from various sources, and phage activity, resistance and safety in vivo on non-pulmonary and pulmonary host models infected by various laboratory and clinical CF PA strains.

20.3.1 Methods Used

In our previous published argumentative review (Rossitto et al. 2018), using as keywords "phage lytic activity in vitro" and "in vivo" in treating CF PA, we searched systematically in PubMed and in a meta-engine (TRIPDATABASE, https://www.tripdatabase.com/), without language, country, and time limits by November 2016. In this chapter, from the retrieved 125 publications, previously screened by title and abstract, and having excluded 94 irrelevant publications, we reassessed 33 full texts, excluded two studies previously included, one irrelevant for resembling CF conditions (Lim et al. 2016), and another reporting the use of a lysogenic phage unsuitable in human trials (Hanlon et al. 2001), and finally included only 22 studies eligible for analyzing and summarizing data on lytic phages tested in vitro and in vivo in various host models infected by laboratory and clinical CF PA strains. To better clarify overlapping information from studies that reported data from both in vitro and in vivo experiments, we agreed on mapping and coding information from the papers in two separate tables using the same author terms, and reported 8 items in in vitro studies (Table 20.2), and 11 items in in vivo studies (Table 20.3), according to our experience, and published suggestions on what research should report (Abedon 2017a). Therefore, the items identified included lines of arguments in key results, problems, limitations, and appraised key results thus assessing upsides and downsides in study findings. After discussion between the threee authors, including the information from the 22 appraised papers and recent relevant studies, a final synthesis was then produced to gain major information that we deemed essential for planning and advancing future clinical research in CF (Fig. 20.1).

Table 20.2 Line of arguments from 18 studies included in this chapter investigating in vitro lytic bacteriophage (phage) effects against laboratory *Pseudomonas aeruginosa* (PA) strains or non-cystic fibrosis (CF) strains or PA strain hosts isolated from patients' wounds, patients with diseases other than CF, and patients with CF (studies reported alphabetically according to the first authors' surnames)

First author, year (Country)	Phage taxonomy (family) [phage bank and genome sequence accession numbers][a]	*Pseudomonas aeruginosa* (PA) strain hosts	Methods used to assess phage host range against PA	Key results from experiments after assessing lytic phage efficacy against PA	Problems reported on safety and efficacy	Limitations to overcome in future research on CF PA	Reliable and repeatable findings useful for planning future CF clinical research in vitro
1. Alemayehu, 2012 (Ireland)[b]	Two newly-isolated φNH-4 (*Myoviridae*) and φMR299-2 (*Podoviridae*) [GenBank numbers: JN254800, φNH-4; JN254801, MR299-2]	*Lux*-tagged PA strains NH57388A (mucoid) and MR299 (non-mucoid) isolated from patients with CF	Plaque assay to determine host range against 10 CF PA strains from University College Cork, Cork University Hospital, Alimentary Health Ltd. (Cork, Ireland)	A mix of two phages reduced bacterial cell numbers in PA biofilms grown on a bronchial epithelial cell monolayer from patients with cystic fibrosis. The phage-mix decreased the risk of resistant mucoid and non-mucoid PA colonies	None	Only two phages tested	Studies *in vitro* need to test phage cocktail therapy in mucoid and non-mucoid *lux*-tagged PA biofilms grown on a bronchial epithelial cell monolayer from patients with CF
2. Alves, 2015 (United Kingdom)	Six newly-isolated phages: DL52, DL60, DL68 (Myoviridae), DL54, DL62 and DL64 (*Podoviridae*) [GenBank numbers: KR054028, DL52; KR054030, DL60; KR054033, DL68; KR054029, DL54; KR054031, DL62; KR054032, DL64]	PAO1 strain isolated from a patient's wound	Spot test	A phage cocktail was more effective than a single phage in decreasing planktonic PAO1 culture and biomass in biofilm models	None	The phage cocktail was tested only on laboratory PAO1 strain	To mimic a more realistic infection situation active phage replication on PAO1 biofilm should be checked under dynamic flow conditions
				In the planktonic PAO1 culture, the phage cocktail exhibited a more positive and rapid effect in eliminating the PA load in broth cultures than in PA biofilms			
				Under static conditions, the bacterial load was disrupted at a lower rate than the load in the planktonic PAO1 culture, and biofilm cells regrew			
				Biofilms took longer to disrupt under dynamic flow conditions than under static conditions, and no PA regrowth appeared within the timeframe of the experiment			

3. Betts, 2013 (France)	PEV2, LUZ7, LKD16 (*Podoviridae*), 14/1 (*Myoviridae*) [GenBank numbers: KU948710.1, PEV2; FN422398, LUZ7; AM265638.1, LKD16; data for phage 14/1 unreported and irretrievable]	PAO1 strain isolated from a patient's wound	No host range reported	Phage activity improved through a standard training process (at least 5–6 phage passages on the host tested)	Resistant PA strains developed	PA resistance was evaluated only after the first of the serial passages used for the standard training procedure	PA bacterial resistance and cross-resistance to other untested phages should be evaluated when phage standard training ends
				During phage training, bacteria became resistant to ancestor phages and cross-resistance developed to other untested (foreign) phages			
4. Coulter, 2014 (Texas, USA)	PB-1 (*Myoviridae*) [The study reported ATCC® number 15692-B3™; GenBank number EU716414.1]	PAO1 strain isolated from a patient's wound	Host range evaluated with an unspecified test	Tobramycin and PB-1 given simultaneously was as effective as tobramycin alone in decreasing PA biofilm mass	None	The bactericidal antibiotic tobramycin, given simultaneously to the phage, inhibited PA protein synthesis, thus interfering with phage production	Potential phage-antibiotic interactions should be tested after assessing useful predefined tobramycin and phage concentrations. In future studies tobramycin and phage therapy treatments should be given sequentially to avoid the tobramycin bactericidal effect induced by simultaneous administration with phages
				Phage and antibiotic treatment combined reduced PA resistance			

(continued)

Table 20.2 (continued)

First author, year (Country)	Phage taxonomy (family) [phage bank and genome sequence accession numbers][a]	*Pseudomonas aeruginosa* (PA) strain hosts	Methods used to assess phage host range against PA	Key results from experiments after assessing lytic phage efficacy against PA	Problems reported on safety and efficacy	Limitations to overcome in future research on CF PA	Reliable and repeatable findings useful for planning future CF clinical research in vitro
5. Danis-Wlodarczyk, 2015 (Poland)	Two newly-isolated KT28 and KTN6 (*Myoviridae*) [GenBank numbers: KP340287, KT28; KP340288, KTN6]	PAO1 and PA0038 strains isolated from a patient's wound, and PA708 isolated from a patient with CF	Host range evaluated with an unspecified test	The various tests used showed that the phage cocktail effectively stimulated PA biofilm eradication activities. Crystal violet (CV) staining and colony-forming unit (CFU) counts detected a 70–90 bacterial cell reduction in 24–72 h-old biofilm; spectrophotometry (SPM) and fluorometry (FM) detected a reduction in pyocyanin and pyoverdine secretions, derived from PA biofilm degradation; laser interferometry (IFM) and goniometry (GM) after phage application detected in PA01 strain an increase in biofilm structure disruption products	After active KT28 and KTN6 phage treatment for one day, more than 90% of persistent PA cells became insensitive to both phages, and cross-resistance developed	No information on whether PA 708, isolated from a patient with CF, was a mucoid or non-mucoid phenotype or the patient with CF had a chronic PA infection	A composite standard reference test (including CV, CFU, SPM, FM, IFM and GM) should be used to detect phage cocktail efficacy on PA biofilm eradication
				No phage mutants appeared in PA-resistant host variants			Phage cocktail treatment on PA hosts must be prolonged over 24 h to detect possible PA cells insensitive to phages, and cross-resistance

6. Danis-Wlodarczyk, 2016 (Poland)[b]	One newly-isolated KTN4 (*Myoviridae*) [GenBank number: KU521356]	PAO1 and non-CF 0038 PA strains isolated from a patient's wound, and the small colony variant PA strain CF708 isolated from a patient with CF	Host range evaluated with an unspecified test against 58 clinical strains from Military Hospital Neder-Over-Heembeek, Brussels, Belgium	The KTN4 phage had a strong bactericidal effect against PA strains, as tested in a gentamicin exclusion assay on airway surface liquid (ASL) models	In a CV binding assay phage treatment increased the biofilm biomass	Colistin caused PA cell deaths by destabilizing cell membrane, thus limiting phage propagation	CV and several well-controlled standard reference tests, and standardized PA host conditions, mimicking normal and CF lung environments, should test phage effect treatment combined with antibiotics on PA biofilms
				CV and several well-controlled, standard reference tests showed that pigmented signaling molecules secreted by the PA hosts decreased significantly, proving that the KNT4 phage was strongly potent in PA biofilm treatment			To design experimental *in vivo* studies, KTN4 phage bactericidal effect combined with antibiotics, and PA host-pathogen interactions should be tested in vitro by using the gentamicin exclusion assay on an ASL in vitro model that is flexible, and generates reproducible data on non-CF and CF epithelial cell lines
				The KTN4 phage freely diffused and gained access to the PA hosts in ASL models, but treatment success depended strongly on the PA strain features			To avoid a colistin bactericidal effect, *in vitro* experiments combining phage and colistin, the antibiotic should be given sequentially and not simultaneously
				Colistin and phage, given simultaneously, had no synergistic effects			

(continued)

Table 20.2 (continued)

First author, year (Country)	Phage taxonomy (family) [phage bank and genome sequence accession numbers][a]	*Pseudomonas aeruginosa* (PA) strain hosts	Methods used to assess phage host range against PA	Key results from experiments after assessing lytic phage efficacy against PA	Problems reported on safety and efficacy	Limitations to overcome in future research on CF PA	Reliable and repeatable findings useful for planning future CF clinical research in vitro
7. Essoh, 2013 (France)	6 Pyophage-derived phages: P1-14_{pyo} (Myoviridae), P1-15_{pyo}, P8-13_{pyo},, P2-10_{pyo}, P3-20_{pyo}, PTr$60_{pyo.}$ (*Podoviridae*) [data unreported and irretrievable]	Forty-seven PA strains from different clusters isolated from patients with CF	Spot test	Efficient cocktails contained several phage genera, but a phage cocktail that efficiently reduced all PA strains was difficult to prepare	The combined pyophage preparation or each newly-isolated phage individually failed to lyse 13 of the 47 PA strains	Bacteria sensitivity and phage host range determined by phage spotting on a lawn of stationary growing bacteria. This metabolic bacterial status could reduce phage lytic activity as others have shown (Abedon and Yin 2009; Middelboe 2000; Sillankorva et al. 2004)	Experiments to test phage cocktail efficiency, should evaluate the presence of the PA resistance mechanism CRISPR-Cas in several PA-resistant strains
	Newly-isolated phages: P1-14_{Or01} (Podoviridae) and P2-10_{Ab01} (*Myoviridae*) [EMBL-EBI numbers: HE983844, P1-$14_{Or01;}$ HE983845, P2-10_{Ab01}]			Phages behaved similarly alone, or in a cocktail			Care is needed to avoid the metabolic PA status owing to a lawn of stationary growing bacteria that might reduce phage lytic activity
				None of the 13 phage-resistant PA strains contained the clustered, regularly interspaced, short palindromic repeat (CRISPR)-Cas system			

8. Friman, 2016, (United Kingdom)	14/1, phiKZ, PT7 (*Myoviridae*) and PNM (*Podoviridae*) [GenBank number: NC_004629.1, phiKZ; data unreported and irretrievable for phages 14/1, PT7 and PNM]	Ten PA strains isolated from patients with CF (five with intermittent, and five with chronic infections)	Host range not evaluated	Trained (evolved) phages were more efficient than ancestral phages in reducing PA bacterial densities, whereas this effect was greater and more consistent when bacteria strains originated from chronic than from intermittent infections	No difference was found among mean PA-resistance levels tested in vitro with the various trained phages	Limited number of PA strains used despite the considerable existing strain variation within and between individual patients with CF	Trained (evolved) phages should be selected by estimating the decrease in phage-resistant PA growth (fitness cost), especially comparing intermittent and chronic PA infections
				Phage selection, during *in vitro* phage bacterial coevolution, led to decreased bacterial growth (fitness cost) measured in the absence of phages only for chronic isolates	The phage effect differed considerably between different PA isolates. This variation was more evident within intermittent than in chronic PA isolates		Experiments should evaluate PA resistance to ancestral and trained phages when phage phage-PA co-evolution ends
				PA became resistant to ancestral and evolved phages at the end of the in vitro phage-PA co-evolution procedure (5 days)			
9. Garbe, 2010 (Germany)	Newly-isolated JG024 (*Myoviridae*) [GenBank number: GU815091]	PAO1, PA mucA mutant strain (mucoid variant of PAO1) and mucoid BT73 strain (isolated from a patient with CF)	Spot test to determine host range against 19 clinical isolates from CF patients, and from urinary tract infections as well as a collection of 100 environmental strains	The BT73 CF mucoid strain infected by phage JG024 *in vitro* showed lower susceptibility to phage infection than PAO1 and PA mucA mutant strains	None	Phage activity in ASM was tested against a single CF strain	In experiments testing phages and mucoid PA hosts in LB broth, alginate should be added, to provide evidence that alginate overproduction can influence phage efficiency in intermittent and in chronic PA infections
				An artificial sputum medium (ASM) mimicking CF lung infection conditions was a suitable method for investigating whether phages lysed PA strains		No information on whether the BT73 CF mucoid strain had been isolated from a CF patient with a chronic PA infection	
				In accordance with the hypothesis that alginate produced by mucoid strains can reduce phage activity, adding alginate to Luria Bertani (LB) broth reduced phage infection efficiency			

(continued)

Table 20.2 (continued)

First author, year (Country)	Phage taxonomy (family) [phage bank and genome sequence accession numbers][a]	*Pseudomonas aeruginosa* (PA) strain hosts	Methods used to assess phage host range against PA	Key results from experiments after assessing lytic phage efficacy against PA	Problems reported on safety and efficacy	Limitations to overcome in future research on CF PA	Reliable and repeatable findings useful for planning future CF clinical research in vitro
10. Hall, 2012 (United Kingdom)[d]	14/1, phiKZ, PT7 (*Myoviridae*) and PNM (*Podoviridae*) [GenBank number: NC_004629.1, phiKZ; data unreported and irretrievable for phages 14/1, PT7, and PNM]	PAO1 strain isolated from a patient's wound	Host range not evaluated	Multi-phage therapy (simultaneous application) was more successful than single-phage therapy (sequential application) for reducing PA bacterial densities (estimated by optical density measurements), and caused no significant increase in the frequency of multi-resistance. Phage-resistant PA emerged in all experimental treatments, and incurred significant fitness costs, expressed as reduced growth rate in the absence of phages	Phage-resistant PA emerged	PA bacterial densities estimated by optical density measurements were probably altered by phage-resistant PA genotypes overproducing alginate or extracellular polymeric substances	To estimate simultaneous application of multi-phage therapy efficiency, phages should be selected by estimating the decrease in phage-resistant PA growth (fitness cost in CF PA clinical strains
11. Henry, 2013 (France)[b]	Nine newly-isolated phages: PAK_P1, PAK_P2, PAK_P3, PAK_P4, PAK_P5 (*Myoviridae*) hosted on the PAK strain; PhiKZ (*Myoviridae*), and LUZ19 (*Podoviridae*) amplified on PAO1; CHA_P1 hosted on the PA CHA strain; LBL3 (*Myoviridae*) hosted on the Aa245 strain [GenBank numbers: KC862297, PAK_P1; KC862298, PAK_P2; KC862299, PAK_P3; KC862300, PAK_P4; KC862301, PAK_P5; NC_004629.1, PhiKZ; NC_010326.1, LUZ19; KC862295, CHA_P1; NC_011165.1, LBL3]	The laboratory PAK, PAK-lumi (PAK bioluminescent version) and PAO1 strains, the multidrug-resistant-mucoid PA CHA strain isolated from a patient with CF, and the Aa245 strain isolated from a burn wound	Host range not evaluated	Efficiency of plating (EOP)[e] used to evaluate phage lytic efficacy on PAK lumi strain in vitro predicted the *in vivo* efficacy of therapeutic phages against PA infections, and in vitro and in vivo EOP results showed a good correlation	None	The experiments testing phage activity by EOP were conducted using the laboratory PAK-lumi strain and not a CF strain host	Experiments designed to compare the lytic activity of each phage with the activity observed on PA hosts should use the EOP test in vitro rather than simple spot tests

12. Larché, 2012 (France)	Three newly-isolated phages: A, B, C (NA) [data unreported and irretrievable]	Forty-four multi-drug-resistant (MDR) and extensively-drug-resistant (XDR) PA strains belonging to different clonal complexes from patients with unspecified pathologies	Spot test to determine host range against 30 PA isolates from Narbonne hospital and 8 from Percy hospital (France), and 6 from Korea	Forty-two MDR and XDR PA strains (95.4%) displayed high susceptibility to at least one phage, as well as to the phage cocktail	Two isolates belonging to the same lineage (representing alone 50% of all MDR isolates), showed resistance to all three phages, both individually and in cocktail	No complete information on DNA analysis to define whether the phages tested were lytic MDR and XDR PA strains of unspecified origin PA strain resistance evaluated from different laboratories from different countries, thus reducing generalizability of the results	Phage cocktail efficacy should be tested with MDR and XDR PA strain hosts from patients with CF
13. Lehman, 2016 (United Kingdom)[b,c†]	Four newly-isolated phages combined in the cocktail AB-PA01 (NA) [data unreported and irretrievable]	Three-hundred sixty-nine PA strains from patients with CF, and 60 PA strains from non-CF patients, collected from 2007 to 2015. PA isolates included both antibiotic susceptible/resistant and mucoid/non-mucoid strains	Spot test to determine host range against 67 distinct CF strains	The phage-mix infected both antibiotic susceptible or resistant, and mucoid or non-mucoid CF PA strain hosts, showing that AB-PA01 had a broad range of activity on PA strains	A total 12.2% CF PA strains were insensitive to the phage cocktail	PA susceptibility to phages tested only by spotting phages on a lawn of PA in an unspecified metabolic status	Because the cocktail AB-PA01 has a broad range of activity on PA strain hosts it should be tested under dynamic metabolic conditions on mucoid or non-mucoid CF PA strains susceptible or resistant to antibiotics

(continued)

Table 20.2 (continued)

First author, year (Country)	Phage taxonomy (family) [phage bank and genome sequence accession numbers][a]	*Pseudomonas aeruginosa* (PA) strain hosts	Methods used to assess phage host range against PA	Key results from experiments after assessing lytic phage efficacy against PA	Problems reported on safety and efficacy	Limitations to overcome in future research on CF PA	Reliable and repeatable findings useful for planning future CF clinical research in vitro
14. Olszak, 2015 (Poland)[b]	28 newly-isolated phages, 2 of which were characterized: PA5oct and KT28 (*Myoviridae*) [data unreported and irretrievable]	A total 121 PA strains (including 29 mucoid strains differing in virulence) isolated from patients with CF, PAO1 and a clinical non-CF strain	Spot test to determine host range against all 121 CF strains tested	PA host features influencing phage activity were slow growth rate, low biofilm production, reduced twitching motility and chemical composition	None	Unspecified PA infection stage	To determine whether PA host features (including slow growth rate, low biofilm production, reduced twitching motility and chemical composition) influence phage activity, these features should be determined in CF and non-CF PA strain hosts isolated from PA colonized patients

15. Pires, 2011 (Portugal)	Newly-isolated: phiIBB-PAA2, phiIBB-PAC23, phiIBB-PACL12, phiIBB-PAP21 (*Podoviridae*) [data unreported and irretrievable]	PAO1 strain isolated from a patient's wound, and an ATCC 10145 strain (unspecified origin)	Spot test to determine host range against 35 strains of PA	When treated with phages, PA biofilm initially reduced biomass and then regrew	Phages isolated from clinical PA strains had a reduced host range	The origin of the ATCC 10145 strain remained unspecified	To evaluate phage resistance, PA planktonic cultures and biofilm should be tested after incubating phage cocktails for 6–10 h
				Phages were equally efficient towards stationary and exponential phase PA cells	Some broad lytic phages failed to infect planktonic cultures		Avoid using the phiIBB-PAA2 rapidly entrapped in a PA biofilm
					PA planktonic cultures and biofilm could become phage-resistant after the first 6–10 h		To demonstrate their presence and active replication during treatments, phage levels in PA biofilm should be checked 2-h after infection
					Higher amounts of the phiIBB-PAA2 than the phiIBB-PaP21 phage became entrapped in the PA biofilms after 2-h infection		

(continued)

Table 20.2 (continued)

First author, year (Country)	Phage taxonomy (family) [phage bank and genome sequence accession numbers][a]	*Pseudomonas aeruginosa* (PA) strain hosts	Methods used to assess phage host range against PA	Key results from experiments after assessing lytic phage efficacy against PA	Problems reported on safety and efficacy	Limitations to overcome in future research on CF PA	Reliable and repeatable findings useful for planning future CF clinical research in vitro
16. Saussereau, 2014 (France)	PAK_P1, PAK_P2, PAK_P3, PAK_P4, PAK_P5, P3_CHA, CHA_P1, phiKZ, LBL3 (*Myoviridae*) and LUZ19 (*Podoviridae*) [GenBank numbers: KC862297, PAK_P1; KC862298, PAK_P2; KC862299, PAK_P3; KC862300, PAK_P4; KC862301, PAK_P5; HM173081, P3-CHA; KC862295, CHA_P1; NC_004629.1, phiKZ; NC_011165.1, LBL3; NC_010326.1, LUZ19]	Twenty small or large PA-colonies, including mucoid and non-mucoid PA strains, from four aliquots collected from 48 sputum samples from patients with CF in three French hospitals	Host range not evaluated	None of the clinical parameters, retrospectively checked in patients with CF, were associated, either positively or negatively, with phage efficacy on their sputum samples infected with PA strains. Hence the microenvironment in the lungs of patients with CF had no influence on phage efficacy ("patient-independent efficacy")	None	Inadequate information on phage lytic activity evaluated on mucoid and non-mucoid PA collected from CF patients' sputum in a selective medium	To test phage efficacy the correlation with clinical criteria should be avoided
				The most efficient phages *in vivo* were those isolated 'deliberately', using patients' PA strains, supporting the use of a personalized approach to achieve optimal treatment			The most efficient technique to yield evidence on formulating new phage cocktails efficient for CF is that deliberately isolating phages on PA patients' strains (personalized approach)

17. Torres-Barceló, 2014 (France)	LUZ7 (*Podoviridae*) [GenBank number: FN422398]	PAO1 isolated from a patient's wound	Host range not evaluated	Phage at a multiplicity of infection (MOI)[c] unspecified and streptomycin treatments (at the dose100 or 240 μg/mL) were tested on PAO1 at different time points: simultaneously, or streptomycin at 12 and 24 h after phage administration	None	MOI unspecified	In in vitro studies studying independent and synergistic effect of combined phages and antibiotics should predefine sub-lethal and MIC antibiotic concentrations for PA hosts
				Phage and streptomycin treatment combined acted synergistically, given that after adding the two treatments singly, combined treatment lowered bacterial density more than expected			To obtain the optimal treatment window, antibiotics should be given 12 h after phages
				The key for minimizing PA resistance levels depended not on the antibiotic dose but on applying antibiotics at the peak phage efficacy, and not simultaneously			To detect phage-resistance PA hosts bacteria should be exposed to prolonged treatments for 70 h
				Combined treatment provided no evidence for a trade-off between antibiotic resistance and phage resistance. Higher phage resistant levels were associated with higher antibiotic resistance			Prediction on phage application should be determined on a MOI-based finding

(continued)

Table 20.3 (continued)

First author, year (Country)	Phage taxonomy (family) [phage bank and genome sequence accession numbers][a]	*Pseudomonas aeruginosa* (PA) strain hosts	Methods used to assess phage host range against PA	Key results from experiments after assessing lytic phage efficacy against PA	Problems reported on safety and efficacy	Limitations to overcome in future research on CF PA	Reliable and repeatable findings useful for planning future CF clinical research in vitro
18. Uchiyama, 2016 (Japan)	Newly-isolated KPP22 (*Myoviridae*) [GenBank number: LC105987]	PAO1 isolated from a patient's wound	Host range not evaluated	EOPs of ancestral KPP22 phage on the KPP22-PAO1 resistant clones were significantly lower (ca. 10^{-4} to 10^{-5} times) than that on the laboratory PAO1 host strain. In contrast, the EOPs of the three mutant KPP22 phages on the KPP22-resistant PAO1 clones showed the same efficacy expressed against the laboratory PAO1 host strain	None	The encouraging results obtained by testing mutant phages need to be replicated in CF PA from chronic pulmonary infections	Phage infectivity levels should be examined by comparing EOP in ancestral and mutant phages The best mutant phages efficient against phage-resistant PA hosts should be detected and reported by genetic analysis and EOP

Note: American type culture collection ATCC® (1925) https://www.lgcstandards-atcc.org/About/About_ATCC/Who_We_Are.aspx Cited 2016; *CF* cystic fibrosis; European molecular biology laboratory-European bioinformatics institute EMBL-EBI (1994). Available at: https://www.ebi.ac.uk/. Cited December 2019; GenBank: Genetic sequence collection of the National Institute of Health of all publicly available DNA sequences (2013) Available at: https://www.ncbi.nlm.nih.gov/genbank. Cited January 2013

[a]In accordance with the International Committee on Taxonomy of Viruses (ICTV) (2019). Available at: https://talk.ictvonline.org/taxonomy/. Cited December 2019; phage bank and genome sequence accession numbers for 6 papers (Betts et al. 2013; Coulter et al. 2014; Friman et al. 2016; Hall et al. 2012; Saussereau et al. 2014; Torres-Barceló et al. 2014) retrieved by the review authors; [b]Results in vivo reported in Table 20.3; [c]*MOI* multiplicity of infection = the ratio between the plaque-forming units (PFU) and the colony-forming units (CFU); *NA* data not available; [d]The in vivo results from this study were excluded because they exclusively referred to PAO1 infecting a non-pulmonary animal model; [e]*EOP* efficiency of plating = the ratio between the average PFU on target bacteria and average PFU on host bacteria; [†]Poster presented at the European Congress of Clinical Microbiology and Infectious Diseases 2016 (no published results)

Table 20.3 Line of arguments from 9 studies included in this chapter investigating in vivo lytic bacteriophage (phage) effects against laboratory *Pseudomonas aeruginosa* (PA) strains or non-cystic fibrosis (CF) strains or PA strain hosts isolated from patients' wounds, patients with diseases other than CF, and patients with CF in pulmonary and non-pulmonary host models (PA-infected animal models) (studies reported alphabetically according to the first authors' surnames)

First author, year (country)	Phage taxonomy (family) [Phage bank and genome sequence accession numbers][a]	PA strain(s) used in the experiments	Animal host models tested	Phage host range methods used to assess lytic activity against PA	Products used for suspending phages	Control group treatments	Key results	Problems reported in testing safety and efficacy	Limitations to overcome in future research on CF	Reliable and repeatable findings useful for planning future CF clinical research in vivo
1. Alemayehu, 2012 (Ireland)[b]	Two newly-isolated φNH-4 (*Myoviridae*) and φMR299-2 (*Podoviridae*) [GenBank numbers: JN254800, φNH-4; JN254801, MR299-2]	*Lux*-tagged PA strains NH57388A (mucoid) and MR299 (non-mucoid) isolated from patients with CF	Mice	Phage plaque assay on 10 CF strains from University College Cork, Cork University Hospital, Alimentary Health Ltd. (Cork, Ireland)	PBS solution	Infected control group received phosphate-buffered saline (PBS) instead of phage mix	The two-phage mix is effective in killing *lux*-tagged PA in the lungs of infected mice	None reported	Lack of a negative control group. The study tested only a two-phage cocktail	Studies in animal models need to test phage cocktail efficiency on mucoid and non-mucoid *lux*-tagged CF PA strains

(continued)

Table 20.3 (continued)

First author, year (country)	Phage taxonomy (family) [Phage bank and genome sequence accession numbers][a]	PA strain(s) used in the experiments	Animal host models tested	Phage host range methods used to assess lytic activity against PA	Products used for suspending phages	Control group treatments	Key results	Problems reported in testing safety and efficacy	Limitations to overcome in future research on CF	Reliable and repeatable findings useful for planning future CF clinical research in vivo
2. Beeton, 2015 (United Kingdom)	Newly-isolated DL52, DL60, DL68 (*Myoviridae*). DL54, DL62, DL64 (*Podoviridae*) [GenBank numbers: KR054028, DL52; KR054030, DL60; KR054033, DL68; KR054029, DL54; KR054031, DL62; KR054032, DL64]	PAO1, PA45291 (isolated from bacteremia) and BC09007 (isolated from a patient with CF)	*Galleria mellonella* larvae	No tests used	Not reported	Three control groups, two negative (one of PA uninfected-phosphate-buffered saline-PBS injected *Galleria mellonella* larvae, to assess the impact of any negative effect from the injection procedure; and one of PA uninfected phage-treated larvae to assess the toxicity of the phage cocktail), and one positive (PA infected larvae treated with PBS solution)	A 6-phage cocktail used in the preventive and curative approaches, prolonged survival of infected larvae in a dose-dependent manner This study validates the use of *Galleria mellonella* as a simple, robust and cost-effective model for initial *in vivo* examination of PA-targeted phage therapy The investigators suggest that continual PAO1 survival in the presence of a high phage titer could be caused by intracellular localization of PA, having ruled out phage resistance developing within the larvae The efficacy of phage treatment is multiplicity of infection (MOI)[d] related (the higher the MOI the better the result) Susceptibility reassessed in re-isolated PAO1	Phage-treated *Galleria mellonella* infected by PAO1 died at 30 h post-infection and larvae infected by clinical strains PA45291 and BC09007 eventually succumbed at 40 h	The authors failed to provide information on why they used phage suspensions at MOI 100 only for the preventive model, but presumably owing to experimental risks including killing the larvae	Reassessing susceptibility in re-isolated PAO1 will rule out possible PAO1 phage resistance during *in vivo* infection

3. Danis-Wlodarczyk, 2016 (Poland)[b]	Newly-isolated KTN4 (*Myoviridae*) [GenBank number: KU521356]	PAO1, non-CF0038 strains isolated from a patient's wound, and the small colony variant CF708 strain isolated from a patient with CF	*Galleria mellonella* larvae	Unspecified phage efficacy test on 58 clinical PA strains (from the Military Hospital Neder-Over-Heembeek, Brussels, Belgium)	Not reported	The controls consisted of PA uninfected *G. mellonella* larvae receiving phage lysate only (negative), and larvae infected with bacterial lethal dose (positive)	The KTN4 phage has an antibacterial strain-dependent efficacy against clinical isolates	None reported	Because it grows slowly and leads to slow biofilm formation, low type IV pili expression, and lower virulence than PAO1 and non-CF0038, the small colony variant CF 708 strain is not a representative PA strain colonizing patients with CF	Studies on slowly growing PA strains should incubate larvae for at least 120 h
							Infected larvae were incubated for up to 120 h			
4. Debarbieux, 2010 (France)	Newly-isolated PAK-P1 (*Myoviridae*) [GenBank number: KC862297]	Bioluminescent laboratory PAK (PAK lumi) strain	Mice	Efficiency of plating (EOP)[c] test for PAK P1 phage on 10 CF PA strains from primary, and 10 CF PA from chronic colonization (French CF strain collection center)	PBS solution	For curative treatment: PA uninfected phage-treated mice (negative), PA infected-phosphate-buffered saline-treated (PBS)-mice (positive). For preventive treatment, PBS-pre-treated PA infected mice (positive)	Phage treatment is effective in saving mice from lethal infection in the curative model, and prevents lung infection when given 24 h before bacterial infection	None reported	The PAK P1 environmental phage effectively lysed in plates only 10% of PA from patients with chronic infection, and was effective in lysing 50% of clinical strains isolated from patients with primary colonization	Phage host range needs to be evaluated by EOP Studies on preventing PA infection should use non-invasive techniques (i.e. bioluminescence)
							Giving a bioluminescent PA intra-nasally quantifies efficacy by phage multiplication in the respiratory tract. Phages tend to persist in the lungs		Owing to high phage variability, completely annotating a phage genome requires in-depth bioinformatic analysis	To evaluate the safety of a high phage treatment dose, mice should be monitored for behavior, fur and weight for 10 days

(continued)

Table 20.3 (continued)

First author, year (country)	Phage taxonomy (family) [Phage bank and genome sequence accession numbers][a]	PA strain(s) used in the experiments	Animal host models tested	Phage host range methods used to assess lytic activity against PA	Products used for suspending phages	Control group treatments	Key results	Problems reported in testing safety and efficacy	Limitations to overcome in future research on CF	Reliable and repeatable findings useful for planning future CF clinical research in vivo
							Active phage efficacy in PA infected mice is dose and time-dependent, and a phage-to-bacterium ratio of 10:1 is harmless to mice. A solution of heat-killed PAK-P1 phage, given 2 h after PA infection, killed similar mouse numbers in the negative and positive control groups		In uninfected PA mice, phage replication showed a 2-log decrease in 24 h	In PA-infected mice, the maximum possible delay in phage treatment, for maintaining a 100% survival, is 2 h Studies assessing phage safety in infected PA mice should ensure a phage-to-bacterium ratio that is harmless to mice
							To monitor phage treatment harmless to mice treated with MOI 100, 10 times higher than a standard dose, mice behaviour, fur and weight were monitored for 10 days.			Measuring lactate dehydrogenase enzyme levels (released from lung damaged cells) in mouse BALs at 6 h after PA infection will identify the time in which bacteria multiply fastest and bacterial phage susceptibility is highest
							To evaluate mouse lung infection status in PA phage-treated infected mice and in mouse controls lactate dehydrogenase enzyme levels were measured in BALs 6 h after the infection			

5. Henry, 2013 (France)[b]	Nine newly-isolated phages: PAK_P1, PAK_P2, PAK_P3, PAK_P4, PAK_P5 (*Myoviridae*) hosted on the PAK strain; PhiKZ (*Myoviridae*), and LUZ19 (*Podoviridae*) amplified on PAO1; CHA_P1 hosted on the PA CHA strain; LBL3 (*Myoviridae*) hosted on the Aa245 strain [GenBank numbers: KC862297, PAK_P1; KC862298, PAK_P2; KC862299, PAK_P3; KC862300, PAK_P4; KC862301, PAK_P5; NC_004629.1, PhiKZ; NC_010326.1, LUZ19; KC862295, CHA_P1; NC_011165.1, LBL3]	A bioluminescent laboratory PAK strain (PAK-lumi)	Mice	EOP test for 9 phages on PAK lumi strain	Not reported	Infected mice treated with PAK_P1 (MOI 0.1) or phosphate-buffered saline (PBS)	The good correlation between *in vitro* results and *in vivo* efficacy for 7 phages, gives a reliable index to predict phage efficacy. *In vitro* efficacy is insufficient to ensure *in vivo* efficacy. The optimal efficacy achieved with the 5 phages isolated on the PA strain used in the mouse model highlights a possible effect of the bacterial host used for isolation on the efficacy of the treatment.	CHA_P1 phage (amplified on a clinical PA strain CHA) was unable to cure animals infected with the PAK-lumi strain, despite being genetically closely related to PAK_P3 and PAK_P5. In addition, an attempt to adapt this phage to the PAK strain failed	The experiments testing phage activity by EOP conducted using the laboratory PAK-lumi strain and not a CF strain as the phage host	Phage host range needs to be evaluated by EOP.
							Bioluminescence in mice was recorded by a real-time imaging system (specialized living image software) to monitor infection in the whole lung		Using phages isolated on patients' PA strains is time consuming	To validate results from studies *in vitro*, studies *in vivo* need to test cocktail phage therapy isolated on the same PA strains (i.e. personalized) used in mouse model treatment, and measure efficacy with real-time imaging system software

(continued)

Table 20.3 (continued)

First author, year (country)	Phage taxonomy (family) [Phage bank and genome sequence accession numbers][a]	PA strain(s) used in the experiments	Animal host models tested	Phage host range methods used to assess lytic activity against PA	Products used for suspending phages	Control group treatments	Key results	Problems reported in testing safety and efficacy	Limitations to overcome in future research on CF	Reliable and repeatable findings useful for planning future CF clinical research in vivo
6. Lehman, 2016 (United Kingdom)[b,e]	Four newly-isolated phages combined in the cocktail AB-PA01 (NA) [data ureported and irretrievable]	Three hundred and 69 PA strains from patients with CF and 60 PA strains from non-CF patients collected between 2007 and 2015. Isolates included both antibiotic susceptible or resistant and mucoid or non-mucoid strains	Mice	Unspecified phage efficacy test on 67 PA strains from CF patients (unspecified collection center)	Not reported	Unspecified non-CF or CF PA-infected mice treated with the phage diluent	The study confirmed the usability of AB-PA01 for clinical use (exclusively lytic, efficacious *in vivo*), nebulization (no significant decreases in titers were observed) and long-term phage stability (GMP manufacturing current process optimization)	None reported in the poster	The PA strains used to infect mice were not specified. No negative control group reported Failed to specify where the PA strains used for testing phage efficacy were collected	Studies on antibiotic susceptible or resistant and mucoid or non-mucoid PA strains need to test phage efficacy on several PA strains taken from patients with CF
7. Morello, 2011 (France)	P3-CHA derived and trained (evolved) from PAK-P3 (*Myoviridae*) [GenBank: HM173081]	Multidrug-resistant (MDR) and mucoid PA CHA strain isolated from a patient with CF	Mice	EOP test for PAK-P3 and P3-CHA on 10 CF PA strains from primary, and 10 CF PA from chronic colonization (French CF strain collection center)	PBS solution	Curative treatment involved PA CHA infected- phosphate-buffered saline (PBS) treated mice (positive control) and PA CHA uninfected heat-killed P3-CHA treated mice (negative control).	Curative phage treatment acts cooperatively with the mice immune response to eliminate acute lung infection caused by MDR PA CHA. Preventive phage treatment requires non-heat-killed active phages	None reported	For curative treatment, phage concentration data reported in the table and in the results showed major discrepancies, thus determining possible reporting bias	Phage host range needs to be evaluated by EOP Studies testing in vivo trained phage therapy on CF MDR PA strains need to adhere to international MDR bacteria definitions
						Preventive phage treatment involved heat-killed P3-CHA pre-treated PA CHA infected mice (positive control) and P3-CHA pre-treated uninfected mice (negative control)	Training (evolved) in vitro a phage towards a CF multidrug resistant clinical strain improves its efficacy in curative and preventive experiments *in vivo*		The reported MDR PA strain used in the experiment, failed to adhere to the international MDR bacteria definitions (Sievert et al. 2013)	Training (evolved) phage (using an endotoxin-free phage solution) in vitro, could avoid stimulating a host immune response that could mask the effects of phage curative and preventive treatments in vivo

8. Olszak, 2015 (Poland)[b]	28 newly-isolated phages (only two characterized): PA5oct and KT28 (*Myoviridae*) [data ureported and irretrievable]	PAO1, non-CF0038 strain and 4 clinical CF strains (CF217, CF708, CF532, CF832) with diverse degrees of virulence	*Galleria mellonella* larvae	Phage spot test on 123 PA isolates (PAO1, 121 clinical CF PA from Prague CF Center collection, non-CF0038 clinical PA strain from the Institute of Genetics and Microbiology collection, University of Wroclaw, Poland)	Not reported	Uninfected *Galleria mellonella* larvae, sham-infected larvae, larvae receiving phage lysate only (negative), PA infected-untreated larvae, and infected larvae treated with ultra-violet (UV)-inactivated phages (positive)	The efficacy of single phages or cocktails is strain related	Phage failure to rescue larvae infected with the 3 weakly virulent CF isolates given at a high load able to cause larva infections could depend on the toxic compounds released during massive bacterial cell lysis after phage propagation, thus causing greater mortality in treated than in untreated larvae	Phages chosen for cocktail preparations were not selected for different host receptor affinity, hence they competed for bacterial receptors	Phage cocktail formulations need to be chosen on the basis of phages not competing for bacterial receptors
							Larval survival reflects phage lytic activity rather than host immune stimulation		Because the mixture was less effective in lysing PAO1 and non-CF0038 than single lytic phage injection, phages need to be selected rigorously	Active phages rather than UV-inactivated phages need to be used in cocktails
									Because weakly virulent PA strains given at a high load release toxic compounds by phage lysis, PA strains need to be carefully selected to avoid severe systemic inflammatory response and death in the animal hosts	Unless further research describes developments reducing toxic compound release, studies on weakly virulent PA strains at a high load need to be avoided

(continued)

Table 20.3 (continued)

First author, year (country)	Phage taxonomy (family) [Phage bank and genome sequence accession numbers][a]	PA strain(s) used in the experiments	Animal host models tested	Phage host range methods used to assess lytic activity against PA	Products used for suspending phages	Control group treatments	Key results	Problems reported in testing safety and efficacy	Limitations to overcome in future research on CF	Reliable and repeatable findings useful for planning future CF clinical research in vivo
9. Pabary, 2016 (United Kingdom)	Phage cocktail PA 24, PA 25, PA 7 (NA) [data ureported and irretrievable]	PAO1 and PA 12B-4973 (isolated from a patient with CF)	Mice	Phage cocktail spot test on PAO1 and five CF PA strains from adult inpatient sputa (patients attending the Royal Brompton Hospital, London)	Not reported	Infected-sodium magnesium buffer-treated mice (positive)	No evidence of murine toxicity following rapid phage-induced lysis of PAO1 and CF PA strains	PA 12B-4973 was unusable in the delayed and prophylactic approaches owing to its virulence	No *in vitro* experiments were done to retest PA colonies recovered from mice receiving delayed or prophylactic phage doses, hence phage susceptibility and phage resistant PA strains remain unknown	Future human clinical trials testing phage cocktail efficacy in patients with CF should measure PA strain clearance in BALs
							Evidence on the phage cocktail treatment benefits on PAO1 and CF PA infected mice. Giving phages before infecting mice with PA strains results in BAL clearance, and reduces neutrophilic inflammation		Only one virulent CF strain tested in a single experimental condition owing to problems related to virulence	Evaluating inflammatory cell levels in mouse BAL and spleen cultures, provides reliable evidence on PA infection and dissemination
									Lack of negative controls	

Note: *CF* cystic fibrosis, *NA* data not available, *PBS* phosphate-buffered saline, *BAL* bronchoalveolar lavage

[a]In accordance with the International Committee on Taxonomy of Viruses (ICTV)(2019). Available at: https://talk.ictvonline.org/taxonomy/. Cited December 2019; [b]Results in vitro reported in Table 20.2; [c]*EOP* efficiency of plating = the ratio between the average plaque-forming units (PFU) on target bacteria and average PFU on host bacteria – is the standard method for testing lytic phage activity (in accordance with Mirzaei et al. 2016); [d]*MOI* multiplicity of infection = the ratio between the PFU and the colony-forming units (CFU); [e]Poster presented at the European Congress of Clinical Microbiology and Infectious Diseases 2016 (no published results)

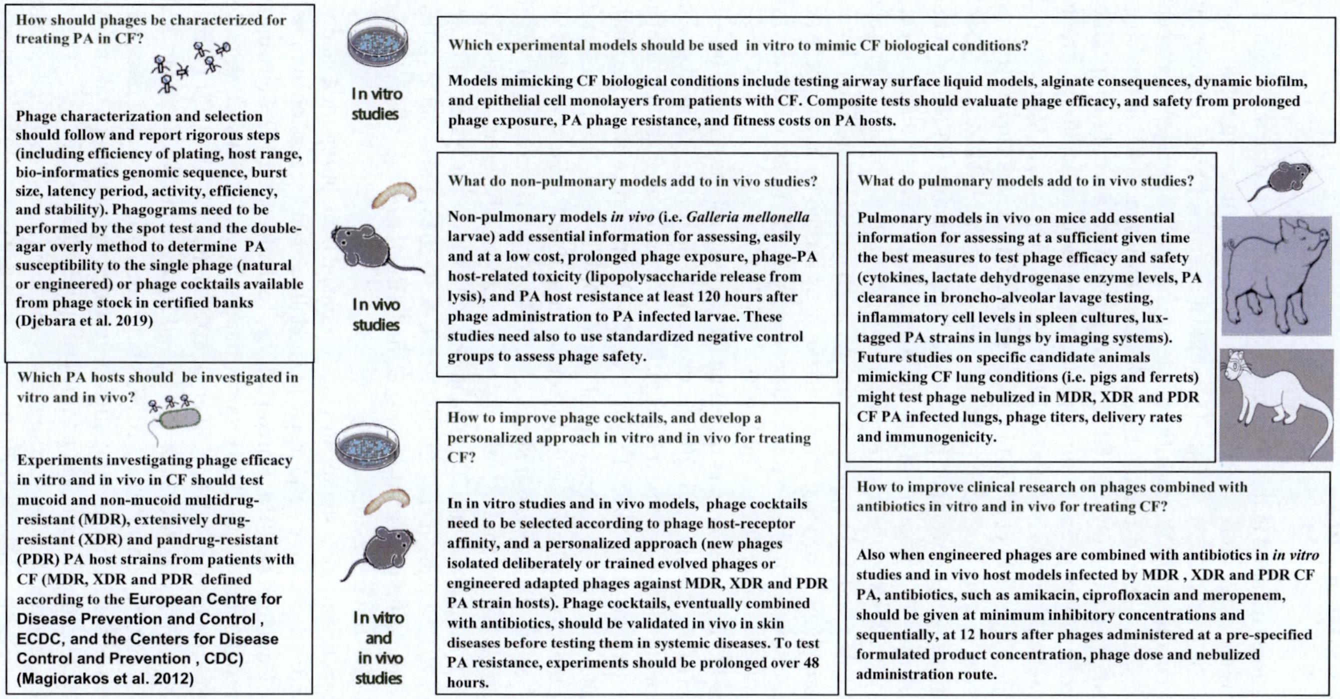

Fig. 20.1 Information from the 22 studies included in this chapter, and other recent studies answering major open questions essential for planning future studies in vitro and in vivo on phage therapy to treat *Pseudomonas aeruginosa* (PA) lung infections in patients with cystic fibrosis (CF)

20.3.2 General Results on Lytic Phages in In Vitro and In Vivo Studies

Of the 22 publications included in this chapter 13 referred to experiments in vitro, 4 to studies in vivo, and 5 to studies in vitro and in vivo (18 studies in vitro, and 9 in vivo) (Tables 20.1 and 20.2). Of the18 publications describing experiments in vitro, 13 were conducted in Western European countries, 3 in Eastern European countries, 1 in the Far East, and 1 in the USA. All the studies included in this chapter were published in the past 16 years. Of the 22 papers reviewed, only 10 specified the criteria used to justify phage selection, 6 selected them according to their range of activity (Alves et al. 2015; Garbe et al. 2010; Lehman et al. 2016; Morello et al. 2011; Olszak et al. 2015; Pires et al. 2011), 1 according to previous studies (Friman et al. 2016), 1 using custom-developed phages against resistant PA strains (Essoh et al. 2013), 1 using custom-developed phages against a laboratory PA (Henry et al. 2013), and 1 according to phage ability to delay and inhibit phage-resistant cells (Pabary et al. 2016). Our argumentative review reported in this chapter, therefore, failed to identify a standardized method for justifying phage selection. Only 3 studies in vitro reported phage functional characterization estimated by latent phase and burst size (the mean number of phages released per bacterial cell) (Danis-Wlodarczyk et al. 2015, 2016; Garbe et al. 2010) (Tables 20.1 and 20.2).

Only 2 studies in vitro selected and tested phage activity and host range by efficiency of plating (EOP) (Henry et al. 2013; Uchiyama et al. 2016), 11 used plaque assays, spot test or streak test (Alemayehu et al. 2012; Alves et al. 2015; Betts et al. 2013; Danis-Wlodarczyk et al. 2015; Essoh et al. 2013; Friman et al. 2016; Garbe et al. 2010; Larché et al. 2012; Lehman et al. 2016; Olszak et al. 201; Pires et al. 2011), and 5 left the test used unspecified (Coulter et al. 2014; Danis-Wlodarczyk et al. 2016; Hall et al. 2012; Saussereau et al. 2014; Torres-Barceló et al. 2014) (Table 20.2). Of the 9 in vivo studies, only 3 used the EOP (Debarbieux et al. 2010; Henry et al. 2013; Morello et al. 2011), and 4 used plaque assays or spot test (Alemayehu et al. 2012; Lehman et al. 2016; Olszak et al. 2015; Pabary et al. 2016). Two further studies left the test used unspecified (Beeton et al. 2015; Danis-Wlodarczyk et al. 2016) (Table 20.3).

Of the 22 papers reviewed, 6 studies in vitro left phage bank unreported, and we retrieved and reported genomic sequences from online bank accession numbers (Betts et al. 2013; Coulter et al. 2014; Friman et al. 2016; Hall et al. 2012; Saussereau et al. 2014; Torres-Barceló et al. 2014). For 8 papers in vitro (Betts et al. 2013; Essoh et al. 2013; Friman et al. 2016; Hall et al. 2012; Larché et al. 2012; Lehman et al. 2016; Olszak et al. 2015; Pires et al. 2011), and 3 papers in vivo (Lehman et al. 2016; Olszak et al. 2015; Pabary et al. 2016) phage genomic sequence data were unreported or irretrievable. Only one investigator reported using a phage taken from a central certified collection (Coulter et al. 2014) (Tables 20.1 and 20.2).

Only 3 studies reported phage stability assessment on newly-isolated phages alone (Danis-Wlodarczyk et al. 2015, 2016) and in a phage cocktail (Lehman et al. 2016) (Table 20.2 and 20.3). Of the in vivo 9 studies, 6 in PA-infected mouse models instilled phages intranasally, whereas the 3 in larva models delivered phages into the larval hemolymph. Only 3 of the 9 in vivo studies reported the formulated products used for administering the phage suspension to host animal models (Alemayehu et al. 2012; Debarbieux et al. 2010; Morello et al. 2011) (Table 20.3). Few papers, therefore, gave essential information on phage functional characterization, selection, activity, genomic sequence, stability, and the formulated products used in phage suspensions.

Several investigators reported in vitro newly-isolated or trained phage lytic efficacy alone or in cocktails in disrupting the biofilm matrix (including dynamic biofilm models), reducing biomass and planktonic cell numbers in laboratory PA, non-CF and CF mucoid or non-mucoid PA strains, and MDR PA (Alemayehu et al. 2012; Alves et al. 2015; Coulter et al. 2014; Danis-Wlodarczyk et al. 2015, 2016; Larché et al. 2012; Lehman et al. 2016; Morello et al. 2011; Pires et al. 2011) Even though phages penetrated alginate, its overproduction in mucoid CF and non-CF PA strains, under simulated CF lung conditions, reduced phage infection efficiency (Garbe et al. 2010) (Table 20.2). These findings underlined the risk that phage efficacy could be influenced by alginate overproduction especially important in patients with CF.

During phage treatment against various PA strains, 4 studies reported immediate phage resistance to ancestral or newly-isolated or trained single phages (including cross-resistance to new phages) within 1 and 5 days (Betts et al. 2013; Danis-Wlodarczyk et al. 2015; Friman et al. 2016; Pires et al. 2011), and only 1 study reported the right time to test resistance or cross-resistance to trained phages when standard phage training ends (Betts et al. 2013). In 1 of these studies, trained phages halved the percentage of resistant bacterial cells, leading to lower bacterial growth (fitness cost) in chronic than in intermittent PA isolates (Friman et al. 2016) (Table 20.2). Another 4 in vitro studies investigated PA resistance to phage cocktails (Alves et al. 2015; Essoh et al. 2013; Hall et al. 2012; Saussereau et al. 2014). Although phage cocktails were tested in vitro for no more than 48 h, phage cocktails showed less PA resistance than phages used alone (Alves et al. 2015; Hall et al. 2012). One study, using three phage cocktails inducing similar resistance in several CF PA strains, failed to identify the clustered regularly interspaced short palindromic repeat (CRISPR)-Cas protein system (Essoh et al. 2013). In another study, designed as an ex vivo study mimicking a clinical condition, Saussereau et al. (2014) reported that CF PA colonies resistant to a phage cocktail were not resistant to phages per se (Table 20.2). Only 2 studies in vitro addressed phage resistance in MDR and XDR PA (some from CF or unspecified origin) (Larché et al. 2012; Lehman et al. 2016). Another in vitro study reported that combining simultaneously a single phage with a bactericidal antibiotic (tobramycin), PA resistance decreased (Coulter et al. 2014). No study investigated resistance to phages combined with antibiotics for CF MDR, XDR or PDR PA infections (Table 20.2 and Fig. 20.1). Collectively, few studies reported CF PA phage resistance on phages alone or in cocktails, which have high PA susceptibility. These findings gave no help on selecting, and personalizing phages to use in cocktails to combat chronic CF PA infections.

Of the 9 papers describing phage efficiency in in vivo studies, 3 used non-pulmonary host larvae models (*Galleria mellonella*) (Beeton et al. 2015; Danis-Wlodarczyk et al. 2016; Olszak et al. 2015), and 6 pulmonary models (mice) (Alemayehu et al. 2012; Debarbieux et al. 2010; Henry et al. 2013; Lehman et al. 2016; Morello et al. 2011; Pabary et al. 2016) (Table 20.3 and Fig. 20.1). Only 1 of the 3 studies on non-pulmonary models, tested both single phages and a phage cocktail in non-CF and CF PA-infected larvae, and assessed adverse events from phage injection testing a standardized negative control group, PA uninfected larvae (Olszak et al. 2015). All these studies highlighted the need to prolong PA-infected larvae exposure to phages, given singly or in cocktails, for up to 120 h.

Few studies on in vivo non-pulmonary models provided reliable findings on phage efficiency, adverse events and host survival, which PA should be selected to avoid massive toxic compound release, or which standardized negative control groups should be assessed to clarify safety from phage injection (Table 20.3). The 6 papers testing phages in vivo on mouse PA-infected host models mimicked pulmonary conditions (Alemayehu et al. 2012; Debarbieux et al. 2010; Henry et al. 2013; Lehman et al. 2016; Morello et al. 2011; Pabary et al. 2016) (Table 20.3).

In 3 studies, the investigators proved phage efficacy in preventive and curative mouse models (Debarbieux et al. 2010; Morello et al. 2011; Pabary et al. 2016). In the preventive model, they showed that single ancestral and trained phages prevented PA lung infection. In the curative model, by quantifying imaging (PAKlumi) and bronchoalveolar lavage (BAL) PA counts, they showed that PA loads decreased, rescuing 100% of mice (Debarbieux et al. 2010; Morello et al. 2011). Of the 6 studies using mouse host models, 4 administering phage-cocktails (two or more phages) to mouse lung infected by various PA strains, all reported that phage cocktails reduced PA loads (Alemayehu et al. 2012; Henry et al. 2013; Lehman et al. 2016; Pabary et al. 2016) and one, comparing PA-infected phage-treated mice with a group of PA-infected mice alone, also showed reduced inflammatory markers in BALs (Pabary et al. 2016). One of these 4 studies, provided an index to predict phage efficacy reliably especially for phages isolated with a personalized approach (Henry et al. 2013). No study tested in vivo phages in cocktails on MDR CF PA, or PT in other animals (pigs and ferrets), possibly better candidates than mice to mimick pulmonary models. Collectively the 4 studies testing efficiency for 2–5 phage cocktails in various CF and non-CF PA strains gave inconclusive results (Table 20.3).

20.3.3 Results on Phages Combined with Antibiotics in In Vitro Studies

Only three studies in vitro and no studies in vivo investigated whether a single phage combined with antibiotics (tobramycin, colistin, and streptomycin) reduced a laboratory PA biomass or planktonic PA cells (Coulter et al. 2014; Danis-Wlodarczyk et al. 2016; Torres-Barceló et al. 2014). Although Coulter et al. and Torres-Barceló et al. reported that combined therapy was as effective as the tested antibiotics alone, the combined treatment reduced the emergence of phage PA-resistant cells. In their study, Torres-Barceló et al. also recommended ensuring maximum antibacterial efficacy, by giving antibiotics 12 h after phages rather than simultaneously, and obtained a synergistic effect in decreasing laboratory PA biofilm biomass or density. Testing a single phage on several strains, Danis-Wlodarczyk et al. demonstrated colistin antagonism to the phage. No study tested phage cocktails combined with antibiotics (Table 20.2). In defining experimental efficiency for phages and antibiotics combined, no study mentioned essential information such as host receptor affinity, personalized approaches, well-selected phage cocktails, antibiotics selected by minimal inhibitory concentration (MIC) in vitro on MDR and XDR PA, or in vivo on PA-infected animal host models (Fig. 20.1).

20.3.4 *Issues Related to Phage Efficacy and Safety in In Vitro and In Vivo Studies*

Of the 22 studies, only 13 reported experimental issues related to phage efficacy and safety (9 in vitro and 4 in vivo). Three investigators reported in vitro problems related to PA biofilm: bacterial regrowth (Pires et al. 2011), increased PA biomass in the phage-treated group (Alves et al. 2015), and PAO1 phage insensitivity (Danis-Wlodarczyk et al. 2015). Although Pires et al. (2011) suggested to overcome these issues by using short-term PT, preferably in cocktails to avoid PA resistance, no other studies mentioned possible solutions (Tables 20.2 and 20.3).

Four groups reported in vivo phage inefficacy including genetically-related phage inability to lyse the same PA strain (Henry et al. 2013), *Galleria mellonella* larval death several hours after phage infection (Beeton et al. 2015; Olszak et al. 2015), and phage inefficacy in the delayed and prophylactic approaches (Pabary et al. 2016). In their in vivo study in a larva model, Olszak et al., to infect larvae before phage infection, gave weakly virulent PA strains at high loads, and consequent PA lysis caused massive toxic compound release followed by larvae death at 96 h. Although no study suggested how to overcome problems in in vivo PT, Olszak et al. (2015) recommended carefully selecting phages (Table 20.3).

Finally, although Torres-Barcelò et al. (2014) found evidence of a synergistic effect using a single phage combined with an antibiotic (streptomycin), Danis-Wlodarczyk et al. (2016) detected colistin and phage antagonism. Few investigators suggested how to overcome issues related to efficacy and safety in phage treatment or by using a phage combined with antibiotics in CF PA. None of the studies we reviewed compared phage efficacy in vitro or in vivo, according to administration routes (Tables 20.2 and 20.3).

20.3.5 *Critical Appraisal of the Information Provided by Our Argumentative Review on Lytic Phages in Cystic Fibrosis (CF). Current Challenges*

Our argumentative review reported in this chapter provided critically appraised, up-to-date previously unavailable information for, and against, in vitro and in vivo lytic phage efficacy in various CF PA strains, useful for planning future clinical research into PT for patients with CF (Tables 20.2, 20.3 and Fig. 20.1). Although our multiple search strategy avoided language and time limits, the most promising and relevant information came mainly from studies published by renowned international research groups over the past seven years, presumably owing to MDR PA spread, and disinterest in producing new antibacterial drugs (Salmond and Fineran 2015; Krylov et al. 2016).

By collecting information from 22 relevant publications describing lytic phage activity and efficacy in vitro, and efficacy and safety in vivo experiments on pulmonary and non-pulmonary host models infected with PA from various sources, the reviewed findings reported in this chapter filled some major knowledge gaps on phages that can effectively combat PA strains infecting lungs in patients with CF. Our findings also provided new insights that will prompt renewed clinical research in CF PA to test well-selected (ancestral or trained) lytic phages to include in cocktails, develop personalized PT, address phage-antibiotic combinations, and envisage even genomic engineered phages combined with antibiotics to treat MDR, XDR and PDR PA in patients with CF (Fig. 20.1).

Among research that especially takes the field ahead in phage treatment for CF PA, our review provides helpful information on the various methods used for testing host range and lytic phage activity against a variety of target bacteria (spot tests, plaque assays, and EOP). Ample evidence suggests that the essential, reliable step for confirming lytic phage cocktail efficiency as preliminarily tested by spot test or plaque assay, is to test single phages for effectiveness with EOP (Debarbieux et al. 2010; Morello et al. 2011; Henry et al. 2013; Uchiyama et al. 2016) (Tables 20.2, 20.3 and Fig. 20.1). Leaving EOP in vitro unassessed or using only spots tests that simply reflect a bactericidal effect, as many papers in our review did (Beeton et al. 2015; Danis-Wlodarczyk et al. 2016; Lehman et al. 2016; Alemayehu et al. 2012; Olszak et al. 2015; Pabary et al. 2016), risks selecting phages that despite their broad lytic activity, as assessed in vitro by other methods (spot tests or plaque assays), fail to infect CF PA planktonic cells (Pires et al. 2011). These findings were later confirmed by two investigators who concluded that EOP is the standard method also for confirming lytic phage activity against other Gram-negative bacteria, namely *Escherichia coli* and *Salmonella* reference collections (Mirzaei and Nilsson 2015).

Another major in vitro and in vivo experimental requirement related to formulating efficient newly-isolated phage cocktails to combat PA lung infection in patients with CF is to select highly virulent lytic phages having good, varying target host-receptor affinity (Fig. 20.1). Choosing phages that differ in host receptor affinity

avoids competition in absorption to the bacterial surface, and improves synergic phage cocktail efficacy (Drulis-Kawa et al. 2012). Hence, surprisingly, among the 22 papers on CF PA reviewed, only one underlines the need to select phages, according to their host receptor affinity, to avoid antagonism, and ensure a phage synergistic effect (Essoh et al. 2013), whereas in Olszak et al. (2015), the phage mixture probably competed for bacterial receptors (Tables 20.2 and 20.3).

A key problem in ensuring the efficacy of newly-isolated lytic phages in cocktails is selecting and mixing specific phages having high lytic activity that can also disrupt and reduce the CF PA biofilm matrix, by producing or inducing biofilm matrix-degrading enzymes (Harper et al. 2014; Pires et al. 2016b). In biofilm aggregates, enzymes produced by the CF PA community breakdown the polysaccharides that hold the biofilm together, thus actively releasing surface bacteria (planktonic cells) that colonize fresh substrates. In our review, two studies reported that two different 2-phage cocktails disrupt and reduce, though by only 2-log (Alemayehu et al. 2012) the CF PA biofilm, also showing age-independent biofilm efficacy (Danis-Wlodarczyk et al. 2015). Phage cocktails, probably by mixing specific biofilm-exopolysaccharide-degrading enzymes, increase lytic phage efficacy by reducing bacterial population densities (Hall et al. 2012), and preventing bacterial resistance (Alves et al. 2015) (Table 20.2). The lytic phages described in our review, and in current practice, often lack detailed criteria for justifying phages chosen for experiments, their genomic sequences, and database accession numbers. These issues reduce the chances for selecting an optimal set of lytic phages to mix in a cocktail, impede experimental repeatability, and make it hard to generalize the results (Abedon 2017b; Adriaenssens and Brister 2017) (Tables 20.2, 20.3 and Fig. 20.1). Overall, these findings imply that regulating and selecting phages for cocktail therapy might be useful in preventing PA lung colonization, and in reducing pulmonary exacerbations in patients with CF (Anderson 2012), as recently tested in preventing and treating PA pulmonary and bacteremic infections in animal models (Forti et al. 2018). Accordingly, new evidence confirming that phage cocktail efficacy depends on selecting the correct phage-degrading-enzyme mixtures arises also from those who claim that in clinical practice PT will probably have restricted usefulness in eliminating bacterial infections caused by biofilm aggregates, i.e. microbial communities inhabiting chronic infections. Conversely, PT could be of great use in decreasing cells that detach themselves from the biofilm and revert to the virulent planktonic phenotype (Criscuolo et al. 2017; Darch et al. 2017). Equally important, new very promising recent studies on tailocins, or phage tail-like bacteriocins from "headless" phages (Ghequire and De Mot 2015), or on lysins selected from phages (Raz et al. 2019) report promising foreground information that can drive research on phage-related therapeutics (Young and Gill 2015). In their study, Ghequire and De Mot (2015), illustrating the "daedalian capacity" of germs to modify exogenous genetic elements for their own benefit, theorized that tailocins could be used as "precision weapons for interbacterial warfare", whileRaz et al. (2019) in their study, selecting two lysins from 16 characterized phage lysins, found evidence that these lysins exhibited >5-log killing activity against PA, thus protecting mice in a lung infection model. Therefore, these lysins are potential drug candidates to treat Gram-negative bacterial and PA infections (Young and Gill 2015).

Whether lytic phages can be used to prevent as well as treat PA infections in patients with CF remains unclear until we have long-term results on using pulmonary models in the best animal candidates. Of the 22 studies reviewed, two showed that preventive treatment requires not heat-killed but active phages (Debarbieux et al. 2010; Morello et al. 2011), thus providing evidence that phage efficiency depends on phage life cycles. Of the 9 in vivo studies we reviewed, 4 investigated phage treatments for preventing PA infections (Beeton et al. 2015; Debarbieux et al. 2010; Morello et al. 2011; Pabary et al. 2016). Despite the impressive efficacy reported, some investigators suggest that the preventive approach is of minor clinical interest because it cannot predict the specific PA strain eventually infecting the patient (Morello et al. 2011). Another open question is whether PT can eradicate MDR, XDR and PDR PA from the airways in patients with CF. Findings in recent years provide evidence that PT can prevent the high-density MDR PA increase and antibiotic treatment failure (Chan et al. 2016; Torres-Barcelò and Hochberg 2016; Chaudhry et al. 2017). A significant and sustained reduction in the number and activity of infecting PA could undoubtedly improve clinical conditions in patients with CF (Krylov 2014).

More information advancing clinical microbiological research in CF, and optimizing phage-cocktail efficacy in treating PA concerns the need to personalize PT by using a standardized approach. Two papers in our review underlined that the most efficacious phages are those isolated on purpose (personalized phage cocktails) against the patients' infecting, and eventually antibiotic-resistant PA strains (sur-mesure, patient-tailored therapy) (Henry et al. 2013; Saussereau et al. 2014). Yet, despite its benefits, the personalized approach has several drawbacks because it is time consuming (phages need to be isolated, purified and completely characterized). Difficulties also arise in assessing phage effectiveness on PA strains isolated in certain medical facilities, and from genetically poorly-differentiated local clones, thus introducing subjectivity in assessing phage lytic activity (Krylov et al. 2016).

Our review also provided new insights into CF PA resistance, including defense mechanisms that allow phages to adapt to changing host systems, thus incurring significant fitness costs. One study in our review emphasizes that a phage-cocktail resistant PAO1 emerges in experiments investigating simultaneous and sequential phage-cocktail exposure. In this study, fitness costs probably arose when phage-cocktail resistant PA overproduced alginate or extracellular polymeric substances (EPS) (Hall et al. 2012). Although how PA phage resistance develops in CF or how PA naturally alters its genotype or phenotype to combat phage virulence is still debatable (Mendes et al. 2014; Krylov et al. 2016), a recent study, supports the hypothesis that a mutant PA, resistant to phage predation, can exhibit a decreased biofilm production, suggesting a fitness cost associated with this resistance (Li et al. 2018). Although a study conducted in recent years suggests that many clinical PA isolates develop the CRISPR-Cas system (Cady et al. 2012), in our review Essoh et al. (2013) observed in the 13 phage-resistant PA strains studied from different clusters isolated from patients with CF no CRISPR-Cas system (Table 20.2).

Although Bondy-Denomy et al. (2013) underline that specific phage genes inactivate the CRISPR-Cas system, our review on CF PA lacks information on phage-borne CRISPR inactivation systems. To advance these findings, a recent major study reported the discrepancies between phage-receptor mutations and CRISPR in PA trade off in vitro and in natural environments. Interestingly, in this study the authors explained the differences between in vitro mutations and natural environments with the biotic complexity of other microbial communities in nature, thus concluding that, by identifying the drivers and consequences of CRISPR-resistance evolution, we might predict and manipulate bacteria-phage interactions in natural and clinical settings (Alseth et al. 2019).

Besides using phages in cocktails to treat CF PA, another strategy for overcoming PA resistance in patients with CF includes phage training. Only one in vivo study (Morello et al. 2011) reported that trained (man-guided evolved) phages were more efficient than untrained (ancestral) phages in killing CF MDR mucoid PA lung-infected mice, and in reducing bacterial resistance in vivo. They also observed that in the curative phage treatment model, trained phages and ancestral phages act cooperatively with the mouse immune response to eliminate MDR PA acute lung infections. Others subsequently highlighted contrasting results, and possible drawbacks in using trained phages in vitro (Betts et al. 2013; Friman et al. 2016), thus reducing their therapeutic possible use (Tables 20.2 and 20.3). On this side, also Henry et al. (2013) demonstrated that phage training is not always feasible, because in their in vivo study they were unable to evolve a highly-virulent phage. Insofar as phage optimization by only two single nucleotide mutations extends host range and is sufficient to improve phage virulence towards CF MDR PA clinical strains (Morello et al. 2011), trained phages possess broad lytic activity and virulence, and might be useful in advancing future research on phage genetic changes (Table 20.3 and Fig. 20.1).

Another problem that has in the past discouraged research on PT in general clinical use, is how to define doses for testing phage safety. Our review findings on reviewing in vivo studies, gave up-to-date information for designing future CF PA clinical studies possibly avoiding risky phage-induced reactions and phage-PA interactions. Many phages collected in libraries and several phages in the studies included in our review came from human waste (sewage). As human beings, we normally encounter phages throughout our lifetime, owing to the complex interactions between bacteria and phage in our colon, upper respiratory system, and on our skin (Merril et al. 2003). In assessing safety in vivo, to investigate possible adverse events we therefore need to include standard negative control groups (PA-uninfected or phage-uninfected animals). For example, experiments testing sham solution injected or nebulized in PA-and-phage-uninfected animals can detect injection-related adverse events, whereas those testing PA-uninfected and phage-treated animals investigate potential phage virulence. Surprisingly, of the 9 in vivo studies examined in this review 6 used unstandardized negative control groups, and reported no safety problems, and only one study assessed safety with standardized negative control groups (Olszak et al. 2015). In this study, Olszak et al. concluded that larval survival was entirely due to phage lytic activity rather than to host immune stimulation (Table 20.3). In another study, Debarbieux et al. (2010) gave bioluminescent-non-mucoid non-CF PAK-infected mice a high phage concentration, monitored their behavior for 10 days, and again reported no adverse effects (Table 20.3). Similar results came from a study conducted by Wang and colleagues in bacteremic mice infected with an imipenem-resistant PA strain, who concluded that inoculating phages at high doses produced no adverse effects in mice (Wang et al. 2006). In the same experiment, Wang et al. underline that to avoid severe adverse effects potentially caused by microbial debris, including the endotoxin LPS released during phage propagation, phages prepared for therapeutic applications must be highly purified (Biswas et al. 2002; Chan et al. 2013). A possible solution to avoid LPS-related mortality, rather than using phages having a high burst size (the average number of phages produced by infected cells) thus producing many virions guaranteeing that sufficient phage numbers reach the infection site (Mirzaei and Nilsson 2015), in patients with CF we suggest selecting a poorly virulent phage cocktail, with a low burst size nebulized directly to the infection pulmonary site (Fig. 20.1). Giving patients nebulized drugs to treat lung infections is a standardized procedure in routine clinical practice and, as so, it is a well-accepted route for patients with CF. Ample evidence shows that phages can be delivered as a dry powder or nebulized (Matinkhoo et al. 2011; Sahota et al. 2015, Lin et al. 2018). If inhaled phages reach the infection site directly, possible characterized and selected phage candidates might include those having a low burst size (causing the infection to progress at a slower rate), hence probably diluting LPS release over time (Fig. 20.1).

Another concern about using vital virus to inhibit PA strains (Morello et al. 2011; Olszak et al. 2015), is the need to assess whether treatment remains effective and safe over time. For example, it should avoid complications related to phage-resistant hosts, bacteria whose pathogenicity has been altered by transduction phenomena, and phage-host genetic rearrangements (Krylov 2014). Although these deleterious events are extremely rare, they might develop especially during the prolonged PT needed for a chronic disease (Waddell et al. 2009). A crucial finding on the issue of using vital virus has been highlighted by a recent study on the importance of an illegitimate phage genome recombination and bacterial chromosome rearrangements that took place during the maintenance of long-term complex host/phage association (Latino et al. 2019). Therefore, there is the need to widen the phage mixtures use delivered from certified reserve storage (phage banks or libraries) including characterized phages having a wide host range, and differing in PA receptor affinity (Kutateladze and Adamia 2008; Domingo-Calap et al. 2016; Górski et al. 2018; Djebara et al. 2019). These phage mixtures could become an extremely valuable clinical resource for ensuring safety, should help laboratory clinicians rapidly to select appropriate phages to include in cocktails, and possibly adapt cocktails to the MDR, XDR and PDR PA strains isolated during CF pulmonary exacerbations (Krylov et al. 2016; Djebara et al. 2019) (Fig. 20.1). Even though our review included only one study that specified the use of a phage from a certified bank (Coulter et al. 2014), and although phage banks lack international institutional regulations, our review identified no information or data on a promising new research direction for clinical PT envisaging combining phage cocktails from individual researchers' phage banks or phage libraries, and a personalized approach using phages isolated on CF PA strains (Tables 20.2 and 20.3). An alternative solution to phage libraries for testing and selecting personalized phage-cocktails, considering the extremely high PA genome plasticity (Table 20.2), might be to set up an international reference library based on clinical CF PA strains with various characteristics (Kutateladze and Adamia 2008; Górski et al. 2018; Riou et al. 2010; Shen et al. 2006; Krylov 2014; Djebara et al. 2019). Although disseminating detailed unpublished research data could be a challenging matter, phage libraries are expensive, need to be continuously updated to account for variability in target bacteria, and can vary from region to region (Kutateladze and Adamia 2008; Chan et al. 2013; Górski et al. 2018; Pirnay et al. 2018; Djebara et al. 2019).

Our review missed data on whether and to what extent phage efficacy depends on administration routes. No studies have compared the various phage administration routes in treating MDR PA or CF PA in vitro or in vivo models. Although the best way to reach PA-infected host lungs may be by delivering phages as a liquid or a dry powder (Carrigy et al. 2017; Chang et al. 2017), the murine CF PA-lung-infected models we reviewed only used intranasal instillation (Alemayehu et al. 2012; Debarbieux et al. 2010; Henry et al. 2013; Lehman et al. 2016; Morello et al. 2011; Pabary et al. 2016). New studies need to address the potential benefits from giving phage by inhalation instead of intranasally. What we mainly need to know is which device to choose for phage inhalation and the right animal models for reproducing human CF airways anatomically and physiologically (Lavelle et al. 2016). We found only one study that in vitro has compared two nebulizers and one inhaler to define the best active liquid phage delivery in treating Mycobacterium tuberculosis. The investigators concluded that delivering phages with a vibrating mesh nebulizer is significantly more effective than using a jet nebulizer ($p < 0.01$), whereas a soft mist inhaler may be useful for self-administration of a phage aerosol (Carrigy et al. 2017). Conversely, comparing efficiency between an air-jet versus a vibrating mesh nebulizer, another recent in vitro study investigating the synergistic antibacterial effect of phage PEV20 in combination with various antibiotics (amikacin, aztreonam, ciprofloxacin, colistin, and tobramycin for inhaled administration) on PA growth derived from patients with CF, reported that the air-jet nebulizer may be more suitable to nebulize phage-antibiotic combination, and observed a completely PA growth inhibition only when ciprofloxacin was used (Lin et al. 2018). Apart from the type of device, another important question to address is the time for nebulizing phages. Known mathematical models can help to assess the best time for nebulizing phages, thus increasing PA clearance from the airways (Nadithe et al. 2003). Although again in a murine model, new information in a non-comparative study on an MDR PA-neutropenic lung-infected mouse model suggests that a dry powder phage formulation might have advantages over a liquid formulation, and could be applied also in CF MDR PA-infected animals (Chang et al. 2017). New promising information advancing in vivo experiments overcoming murine model limitations (no mucus hypersecretion in mice, and the cystic fibrosis transport regulator, CFTR, gene is expressed only in the proximal trachea) envisages specific animal candidates (pigs and ferrets) (Lavelle et al. 2016). Like humans, these animals exhibit the same high CFTR expression in serous cells of submucosal glands in the cartilagineous airways, thus mimicking CF lungs (Fig. 20.1).

By using these experimental procedures in in vivo models, to deliver the best sufficient active phage titers for targeting MDR CF PA in the lungs, we could also address phage pharmacokinetics (Trend et al. 2017). Similarly, by using various phage administration routes to address efficacy on combating MDR, XDR and PDR CF PA lung infections, and negative or adverse phage effects, we could also investigate another unaddressed problem, phage pharmacodynamics (Abedon 2017a). Pharmacokinetics and pharmacodynamics can help transfer research results into clinical care. Even more important, we need to test in vivo phage adverse effects related to immunogenicity in pig and ferret models in studies with a follow-up long enough to allow chronic pulmonary inflammation determined by bacterial persistence to develop (Lavelle et al. 2016; Lin et al. 2017; Trend et al. 2017). Pig and ferret models can also address research on nebulized phage treatment to test what might happen in immune interference from innate immune responses, neutrophil elastase, secretory IgA and macrophages engulfing phage particles after repeated phage administration in CF lungs with PA and other microbial communities (Trend et al. 2017; O'Toole 2018) (Fig. 20.1).

Considering the known resistance to a three-phage cocktail in MDR PA and XDR PA (Larché et al. 2012), another promising approach that helps in reducing MDR PA biofilms, as well as the PA planktonic cells responsible for exacerbations in CF, is to improve treatment effectiveness by combining PT and antibiotics (Fig. 20.1). Although combined phage-antibiotic treatment started when the antibiotic era began (Himmelweit 1945), questions remain on whether to use sequential treatment, optimal phage administration routes, dosing, suspended formulated products, and treatment duration. Surprisingly, only three studies in our review investigated PT combined with tobramycin, streptomycin and colistin on laboratory PAO1 (Coulter et al. 2014; Torres-Barcelò et al. 2014; Danis-Wlodarczyk et al. 2016). The best strategy to ensure phage efficacy, allow a wide antibiotic choice, and minimize possible antibiotic antagonism and PA resistance, is to apply antibiotics at peak phage efficacy (within about 12 h) sequentially rather than simultaneously (Torres-Barcelò et al. 2014), like recently has been confirmed (Chaudhry et al. 2017) (Table 20.2 and Fig. 20.1). We retrieved no studies testing other antibiotics (aminoglycosides, beta-lactams, fluoroquinolones, carbapenems or macrolides) combined with phage against CF and non-CF PA strains, and none investigated PT combined with antibiotics for CF XDR or PDR PA strain infections in vitro or in vivo. We remain skeptical about why Danis-Wlodarczyk et al. (2016) combined PT with colistin, a drug known to alter the bacterial membrane thereby removing phage bacterial receptors (Martis et al. 2014). No comparative head-to-head studies reported antibiotic therapy and trained phages as alternative treatments in CF MDR, XDR or PDR PA strains. The recent promising results of a synergistic antibacterial effect of phage PEV20 in combination with various antibiotics, mainly inhaled ciprofloxacin, support the prospectively clinical use of this combination (Lin et al. 2018). Besides, this finding supports our ongoing preliminary unpublished results on a combination of phages and fluoroquinolones to inhibit MDR PA growth in CF patients. The evolutionary rationale for combining phages with antibiotics provides support for designing pragmatic ecological clinical trials using combined phage-antibiotic therapy possibly in skin infections, such as diabetic ulcers, before applying the best results to systemic diseases in humans (Torres-Barcelò and Hochberg 2016; Chaudhry et al. 2017). Overall, the promising findings this review highlights on trained (man-guided) phages and on how certain antibiotics effectively combat CF MDR, XDR and PDR PA and new preliminary information suggesting that the amikacin-phage combination could have potentially more benefits on PA biofilms than meropenem (Nouraldin et al. 2016) lead us to conclude that trained or even engineered phages combined with inhaled amikacin, ciprofloxacin and meropenem at sub-inhibitory concentrations might reduce PA resistance, thus boosting interest in PT (Torres-Barcelò et al. 2014; Pires et al. 2016a; Lin et al. 2018) (Fig. 20.1).

New evidence implies that genetically engineered phages that might act against MDR, XDR and PDR PA biofilms better than newly-isolated phages, expand their host range and, by so far undefined co-evolutionary phage-bacteria interactions, possibly sensitize MDR, XDR and PDR PA antibiotic efficacy (Magiorakos et al. 2012; Pires et al. 2015, 2016a; Salmond and Fineran 2015; Chan et al. 2016; Torres-Barcelò and Hochberg 2016). Again, underlining the need for combining phage with antibiotics in treating CF MDR and XDR PA, a future crucial research step, before undertaking in vivo therapy, must include pre-testing antibiotics to assess the standardized MIC for MDR, XDR and PDR PA isolates. Other essential stages include characterizing and selecting phages assessed for host range by EOP, by host receptor affinity, and by a personalized approach (by phagograms) (Fauconnier 2019). Final steps include choosing phages also from phage libraries for formulating well-designed phage cocktails, and conducting research on antibiotic and phage dosing, timing, and administration routes (Merabishvili et al. 2009; Djebara et al. 2019) (Fig. 20.1).

20.4 Future Perspectives on Phage Therapy (PT) Against Multidrug-Resistant (MDR), Extensively Drug-Resistant (XDR) and Pandrug-Resistant (PDR) PA in Patients with Cystic Fibrosis (CF)

Despite the promising research highlighted by the argumentative review reported in this chapter on PT against CF PA, clinical trials for PT in patients with CF seem a long way off in Western countries (Parracho et al. 2012; Reardon 2014; Nilsson 2019). Even though the alarming evidence on growing resistance to antibiotics hugely encourage efforts to exploit the promising PT for treating MDR, XDR and PDR PA lung infections in patients with CF (Djebara et al. 2019; Horcajada et al. 2019), researchers responsible for designing future clinical trials for PT in humans neither have defined the right strategy for experiments in vitro and in vivo models on biofilm-producing mucoid PA (Costerton et al. 2003), nor international bodies have issued regulations for defining phage products and study designs to test PT (see Sect. 20.5).

As a first step, clinical researchers should preferably test in vitro a mono-microbial-biofilm-forming mucoid PA model or a non-biofilm-forming PA incorporated in biofilm-producing isolates mimicking a CF lung condition (Deligianni et al. 2010). These biofilm and non-biofilm forming PA often coexist in poly-microbial communities in CF lungs thus determining an evolving microbial community difficult to test (Danis-Wlodarczyk et al. 2015, 2016; O'Toole 2018). Equally important, although phages focus their action specifically on the target pathogen (Loc-Carrillo and Abedon 2011), and although Rogers et al. (2005) state that the various bacteria persist in an active form without interfering with the lung disease or influencing patients' prognosis, others observed that other various microbial species contemporaneously existing in the biofilm can alter phage-induced responses, influencing bacterial and phage densities, and determining how bacterial resistance evolves (Mumford and Friman 2017; O'Toole 2018; Alseth et al. 2019). What we especially need are dynamic models closely resembling physiological conditions (Alves et al. 2015; O'Toole 2018; Alseth et al. 2019), or models mimicking PA growth in CF lungs including the mucus barrier, such as the airway surface liquid model used on epithelial cell lines (Danis-Wlodarczyk et al. 2016), artificial sputum medium (Garbe et al. 2010) or eventually patients' sputum (Saussereau et al. 2014), and test in vivo specific animal candidate models (pig and ferret) (Trend et al. 2017) (Fig. 20.1).

Another crucial step is to characterize selected phages functionally (latency and burst size) and to detect those carrying undesirable genes coding for toxins and antibiotic resistance by combining information from new mathematical modelling for advancing tests in vivo pharmacokinetic and pharmacodynamics processes quantitatively assessed (Loc-Carrillo and Abedon 2011; Nilsson 2019). Some evidence suggests that safety can be efficiently and reliably assessed by separating lytic phages from lysogenic (temperate) phages, characterizing them by genome sequencing and analysis, and by using standardized control groups in in vivo studies (Brüssow 2012; Chan et al. 2013; Mattila et al. 2015; Forti et al. 2018). After having identified the target and the strain to treat, clinical researchers need to select the most appropriate phages to combine in a cocktail (according to their EOP), and select their affinity for the various PA receptors by testing phagograms (Barbu et al. 2016; Kakabadze et al. 2018; Djebara et al. 2019). Researchers need also to test phage ability to degrade the biofilm matrix, and phage low burst size (Forti et al. 2018). Future research on phage administration routes will consider nebulizing phages, suspended them in pre-specified material or giving them as a dry powder (Matinkhoo et al. 2011; Sahota et al. 2015, Lin et al. 2018) (Fig. 20.1). Future directions should include research to investigate in pragmatic trials effective trained phages in cocktails (Domingo-Calap et al. 2016) or engineered phages for formulating new phage cocktails, even combined with antibiotics given also by inhaled route (Lu and Collins 2007; Pires et al. 2016a; Lin et al. 2018). Phage cocktail efficacy and safety, and phages eventually combined with antibiotics, should then be validated in vivo in skin diseases before testing them in systemic diseases easily and at a low cost in non-pulmonary models (Jault et al. 2019) (Fig. 20.1).

Because human immunological responses to phages differ significantly during chronic infections (Li et al. 2005; Lin et al. 2017; Waters et al. 2017), we also need to test outcomes in ecologic pragmatic studies investigating long-term PA infections and biofilm forming PA (O'Toole 2018; Horcajada et al. 2019; Latino et al. 2019). All experimental procedures should include standardized negative control groups (Olszak et al. 2015) (Fig. 20.1). Future promises and perspectives in PT for treating MDR, XDR and PDR PA lung infections in CF include models designed to apply antibiotics to phages at peak phage efficacy (within about 12 h) sequentially rather than simultaneously (Torres-Barcelò et al. 2014) (Fig. 20.1). We envisage future head-to-head studies comparing antibiotics given at sub-inhibitory concentrations combined with trained or engineered phages and antibiotics given alone or in cocktails from certified banks (Kutateladze and Adamia 2008; Merabishvili et al. 2009; Nouraldin et al. 2016; Pires et al. 2016a; Lin et al. 2017, 2018; Górski et al. 2018; Djebara et al. 2019). These research advances might reduce PA resistance, and boost interest in PT for treating patients with CF (Fig. 20.1).

A hypothesized modular approach proposed in 2007 (Lu and Collins 2007) and reported also by Krylov et al. (2016) entails formulating phage mixtures (monospecies or hetero-species), enhancing their activity by introducing newly characterized engineered phages to express the most effective EPS-degrading enzymes specific to the target biofilm, and combining them in a personalized way to treat PA infection in patients with CF (Pires et al. 2016a, b). These phage mixtures included in certified phage libraries, such as the Belgian phage bank (Pirnay et al. 2018) could allow different research groups to cooperate in developing publicly free phage libraries, to ensure distributing phage mix on call, thus encouraging the use of the PT in untreatable CF MDR, XDR and PDR PA pulmonary infections (Krylov et al. 2016; Sybesma et al. 2018; Djebara et al. 2019) (Fig. 20.1). Gaining crucial information on phage pedigree properties, standardization and integration into development models (Abedon 2017a, b; Forti et al. 2018), and on potential PT in CF (Chan et al. 2013) from reliable and repeatable experiments on PT against PA, MDR, XDR and PDR PA in CF patients, could permit to pilot studies for testing phage efficacy and safety in hospital settings, hence avoiding any serious immunological reactions, and therefore optimize rather than maximize PT in CF (Djebara et al. 2019; Horcajada et al. 2019; Nilsson 2019).

20.5 Phages in International Regulations. Current Use and Perspectives of Phage Therapy (PT) in Patients with Cystic Fibrosis (CF)

Admittedly, PT faces two unsolved major issues that plague the still confused international regulations. The first issue concerns how to consider phages in therapeutics. Phages, as unconventional active self-evolving viruses with complex dynamics in phage-bacteria coevolution, can be unlikely defined as canonical drugs, or investigated by using standard pharmacokinetic models (Fauconnier 2019). Conversely, investigating phages pragmatically, during their ecological processes in microbial communities (Koskella and Brockhurst 2014; Torres-Barcelò 2018), could reveal their full dimension, and their potential to develop therapeutic strategies in chronic lung infections and in MDR, XDR and PDR PA CF infections (O'Toole 2018). The second concern, when considering phages as drugs, is the huge research expenditure that would be required by testing phages in clinical trials (Fauconnier 2019). As such, in clinical trials costly regulatory framework of Good Manufacturing Practices (GMPs) need to be applied, in spite of the standardized clinical trial design unsuitable for analyzing phages that are active and self-replicating entities co-evolving in ecological microbial community (Young and Gill 2015). Equally important, in Europe, standard regulations on trials after the thalidomide drama in the sixties, had been issued. The specific regulation on PT in the European Directive 2001/83/EC (European Commission 2001) was assimilated to medicinal products (PTMPs), namely "medicinal products for human use intended to be placed on the market in Member States and either prepared industrially or manufactured by a method involving an industrial process". This implied that phages could not be included in the regulation of magistral formulas (Pirnay et al. 2018), whereas the specific features of trained or engineered phages or phage cocktails for personalized approaches, fit well within magistral/galenic mixtures prepared in local hospitals (Pires et al. 2016a; Pirnay et al. 2018; Fauconnier 2019). The still ongoing European regulatory debate on the definition and application of phage therapeutics as PTMPs ready-prepared medicines (European Commission 2001) or magistral formulas (Pirnay et al. 2018; Fauconnier 2019) show similarities with other biological entities (i.e. fecal microbiota transplant, FMT) (Vyas et al. 2015). These challenges remain unsolved along the years. To overcome this issue, the European legislation coined the term advanced therapy medicinal products (ATMPs), and designed a tailored regulatory framework to address ATMPs as customized hospital-made or possibly engineered products for personalized approaches (Leyens et al. 2015; Pires et al. 2016a). Even though ATMPs have some similarity with phages, phages as ATMPs cannot progress without expensive industrially-based development. For example, in the US although two phase I/II trials tested safety, efficacy and tolerability of a single phage in skin lesions or a phage cocktail to treat burn wounds infected by PA (Jault et al. 2019), no phage products were licensed or are currently marketed. So, to advance pharmaceutical research and development (R&D) on PT, the US issued a regulation and granted PT for an "expanded access" to unauthorized drugs outside clinical trials (FDA 1998; Brüssow 2012; FDA 2014; LaVergne et al. 2018), whereas in Europe, although some flexibility in the Regulation 536/2014, the boundaries of the GMP are still in place for unauthorized drugs (European Commission 2014).

By the way, in Europe, to treat CF with PT, we could rely on other various national regulations. For instance, in Eastern Europe during decades of clinical experience inside official health institutions (Georgia, Poland, Russia), patients with CF are being treated with PT (Kutateladze and Adamia 2010; Krylov et al. 2016), whereas in Western Europe, international regulatory bodies, unable to solve ethical concerns to advance clinical trials involving MDR, XDR or PDR PA strain infecting CF lungs or to pilot possible clinical case-control studies (Verbeken et al.

2007; Djebara et al. 2019), PT in clinical practice is not permitted unless for compassionate use, according to specific "adaptive pathways" regulating new unproven drugs (European Commission Directive 83/2001; European Commission Regulation 726/2004; European Commission Regulation 536/2014; European Medicines Agency 2015). Similarly, the US regulated the use of PT in clinical practice as a drug under the "expanded access" to unauthorized drugs if any other treatment exists (FDA 1998, 2014; Pirnay et al. 2018).

20.5.1 A Call for Developing a Framework for Advanced Phage Biological Therapy in Cystic Fibrosis (CF)

Although unheard by international regulatory bodies, clinical researchers globally and repeatedly demand PT when there are no available therapeutic protocols for patients with chronic diseases and CF, and use PTMPs in PT R&D in their clinical routine (Pelfrene et al. 2016; Furfaro et al. 2018; Górski et al. 2018; Pelfrene et al. 2019). To bypass the challenges of pharmaceutical regulations, we claim to consider phages as biological products and, while envisaging R&D to assure for PT a strong patent protection, we wish also to highlight the potential huge cost of possible useless clinical trials, owing to the specific microbial and ecological features of phages (Todd 2019). Therefore, we envisage clinical experiments on advanced phage biological therapy (APBT) for patients with CF. Potent phages can be prepared by synthetic biology (Barbu et al. 2016; Pires et al. 2016a) thus encouraging the use of APBT when MDR, XDR and PDR PA strain infections have the antibiogram evidence of no effective antibiotic therapy (World Medical Association 2013).[1] Since 1902, when the Biologics Control Act stated the federal regulations for producing vaccines and other biological products, biologicals prepared from human or animal materials or from microbiologic origin (such as vaccines, insulin or blood cells) represented one of the major progresses in public health (Baylor and Midthun 2008). As well as happened with vaccines, we need to progress in phage regulation, R&D, so as to define GMP procedures for APBT products. Besides, pharmacovigilance and surveillance can advance phage clinical research rapidly and globally (Merabishvili et al. 2009; Moelling et al. 2018; Pirnay et al. 2018; Sybesma et al.

[1] The World Medical Association (WMA) in 1964 developed the Declaration of Helsinki as a statement of ethical principles for medical research involving human subjects. In the Article 37 the WMA states: "In the treatment of an individual patient, where proven interventions do not exist or other known interventions have been ineffective, the physician, after seeking expert advice, with informed consent from the patient or a legally authorised representative, may use an unproven intervention if, in the physician's judgement, it offers hope of saving life, re-establishing health, or alleviating suffering. This intervention should subsequently be made the object of research, designed to evaluate its safety and efficacy. In all cases, new information must be recorded and, where appropriate, made publicly available". Available at: https://www.wma.net/policies-post/wma-declaration-of-helsinki-ethical-principles-for-medical-research-involving-human-subjects/ Cited October 2013.

2018). To do so, to overcome clinical trials in PT, other study designs should be considered, such as N-of-1, i.e. single subject clinical trials where a patient embodies the study, or basket trials, that allow a patient selection before the enrollment based on their bacteria sensitivity to phage treatment (phagogram) defined by the spot test and the double-agar overlay method (Barbu et al. 2016; Kakabadze et al. 2018; Moelling et al. 2018; Sybesma et al. 2018). Admittedly, one of the main reasons why Big Pharma neglect R&D on PT is because drug companies cannot get intellectual property protection on PT or even make significant profit out of PT, because clinical researchers are repeatedly using phages for years (Parfitt 2005). Therefore, if regulatory bodies insist in considering phages as drugs, PT should have to be proven even by expensive phase I clinical trials, in spite of the ever-increasing drug companies interests in developing drugs for chronic ailments or depression rather than for researching an efficient one-off treatment for a resistant bacterial infection (Parfitt 2005).

Fortunately, despite Big Pharma and European regulatory bodies reluctance in providing effective pragmatic solutions to advance PT, we deem essential the breakthrough made since 2007 at the Queen Astrid Military Hospital (QAMH) in Brussels, Belgium (Djebara et al. 2019). At the QAMH, Pirnay and his colleagues, aiming to overcome PT current definitions and regulatory barriers, under the umbrella of the Art. 37 of the Declaration of Helsinki, issued a monograph with written standards to assess the quality of phages as active substances thus advancing the preparation of magistral formulas. They tested the phage stock, certified the phage material, according to the current technical R&D for preparing customized products based on the medical prescriptions, and accelerated the production of new potent phages by synthetic biology (Barbu et al. 2016; Pirnay et al. 2018). Since 2007, the QAMH gathered 260 requests from the Northern and Southern European countries, and the most frequent requests were for treating MDR, XDR and PDR PA strain infections in lower respiratory tracts and CF (Djebara et al. 2019). Unfortunately, owing to the lack of a specific single phage or phage cocktail to treat the infection, about a half of the requests had to be dismissed (Djebara et al. 2019).

In summary, although a desirable phage selection on an internationally available PT reference set still lacks international regulations, we claim specific rules to use ABPT containing well-combined and continuously updated newly-isolated lytic phages given by publicly available certified banks, like the Belgian laboratory nationally approved (Pirnay et al. 2018). This new specific rules should permit a regulatory framework for phage production and its therapeutic use as "magistral formulas", namely patient-tailored galenic mixtures prepared by pharmacists from ABPT products provided by the Belgian certified bank or other similar certified banks from Georgia, Poland or Russia (Kutateladze and Adamia 2008; Górski et al. 2018; Djebara et al. 2019). W

hat we need now is a committee held by European regulators and public health institutions involved in PT to urgently define PT as ABPT products thus advancing PT to treat MDR, XDR and PDR PA strains in CF (Moelling et al. 2018; Pirnay et al. 2018; Djebara et al. 2019; Fauconnier 2019).

20.6 Conclusions

In conclusion, although the results reported as an argumentative review in this chapter addressed practical in vivo models for advancing clinical research on PT to design protocols for pivotal clinical trials, instead we claim to test ABPT against MDR, XDR and PDR PA strains in CF by pragmatic design studies in close coordination among central certified phage banks, microbiologists and clinicians caring for patients with CF (Fig. 20.1). Our claim also involves implementing dialogues between the medical community and the CF community reference groups thus helping to overcome complex ethical dilemmas (Caplan 2015; Trend et al. 2017; Djebara et al. 2019; Fauconnier 2019). The approach of considering PT as ABPT promises clinical advantages including reducing costs, speeding phage cocktail delivery, analyzing dose-response curves, and "weeding out" phages that could show incompatibility (Chan and Abedon 2012; Alseth et al. 2019; Djebara et al. 2019). Although no research has yet defined the kinetic mechanisms responsible for phage actions especially against CF PA biofilms, to simplify ABPT R&D we agree with others (Chan and Abedon 2012; Domingo-Calap et al. 2016; O'Toole 2018; Alseth et al. 2019) on the need to design pragmatic clinical protocols investigating a minimum number of active phages (2-3 phages) via a nebulizer or by an inhaler delivered directly to the CF lungs (Carrigy et al. 2017; Trend et al. 2017). This standard way is effective and meets with patients' approval. Given the importance of simulating in vivo a condition resembling that in CF patients' lungs (O'Toole 2018), overall the findings from our review suggest that research needs also to develop models in vivo able to detect microbial communities in chronic PA infections in CF, and to investigate the most important concerns on MDR, XDR and PDR PA phage co-evolutionary interactions and immunogenicity (O'Toole 2018). Future studies on immunogenicity in animals (pigs and ferrets) that exhibit a high CFTR expression in lungs, likely CF, will show whether mucus can stimulate phage adhesion to the lungs, prolong phage therapeutic activity, and how phage particles can overcome immune interference (Barr et al. 2013; Trend et al. 2017; O'Toole 2018; Alseth et al. 2019). A crucial step is to use phage genome sequencing to identify possible virulence factors or prophages from host strains, thus avoiding to transfer possible contamination in phage stocks (Abedon 2017a; Fig. 20.1).

Convincing research from synthetic biology and nanotechnology techniques already exploits new engineered phages that display selected proteins binding specific CF PA targets isolated and re-amplified from commercially available certified phage libraries (Smith 1985; Kaur et al. 2012; Henry et al. 2015; Barbu et al. 2016; Pires et al. 2016a; Criscuolo et al. 2017; Pirnay et al. 2018; Djebara et al. 2019). Using phages from certified libraries and considering phages as biological products, namely ABPT, will help to overcome concerns requiring international regulatory agencies to approve or re-approve phage cocktails for each individual possible clinical trial (Parracho et al. 2012; Sarhan and Azzazy 2015; Salmond and Fineran 2015; Criscuolo et al. 2017). Even though obtaining a phage set that effectively lyses all variants in each pathogen is a challenging problem (Pirnay et al. 2012; Chan et al.

2013; Djebara et al. 2019), our findings suggest that successful ABPT on a larger scale in patients with CF cannot depend on involving the pharmaceutical industry in preclinical development and formulation before undertaking clinical trials (Parfitt 2005). It is unlikely that preclinical industry-funded phage formulation could be developed thus creating the claimed first-in-human advisory system (Kimmelman and Federico 2017). Conversely, this regulatory board system could critically match preclinical evidence and clinical scenarios for ABPT in CF PA, and envisage solutions on how to best deal with the high investment needed for creating an advisory board regulatory infrastructure. This infrastructure is essential to justify first-in-human trials for PT, thus advancing PT to treat even CF MDR, XDR and PDR PA. Hence, the new ABPT approach we claim, will improve laboratory techniques intended to develop ABPT combined with antibiotics so as to combat MDR, XDR and PDR PA strains effectively(Chan et al. 2013; Domingo-Calap et al. 2016; Torres-Barcelò and Hochberg 2016; Lin et al. 2017; Moelling et al. 2018; Djebara et al. 2019). Eventually, combining ABPT with antibiotics might be more appealing to pharmaceutical companies because it could give them opportunities for new patents and profits.

Collectively, our findings argue that, despite the continuing need to improve phage efficacy and safety in vivo and solve ethical and regulatory concerns, research into lytic ABPT in CF PA strains holds promise. Progress over the past 20 years especially intended to combat increasing antibiotic resistance, including hospital studies, already suggests that lytic ABPT of CF MDR, XDR and PDR PA infections will become a useful, safe medical procedure (Moelling et al. 2018; Pirnay et al. 2018; Djebara et al. 2019). Though ABPT will never completely replace antibiotics in patients with CF, pharmaceutical industries should eventually combine ABPT with one or two antibiotics scaled by MIC to pre-test their sensitivity, and also the phagogram, to treat CF MDR, XDR and PDR PA (Djebara et al. 2019; Fauconnier 2019). Advances that should encourage and simplify phage therapeutic use in CF MDR, XDR and PDR PA include regularly updating cocktail preparations in certified publicly available phage libraries for personalized ABPT, preferably using genomic and patentable engineered phages so as to ensure safety certified by international regulatory agencies (Verbeken et al. 2007; Pires et al. 2016a; Pirnay et al. 2018; Djebara et al. 2019; Fauconnier 2019). Researchers should release all available ABPT results to the medical society so that procedures could be corrected and adjusted accordingly. Owing to easy phage isolation, sequence-specific targeting and replication on specific bacterial hosts, active penetration into bacterial biofilms, low production costs, safety, lack of known immunogenicity, genomic engineered phages from patented phage libraries regulated to permit a promising clinical strategy in CF, and Big Pharma disinterest in producing new antibiotics, introducing ABPT in clinical practice by efficient personalized cocktails combined with existing PA-sensitive antibiotics in treating MDR, XDR and PDR PA in CF appears a promising development over time.

References

Abedon ST (2017a) Information phage therapy research should report. Pharmaceuticals 10:43. https://doi.org/10.3390/ph10020043

Abedon ST (2017b) Phage name check. http://www.phage.org/phage_name_check.html. Cited 26 Apr 2017

Ackermann HW (2011) Bacteriophage taxonomy. Microbiol Aust 32:90–94

Adriaenssens E, Brister JR (2017) How to name and classify your phage: an informal guide. Viruses 9:70. https://doi.org/10.3390/v9040070

Alemayehu D, Casey PG, McAuliffe O, Guinane CM, Martin JG, Shanahan F, Coffey A, Ross RP, Hill C (2012) Bacteriophages ΦMR299-2 and ΦNH-4 can eliminate *Pseudomonas aeruginosa* in the murine lung and on cystic fibrosis lung airway cells. M Bio 3:e00029–e00012. https://doi.org/10.1128/mBio.00029-12

Alseth EO, Pursey E, Luján AM, McLeod I, Rollie C, Westra ER (2019) Bacterial biodiversity drives the evolution of CRISPR-based phage resistance. Nature 574:549–552

Alves DR, Perez-Esteban P, Kot W, Bean J, Arnot T, Hansen L, Enright MC, Jenkins ATA (2015) A novel bacteriophage cocktail reduces and disperses *Pseudomonas aeruginosa* biofilms under static and flow conditions. Microb Biotechnol 9:61–74. https://doi.org/10.1111/1751-7915.12316

Anderson GG (2012) *Pseudomonas aeruginosa* biofilm formation in the CF lung and its implications for therapy. In: Sriramulu D (ed) Cystic fibrosis – renewed hopes through research. IntechOpen, London, pp 153–180

Azeredo J, Sutherland IW (2008) The use of phages for the removal of infectious biofilms. Curr Pharm Biotechnol 9:261–266

Bacci G, Paganin P, Lopez L, Vanni C, Dalmastri C, Cantale C, Daddiego L, Perrotta G, Dolce D, Morelli P et al (2016) Pyrosequencing unveils cystic fibrosis lung microbiome differences associated with a severe lung function decline. PLoS One 11:e0156807. https://doi.org/10.1371/journal.pone.0156807

Barbu EM, Cady KC, Hubby B (2016) Phage therapy in the era of synthetic biology. Cold Spring Harb Perspect Biol 8:a023879

Barr JJ, Auro R, Furlan M, Whiteson KL, Erb ML, Pogliano J, Stotland A, Wolkowicz R, Cutting AS, Doran KS et al (2013) Bacteriophage adhering to mucus provide a non–host-derived immunity. Proc Natl Acad Sci U S A 110:10771–10776. https://doi.org/10.1073/pnas.1305923110

Bassetti M, Ginocchio F, Mikulska M (2011) New treatment options against gram-negative organisms. Crit Care 15:215. https://doi.org/10.1186/cc9997

Battán PC, Barnes AI, Albesa I (2004) Resistance to oxidative stress caused by ceftazidime and piperacillin in a biofilm of *Pseudomonas*. Luminescence 19:265–270. https://doi.org/10.1002/bio.779

Baylor NW, Midthun K (2008) Regulation and testing of vaccines. In: Plotkin S, Orenstein W, Offit PA (eds) Vaccines, 5th edn. WB Saunders, Philadelphia, pp 1611–1627

Beeton ML, Alves DR, Enright MC, Jenkins ATA (2015) Assessing phage therapy against *Pseudomonas aeruginosa* using a *Galleria mellonella* infection model. Int J Antimicrob Agents 46:196–200. https://doi.org/10.1016/j.ijantimicag.2015.04.005

Betts A, Vasse M, Kaltz O, Hochberg ME (2013) Back to the future: evolving bacteriophages to increase their effectiveness against the pathogen *Pseudomonas aeruginosa* PAO1. Evol Appl 6:1054–1063. https://doi.org/10.1111/eva.12085

Biswas B, Adhya S, Washart P, Paul B, Trostel AN, Powell B, Carlton R, Merril RC (2002) Bacteriophage therapy rescues mice bacteremic from a clinical isolate of vancomycin-resistant *Enterococcus faecium*. Infect Immun 70:204–210. https://doi.org/10.1128/IAI.70.1.204-210.2002

Boles BR, Thoendel M, Singh PK (2004) Self-generated diversity produces insurance effects in biofilm communities. Proc Natl Acad Sci U S A 101:16630–16635. https://doi.org/10.1073/pnas.0407460101

Bondy-Denomy J, Pawluk A, Maxwell KL, Davidson AR (2013) Bacteriophages genes that inactivate the CRISPR/Cas bacterial immune system. Nature 493:429–432. https://doi.org/10.1038/nature11723

Bradley DE, Robertson D (1968) The structure and infective process of a contractile *Pseudomonas aeruginosa* bacteriophage. J Gen Virol 3:247–254

Brunak S, Danchin A, Hattori M, Nakamura H, Shinozaki K, Matise T, Preuss D (2002) Nucleotide sequence database policies. Science 298:1333–1334

Brüssow H (2012) What is needed for phage therapy to become a reality in western medicine? Virology 434:138–142. https://doi.org/10.1016/j.virol.2012.09.015

Cady KC, Bondy-Denomy J, Heussler GE, Davidson AR, O'Toole GA (2012) The CRISPR/Cas adaptive immune system of *Pseudomonas aeruginosa* mediates resistance to naturally occurring and engineered phages. J Bacteriol 194:5728–5738. https://doi.org/10.1128/JB.01184-12

Caplan A (2015) Morals, microbes, and methods. Lancet 385:e45. https://doi.org/10.1016/S0140-6736(15)60928-X

Carrigy NB, Chang RY, Leung SS, Harrison M, Petrova Z, Pope WH, Hatfull GF, Britton WJ, Chan HK, Sauvageau D et al (2017) Anti-tuberculosis bacteriophage D29 delivery with a vibrating mesh nebulizer, jet nebulizer, and soft mist inhaler. Pharm Res 34:2084–2096. https://doi.org/10.1007/s11095-017-2213-4

Chan BK, Abedon ST (2012) Phage therapy pharmacology: phage cocktails. Adv Appl Microbiol 78:1–23. https://doi.org/10.1016/B978-0-12-394805-2.00001-4

Chan BK, Abedon ST, Loc-Carrillo C (2013) Phage cocktails and the future of phage therapy. Future Microbiol 8:769–783. https://doi.org/10.2217/fmb.13.47

Chan BK, Sistrom M, Wertz JE, Kortright KE, Narayan D, Turner PE (2016) Phage selection restores antibiotic sensitivity in MDR *Pseudomonas aeruginosa*. Sci Rep 6:26717

Chang RYK, Chen K, Wang J, Wallin M, Britton W, Morales S, Kutter E, Li J, Chan HK (2017) Anti-Pseudomonal activity of phage PEV20 in a dry powder formulation. A proof-of-principle study in a murine lung infection model. Antimicrob Agents Chemother AAC-1714. https://doi.org/10.1128/AAC.01714-17

Chaturongakul S, Ounjai P (2014) Phage-host interplay: examples from tailed phages and Gram-negative bacterial pathogens. Front Microbiol 5:442. https://doi.org/10.3389/fmicb.2014.00442

Chaudhry WN, Concepción-Acevedo J, Park T, Andleeb S, Bull JJ, Levin BR (2017) Synergy and order effects of antibiotics and phages in killing *Pseudomonas aeruginosa* biofilms. PLoS One 12:e0168615

Chmiel JF, Davis PB (2003) State of the art: why do the lungs of patients with cystic fibrosis become infected and why can't they clear the infection? Respir Res 4:8. https://doi.org/10.1186/1465-9921-4-8

Clokie MR, Millard AD, Letarov AV, Heaphy S (2011) Phages in nature. Bacteriophage 1:31–45. https://doi.org/10.4161/bact.1.1.14942

Cochran W, Suh SJ, McFeters G, Stewart P (2000) Role of RpoS and AlgT in *Pseudomonas aeruginosa* biofilm resistance to hydrogen peroxide and monochloramine. J Appl Microbiol 88:546–553. https://doi.org/10.1046/j.1365-2672.2000.00995.x

Costerton JW, Stewart PS, Greenberg EP (1999) Bacterial biofilms: a common cause of persistent infections. Science 284:1318–1322. https://doi.org/10.1126/science.284.5418.1318

Costerton W, Veeh R, Shirtliff M, Pasmore M, Post C, Ehrlich G (2003) The application of biofilm science to the study and control of chronic bacterial infections. J Clin Invest 112:1466–1477. https://doi.org/10.1172/JCI200320365

Coulter LB, McLean RJ, Rohde RE, Aron GM (2014) Effect of bacteriophage infection in combination with tobramycin on the emergence of resistance in *Escherichia coli* and *Pseudomonas aeruginosa* biofilms. Viruses 6:3778–3786. https://doi.org/10.3390/v6103778

Criscuolo E, Spadini S, Lamanna J, Ferro M, Burioni R (2017) Bacteriophages and their immunological applications against infectious threats. J Immunol Res 2017:3780697. https://doi.org/10.1155/2017/3780697

D'Hérelle F (1917) Sur un microbe invisible antagoniste des bacilles dysentériques. C R Acad Sci Ser D 165:373–375

Danis-Wlodarczyk K, Olszak T, Arabski M, Wasik S, Majkowska-Skrobek G, Augustyniak D, Gula G, Briers Y, Jang HB, Vandenheuvel D et al (2015) Characterization of the newly isolated lytic bacteriophages KTN6 and KT28 and their efficacy against *Pseudomonas aeruginosa* biofilm. PLoS One 10:e0137015. https://doi.org/10.1371/journal.pone.0127603

Danis-Wlodarczyk K, Vandenheuvel D, Jang HB, Briers Y, Olszak T, Arabski M, Wasik S, Drabik M, Higgins G, Tyrrell J et al (2016) A proposed integrated approach for the preclinical evaluation of phage therapy in *Pseudomonas* infections. Sci Rep 6:28115. https://doi.org/10.1038/srep28115

Darch SE, Kragh KN, Abbott EA, Bjarnsholt T, Bull JJ, Whiteley M (2017) Phage inhibit pathogen dissemination by targeting bacterial migrants in a chronic infection model. M Bio 8:e00240–e00217. https://doi.org/10.1128/mBio.00240-17

Davies KJ, Lloyd D, Boddy L (1989) The effect of oxygen on denitrification in *Paracoccus denitrificans* and *Pseudomonas aeruginosa*. J Gen Microbiol 135:2445–2451. https://doi.org/10.1099/00221287-135-9-2445

Debarbieux L, Leduc D, Maura D, Morello E, Criscuolo A, Grossi O, Balloy V, Touqui L (2010) Bacteriophages can treat and prevent *Pseudomonas aeruginosa* lung infections. J Infect Dis 201:1096–1104. https://doi.org/10.1086/651135

Deligianni E, Pattison S, Berrar D, Ternan NG, Haylock RW, Moore JE, Elborn SJ, Dooley JSG (2010) *Pseudomonas aeruginosa* cystic fibrosis isolates of similar RAPD genotype exhibit diversity in biofilm forming ability in vitro. BMC Microbiol 10:38. https://doi.org/10.1186/1471-2180-10-38

Djebara S, Maussen C, De Vos D, Merabishvili M, Damanet B, Win Pang K, De Leenheer P, Strachinaru I, Soentjens P, Pirnay JP (2019) Processing phage therapy requests in a Brussels military hospital: lessons identified. Viruses 11:265. https://doi.org/10.3390/v11030265

Domingo-Calap P, Georgel P, Bahram S (2016) Back to the future: bacteriophages as promising therapeutic tools. HLA 87:133–140. https://doi.org/10.1111/tan.12742

Donlan RM, Costerton JW (2002) Biofilms: survival mechanisms of clinically relevant microorganisms. Clin Microbiol Rev 15:167–193. https://doi.org/10.1128/CMR.15.2.167-193.2002

Drulis-Kawa Z, Majkowska-Skrobek G, Maciejewska B, Delattre AS, Lavigne R (2012) Learning from bacteriophages-advantages and limitations of phage and phage-encoded protein applications. Curr Protein Pept Sci 13:699–722. https://doi.org/10.2174/138920312804871193

Edgar R, Friedman N, Molshanski-Mor S, Qimron U (2012) Reversing bacterial resistance to antibiotics by phage-mediated delivery of dominant sensitive genes. Appl Environ Microbiol 78:744–751. https://doi.org/10.1128/AEM.05741-11

Essoh C, Blouin Y, Loukou G, Cablanmian A, Lathro S, Kutter E, Thien HV, Vergnaud G, Pourcel C (2013) The susceptibility of *Pseudomonas aeruginosa* strains from cystic fibrosis patients to bacteriophages. PLoS One 8:e60575. https://doi.org/10.1371/journal.pone.0060575

European Commission (2001) Directive (EC) 83/2001 of the European Parliament and of the Council of 6 November 2001 on the Community code relating to medicinal products for human use. Available at: https://eur-lex.europa.eu/legal-content/en/ALL/?uri=CELEX%3A32001L0083. Cited 28 Nov 2001

European Commission (2004) Regulation (EC) 726/2004 of the European Parliament and of the Council of 31 March 2004 laying down Community procedures for the authorisation and supervision of medicinal products for human and veterinary use and establishing a European Medicines Agency. Available at: https://eur-lex.europa.eu/legal-content/EN/TXT/?uri=CELEX%3A32004R0726. Cited 30 Apr 2004

European Commission (2014) Regulation (EC) 536/2014 of the European Parliament and of the Council of 16 April 2014 on clinical trials on medicinal products for human use, and repealing Directive 2001/20/EC. Available at: https://ec.europa.eu/health/sites/health/files/files/eudralex/vol-1/reg_2014_536/reg_2014_536_en.pdf. Cited 16 Apr 2014

European Medicines Agency, EMA (2015) Adaptive pathways. Available at: https://www.ema.europa.eu/en/human-regulatory/research-development/adaptive-pathways. Cited 20 Nov 2015

Farrell PM, Kosorok MR, Rock MJ, Laxova A, Zeng L, Lai HC, Hoffman G, Laessig RH, Splaingard ML (2001) Early diagnosis of cystic fibrosis through neonatal screening prevents severe malnutrition and improves long-term growth. Wisconsin Cystic Fibrosis Neonatal Screening Study Group. Pediatrics 107:1–13

Fauconnier A (2019) Phage therapy regulation: from night to dawn. Viruses 11:352. https://doi.org/10.3390/v11040352

Food and Drug Administration (FDA) (1998) Emergency use of an investigational drug or biologic. In: Investigational new drug (IND) application. Available at: https://www.fda.gov/regulatory-information/search-fda-guidance-documents/emergency-use-investigational-drug-or-biologic. Cited 12 July 2018

Food and Drug Administration (FDA) (2014) Section 503A of the Federal Food, Drug, and Cosmetic Act. In: FDA guidance on pharmacy compounding of human drug products. https://www.fda.gov/drugs/human-drug-compounding/section-503a-federal-food-drug-and-cosmetic-act. Cited 21 June 2018

Forti F, Roach DR, Cafora M, Pasini ME, Horner DS, Fiscarelli EV, Rossitto M, Cariani L, Briani F, Debarbieux et al (2018) Design of a broad-range bacteriophage cocktail that reduces *Pseudomonas aeruginosa* biofilms and treats acute infections in two animal models. Antimicrob Agents Chemother 62:e02573–e02517. https://doi.org/10.1128/AAC.02573-17

Frederiksen B, Koch C, Høiby N (1997) Antibiotic treatment of initial colonization with *Pseudomonas aeruginosa* postpones chronic infection and prevents deterioration of pulmonary function in cystic fibrosis. Pediatr Pulmonol 23:330–335. https://doi.org/10.1002/(SICI)1099-0496(199705)23:5<330:AID-PPUL4>3.0.CO;2-O

Friman VP, Soanes-Brown D, Sierocinski P, Molin S, Johansen HK, Merabishvili M, Pirnay JP, De Vos D, Buckling A (2016) Pre-adapting parasitic phages to a pathogen leads to increased pathogen clearance and lowered resistance evolution with *Pseudomonas aeruginosa* cystic fibrosis bacterial isolates. J Evol Biol 29:188–198. https://doi.org/10.1111/jeb.12774

Furfaro LL, Payne MS, Chang BJ (2018) Bacteriophage therapy: clinical trials and regulatory hurdles. Front Cell Infect Microbiol 8. https://doi.org/10.3389/fcimb.2018.00376

Furukawa S, Kuchma S, O'Toole G (2006) Keeping their options open: acute versus persistent infections. J Bacteriol 188:1211–1217. https://doi.org/10.1128/JB.188.4.1211-1217.2006

Garbe J, Wesche A, Bunk B, Kazmierczak M, Selezska K, Rohde C, Sikorski J, Rohde M, Jahn D, Schobert M (2010) Characterization of JG024, a *Pseudomonas aeruginosa* PB1-like broad host range phage under simulated infection conditions. BMC Microbiol 10:301. https://doi.org/10.1186/1471-2180-10-301

Gellatly SL, Hancock RE (2013) *Pseudomonas aeruginosa*: new insights into pathogenesis and host defenses. Pathog Dis 67:159–173. https://doi.org/10.1111/2049-632X.12033

Ghequire MGK, De Mot R (2015) The tailocin tale: peeling off phage tails. Trends Microbiol 23:587–590. https://doi.org/10.1016/j.tim.2015.07.011

Gibson RL, Burns JL, Ramsey BW (2003) Pathophysiology and management of pulmonary infections in cystic fibrosis. Am J Respir Crit Care Med 168:918–951. https://doi.org/10.1164/rccm.200304-505SO

Gómez MI, Prince A (2007) Opportunistic infections in lung disease: *Pseudomonas* infections in cystic fibrosis. Curr Opin Pharmacol 7:244–251. https://doi.org/10.1016/j.coph.2006.12.005

Gooderham WJ, Hancock RE (2009) Regulation of virulence and antibiotic resistance by two-component regulatory systems in *Pseudomonas aeruginosa*. FEMS Microbiol Rev 33:279–294. https://doi.org/10.1111/j.1574-6976.2008.00135.x

Górski A, Międzybrodzki R, Łobocka M, Głowacka-Rutkowska A, Bednarek A, Borysowski J, Jończyk-Matysiak E, Łusiak-Szelachowska M, Weber-Dąbrowska B, Bagińska N et al (2018) Phage therapy: what have we learned? Viruses 10:288. https://doi.org/10.3390/v10060288

Hall AR, De Vos D, Friman VP, Pirnay JP, Buckling A (2012) Effects of sequential and simultaneous applications of bacteriophages on populations of *Pseudomonas aeruginosa* in vitro and in wax moth larvae. Appl Environ Microbiol 78:5646–5652. https://doi.org/10.1128/AEM.00757-12

Hanlon GW, Denyer SP, Olliff CJ, Ibrahim LJ (2001) Reduction in exopolysaccharide viscosity as an aid to bacteriophage penetration through *Pseudomonas aeruginosa* biofilms. Appl Environ Microbiol 67:2746–2753. https://doi.org/10.1128/AEM.67.6.2746-2753.2001

Harper DR, Parracho HM, Walker J, Sharp R, Hughes G, Werthén M, Lehman S, Morales S (2014) Bacteriophages and biofilms. Antibiotics 3:270–284. https://doi.org/10.3390/antibiotics3030270

Henry M, Lavigne R, Debarbieux L (2013) Predicting in vivo efficacy of therapeutic bacteriophages used to treat pulmonary infections. Antimicrob Agents Chemother 57:5961–5968. https://doi.org/10.1128/AAC.01596-13

Henry KA, Arbabi-Ghahroudi M, Scott JK (2015) Beyond phage display: non-traditional applications of the filamentous bacteriophage as a vaccine carrier, therapeutic biologic, and bioconjugation scaffold. Front Microbiol 6:755. https://doi.org/10.3389/fmicb.2015.00755

Himmelweit F (1945) Combined action of penicillin and bacteriophage on Staphylococci. Lancet 246:104–105. https://doi.org/10.1016/S0140-6736(45)91422-X

Høiby N, Ciofu O, Bjarnsholt T (2010) *Pseudomonas aeruginosa* biofilms in cystic fibrosis. Future Microbiol 5:1663–1674. https://doi.org/10.2217/fmb.10.125

Horcajada JP, Montero M, Oliver A, Sorlí L, Luque S, Gómez-Zorrilla S, Benito N, Grau S (2019) Epidemiology and treatment of multidrug-resistant and extensively drug-resistant *Pseudomonas aeruginosa* infections. Clin Microbiol Rev 32:e00031–e00019

Hraiech S, Brégeon F, Rolain JM (2015) Bacteriophage-based therapy in cystic fibrosis-associated *Pseudomonas aeruginosa* infections: rationale and current status. Drug Des Dev Ther 9:3653–3663. https://doi.org/10.2147/DDDT.S53123

Hurley MN, Cámara M, Smyth AR (2012) Novel approaches to the treatment of *Pseudomonas aeruginosa* infections in cystic fibrosis. Eur Respir J 40:1014–1023. https://doi.org/10.1183/09031936.00042012

International Nucleotide Sequence Database Consortium (INSDC) (2018) The DDBJ/ENA/GenBank Feature Table Definition Version 10.8. Available at: http://www.insdc.org/. Cited December 2018

Jassim SA, Limoges RG (2014) Natural solution to antibiotic resistance: bacteriophages 'the living drugs'. World J Microbiol Biotechnol 30:2153–2170. https://doi.org/10.1007/s11274-014-1655-1657

Jault P, Leclerc T, Jennes S, Pirnay JP, Que YA, Resch G, Rousseau AF, Ravat F, Carsin H, Le Floch R et al (2019) Efficacy and tolerability of a cocktail of bacteriophages to treat burn wounds infected by *Pseudomonas aeruginosa* (PhagoBurn): a randomised, controlled, double-blind phase 1/2 trial. Lancet Infect Dis 19:35–45

Jesaitis AJ, Franklin MJ, Berglund D, Sasaki M, Lord CI, Bleazard JB, Duffy JE, Beyenal H, Lewandowski Z (2003) Compromised host defense on *Pseudomonas aeruginosa* biofilms: characterization of neutrophil and biofilm interactions. J Immunol 171:4329–4339. https://doi.org/10.4049/jimmunol.171.8.4329

Kakabadze E, Makalatia K, Grdzelishvili N, Bakuradze N, Goderdzishvili M, Kusradze I, Phoba MF, Lunguya O, Lood C, Lavigne R et al (2018) Selection of potential therapeutic bacteriophages that lyse a CTX-M-15 extended spectrum β-lactamase producing *Salmonella enterica* serovar typhi strain from the Democratic Republic of the Congo. Viruses 10:172

Kaur T, Nafissi N, Wasfi O, Sheldon K, Wettig S, Slavcev R (2012) Immunocompatibility of bacteriophages as nanomedicines. J Nanotech 2012:ID 247427. https://doi.org/10.1155/2012/247427

Keogh RH, Szczesniak R, Taylor-Robinson D, Bilton D (2018) Up-to-date and projected estimates of survival for people with cystic fibrosis using baseline characteristics: a longitudinal study using UK patient registry data. J Cyst Fibros 17(2):218–227. https://doi.org/10.1016/j.jcf.2017.11.019

Kerem B, Rommens JM, Buchanan JA, Markiewicz D, Cox TK, Chakravarti A, Buchwald M, Tsui LC (1989) Identification of the cystic fibrosis gene: genetic analysis. Science 245(4922):1073–1080

Kimmelman J, Federico C (2017) Consider drug efficacy before first-in-human trials. Nature 542:25–27. https://doi.org/10.1038/542025a

Koskella B, Brockhurst MA (2014) Bacteria-phage coevolution as a driver of ecological and evolutionary processes in microbial communities. FEMS Microbiol Rev 38:916–931

Krylov V (2014) Bacteriophages of *Pseudomonas aeruginosa*: long-term prospects for use in phage therapy. In: Maramorosch K, Murphy FA (eds) Advances in virus research, vol 88. Elsevier, Oxford, pp 227–278

Krylov V, Shaburova O, Pleteneva E, Bourkaltseva M, Krylov S, Kaplan A, Chesnokova E, Kulakov L, Magill D, Polygach O (2016) Modular approach to select bacteriophages targeting *Pseudomonas aeruginosa* for their application to children suffering with cystic fibrosis. Front Microbiol 7:1631. https://doi.org/10.3389/fmicb.2016.01631

Kutateladze M, Adamia R (2008) Phage therapy experience at the Eliava Institute. Med Mal Infect 38(8):426–430. https://doi.org/10.1016/j.medmal.2008.06.023

Kutateladze M, Adamia R (2010) Bacteriophages as potential new therapeutics to replace or supplement antibiotics. Trends Biotechnol 28:591–595. https://doi.org/10.1016/j.tibtech.2010.08.001

Larché J, Pouillot F, Essoh C, Libisch B, Straut M, Lee JC, Soler C, Lamarca R, Gleize E, Gabard J et al (2012) Rapid identification of international multidrug-resistant *Pseudomonas aeruginosa* clones by multiple-locus variable number of tandem repeats analysis and investigation of their susceptibility to lytic bacteriophages. Antimicrob Agents Chemother 56:6175–6180. https://doi.org/10.1128/AAC.01233-12

Latino L, Midoux C, Vergnaud G, Pourcel C (2019) Investigation of *Pseudomonas aeruginosa* strain PcyII-10 variants resisting infection by N4-like phage Ab09 in search for genes involved in phage adsorption. PLoS One 14:e0215456. https://doi.org/10.1371/journal.pone.0215456

Lavelle GM, White MM, Browne N, McElvaney NG, Reeves EP (2016) Animal models of cystic fibrosis pathology: phenotypic parallels and divergences. Biomed Res Int 2016:5258727. https://doi.org/10.1155/2016/5258727

LaVergne S, Hamilton T, Biswas B, Kumaraswamy M, Schooley RT, Wooten D (2018) Phage therapy for a multidrug-resistant *Acinetobacter baumannii* craniectomy site infection. Open Forum Infect Dis 5(4):ofy064. https://doi.org/10.1093/ofid/ofy064

Lehman S, Branston S, Smrekar F, Pabary R, Alton E, Davies J, Morales S (2016) Bacteriophage therapy for the treatment of *P. aeruginosa* infections in cystic fibrosis patients. Abstract. In: Proceedings from the EMBO Conference 2016. Available at: http://pcwww.liv.ac.uk/~hallison/EMBOVoM2016/data/abstracts/abstract_275.html. Cited July 2016

Leyens L, Richer É, Melien Ø, Ballensiefen W, Brand A (2015) Available tools to facilitate early patient access to medicines in the EU and the USA: analysis of conditional approvals and the implications for personalized medicine. Public Health Genomics 18(5):249–259. https://doi.org/10.1159/000437137

Li Z, Kosorok MR, Farrell PM, Laxova A, West SE, Green CG, Collins J, Rock MJ, Splaingard ML (2005) Longitudinal development of mucoid *Pseudomonas aeruginosa* infection and lung disease progression in children with cystic fibrosis. JAMA 293:581–588. https://doi.org/10.1001/jama.293.5.581

Li G, Shen M, Yang Y, Le S, Li M, Wang J, Lu S (2018) Adaptation of *Pseudomonas aeruginosa* to phage PaP1 predation via O-antigen polymerase mutation. Front Microbiol 9:1170. https://doi.org/10.3389/fmicb.2018.01170

Lim WS, Phang KK, Tan AHM, Li SFY, Ow DSW (2016) Small colony variants and single nucleotide variations in Pf1 region of PB1 phage-resistant *Pseudomonas aeruginosa*. Front Microbiol 7:282. https://doi.org/10.3389/fmicb.2016.00282

Lin DM, Koskella B, Lin HC (2017) Phage therapy: an alternative to antibiotics in the age of multidrug resistance. World J Gastrointest Pharmacol Ther 8:162–173. https://doi.org/10.4292/wjgpt.v8.i3.162

Lin Y, Chang RYK, Britton WJ, Morales S, Kutter E, Chan HK (2018) Synergy of nebulized phage PEV20 and ciprofloxacin combination against *Pseudomonas aeruginosa*. Int J Pharm 551:158–165. https://doi.org/10.1016/j.ijpharm.2018.09.024

Loc-Carrillo C, Abedon ST (2011) Pros and cons of phage therapy. Bacteriophage 1:111–114. https://doi.org/10.4161/bact.1.2.14590

Lu TK, Collins JJ (2007) Dispersing biofilms with engineered enzymatic bacteriophage. Proc Natl Acad Sci U S A 104:11197–11202. https://doi.org/10.1073/pnas.0704624104

Magiorakos AP, Srinivasan A, Carey RB, Carmeli Y, Falagas ME, Giske CG, Harbarth S, Hindler JF, Kahlmeter G, Olsson-Liljequist B et al (2012) Multidrug-resistant, extensively drug-resistant and pandrug-resistant bacteria: an international expert proposal for interim standard definitions for acquired resistance. Clin Microbiol Infect 18(3):268–281. https://doi.org/10.1111/j.1469-0691.2011.03570.x

Mah TF, O'Toole GA (2001) Mechanisms of biofilm resistance to antimicrobial agents. Trends Microbiol 9:34–39

Martis N, Leroy S, Blanc V (2014) Colistin in multi-drug resistant *Pseudomonas aeruginosa* blood-stream infections: a narrative review for the clinician. J Infect 69:1–12. https://doi.org/10.1016/j.jinf.2014.03.001

Matinkhoo S, Lynch KH, Dennis JJ, Finlay WH, Vehring R (2011) Spray-dried respirable powders containing bacteriophages for the treatment of pulmonary infections. J Pharm Sci 100:5197–5205. https://doi.org/10.1002/jps.22715

Mattila S, Ruotsalainen P, Jalasvuori M (2015) On-demand isolation of bacteriophages against drug-resistant bacteria for personalized phage therapy. Front Microbiol 6:1271. https://doi.org/10.3389/fmicb.2015.01271

Mendes JJ, Leandro C, Mottola C, Barbosa R, Silva FA, Oliveira M, Vilela CL, Melo-Cristino J, Górski A, Pimentel M (2014) In vitro design of a novel lytic bacteriophage cocktail with therapeutic potential against organisms causing diabetic foot infections. J Med Microbiol 63:1055–1065. https://doi.org/10.1099/jmm.0.071753-0

Merabishvili M, Pirnay JP, Verbeken G, Chanishvili N, Tediashvili M, Lashkhi N, Glonti T, Krylov V, Mast J, Van Parys L et al (2009) Quality-controlled small-scale production of a well-defined bacteriophage cocktail for use in human clinical trials. PLoS One 4:e4944

Merelle ME, Schouten JP, Gerritsen J, Dankert-Roelse JE (2001) Influence of neonatal screening and centralized treatment on long-term clinical outcome and survival of CF patients. Eur Respir J 18:306–315

Merril CR, Scholl D, Adhya SL (2003) The prospect for bacteriophage therapy in Western medicine. Nat Rev Drug Discov 2:489–497. https://doi.org/10.1038/nrd1111

Mirzaei MK, Nilsson AS (2015) Isolation of phages for phage therapy: a comparison of spot tests and efficiency of plating analyses for determination of host range and efficacy. PLoS One 10:e0118557. https://doi.org/10.1371/journal.pone.0118557

Moelling K, Broecker F, Willy C (2018) Wake-up call: we need phage therapy now. Viruses 5;10(12). https://doi.org/10.3390/v10120688

Moreau-Marquis S, Stantona BA, OToole GA (2008) *Pseudomonas aeruginosa* biofilm formation in the cystic fibrosis airway: a short review. Pulm Pharmacol Ther 21:595–599. https://doi.org/10.1016/j.pupt.2007.12.001

Morello E, Saussereau E, Maura D, Huerre M, Touqui L, Debarbieux L (2011) Pulmonary bacteriophage therapy on *Pseudomonas aeruginosa* cystic fibrosis strains: first steps towards treatment and prevention. PLoS One 6:e16963. https://doi.org/10.1371/journal.pone.0016963

Moretti S, Renga G, Oikonomou V, Galosi C, Pariano M, Iannitti RG, Borghi M, Puccetti M, De Zuani M, Pucillo CE et al (2017) A mast cell-ILC2-Th9 pathway promotes lung inflammation in cystic fibrosis. Nat Commun 8:14017. https://doi.org/10.1038/ncomms14017

Moskowitz SM, Foster JM, Emerson J, Burns JL (2004) Clinically feasible biofilm susceptibility assay for isolates of *Pseudomonas aeruginosa* from patients with cystic fibrosis. J Clin Microbiol 42:1915–1922. https://doi.org/10.1128/JCM.42.5.1915-1922.2004

Mumford R, Friman VP (2017) Bacterial competition and quorum-sensing signalling shape the eco-evolutionary outcomes of model in vitro phage therapy. Evol Appl 10:161–169. https://doi.org/10.1111/eva.12435

Murray TS, Egan M, Kazmierczak BI (2007) *Pseudomonas aeruginosa* chronic colonization in cystic fibrosis patients. Curr Opin Pediatr 19:83–88. https://doi.org/10.1097/MOP.0b013e3280123a5d

Nadithe V, Rahamatalla M, Finlay WH, Mercer JR, Samuel J (2003) Evaluation of nose-only aerosol inhalation chamber and comparison of experimental results with mathematical simulation of aerosol deposition in mouse lungs. J Pharm Sci 92:1066–1076. https://doi.org/10.1002/jps.10379

National Heart, Lung, and Blood Institute (NHLBI) Information Center, The Cystic Fibrosis Foundation (1995) Cystic fibrosis. Bethesda, MD. Available at: https://www.cdc.gov/scienceambassador/documents/cystic-fibrosis-fact-sheet.pdf. Cited November 1995

Nilsson AS (2019) Pharmacological limitations of phage therapy. Ups J Med Sci 14:1–10. https://doi.org/10.1080/03009734.2019.1688433

Nouraldin AAM, Baddour MM, Harfoush RAH, Essa SAM (2016) Bacteriophage-antibiotic synergism to control planktonic and biofilm producing clinical isolates of *Pseudomonas aeruginosa*. Alex Med J 52:99–105. https://doi.org/10.1016/j.ajme.2015.05.002

Olszak T, Zarnowiec P, Kaca W, Danis-Wlodarczyk K, Augustyniak D, Drevinek P, de Soyza A, McClean S, Drulis-Kawa Z (2015) In vitro and in vivo antibacterial activity of environmental bacteriophages against *Pseudomonas aeruginosa* strains from cystic fibrosis patients. Appl Microbiol Biotechnol 99:6021–6033. https://doi.org/10.1007/s00253-015-6492-6

O'Toole GA (2018) Cystic fibrosis airway microbiome: overturning the old, opening the way for the new. J Bacteriol 200:e00561–e00517. https://doi.org/10.1128/JB.00561-17

Pabary R, Singh C, Morales S, Bush A, Alshafi K, Bilton D, Alton EW, Smithyman A, Davies JC (2016) Anti-pseudomonal bacteriophage reduces infective burden and inflammatory response in murine lung. Antimicrob Agents Chemother 60:744–751. https://doi.org/10.1128/AAC.01426-15

Parfitt T (2005) Georgia: an unlikely stronghold for bacteriophage therapy. Lancet 9478:2166–2167

Parks QM, Young RL, Poch KR, Malcolm KC, Vasil ML, Nick JA (2009) Neutrophil enhancement of *Pseudomonas aeruginosa* biofilm development: human F-actin and DNA as targets for therapy. J Med Microbiol 58:492–502. https://doi.org/10.1099/jmm.0.005728-0

Parracho HMRT, Burrowes BH, Enright MC, McConville ML, Harper DR (2012) The role of regulated clinical trials in the development of bacteriophage therapeutics. J Mol Genet Med 6:279–286

Pei R, Lamas-Samanamud GR (2014) Inhibition of biofilm formation by T7 bacteriophages producing quorum-quenching enzymes. Appl Environ Microbiol 80:5340–5348. https://doi.org/10.1128/AEM.01434-14

Pelfrene E, Willebrand E, Cavaleiro Sanches A, Sebris Z, Cavaleri M (2016) Bacteriophage therapy: a regulatory perspective. J Antimicrob Chemother 71:2071–2074. https://doi.org/10.1093/jac/dkw083

Pelfrene E, Sebris Z, Cavaleri M (2019) Comment on Fauconnier A. Phage therapy regulation: from night to dawn. Viruses 11:771. https://doi.org/10.3390/v11090771

Pires D, Sillankorva S, Faustino A, Azeredo J (2011) Use of newly isolated phages for control of *Pseudomonas aeruginosa* PAO1 and ATCC 10145 biofilms. Res Microbiol 162:798–806. https://doi.org/10.1016/j.resmic.2011.06.010

Pires DP, Boas DV, Sillankorva S, Azeredo J (2015) Phage therapy: a step forward in the treatment of *Pseudomonas aeruginosa* infections. J Virol 89:7449–7456. https://doi.org/10.1128/JVI.00385-15

Pires DP, Cleto S, Sillankorva S, Azeredo J, Lu TK (2016a) Genetically engineered phages: a review of advances over the last decade. Microbiol Mol Biol Rev 80:523–543. https://doi.org/10.1128/MMBR.00069-15

Pires DP, Oliveira H, Melo LD, Sillankorva S, Azeredo J (2016b) Bacteriophage-encoded depolymerases: their diversity and biotechnological applications. Appl Microbiol Biotechnol 100:2141–2151. https://doi.org/10.1007/s00253-015-7247-0

Pirnay JP, De Vos D, Verbeken G, Merabishvili M, Chanishvili N, Vaneechoutte M, Van den Mooter G (2011) The phage therapy paradigm: pret-a-porter or sur-mesure? Pharm Res 28:934–937. https://doi.org/10.1007/s11095-010-0313-5

Pirnay JP, Verbeken G, Rose T, Jennes S, Zizi M, Huys I, Lavigne R, Merabishvili M, Vaneechoutte M, Buckling A et al (2012) Introducing yesterday's phage therapy in today's medicine. Future Virol 7:379–390. https://doi.org/10.2217/fvl.12.24

Pirnay JP, Verbeken G, Ceyssens PJ, Huys I, De Vos D, Ameloot C, Fauconnier A (2018) The magistral phage. Viruses 10:64. https://doi.org/10.3390/v10020064

Pollini S, Fiscarelli E, Mugnaioli C, Di Pilato V, Ricciotti G, Neri AS, Rossolini GM (2011) *Pseudomonas aeruginosa* infection in cystic fibrosis caused by an epidemic metallo-β-lactamase-producing clone with a heterogeneous carbapenem resistance phenotype. Clin Microbiol Infect 17:1272–1275. https://doi.org/10.1111/j.1469-0691.2011.03466.x

Rasamiravaka T, Labtani Q, Duez P, El Jaziri M (2015) The formation of biofilms by *Pseudomonas aeruginosa*: a review of the natural and synthetic compounds interfering with control mechanisms. Biomed Res Int 2015:17. https://doi.org/10.1155/2015/759348

Raz A, Serrano A, Hernandez A, Euler CW, Fischetti VA (2019) Isolation of phage lysins that effectively kill *Pseudomonas aeruginosa* in mouse models of lung and skin infection. Antimicrob Agents Chemother 63:e00024-19. https://doi.org/10.1128/AAC.00024-19

Reardon S (2014) Phage therapy gets revitalized. Nature 510:15–16. https://doi.org/10.1038/510015a

Riou M, Carbonnelle S, Avrain L, Mesaros N, Pirnay JP, Bilocq F, De Vos D, Simon A, Piérard D, Jacobs F et al (2010) In vivo development of antimicrobial resistance in *Pseudomonas aeruginosa* strains isolated from the lower respiratory tract of intensive care unit patients with nosocomial pneumonia and receiving antipseudomonal therapy. Int J Antimicrob Agents 36:513–522. https://doi.org/10.1016/j.ijantimicag.2010.08.005

Rogers GB, Carroll MP, Serisier DJ, Hockey PM, Kehagia V, Jones GR, Bruce KD (2005) Bacterial activity in cystic fibrosis lung infections. Respir Res 6:49. https://doi.org/10.1186/1465-9921-6-49

Rossitto M, Fiscarelli EV, Rosati P (2018) Challenges and promises for planning future clinical research into bacteriophage therapy against *Pseudomonas aeruginosa* in cystic fibrosis. An argumentative review. Front Microbiol 9:775. https://doi.org/10.3389/fmicb.2018.00775

Ryan EM, Gorman SP, Donnelly RF, Gilmore BF (2011) Recent advances in bacteriophage therapy: how delivery routes, formulation, concentration and timing influence the success of phage therapy. J Pharm Pharmacol 63:1253–1264. https://doi.org/10.1111/j.2042-7158.2011.01324.x

Sahota JS, Smith CM, Radhakrishnan P, Winstanley C, Goderdzishvili M, Chanishvili N, Kadioglu A, O'Callaghan C, Clokie MR (2015) Bacteriophage delivery by nebulization and efficacy against phenotypically diverse *Pseudomonas aeruginosa* from cystic fibrosis patients. J Aerosol Med Pulm Drug Deliv 28:353–360. https://doi.org/10.1089/jamp.2014.1172

Saiman L, Siegel JD, LiPuma JJ, Brown RF, Bryson EA, Chambers MJ, Downer VS, Fliege J, Hazle LA, Jain M et al (2014) Infection prevention and control guideline for cystic fibrosis: 2013 update. Infect Control Hosp Epidemiol 35:S1–S67. https://doi.org/10.1086/676882

Salmond GP, Fineran PC (2015) A century of the phage: past, present and future. Nat Rev Microbiol 13:777–786. https://doi.org/10.1038/nrmicro3564

Sarhan WA, Azzazy HME (2015) Phage approved in food not as a therapeutic? Expert Rev Anti Infect Ther 13:91–101. https://doi.org/10.1586/14787210.2015.990383

Saussereau E, Vachier I, Chiron R, Godbert B, Sermet I, Dufour N, Pirnay JP, De Vos D, Carrié F, Molinari N et al (2014) Effectiveness of bacteriophages in the sputum of cystic fibrosis patients. Clin Microbiol Infect 20:O983–O990. https://doi.org/10.1111/1469-0691.12712

Shen K, Sayeed S, Antalis P, Gladitz J, Ahmed A, Dice B, Janto B, Dopico R, Keefe R, Hayes J et al (2006) Extensive genomic plasticity in *Pseudomonas aeruginosa* revealed by identification and distribution studies of novel genes among clinical isolates. Infect Immun 74:5272–5283. https://doi.org/10.1128/IAI.00546-06

Siegel JD, Rhinehart E, Jackson M, Chiarello L, The Healthcare Infection Control Practices Advisory Committee (2017) Multidrug-resistant organisms (MDRO) Management. Update of the management of multidrug-resistant organisms in healthcare settings, 2006. Available at: https://www.cdc.gov/infectioncontrol/pdf/guidelines/mdro-guidelines.pdf. Cited 15 Feb 2017

Sievert DM, Ricks P, Edwards JR, Schneider A, Patel J, Srinivasan A, Kallen A, Limbago B, Fridkin S (2013) Antimicrobial-resistant pathogens associated with healthcare-associated infections summary of data reported to the National Healthcare Safety Network at the Centers for Disease Control and Prevention 2009–2010. Infect Control Hosp Epidemiol 34:1–14. https://doi.org/10.1086/668770

Smith GP (1985) Filamentous fusion phage: novel expression vectors that display cloned antigens on the virion surface. Science 228:1315–1317

Smith EE, Buckley DG, Wu Z, Saenphimmachak C, Hoffman LR, DArgenio DA, Miller SI, Ramsey BW, Speert DP, Moskowitz SM et al (2006) Genetic adaptation by *Pseudomonas aeruginosa* to the airways of cystic fibrosis patients. Proc Natl Acad Sci U S A 103:8487–8492. https://doi.org/10.1073/pnas.0602138103

Starkey M, Hickman JH, Ma L, Zhang N, De Long S, Hinz A, Manoil C, Kirisits MJ, Starner TD, Wozniak DJ et al (2009) *Pseudomonas aeruginosa* rugose small-colony variants have adaptations that likely promote persistence in the cystic fibrosis lung. J Bacteriol 191:3492–3503. https://doi.org/10.1128/JB.00119-09

Stuart B, Lin JH, Mogayzel PJ (2010) Early eradication of *Pseudomonas aeruginosa* in patients with cystic fibrosis. Paediatr Respir Rev 11:177–184. https://doi.org/10.1016/j.prrv.2010.05.003

Sulakvelidze A, Alavidze Z, Morris JG (2001) Bacteriophage therapy. Antimicrob Agents Chemother 45:649–659. https://doi.org/10.1128/AAC.45.3.649-659.2001

Sybesma W, Rohde C, Bardy P, Pirnay JP, Cooper I, Caplin J, Chanishvili N, Coffey A, De Vos D, Scholz AH et al (2018) Silk route to the acceptance and re-implementation of bacteriophage therapy – Part II. Antibiotics 7:35. https://doi.org/10.3390/antibiotics7020035

Theuretzbacher U (2009) Future antibiotics scenarios: is the tide starting to turn? Int J Antimicrob Agents 34:15–20. https://doi.org/10.1016/j.ijantimicag.2009.02.005

Thiel K (2004) Old dogma, new tricks 21st century phage therapy. Nat Biotechnol 22:31–36. https://doi.org/10.1038/nbt0104-31

Todd K (2019) The promising viral threat to bacterial resistance: the uncertain patentability of phage therapeutics and the necessity of alternative incentives. Duke Law J 68:767–805

Torres-Barcelò C (2018) Phage therapy faces evolutionary challenges. Viruses 10:323

Torres-Barceló C, Hochberg ME (2016) Evolutionary rationale for phages as complements of antibiotics. Trends Microbiol 24:249–256. https://doi.org/10.1016/j.tim.2015.12.011

Torres-Barceló C, Arias-Sánchez FI, Vasse M, Ramsayer J, Kaltz O, Hochberg ME (2014) A window of opportunity to control the bacterial pathogen *Pseudomonas aeruginosa* combining antibiotics and phages. PLoS One 9:e106628. https://doi.org/10.1371/journal.pone.0106628

Trend S, Fonceca AM, Ditcham WG, Kicic A (2017) The potential of phage therapy in cystic fibrosis: essential human-bacterial-phage interactions and delivery considerations for use in *Pseudomonas aeruginosa*-infected airways. J Cyst Fibros 16:663–670. https://doi.org/10.1016/j.jcf.2017.06.012

Tsui LC (1995) The cystic fibrosis transmembrane conductance regulator gene. Am J Respir Crit Care Med 151:S47

Twort FW (1915) An investigation on the nature of ultra-microscopic viruses. Lancet 186:1241–1243. https://doi.org/10.1016/S0140-6736(01)20383-3

Uchiyama J, Suzuki M, Nishifuji K, Kato SI, Miyata R, Nasukawa T, Yamaguchi K, Takemura-Uchiyama I, Ujihara T, Shimakura H et al (2016) Analyses of short-term antagonistic evolution between *Pseudomonas aeruginosa* strain PAO1 and phage KPP22 belonging to the family Myoviridae genus PB1-like viruses. Appl Environ Microbiol 82:4482–4491. https://doi.org/10.1128/AEM.00090-16

Verbeken G, De Vos D, Vaneechoutte M, Merabishvili M, Zizi M, Pirnay JP (2007) European regulatory conundrum of phage therapy. Future Microbiol 2:485–491. https://doi.org/10.2217/17460913.2.5.485

Vyas D, Aekka A, Vyas A (2015) Fecal transplant policy and legislation. World J Gastroenterol 21(1):6–11

Waddell TE, Franklin K, Mazzocco A, Kropinski AM, Johnson RP (2009) Generalized transduction by lytic bacteriophages. In: MRJ C, Kropinski AM (eds) Bacteriophages: methods and protocols, vol 1. Humana Press, Hertfordshire, pp 293–303

Walker TS, Tomlin KL, Worthen GS, Poch KR, Lieber JG, Saavedra MT, Fessler MB, Malcolm KC, Vasil ML, Nick JA (2005) Enhanced *Pseudomonas aeruginosa* biofilm development mediated by human neutrophils. Infect Immun 73:3693–3701. https://doi.org/10.1128/IAI.73.6.3693-3701.2005

Walters MC, Roe F, Bugnicourt A, Franklin MJ, Stewart PS (2003) Contributions of antibiotic penetration, oxygen limitation, and low metabolic activity to tolerance of *Pseudomonas aeruginosa* biofilms to ciprofloxacin and tobramycin. Antimicrob Agents Chemother 47:317–323. https://doi.org/10.1128/AAC.47.1.317-323.2003

Wang J, Hu B, Xu M, Yan Q, Liu S, Zhu X, Sun Z, Reed E, Ding L, Gong J et al (2006) Use of bacteriophage in the treatment of experimental animal bacteremia from imipenem-resistant *Pseudomonas aeruginosa*. Int J Mol Med 17:309–317. https://doi.org/10.3892/ijmm.17.2.309

Waters EM, Neill DR, Kaman B, Sahota JS, Clokie MR, Winstanley C, Kadioglu A (2017) Phage therapy is highly effective against chronic lung infections with *Pseudomonas aeruginosa*. Thorax 72:666–667. https://doi.org/10.1136/thoraxjnl-2016-209265

Weber-Dąbrowska B, Jończyk-Matysiak E, Żaczek M, Łobocka M, Łusiak-Szelachowska M, Górski A (2016) Bacteriophage procurement for therapeutic purposes. Front Microbiol 7:1177. https://doi.org/10.3389/fmicb.2016.01177

Wei Q, Ma LZ (2013) Biofilm matrix and its regulation in *Pseudomonas aeruginosa*. Int J Mol Sci 14:20983–21005. https://doi.org/10.3390/ijms141020983

Winstanley C, O'Brien S, Brockhurst MA (2016) *Pseudomonas aeruginosa* evolutionary adaptation and diversification in cystic fibrosis chronic lung infections. Trends Microbiol 24:327–337. https://doi.org/10.1016/j.tim.2016.01.008

World Medical Association (WMA) (2013) Declaration of Helsinki: ethical principles for medical research involving human subjects. JAMA 10:2191–2194

Wright A, Hawkins C, Änggård E, Harper D (2009) A controlled clinical trial of a therapeutic bacteriophage preparation in chronic otitis due to antibiotic-resistant *Pseudomonas aeruginosa*; a preliminary report of efficacy. Clin Otolaryngol 34:349–357. https://doi.org/10.1111/j.1749-4486.2009.01973.x

Xu KD, McFeters GA, Stewart PS (2000) Biofilm resistance to antimicrobial agents. Microbiology 146:547–549. https://doi.org/10.1099/00221287-146-3-547

Yoon SS, Hassett DJ (2004) Chronic *Pseudomonas aeruginosa* infection in cystic fibrosis airway disease: metabolic changes that unravel novel drug targets. Expert Rev Anti Infect Ther 2:611–623. https://doi.org/10.1586/14787210.2.4.611

Young R, Gill JJ (2015) Phage therapy redux – what is to be done? Science 350(6265):1163–1164. https://doi.org/10.1126/science.aad6791

Chapter 21
Bacteriophage Applications for Food Safety

Ayman El-Shibiny and Alyaa Dawoud

Abstract Ingestion of food and water contaminated with pathogenic bacteria may cause serious diseases. Many of those pathogenic bacteria are considered normal microbiota of many animals and poultry and they usually infect human through cross contamination. With the increase resistance of bacteria to most available antibiotics, the development of an alternative and effective method to control bacterial infections is becoming an urgent solution. Bacteriophage therapy is considered an alternative to antibiotics to control bacterial contamination of food especially after the U.S. Food and Drug Administration (FDA) approval in 2006 to use phages as a safe food additive. It is very important to use virulent phages and to identify the best cocktail that shows the highest efficacy against their target bacterial host before biocontrol applications. In this chapter, we will discuss the successful applications of phages to decontaminate different types of food such as; meat, poultry, eggs, fish, fruits, vegetables, food animals and drinking water during processing stages to control bacterial infections. In addition, we will discuss the developments in phage delivery systems and the application of nano-based bio-packing material and edible anti-microbial coating to enhance the effectiveness of phages as a biocontrol agent.

21.1 Introduction

Food and water are considered sources of widespread infections. To minimize the risk potential, different control measures must be in place. In industry, a Hazard Analysis and Critical Control Point (HACCP) system is implemented to safeguard food from contamination with harmful microbes such as animal's intestinal contents and feces. Food handling, preparation and serving practices may contribute directly to foodborne diseases and identifying those points of contamination along food chain can reduce the contamination of food and accordingly reduce foodborne diseases. In general, microbes don't intend to cause disease but they try to get food and

A. El-Shibiny (✉) · A. Dawoud
Center for Microbiology and Phage Therapy, Zewail City of Science and Technology, Giza, Egypt
e-mail: aelshinbiny@zewailcity.edu.eg

G. Witzany (ed.), *Biocommunication of Phages*,
https://doi.org/10.1007/978-3-030-45885-0_21

defend themselves. However, their presence induce the symptoms in their hosts. Identifying the source of an individual infection is very rare and difficult. However, any identified foodborne disease outbreak will provide useful information about the pathogens that cause this outbreak and their route of transmission. Lytic bacteriophages are used in food industry as an effective and inexpensive agent to control foodborne pathogenic bacteria and extend the shelf life of food products. Because of its specificity, phages can infect pathogenic bacteria and decrease their numbers without affecting the normal microbiota of foods.

21.2 Food Safety

Most foods provide the required nutrients and other growth factors for the growth of bacteria and other microorganisms such as fungi. The majority of fresh food, including fresh vegetables and fruits, are susceptible to be contaminated with food spoilage bacteria and sometimes with pathogens such as *Salmonella*, *Escherichia coli*, *Campylobacter*, *Staphylococcus aureus*, *Listeria monocytogenes*, and *Clostridium perfringens* that may cause infections with zoonotic pathogens due to improper food handling or cross contamination (Gould et al. 2011; Lynch et al. 2009). Infection through contaminated water with bacterial pathogens such as; *Vibrio cholera*, *E. coli*, *Salmonella enterica* serovar Typhi, and *Pseudomonas aeruginosa* may cause waterborne diseases (Mousavi et al. 2009). These infections could be transmitted through contaminated drinking water, or from contaminated water that may be used in food preparation or in swimming pools. Other bacteria such as *Campylobacter* and *Salmonella* are considered normal microbiota of many animals and poultry and they usually contaminate foods during slaughter, handling and/or carcass processing and subsequently increase the risk of human infections with those pathogens. Infection usually occurs from the ingestion of food contaminated with pathogenic bacteria and/or their toxins that results in disease. The effectiveness of available antibiotics worldwide is affected by the increasing resistance by bacteria and most of available antibiotics are now not effective to control many bacterial infections over time (Akinkunmi et al. 2014; Campos et al. 2015; World Health Organization 2014). Unfortunately, the increase of resistance is mainly because of the misuse of antibiotics in health and agriculture especially in developing countries which will ultimately increase the rate of mortality over time (Fortini et al. 2011; Nagel et al. 2016; Pavlickova et al. 2015). Antibiotics use in livestock as growth promoters has become a major concern because of their contribution to the increase rate of resistance to antibiotics in humans (Smith et al. 2002). However, the continuous improvements in public health, access to safe food, and improvements in sewage water treatment have reduced the exposure chances to pathogenic bacteria. On the other hand, any alterations in the pathogen, like mutations or the surrounding environment, will contribute to the emerging of new diseases which may increase the morbidity and mortality rate.

Therefore, the development of a new effective methodology to control bacterial infections is a must particularly with the high incidence of antibiotic resistance. Such methodology should not have any adverse effect on the quality of food or human health (Loretz et al. 2010). Bacteriophages (phages in short) are considered one of the best alternatives to antibiotics to detect and control bacterial contamination of food (Nagel et al. 2016). They can be used alone or in combination with antibiotics or disinfectants to control bacterial infections (Kutateladze and Adamia 2010). To be used in food, phage applications in food must be in compliance with legislation. The U.S. Food and Drug Administration (FDA) has approved the use of phages as a safe food additive on ready-to-eat products and in veterinary medicine (Bren 2007; Monk et al. 2010).

21.3 Biocontrol of Common Foodborne and Waterborne Bacterial Pathogens

The applications of phage in foods to combat pathogens and control their entry into the food chain has increased during the last few years because of the rise in antibiotic resistance among pathogenic bacteria, as described above. The application of the appropriate phages to control foodborne diseases is very promising and phage dependent strategy. In general, phages can be easily prepared and applied in food without any harm to humans (Bruttin and Brüssow 2005). Before conducting any phage-based biocontrol treatment, it is important to test the virulent phages against their target bacterial host at different conditions in vitro to identify the best cocktail and conditions that can lead to the highest efficacy. For effective biocontrol applications, phages or phage cocktails should be active against a wide host range (Greer and Dilts 1990) and should be applied in high concentrations to the surface of food or packaging materials to increase their ability to make a contact with their host bacterial cells (Bigwood et al. 2008; Turki et al. 2012). However, phage efficiency may be affected by the characteristics of foodstuff. For example, the high fat content of chicken skin and its feather follicles, may support the survival of bacterial cells and reduce the ability of phages to bind to their specific receptors (El-Shibiny et al. 2017).

Different bacteria can be acquired from different foods especially undercooked or raw food. Before thinking of using phages to control foodborne pathogens, it is important to understand the epidemiology of each pathogen to identify the best effective application to control bacterial infection along the food chain (Stone 2002). Phages are used successfully to decontaminate the surface of meat, poultry carcasses, eggs, fish, fruits, and vegetables, during different processing stages to control pathogens (El-Shibiny et al. 2017; Leverentz et al. 2001; O'Flynn et al. 2004; Viazis et al. 2011; Yang et al. 2019). *Salmonella*, a human pathogen that causes salmonellosis, is frequently isolated from contaminated food or water and is considered a big cause of food-borne disease worldwide. The health burdens of this pathogen made it an obvious target for phage biocontrol applications. Phage application successfully reduced the numbers of *Salmonella Typhimurium* in chocolate milk and turkey deli meat by 5 $\log_{10}$ (Guenther et al. 2012) in cheddar cheese (Modi et al. 2001), chicken skin (Pao et al. 2006), chicken breasts and pig skin (Spricigo et al. 2013), energy drinks, skimmed milk, and apple juice (Zinno et al. 2014), alfalfa seeds (Kocharunchitt et al. 2009), and sprouts (Leverentz et al. 2001). Leverentz et al. succeeded to reduce the numbers of *Salmonella* on fresh-cut melon stored at 5 and 10 °C by approximately 3.5 $\log_{10}$ CFU (B. Leverentz et al. 2001). Campylobacteriosis is associated with the consumption food contaminated with campylobacters. Broiler chickens and poultry products are a leading source of campylobacters since *Campylobacter* is considered a part of the normal microbiota of the poultry's gastrointestinal tract (GIT). Phages application successfully reduced the numbers of campylobacters in cooked and raw beef by 2 $\log_{10}$ and the surface area of poultry meat (Atterbury et al. 2003; Bigwood et al. 2008; Goode et al. 2003; Rosenquist et al. 2003). *E. coli* O157:H7 infection, that causes acute hemorrhagic diarrhea, is usually accompanied with the consumption of undercooked beef or via direct contact with animals (Belongia et al. 1991). Phages showed a significant decrease in the numbers of *E. coli* on meat and tomatoes surfaces to below the limit of detection (El-Shibiny et al. 2017). Other studies indicated that, phages significantly reduced the numbers of *L. monocytogenes* on pear, melon, apple, juice, cheese and dairy products (Carlton et al. 2005; Oliveira et al. 2014; Schellekens et al. 2007). It also can be immobilized on cellulose membranes to control the growth *E. coli* and *L. monocytogenes* on experimental meat (Anany et al. 2011).

21.4 Use of Phages as Biocontrol Agents in Food Animals

Food animals which are considered main sources of white meat (i.e poultry) and red meat (i.e sheep, cattle and pig) are major sources of foodborne pathogens especially in pre-harvest stages in food industry processes, particularly, during slaughter when they are able to contaminate the raw meet by their intestinal and fecal content. Therefore using the phage therapy as a biocontrol agent is a worthy solution for this dilemma. Additionally, phages can be used as a prophylactic therapy for food animals to reduce the risk of infection and improve the quality of their products. The first known phage application was done by d'Herelle in chickens after the phage discovery in 1917 (d'Herelle 1917). Over time, phages showed high efficiency to control different bacterial infections, like *Clostridium difficile* (Ramesh et al. 1999), *E. coli* (Huff et al. 2002), *Campylobacter* (Loc Carrillo et al. 2007), *Staphylococcus aureus* (Wills et al. 2005), *Salmonella* (Andreatti Filho et al. 2007)) and *Pseudomonas aeruginosa* (McVay et al. 2007) in large animals and poultry. However, some studies, which focused on using the phage in pre-harvest stages, have highlighted the increase of incidence of the phage resistance by bacteria when they are administered as a single phage. Conversely, if phages are administered as a cocktail, the probability of developing a resistance for a single phage is very low (El-Shibiny et al. 2009;

Goodridge and Bisha 2011; Tanji et al. 2004). However, it is proved that the more the number of phages exists in the cocktail and more properly they are selected, the least phages resistance will be obtained (Raya et al. 2011). Moreover, a study done by Tanji et al. has showed that the oral administration of a cocktail of phages to treat *E. coli* O157:H7 infection in the GIT of mice reduced the colonization rate of *E. coli* significantly (Tanji et al. 2005). That's why it's highly recommended to use phages in a cocktail whether consists of a mixture of different phages or consists of a combination with other chemical compounds. Phages have also showed a prophylactic activity for food animals when they are administered for non-infected animals to protect them from certain pathogens infection. In a study done by Waddell et al. (2002) on a group of calves which were being treated by a 6 phages cocktail for a weak, then, infected with *E. coli* O157:H7. Upon fecal analysis it was found that treated calves have stopped shedding *E. coli* O157:H7 after 8 days in comparison to the untreated group that kept on shedding *E. coli* O157:H7 for at least 12–16 days (Goodridge and Bisha 2011; Waddell et al. 2002).

Regarding the route of administration, virulent phages could be administered to food animals orally or rectally. However, Rozema et al. has shown that oral route is more efficient in a study done on steers to compare both routes of administration (Rozema et al. 2009) since *E. coli* O157:H7 count was the lowest in the group treated with phage orally than the group that treated with phage rectally or the group that treated through both routes; orally and rectally. Furthermore, dose-formulation also affects the efficiency of phage therapy. Researchers pointed out that the importance of phage encapsulation for pages that can be administrated orally to animals. Encapsulation of phages leads to increasing their stability in the acidic stomach condition and enhance their intestinal retention. Colom et al. study has shown that encapsulation of a phage cocktail in liposomes for the sake of their administration to broilers, encapsulated phages where detected in 38.1% of animals while the non-encapsulated phages detected in only 9.5%, after 72 h of treatment (Colom et al. 2015). Moreover, for two boilers groups which had been treated 1 day before the infection onset with encapsulated and non-encapsulated phages, respectively, then experimentally infected with *Salmonella*, and then both groups were given similar dose of both forms of phages daily for 6 days only, encapsulated and non-encapsulated phages have shown the same protection in both groups. However, the protection was preserved in group one for at least 1 week after the end of phage administration prolonging the protection period (Colom et al. 2015).

21.5 Ingested Phages Inside the Gut and Body Compartments

Phages, which are the most abundant entity on the planet, have been found to be the most copious group of human virome, even greater than eukaryotic viruses (Navarro and Muniesa 2017). Phages have been isolated from different body compartments.

They were found as a part of microbiome in blood stream (phagemia), urine, lung, gut, and ascetic fluids consisting, what is called, the human-body phagome (Górski et al. 2006). However, intestine phagome plays an important role in keeping the homeostasis of the intestine microbiome and shaping its bacterial genetic material and thus, contributes in the evolution of the bacterial population. On the other hand, phages disturb the results of some clinically diagnostic tests leading to unreliable results (Brown-Jaque et al. 2016; Manrique et al. 2017; Modi et al. 2014). So, using of phages in food products and as anti-bacterial agent for intestinal bacterial infection should be well-investigated to obtain as much as possible of its benefits.

To study the viability of ingested phages, whether with food or as pure therapy, and their interaction with human immune system *in vivo*, intensive studies on animal models was done to compare different body responses to different dose amount, regimen and route of administration. In an experiment done on mice given T4 phages in drinking water to simulate the oral route of administration. Upon fecal analysis, viable phages in feces present only at low level of IgA, which increased to a significant level on the 79th day of continuous administration of high concentration of the phage. However, active phages recovered again when IgA returned to the low level. Nevertheless, at re-administration of T4 phages, the IgA level increased faster than the first time (Majewska et al. 2015). However, the doses which have been used already for medication and administered orally for the patients, in tablet or liquid formulation forms, were in much lower concentrations that didn't induce the production of antibodies. So, they weren't observed in the safety tests (Górski et al. 2016).

Throughout all studies that were done on phages and their efficiency in fighting back the antimicrobial drug resistant bacteria strains, the context of the medication demanded an answer for a question to be more aware about phages pharmacokinetics; Dose phages translocate from the GIT to the lymph or to the blood stream? To find an answer for this very specific question, animal studies were done, then, followed by clinical trials on human (Table 21.1). All studies proved phage translocation to the lymph or to the blood stream at high concentration of orally administered phage (Górski et al. 2006). Furthermore, there were evidences of the ability of the naked DNA of M13 phage for translocation (Schubbert et al. 1994). However, fluctuation and variability in the translocated phage numbers were the most dominant observation in all the studies. Therefore, well-planned studies with more defined factors like phage titer, health of gut and its permeability, size of the phage, proteins expressed on phages capsid surface, previous neutralization for gastric juice or encapsulation for the phage before administration, and the route of administration as well should be considered.

Furthermore, due to the ability of phages to translocate in human body, it's common to see phage lysis plaques on agar plate during microbiological clinical tests for different body fluids which in turn may interfere with the results of these tests: In experiments done for the investigation of the possibility of phage interference in clinical analysis, it was found that when the inoculum for the antibiogram assay was taken from an agar plate contains lysis plaque, the continuous growth of the bacteria was inhibited by the phage, unless the colony was taken from phage-free area,

Table 21.1 Different phages supplied to animal models and human orally to prove phage translocation to different body tissue from the GIT

Phage name	Specie applied to	Route of administration	Time of detection	Site of detection	References
Bacillus megathericum phages	Mice	Oral	5 min	Blood	Keller and Engley (2013)
T1 coliphages	Rats	Oral	–	Lymph, then prepheral blood	Hildebrand and Wolochow (1962)
T3 coliphages	Mice	Oral	15 min	Blood	Hoffmann (1965)
CF 0103 coliphages	Rabbits	Oral	4th day	Erythrocytes	Reynaud et al. (1992)
T4 phage	Humsn	Oral	Not observed	Not observed	Bruttin and Brüssow (2005)
Ps/68 *Pseudomonas* phage	Humsn	Oral	5th day	Serum	Górski et al. (2006)
Staphylococcus phi 131 phage	Humsn	Oral	5th day	Serum	Górski et al. (2006)
676/Forys *Staphylococcus* phage	Humsn	Oral	5th day	Serum	Górski et al. (2006)

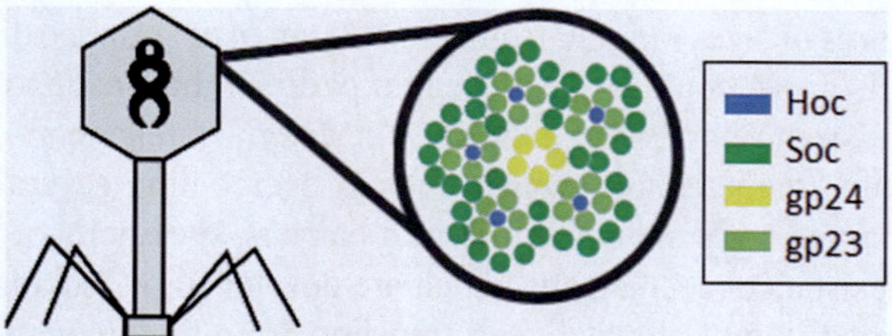

Fig. 21.1 Illustration for composition of T4 phage capsid protein

leading to unreliable antibiogram results. Additionally, when viable infectious phages present in high concentration in urine sample, phages inhibit the growth of Gram-negative bacilli in the sample when cultured on agar plate giving false negative results (Brown-Jaque et al. 2016). Since it's usual to use liquid enrichment culture media, especially for liquid samples like blood or ascitic fluid, for increasing the volume of the collected sample for good measurements' sensitivity or for the sake of selectivity of specific substance or organism, the used media could provide a good environment for the in-sample phage to infect the bacteria interfering the measurement and giving unreliable results (Navarro and Muniesa 2017).

Nevertheless, ingested phages have a very important benefit, which still under investigation, to be used as a method of vaccination. As one of recent application for ingested T4 phage is to be used in vaccination for Ebola virus by fusing the oligopeptide EB1 antigen to Hoc protein (Fig. 21.1) which present on the protein capsid of the phage. EB1-antigene presented on T4 phage was administered to the mice orally in drinking water. Then, mice were tested for specific anti-EB1 IgG and IgA. Both IgG and IgA presented. However, secretory IgA specific to EB1 was detected in the gut samples at significant level (Majewska et al. 2015; Rao and Black 2010).

21.6 Nano-based Bio-packing Material for Phage Delivery into Food Products

Long time ago, especially for Russian countries, research focused on using phages as natural agent for many purposes: natural preservative agent in food material, detecting agent for food pathogens and as an alternative antibiotic for animal and human (Bai et al. 2016; García et al. 2008; Morozova et al. 2018; Summers 1999). However, the more uses in which the phages are exploited, the more challenges are faced and more development is required. Thus, new methods of encapsulation and specific materials have been utilized to allow maximum protection of the phage and most probable target specificity (Abdelsattar et al. 2019; Choińska-Pulit et al. 2015). Furthermore, to guarantee more specific distribution and better body compatibility, nanotechnology has been merged with bio-packing for phages under the umbrella of what's known as nanomedicine to provide the scientists by their research objectives. Nevertheless, the dilemma of drug acceptance by patients turned to be the point of interest by researchers and on which they focus their investigation.

According to National Nanotechnology Initiative (NNI), nanoparticles are structures of sizes ranging from 1 to 100 nm in at least one dimension. However, the term of "nano" could also be given for particles that are up to several hundred nanometers in size (Wilczewska et al. 2012). Materials in nano size exhibit different properties and characteristics which help in overcoming several challenges throughout the human body including transport barriers, specificity of the target site, multiple drug resistance mechanisms which are developed by pathological cells, the harsh acidic condition of stomach, and transport from the bloodstream to target bacterial cells (Hubbell and Chilkoti 2012; Jain and Stylianopoulos 2010). Therefore, many biological and non-biological, or sometimes hybrid, materials are used for drug delivery whether by encapsulating the drug within this nano-sized material or by attaching the drug on its surface by adsorption or covelent bonding to get the benefits of its new properties in such size (Pelaz et al. 2017).

Furthermore, biomaterial is a term given by the American National Institute of Health for any substance or combination of substances, other than drugs, synthetic or natural in origin, which can be used for any period of time inside the body with no or minimum adverse interaction, sometimes exist in polymer structure (Bergmann and Stumpf 2013; Gbassi and Vandamme 2012). Encapsulation of phages in biomaterial is commonly called microcapsulation due to the size of the capsules. However, ethylcellulose, carboxymethyl cellulose, hygroxypropylmethylcellulose, alginate, chitosan, gelatin, carob bean gum, starch, whey proteins and polyesters are the most coomon biomaterials used in phages encapsulation (Choińska-Pulit et al. 2015).

For so long time, many kind of food have been used as a carrier for probiotics: products that contain an adequate calculated dose of living microbes that have been documented in target-host studies to confer a health benefit mainly for the aim of improving the balance of the intestinal microbial profile resulting in several therapeutic benefits (Sanders 2008). However, there are a lot of probiotic products in the world market including Denmark, Germany, France, USSR, Czechoslovakia, Japan, Norway, Sweden, Italy, Bauer, Switzerland, Netherlands, and Egypt. Yogurt and milk products, as dairy foods, have been the most famous carrier for probiotics in all time while the most popular microorganism was the viable *Lactobacillus acidophilus* and *Bifidobacteriu bifidum* cells (Lourens-Hattingh and Viljoen 2001).

For food that carries bacteria, it had many benefits: it provides a synergistic effect with probiotics through keeping a suitable media for probiotic growth, and contains bioactive components that could interact with the probiotic. Additionally, food helps in buffering the probiotic throughout the GIT keeping it in homogenous distribution (Ranadheera et al. 2010). On the other hand, when phages are encapsulated, they seldom react with the food component. However, food will provide an easier and more acceptable administration of the phage for patients with gastroenteritis when it caused by bacterial infection in the intestine. Meanwhile, the properties of packing material will allow the control of phage release in an exact site of the GIT. Food, like yogurt or fruit juices, is a very acceptable carrier in which encapsulated phage can be carried for oral administration to treat the bacterial infection of the intestine region. Besides, nano-sized bio-packing material will provide the target specificity and exact release site helping to keep the maximum dose of phage at the infected region (Pelaz et al. 2017).

21.7 Development of Phage-based Edible Anti-microbial Coating for Food Products

Since the FDA approval for the phage use in certain food products to enhance food safety as a safe bio-preservative in 2006, an extensive research on the phage effect on foodborne contaminants has been done to investigate for more phage activity against food pathogens in beef and chicken meat, dairy products, infant formula, and fresh fruits. Many companies like GengaGen, Inc. in USA and Biopharm, Inc. in Georgia have begun to involve phage research in their studies while other focused their whole research on phage-based products like OmniLytics, Inc. and Intralytix, Inc. That's how phage became used in all food industry processes: starting from phage therapy for farm animal (pre-harvest stages) passing through industrial sanitation, bio-preservative for raw-products, and in marketed products that reaches the consumer (post- harvest stages) (Goodridge and Bisha 2011) (Fig. 21.2).

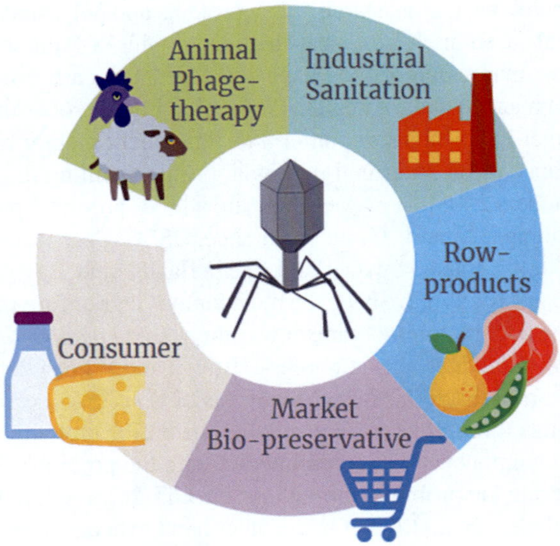

Fig. 21.2 Demonstration of different industrial aspects in which phages can be used to eradicate bacterial pathogens

Biofilm formation that comes after some bacterial contamination in food industry doesn't only cost companies billions of dollars annually, but also causes endemics and cost lives (Brooks and Flint 2008). Throughout history, bacteria were the highly infectious agent found in raw-products and in ready-to-eat food (REF): Bacterial infections were responsible for endemic foodborne botulism resulting from consumption of uncooked aquatic foods, four deaths due to burger eating in US, 1993, more than 65 infected cases by *E. coli* due to drinking of unpasteurized apple juice in USA, 1996, and more than 157 infected cases and one death due to cross-contamination of REF direct and indirect contact with raw meat in UK, 2005 (Fagan et al. 2011; Harper et al. 2014; Pennington 2010).

In a study was done in 2011 by the Centers for Disease Control and Prevention (CDC), it was estimated that approximately 48 million cases of food poisoning occur each year in the United States alone, of which 128,000 result in hospitalization and 3000 in fatalities. Only 31 major pathogens are responsible for 1351 deaths (Endersen et al. 2014; Scallan et al. 2011). All these cases are caused by food contamination by living microorganisms or food spoilage: contamination by the toxins secreted by the microorganisms and is considered as a main irritant in food industry after bacteria killing (Brooks and Flint 2008).

Due to continuous increase of antimicrobial resistance, Phages have displayed the best solution to fight back the bacteria. Namely, phages have an "auto dosing" property through which they multiply increasing their population-number whenever their host is present, and also have the ability to eradicate some biofilm populations whether by producing enzymes that depolymerize the biofilm or by passing through the biofilm and infect the bacteria (Sarhan and Azzazy 2015). Moreover, phages isolation and production is inexpensive and they can be isolated from different sources including the food products themselves.

Lytic Phages as bio-preservatives display two features which most related to and important for food safety. Namely, their safe use as they are harmless to mammalian cells and their high host specificity that allows no disturbance for micro flora (Endersen et al. 2014). Additionally, oral application of phage with the ingested food will affect the viability of phages by their exposure to acidic gastric juice (GJ) during stomach digestion which in turn will decrease the dose of the ingested phages to a negligible level that cannot affect the gut phagome obtaining the advantage of food preservation without disturbing the microflora (Koo et al. 2000). Traditionally, lytic-phage application to food can be done by dipping, spraying, or introduced as a liquid suspension to a large volume of food. However, these methods had drawbacks like being phage-wasteful and contributing to bacterial growth due to the suspension of phages in liquid-nutrient media. Additionally, the probability of phage inactivation by sanitizers used in the operating areas was one of the previous methods' drawbacks. Nevertheless, dry phage was obtained to be used on food products, but phage dilution and concentration was hard to be calculated for this application method leading to more probable phage resistance by the bacteria especially at low concentrations of phage. However, very high concentration can prevent this.

These challenges in free phage utilization as bio-preservative led to the use of immobilization method in which phages are kept on a surface of some material physically or ionically then exposed to the surface that is being treated. Phages have been immobilized on many surfaces like gold-coated biosensors and on polystyrene, but to provide higher efficiency through keeping the orientation of the phage in which phage's tail points outwards, phages were immobilized on cellulose membrane. Furthermore, to make the immobilization process more cost-efficient, the difference of charge distribution on the phage; negatively charged head and positively charged tail in pH above 4, was used and phage were immobilized on cationic polymers, polyvinylamine (Anany et al. 2011).

21.8 Synergistic Combination of Phages and Other Biocontrol Agents

Bacteria are able to develop resistance for phages through different mechanisms including CRISPR-Cas system, blocking the receptor through which phages penetrate the bacterial cell, Super infection exclusion (SIE) systems by which bacteria prevent phage DNA from entry into their cells, and more. However, phages are also very dynamic and when they face an antiviral resistance mechanism, they develop tactic mechanism to abolish these mechanisms in order to survive (Labrie et al. 2010). However, in industry where phages are being used as biocontrol agents, to insure the maximum efficiency and food products protection, phages are no longer used alone. They are used in cocktail of different phages or in combination with other chemical compounds.

In a study done by Leverentz et al. (2003) to test the synergistic effect of bacteriocin combination with phages on freshly-cut vegetables and fruits, it was found that treating the fruits with phage and nisin (bacteriocin) combination provided the most reduction of *Listeria monocytogenes* count on apple and honeydew melon comparing to treating each of them with phage or nicin alone (Table 21.2). The application of phage or phage and nicin was done by spraying and pipetting, however, spraying application reduced *L. monocytogenes* at least as much as pipetting application. Additionally, phage titer was more stable on melon slices than apple (Leverentz et al. 2003).

Synergistic effect was much clear when phages cocktail was combined with *trans*-cinnameldehyde (TC) (Fig. 21.3) in a study was done by Viazis et al. on leafy green vegetables: lettuce and spinach leaves infected with Enterohemorrhagic *E. coli* (EHEC) O157:H7. The study was done using different titers of EHEC (10^4, 10^5, and 10^6 CFU/mL) at different temperatures (4, 8, 23 and 37 °C) and incubated for 24 h. At the lowest titer of EHEC, either phage cocktail or TC showed complete bactericidal activity at both 23 and 37 °C. However, at higher EHEC titer and lower temperature conditions, phages cocktail and TC efficiency, each individually, decreased. Nevertheless, at combination of both, no bacterial cells were detected at all titer concentrations and temperature conditions on both leafy greens, showing the high bactericidal potential of phages cocktail/TC combination and its possible use as preservative in food industry (Viazis et al. 2011).

Synergistic effect between phage SA97, which encodes for endolysin LysSA97, and carvacrol volatile oil (Fig. 21.4) had showed significant antimicrobial activity against eleven *S. aureus* strains that participate in foodborne illness in food industry, including beef and milk, two of the *S. aureus* strains are Methicillin-Resistant *S. aureus* MRSA. Since phage SA97 alone (376 nM) and carvacrol alone (3.33 mM) showed 0.8 ± 0.2 and 1.0 ± 0.0 log CFU/mL reduction in *S. aureus* cell count, respectively. When both combined together at the same concentrations bacterial count decreased by 4.5 ± 0.2 log CFU/mL. However, the combination is affected by

Table 21.2 Summary of an experiment compare the efficiency of nicin bacteriocin, phages, and the synergistic effect of their combination on freshly-cut vegetables and fruits

Treatment *Listeria* method *monocytogenes* infected	Control group (no treatment) 10 °C – 7 days	Phage treatment (spray/pipetting)	Nisin treatment (bacteriocin)	Phage + Nisin treatment (synergism)
Honeydew melon slices	*Listeria* increased significantly	*Listeria* decreased by 2–4.6 log units	*Listeria* decreased by 3.2 log units	*Listeria* decreased by 5.7 log units
Apple slices	*Listeria* survived and increased slightly	*Listeria* reduction is below 0.4 log units	*Listeria* decreased by 2 log units	*Listeria* decreased by 2.3 log units

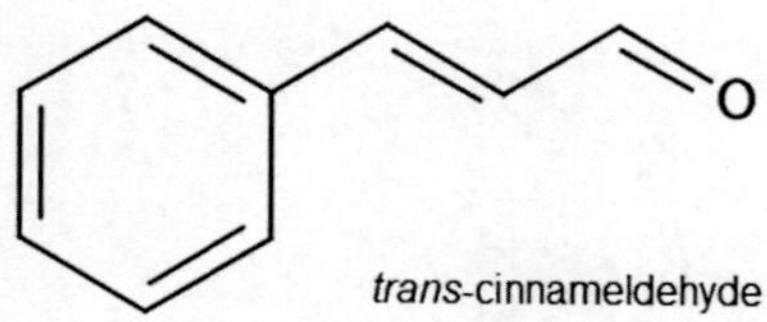

Fig. 21.3 Chemical structure of *trans*-cinnameldehyde (TC) volatile oil that used in combination with phages cocktail producing synergistic antibacterial effect on Enterohemorrhagic *E. coli* (EHEC) *O157*H7

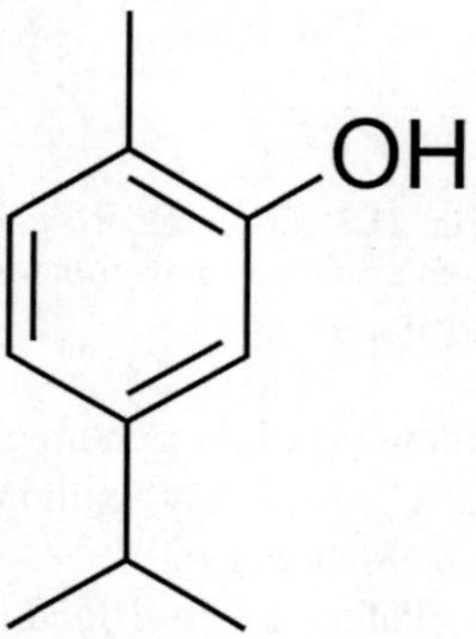

Fig. 21.4 Chemical structure of carvacrol volatile oil that used in combination with endolysin LysSA97-coding phage (phage SA97) producing synergistic antibacterial effect on eleven *S. aureus* strains including two MRSA strains

total lipid content in food. Since the bacterial inactivation in skimmed milk was much significant than that in the whole milk (Chang et al. 2017).

Phage-antibiotic synergy (PAS) has gained a lot of focus recently. In an *in vitro* study done by Ryan et al. to test the effect of T4 phage and cefotaxime (Fig. 21.5) combination on controlling the biofilm produced by *E. coli* ATCC 11303. It was found that minimum biofilm eradication concentration (MBEC) of cefotaxime decreased from 256 to 128 and 32 μg mL^{-1} at phage titer 10^4 PFU mL^{-1} and 10^7 PFU mL^{-1}, respectively, to prove for the first time that combination of phage and conventional antibiotic could enhance in vitro biofilm control. Additionally, T4 phage plaque size and concentration have showed increase when cefotaxime sub-lethal concentration increases at standard plaque assay (Ryan et al. 2012).

21.9 Challenges of Phage Biocontrol

Despite the promising power of phages to solve the antimicrobial-drug-resistance problem, phages still face many challenges to be used in food industry as biocontrol agents (Goodridge and Bisha 2011). Some of the challenges were enumerated by Hagens and Loessner as the following: phages to be used should show broad hast range, be virulent and undergo lytic cycle, allow no transmission of non-phage genetic material (i.e., resistance genes of bacteria), and confirmation of adverse-effects absence that follows oral administration. Moreover, they recommend

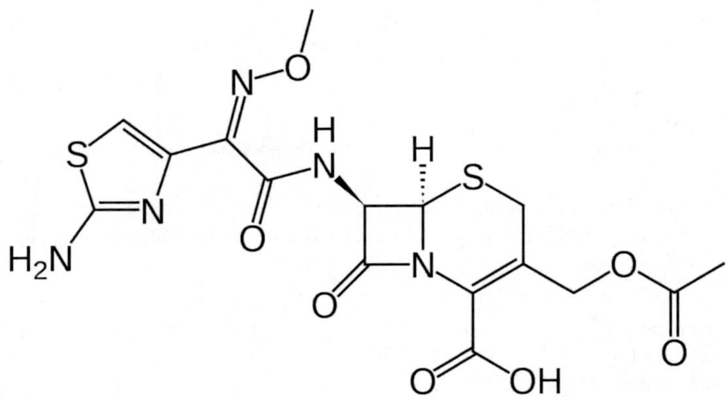

Fig. 21.5 Chemical structure of cefotaxime antibiotic that used with T4 phage to decrease the cefotaxime's minimum biofilm eradication concentration MBEC and increase T4 phage plaque size by synergism

knowing all the genomic information of the used phages and to be sure of complete absence of any pathogenic genes or allergy-inducing proteins (Hagens and Loessner 2010).

Pathogenicity island (PAI) is a term introduced in 1990 and used to descript a cluster of mobile genes code for pathogenicity-contributing genes that have been transferred horizontally (Hacker et al. 1990). However, some investigations have proved phage contribution in transferring two PAI to *E. coli* strain 536 due to presence of phage-related integrase gene (Cheetham and Katz 1995; Dobrindt et al. 2000).

Accordingly, some temperate phages under several environmental condition interconvert from lytic cycle to lysogenic cycle and contribute in transduction of bacterial genes horizontally in a process known as "lysogenic conversion". This change doesn't kill the bacteria and lead to their protection from infection of the same or closely related phages in a process called "superinfection immunity", consequently, increasing the bacterial cell virulence (Endersen et al. 2014).

21.10 Physico-Chemical Factors

It is of upmost importance to consider stability of phage in response to different environmental factors when phages are used in biocontrol of food-borne pathogens. Such factors affect phage indirectly by changing bacterial physiology and subsequently interaction between two of them. Thus, viral usefulness to eradicate pathogen is either improved or deteriorated (i.e. viral replication relies directly on level of bacterial growth). Bacterial host has primary requirements of temperature, pH and salts. Additionally, other factors can directly impact phage stability such as presence of certain chemicals, pH, UV light and heat in which viruses are stored or used for biocontrol. Phage stability is highly variable among different classes and families as its genetic material is always prone to degradation by aforementioned factors (Ishani and Tina 2015; Ly-Chatain 2014).

As previously discussed, *L. monocytogenes was the target of most phage-based preparations produced so far.* Although *L. monocytogenes requires a high concentration of salt, a low temperature and low pH, the bacteria was persistently found in food production (e.g. plants) with inappropriate conditions for optimal growth (Koutsoumanis and Sofos* 2005*; Swaminathan and Gerner-Smidt* 2007*). Phage P100-driven elimination of L. monocytogenes was dramatically affected by the presence of certain chemicals. Although attachment of phage to host remained unaffected, chemicals such as Lutensol AO7, SDS and NaCl caused replication of phage to highly deteriorate. Nonetheless, inclusion of these chemicals into the course of decontamination induced better reduction of bacterial counts, synergistically with phage. Moreover, Temperature is another factor that affects host replication and phage stability before infection, and determines degree of bacterial reduction by phage during decontamination. Bacterial reduction of L. monocytogenes during decontamination by phage P100 was found to be influenced by temperature. At a temperature as low as 4 °C, bacterial count was reduced by* 7 $\log_{10}$ units in 2 week while the *reduction reached* at 20 *°C was only 4–5* $\log_{10}$ units in 1 week. A remarkable phage resistance expressed in an increase in bacterial count was then recorded over a course of 16 weeks in the two temperature conditions (Carlton et al. 2005; Fister et al. 2016; Soni and Nannapaneni 2010).

21.11 Phage Resistance

Bacterial host under stress of phage undergoes several mutations to evade the viral activity, a life-long marathon of evolution that commenced before human race even existed, as phage reactively adapt to invade the host cell again. This dual and highly responsive evolution was investigated in multiple models such as phage Phi2 and *Pseudomonas fluorescens*, phage RIM8 and the marine cyanobacterium *Synechococcus*, and phage PP01 and *E. coli* O157:H7. However, these studies do not accurately reflect the antagonistic evolution that takes place on food surface or *in vivo*, and more investigation is still needed. Although phage preparations against *L. monocytogenes*, for instance, have proven high success in eradication of the pathogen from assorted REF, it was not an utter eradication and viable *L. monocytogenes* were recovered. This is also equivalent to resistance records of *Salmonella* and *E. coli* in food. However, resistance to phage cocktails was less potent as single-phage preparations; as it requires more mutations by the targeted bacteria to overcome all introduced phages (El-Shibiny et al. 2009; Goodridge and Bisha 2011; Tanji et al. 2004). Surprisingly, some contradictory results were also obtained. A significant count of resistant bacteria recovered from food - in case of *L. monocytogenes-* was still susceptible to phage, suggesting the presence of another contributing factor. Moreover, difference in resistance upon phage treatment by

Salmonella-targeting phage in different foods solidified the thought that variations in food matrix affect the distribution of phage virions through it. Additionally, food microbiome can be another limiting factor that affects effectiveness of phage preparation.

On the other hand, some studies of phage resistance gleaned interesting outcomes. Mutations of bacteria delivering phage resistance can lead to loss of bacterial pathogenicity due to consequent changes in some genes modulating virulence (León and Bastías 2015). A good example is cholera-causing bacteria *Vibrio cholera*, isolated from Bangladesh and Haiti that were rendered non-virulent after they became resistant to natural phage invasion. Development of resistance costs *V. cholerae* its disease-causing features and genetic analyses discovered occurrence of mutations in an outer membrane porin expressed by *ompU* gene. Mutations in *ompU* gene lead to failure of phage to recognize the bacterial host, because it is the phage receptor (Attridge et al. 2001). Along the same lines, two pathogens of salmon fish were found to lose their virulence upon adapting to phages. *Flavobacterium psychrophilum* and *F. columnare* had modified surface properties besides loss of motility and eventually rendered non-virulent (León and Bastías 2015). These studies may propose another role of phages to render unaffected hosts to non-virulent forms and further studies are still needed to confirm this hypothesis in food-borne pathogens.

21.12 Multiplicity of Infection

Multiplicity of infection (MOI) is frequently used to define the amount of phages virions introduced to bacterial host cells. This ratio must represent the number of viral particles that were not only attached to their host cells, but also were successful to infect the bacteria. Due to phage resistance approaches of bacteria, not all adsorbed phages can cause infection. The new term "Multiplicity of absorption" was later on used for more accurate depiction (Delbruck 1940).

In food decontamination settings, pollutant bacteria are distributed in low density within food. Maintaining constant MOI levels, rates of bacterial killing by introduced phage was found highly dependent on density of bacterial host cells (Payne and Jansen 2001). Another challenge implicated is the determination of minimum phage density needed to reach a bacterial killing inundation threshold in constant MOI, regardless of phage potential to replicate (Abedon et al. 2011; Payne and Jansen 2001).

21.13 Conclusion

Bacteriophage applications in food industry is started to be used worldwide especially in USA and Europe. Many commercial phage products have been approved to be used to control and detect foodborne pathogens in different food products. The promising results of recent research showed the excessive potential to use phages to improve food safety. Phages can be used in feed and drinking water to reduce pathogenic bacteria in poultry and large animals and can also be added to food packages by immobilization to extend its shelf life (Colom et al. 2015; Endersen et al. 2014). It is important to ensure that the food surface is entirely treated with phage to get the best efficacy. The application of phage cocktails will also reduce the risk of bacterial resistance.

References

Abdelsattar AS, Abdelrahman F, Dawoud A, Connerton IF, El-Shibiny A (2019) Encapsulation of E. coli phage ZCEC5 in chitosan–alginate beads as a delivery system in phage therapy. AMB Express *9*(1):87. https://doi.org/10.1186/s13568-019-0810-9

Abedon ST, Kuhl SJ, Blasdel BG, Kutter EM (2011) Phage treatment of human infections. Bacteriophage 1:66–85. https://doi.org/10.4161/bact.1.2.15845

Akinkunmi EO, Adesunkanmi AR, Lamikanra A (2014) Pattern of pathogens from surgical wound infections in a Nigerian hospital and their antimicrobial susceptibility profiles. Afr Health Sci 14. https://doi.org/10.4314/ahs.v14i4.5

Anany H, Chen W, Pelton R, Griffiths MW (2011) Biocontrol of Listeria monocytogenes and Escherichia coli O157:H7 in meat by using phages immobilized on modified cellulose membranes. Appl Environ Microbiol 77(18):6379–6387. https://doi.org/10.1128/aem.05493-11

Andreatti Filho RL, Higgins JP, Higgins SE, Gaona G, Wolfenden AD, Tellez G, Hargis BM (2007) Ability of bacteriophages isolated from different sources to reduce Salmonella enterica serovar Enteritidis in vitro and in vivo. Poult Sci 86(9):1904–1909

Atterbury RJ, Connerton PL, Dodd CER, Rees CED, Connerton IF (2003) Application of host-specific bacteriophages to the surface of chicken skin leads to a reduction in recovery of campylobacter jejuni. Appl Environ Microbiol 69:6302–6306. https://doi.org/10.1128/AEM.69.10.6302-6306.2003

Attridge SR, Fazeli A, Manning PA, Stroeher UH (2001) Isolation and characterization of bacteriophage-resistant mutants of Vibrio cholerae O139. Microb Pathog 30:237–246. https://doi.org/10.1006/mpat.2000.0426

Bai J, Kim YT, Ryu S, Lee JH (2016) Biocontrol and rapid detection of food-borne pathogens using bacteriophages and endolysins. Front Microbiol 7. https://doi.org/10.3389/fmicb.2016.00474

Belongia EA, MacDonald KL, Parham GL, White KE, Korlath JA, Lobato MN et al (1991) An outbreak of Escherichia coli O157:H7 colitis associated with consumption of precooked meat patties. J Infect Dis 164:338–343. https://doi.org/10.1093/infdis/164.2.338

Bergmann CP, Stumpf A (2013) Dental ceramics: microstructure, properties and degradation. In *Dental ceramics: microstructure, properties and degradation*. https://doi.org/10.1007/978-3-642-38224-6

Bigwood T, Hudson JA, Billington C, Carey-Smith GV, Heinemann JA (2008) Phage inactivation of foodborne pathogens on cooked and raw meat. Food Microbiol 25:400–406. https://doi.org/10.1016/j.fm.2007.11.003

Bren L (2007) Bacteria-eating virus approved as food additive. FDA Consum

Brooks JD, Flint SH (2008) Biofilms in the food industry: problems and potential solutions. Int J Food Sci Technol 43(12):2163–2176. https://doi.org/10.1111/j.1365-2621.2008.01839.x
Brown-Jaque M, Muniesa M, Navarro F (2016) Bacteriophages in clinical samples can interfere with microbiological diagnostic tools. Sci Rep 6(September):1–8. https://doi.org/10.1038/srep33000
Bruttin A, Brüssow H (2005) Human volunteers receiving Escherichia coli phage T4 orally: a safety test of phage therapy. Antimicrob Agents Chemother 49:2874–2878. https://doi.org/10.1128/AAC.49.7.2874-2878.2005
Campos J, Gil J, Mourão J, Peixe L, Antunes P (2015) Ready-to-eat street-vended food as a potential vehicle of bacterial pathogens and antimicrobial resistance: an exploratory study in Porto region, Portugal. Int J Food Microbiol 206:1–6. https://doi.org/10.1016/j.ijfoodmicro.2015.04.016
Carlton RM, Noordman WH, Biswas B, De Meester ED, Loessner MJ (2005) Bacteriophage P100 for control of Listeria monocytogenes in foods: genome sequence, bioinformatic analyses, oral toxicity study, and application. Regul Toxicol Pharmacol 43:301–312. https://doi.org/10.1016/j.yrtph.2005.08.005
Chang Y, Yoon H, Kang DH, Chang PS, Ryu S (2017) Endolysin LysSA97 is synergistic with carvacrol in controlling Staphylococcus aureus in foods. Int J Food Microbiol 244:19–26. https://doi.org/10.1016/j.ijfoodmicro.2016.12.007
Cheetham BF, Katz ME (1995) A role for bacteriophages in the evolution and transfer of bacterial virulence determinants. Mol Microbiol 18(2):201–208. https://doi.org/10.1111/j.1365-2958.1995.mmi_18020201.x
Choińska-Pulit A, Mituła P, Śliwka P, Łaba W, Skaradzińska A (2015) Bacteriophage encapsulation: trends and potential applications. Trends Food Sci Technol 45:212–221. https://doi.org/10.1016/j.tifs.2015.07.001
Colom J, Cano-Sarabia M, Otero J, Cortés P, Maspoch D, Llagostera M (2015) Liposome-encapsulated bacteriophages for enhanced oral phage therapy against Salmonella spp. Appl Environ Microbiol 81(14):4841–4849. https://doi.org/10.1128/AEM.00812-15
d'Herelle F (1917) Sur un microbe invisible antagoniste des bacilles dysentériques. CR Acad. Sci, Paris
Delbruck M (1940) The growth of bacteriophage and lysis of the host. J Gen Physiol 23:643–660. https://doi.org/10.1085/jgp.23.5.643
Dobrindt U, Janke B, Piechaczek K, Nagy G, Ziebuhr W, Fischer G et al (2000) Toxin genes on pathogenicity islands: impact for microbial evolution. Int J Med Microbiol 290(4–5):307–311. https://doi.org/10.1016/S1438-4221(00)80028-4
El-Shibiny A, Scott A, Timms A, Metawea Y, Connerton P, Connerton I (2009) Application of a group II campylobacter bacteriophage to reduce strains of campylobacter jejuni and campylobacter coli colonizing broiler chickens. J Food Prot 72(4):733–740. https://doi.org/10.4315/0362-028X-72.4.733
El-Shibiny A, El-Sahhar S, Adel M (2017) Phage applications for improving food safety and infection control in Egypt. J Appl Microbiol 123:556–567. https://doi.org/10.1111/jam.13500
Endersen L, O'Mahony J, Hill C, Ross RP, McAuliffe O, Coffey A (2014) Phage therapy in the food industry. Annu Rev Food Sci Technol 5(1):327–349. https://doi.org/10.1146/annurev-food-030713-092415
Fagan RP, McLaughlin JB, Castrodale LJ, Gessner BD, Jenkerson SA, Funk EA et al (2011) Endemic foodborne botulism among Alaska native persons-Alaska, 1947-2007. Clin Infect Dis 52(5):585–592. https://doi.org/10.1093/cid/ciq240
Fister S, Robben C, Witte AK, Schoder D, Wagner M, Rossmanith P (2016) Influence of environmental factors on phage-bacteria interaction and on the efficacy and infectivity of phage P100. Frontiers Microbiol 7(JUL):1–13. https://doi.org/10.3389/fmicb.2016.01152
Fortini D, Fashae K, García-Fernández A, Villa L, Carattoli A (2011) Plasmid-mediated quinolone resistance and β-lactamases in Escherichia coli from healthy animals from Nigeria. J Antimicrob Chemother 66:1269–1272. https://doi.org/10.1093/jac/dkr085

García P, Martínez B, Obeso JM, Rodríguez A (2008) Bacteriophages and their application in food safety. Lett Appl Microbiol 47(6):479–485. https://doi.org/10.1111/j.1472-765X.2008.02458.x
Gbassi GK, Vandamme T (2012) Probiotic encapsulation technology: from microencapsulation to release into the gut. Pharmaceutics 4(1):149–163. https://doi.org/10.3390/pharmaceutics4010149
Goode D, Allen VM, Barrow PA (2003) Reduction of experimental Salmonella and Campylobacter contamination of chicken skin by application of lytic bacteriophages. Appl Environ Microbiol 69:5032–5036. https://doi.org/10.1128/AEM.69.8.5032-5036.2003
Goodridge LD, Bisha B (2011) Phage-based biocontrol strategies to reduce foodborne pathogens in foods. Bacteriophage 1(3):130–137. https://doi.org/10.4161/bact.1.3.17629
Górski A, Wazna E, Daṃbrowska BW, Daṃbrowska K, Świtała-Jeleń K, Międzybrodzki R (2006) Bacteriophage translocation. FEMS Immunol Med Microbiol 46(3):313–319. https://doi.org/10.1111/j.1574-695X.2006.00044.x
Górski A, Miedzybrodzki R, Weber-Dabrowska B, Fortuna W, Letkiewicz S, Rogóz P et al (2016) Phage therapy: combating infections with potential for evolving from merely a treatment for complications to targeting diseases. Frontiers Microbiol 7(SEP):1–9. https://doi.org/10.3389/fmicb.2016.01515
Gould LH, Nisler AL, Herman KM, Cole DJ, Williams IT, Mahon BE et al (2011) Surveillance for foodborne disease outbreaks — United States, 2008. Morb Mortal Wkly Rep
Greer GG, Dilts BD (1990) Inability of a bacteriophage pool to control beef spoilage. Int J Food Microbiol 10:331–342. https://doi.org/10.1016/0168-1605(90)90080-O
Guenther S, Herzig O, Fieseler L, Klumpp J, Loessner MJ (2012) Biocontrol of Salmonella Typhimurium in RTE foods with the virulent bacteriophage FO1-E2. Int J Food Microbiol 154:66–72. https://doi.org/10.1016/j.ijfoodmicro.2011.12.023
Hacker J, Bender L, Ott M, Wingender J, Lund B, Marre R, Goebel W (1990) Deletions of chromosomal regions coding for fimbriae and hemolysins occur in vitro and in vivo in various extra intestinal Escherichia coli isolates. Microb Pathog 8(3):213–225. https://doi.org/10.1016/0882-4010(90)90048-U
Hagens S, Loessner (2010) Bacteriophage for biocontrol of foodborne pathogens. Curr Pharm Biotechnol 11(1):58–68. Retrieved from http://e-citations.ethbib.ethz.ch/view/pub:29137
Harper DR, Parracho HMRT, Walker J, Sharp R, Hughes G, Werthén M et al (2014) Bacteriophages and biofilms. Antibiotics 3:270–284. https://doi.org/10.3390/antibiotics3030270
Hildebrand GJ, Wolochow H (1962) Translocation of bacteriophage across the intestinal wall of the rat. Proc Soc Exp Biol Med 109(1):183–185
Hoffmann M (1965) Animal Experiments on Mucosal Passage and Absorption Viraemia of T3 Phages after Oral, Trachéal and Rectal Administration. Zentralblatt Bakteriol Parasitenkd Infekt Hyg 198(4):371–390
Hubbell JA, Chilkoti A (2012) Chemistry. Nanomaterials for drug delivery. Science 337(6092):303–305. https://doi.org/10.1126/science.1219657
Huff WE, Huff GR, Rath NC, Balog JM, Donoghue AM (2002) Prevention of Escherichia coli infection in broiler chickens with a bacteriophage aerosol spray. Poult Sci 81:1486–1491. https://doi.org/10.1093/ps/81.10.1486
Ishani D, Tina S (2015) Bacteriophages as twenty-first century antibacterial tools for food and medicine. J Interdisciplinary Multidisciplinary Res 2(9):92–97. https://doi.org/10.1007/s00253-011-3227-1
Jain RK, Stylianopoulos T (2010) Delivering nanomedicine to solid tumors. Nat Rev Clin Oncol 7:653–664. https://doi.org/10.1038/nrclinonc.2010.139
Keller R, Engley FB (2013) Fate of bacteriophage particles introduced into mice by various routes. Exp Biol Med 98:577–580. https://doi.org/10.3181/00379727-98-24112
Kocharunchitt C, Ross T, McNeil DL (2009) Use of bacteriophages as biocontrol agents to control Salmonella associated with seed sprouts. Int J Food Microbiol 128:453–459. https://doi.org/10.1016/j.ijfoodmicro.2008.10.014

Koo J, DePaola A, Marshall DL (2000) Effect of simulated gastric fluid and bile on survival of Vibrio vulnificus and Vibrio vulnificus phage†. J Food Prot 63:1665–1669. https://doi.org/10.4315/0362-028X-63.12.1665

Koutsoumanis KP, Sofos JN (2005) Effect of inoculum size on the combined temperature, pH and a w limits for growth of Listeria monocytogenes. Int J Food Microbiol 104:83–91. https://doi.org/10.1016/j.ijfoodmicro.2005.01.010

Kutateladze M, Adamia R (2010) Bacteriophages as potential new therapeutics to replace or supplement antibiotics. Trends Biotechnol 28:591–595. https://doi.org/10.1016/j.tibtech.2010.08.001

Labrie SJ, Samson JE, Moineau S (2010) Bacteriophage resistance mechanisms. Nat Rev Microbiol 8:317–327. https://doi.org/10.1038/nrmicro2315

León M, Bastías R (2015) Virulence reduction in bacteriophage resistant bacteria. Front Microbiol 6(APR):1–7. https://doi.org/10.3389/fmicb.2015.00343

Leverentz B, Conway WS, Alavidze Z, Janisiewicz WJ, Fuchs Y, Camp MJ et al (2001) Examination of bacteriophage as a biocontrol method for Salmonella on fresh-cut fruit: A model study. J Food Prot 64:1116–1121. https://doi.org/10.4315/0362-028X-64.8.1116

Leverentz B, Conway WS, Camp MJ, Janisiewicz WJ, Abuladze T, Yang M et al (2003) Biocontrol of Listeria monocytogenes on fresh-cut produce by treatment with lytic bacteriophages and a bacteriocin. Appl Environ Microbiol 69(8):4519–4526

Loc Carrillo CM, Connerton PL, Pearson T, Connerton IF (2007) Free-range layer chickens as a source of campylobacter bacteriophage. Anton Leeuw Int J Gen Mol Microbiol 92:275–284. https://doi.org/10.1007/s10482-007-9156-4

Loretz M, Stephan R, Zweifel C (2010) Antimicrobial activity of decontamination treatments for poultry carcasses: A literature survey. Food Control 21:791–804. https://doi.org/10.1016/j.foodcont.2009.11.007

Lourens-Hattingh A, Viljoen BC (2001) Yogurt as probiotic carrier food. Int Dairy J 11(1–2):1–17. https://doi.org/10.1016/S0958-6946(01)00036-X

Ly-Chatain MH (2014) The factors affecting effectiveness of treatment in phages therapy. Front Microbiol 5. https://doi.org/10.3389/fmicb.2014.00051

Lynch MF, Tauxe RV, Hedberg CW (2009) The growing burden of foodborne outbreaks due to contaminated fresh produce: risks and opportunities. Epidemiol Infect 137:307–315. https://doi.org/10.1017/S0950268808001969

Majewska J, Beta W, Lecion D, Hodyra-Stefaniak K, Kłopot A, Kazmierczak Z et al (2015) Oral application of T4 phage induces weak antibody production in the gut and in the blood. Viruses 7(8):4783–4799. https://doi.org/10.3390/v7082845

Manrique P, Dills M, Young MJ (2017) The human gut phage community and its implications for health and disease. 9–11. https://doi.org/10.3390/v9060141

McVay CS, Velásquez M, Fralick JA (2007) Phage therapy of Pseudomonas aeruginosa infection in a mouse burn wound model. Antimicrob Agents Chemother 51:1934–1938. https://doi.org/10.1128/AAC.01028-06

Modi R, Hirvi Y, Hill A, Griffiths MW (2001) Effect of phage on survival of Salmonella Enteritidis during manufacture and storage of Cheddar cheese made from raw and pasteurized milk. J Food Prot 64:927–933. https://doi.org/10.4315/0362-028X-64.7.927

Modi SR, Lee HH, Spina CS, Collins JJ (2014) Ecological network of the phage metagenome. Nature 499(7457):219–222. https://doi.org/10.1038/nature12212.Antibiotic

Monk AB, Rees CD, Barrow P, Hagens S, Harper DR (2010) Bacteriophage applications: where are we now? Lett Appl Microbiol 51:363–369. https://doi.org/10.1111/j.1472-765X.2010.02916.x

Morozova VV, Vlassov VV, Tikunova NV (2018) Applications of bacteriophages in the treatment of localized infections in humans. Front Microbiol 9(AUG). https://doi.org/10.3389/fmicb.2018.01696

Mousavi, S. L., Rasooli, I., Nazarian, S., & Amani, J. (2009). Simultaneous detection of Escherichia coli O157:H7, toxigenic Vibrio cholerae, and Salmonella typhimurium by multiplex PCR. Iran J Clin Infect Dis. 4(2): 97-103

Nagel TE, Chan BK, De Vos D, El-Shibiny A, Kang'ethe EK, Makumi A, Pirnay JP (2016) The developing world urgently needs phages to combat pathogenic bacteria. Front Microbiol 7. https://doi.org/10.3389/fmicb.2016.00882

Navarro F, Muniesa M (2017) Phages in the human body. Front Microbiol 8(APR):1–7. https://doi.org/10.3389/fmicb.2017.00566

O'Flynn G, Ross RP, Fitzgerald GF, Coffey A (2004) Evaluation of a cocktail of three bacteriophages for biocontrol of Escherichia coli O157:H7. Appl Environ Microbiol 70:3417–3424. https://doi.org/10.1128/AEM.70.6.3417-3424.2004

Oliveira M, Viñas I, Colàs P, Anguera M, Usall J, Abadias M (2014) Effectiveness of a bacteriophage in reducing Listeria monocytogenes on fresh-cut fruits and fruit juices. Food Microbiol 38:137–142. https://doi.org/10.1016/j.fm.2013.08.018

Pao S, Rolph SP, Westbrook EW, Shen H (2006) Use of bacteriophages to control Salmonella in experimentally contaminated sprout seeds. J Food Sci 69:M127–M130. https://doi.org/10.1111/j.1365-2621.2004.tb10720.x

Pavlickova S, Dolezalova M, Holko I (2015) Resistance and virulence factors of Escherichia coli isolated from chicken. J Environ Sci Health, Part B Pestic Food Contam Agric Wastes 50:417–421. https://doi.org/10.1080/03601234.2015.1011959

Payne RJH, Jansen VAA (2001) Understanding bacteriophage therapy as a density-dependent kinetic process. J Theor Biol 208:37–48. https://doi.org/10.1006/jtbi.2000.2198

Pelaz B, Alexiou C, Alvarez-Puebla RA, Alves F, Andrews AM, Ashraf S et al (2017) Diverse applications of nanomedicine. ACS Nano 11(3):2313–2381. https://doi.org/10.1021/acsnano.6b06040

Pennington H (2010) Escherichia coli O157. Lancet 376(9750):1428–1435. https://doi.org/10.1016/S0140-6736(10)60963-4

Ramesh V, Fralick JA, Rolfe RD (1999) Prevention of Clostridium difficile-induced ileocecitis with bacteriophage. Anaerobe 5:69–78. https://doi.org/10.1006/anae.1999.0192

Ranadheera RDCS, Baines SK, Adams MC (2010) Importance of food in probiotic efficacy. Food Res Int 43:1–7. https://doi.org/10.1016/j.foodres.2009.09.009

Rao VB, Black LW (2010) Structure and assembly of bacteriophage T4 head. Virol J 7:1–14. https://doi.org/10.1186/1743-422X-7-356

Raya RR, Oot RA, Moore-Maley B, Wieland S, Callaway TR, Kutter EM, Brabban AD (2011) Naturally resident and exogenously applied T4-like and T5-like bacteriophages can reduce Escherichia coli O157. Bacteriophage 1(1):15–24. https://doi.org/10.4161/bact.1.1.14175

Reynaud A, Cloastre L, Bernard J, Laveran H, Ackermann HW, Licois D, Joly B (1992) Characteristics and diffusion in the rabbit of a phage for Escherichia coli 0103. Attempts to use this phage for therapy. Vet Microbiol 30:203–212. https://doi.org/10.1016/0378-1135(92)90114-9

Rosenquist H, Nielsen NL, Sommer HM, Nørrung B, Christensen BB (2003) Quantitative risk assessment of human campylobacteriosis associated with thermophilic Campylobacter species in chickens. Int J Food Microbiol 83:87–103. https://doi.org/10.1016/S0168-1605(02)00317-3

Rozema EA, Stephens TP, Bach SJ, Okine EK, Johnson RP, Stanford KIM, Mcallister TA (2009) Oral and rectal administration of bacteriophages for control of escherichia coli O157.-H7 in feedlot cattle. J Food Prot 72(2):241–250. https://doi.org/10.4315/0362-028X-72.2.241

Ryan EM, Alkawareek MY, Donnelly RF, Gilmore BF (2012) Synergistic phage-antibiotic combinations for the control of Escherichia coli biofilms in vitro. FEMS Immunol Med Microbiol 65(2):395–398. https://doi.org/10.1111/j.1574-695X.2012.00977.x

Sanders ME (2008) Probiotics: definition, sources, selection, and uses. Clin Infect Dis 46(s2):S58–S61. https://doi.org/10.1086/523341

Sarhan WA, Azzazy HME (2015) Phage approved in food, why not as a therapeutic? Expert Rev Anti-Infect Ther 13(1):91–101. https://doi.org/10.1586/14787210.2015.990383

Scallan E, Hoekstra RM, Angulo FJ, Tauxe RV, Widdowson M, Roy SL, Jones JL, Griffin PM (2011) Foodborne illness acquired in the United States- major pathogens. Emerg Infect Dis 17:7–15. https://doi.org/10.3201/eid1701.P11101

Schellekens MM, Woutersi J, Hagens S, Hugenholtz J (2007) Bacteriophage P100 application to control Listeria monocytogenes on smeared cheese. Milchwissenschaft

Schubbert R, Lettmann C, Doerfler W (1994) Ingested foreign (phage M13) DNA survives transiently in the gastrointestinal tract and enters the bloodstream of mice. MGG Mol Gen Genet 242(5):495–504. https://doi.org/10.1007/BF00285273

Smith DL, Harris AD, Johnson JA, Silbergeld EK, Morris JG (2002) Animal antibiotic use has an early but important impact on the emergence of antibiotic resistance in human commensal bacteria. Proc Natl Acad Sci U S A 99:6434–6439. https://doi.org/10.1073/pnas.082188899

Soni KA, Nannapaneni R (2010) Removal of listeria monocytogenes biofilms with bacteriophage P100. J Food Prot 73:1519–1524. https://doi.org/10.4315/0362-028X-73.8.1519

Spricigo DA, Bardina C, Cortés P, Llagostera M (2013) Use of a bacteriophage cocktail to control Salmonella in food and the food industry. Int J Food Microbiol 165(2):169–174. https://doi.org/10.1016/j.ijfoodmicro.2013.05.009

Stone R (2002) Bacteriophage therapy. Food and agriculture: testing grounds for phage therapy. Science 298:730. https://doi.org/10.1126/science.298.5594.730

Summers WC (1999) Bacteriophage discovered. In Félix d'Herelle and the origins of molecular biology. (p. 230). Retrieved from https://yalebooks.yale.edu/book/9780300071276/felix-dherelle-and-origins-molecular-biology

Swaminathan B, Gerner-Smidt P (2007) The epidemiology of human listeriosis. Microbes Infect 9:1236–1243. https://doi.org/10.1016/j.micinf.2007.05.011

Tanji Y, Shimada T, Yoichi M, Miyanaga K, Hori K, Unno H (2004) Toward rational control of Escherichia coli O157:H7 by a phage cocktail. Appl Microbiol Biotechnol 64(2):270–274. https://doi.org/10.1007/s00253-003-1438-9

Tanji Y, Shimada T, Fukudomi H, Miyanaga K, Nakai Y, Unno H (2005) Therapeutic use of phage cocktail for controlling Escherichia coli O157:H7 in gastrointestinal tract of mice. J Biosci Bioeng 100:280–287. https://doi.org/10.1263/jbb.100.280

Translocation of Bacteriophage Across the Intestinal Wall of the Rat (1962) Proceedings of the Society for Experimental Biology and Medicine. https://doi.org/10.3181/00379727-109-27146

Turki Y, Ouzari H, Mehri I, Ammar AB, Hassen A (2012) Evaluation of a cocktail of three bacteriophages for the biocontrol of Salmonella of wastewater. Food Res Int 45:1099–1105. https://doi.org/10.1016/j.foodres.2011.05.041

Viazis S, Akhtar M, Feirtag J, Diez-Gonzalez F (2011) Reduction of Escherichia coli O157:H7 viability on leafy green vegetables by treatment with a bacteriophage mixture and trans-cinnamaldehyde. Food Microbiol 28:149–157. https://doi.org/10.1016/j.fm.2010.09.009

Waddell TE, Mazzocco A, Pacan J, Ahmed R, Johnson R, Poppe C, Khakhria R (2002) U.S. Patent No. 6,485,902. U.S. Patent and Trademark Office, Washington, DC

Wilczewska AZ, Niemirowicz K, Markiewicz KH (2012) Review Nanoparticles as drug delivery systems

Wills QF, Kerrigan C, Soothill JS (2005) Experimental bacteriophage protection against Staphylococcus aureus abscesses in a rabbit model. Antimicrob Agents Chemother 49:1220–1221. https://doi.org/10.1128/AAC.49.3.1220-1221.2005

World Health Organization (2014) Antimicrobial resistance: global report on surveillance 2014. World Health Organization. doi: 9789241564748

Yang ZQ, Tao XY, Zhang H, Rao SQ, Gao L, Pan ZM, Jiao XA (2019) Isolation and characterization of virulent phages infecting Shewanella baltica and Shewanella putrefaciens, and their application for biopreservation of chilled channel catfish (Ictalurus punctatus). Int J Food Microbiol 292:107–117. https://doi.org/10.1016/j.ijfoodmicro.2018.12.020

Zinno P, Devirgiliis C, Ercolini D, Ongeng D, Mauriello G (2014) Bacteriophage P22 to challsenge Salmonella in foods. Int J Food Microbiol 191:69–74. https://doi.org/10.1016/j.ijfoodmicro.2014.08.037

Chapter 22
Bacteriophages for Environmental Applications: Effect of Trans-organismic Communication on Wastewater Treatments

Soumya Chatterjee and Sonika Sharma

Abstract The emergence of different pathogenic bacteria especially (multi) drug resistant one, is a concern worldwide. Bacteriophage (commonly known as 'phage') based restoration of bacteria contaminated environments has a huge potential. However, phages intricately communicate, both intra- and intercellular levels, for their survival and take decisions for lytic or lysogenised phase of life. This review encompasses recent scientific understanding on bacteriophage communications, infection, and behavioural pattern amongst phages and hosts, and their potential environmental applications.

22.1 Introduction

In the present era, along with the expansion of the global economy, intensification of human population, habitation and related activities have created a critical problem of wastes generation contaminating the environment (Prasad et al. 2019). Similarly, productions of wastewaters from different sources have increased manifold in last few decades polluting soil and water at different proportions. Wastewaters from households, agricultural, industrial, and medical activities have their own typical characteristics in terms of organic and inorganic loads, pH, oxygen demands (both biological- BOD and chemical- COD) and microbial profile. However, mixing of wastewaters in their path creates a more composite nature of wastewater, which creates condition of presence of a number of common microbial or bacterial species (Garner et al. 2019). Several researchers reported that, the composite wastewater may carry a considerable amount of antibiotics which results in development, selection and transmission of antibiotic-resistance bacteria (ARB) &/or

S. Chatterjee (✉) · S. Sharma
Biodegradation Technology Division, Defence Research Laboratory, DRDO, Tezpur, Assam, India
e-mail: schatterjee@drl.drdo.in

G. Witzany (ed.), *Biocommunication of Phages*,
https://doi.org/10.1007/978-3-030-45885-0_22

antibiotic-resistance genes (ARG) in the environment (Ye et al. 2019; Liu K et al. 2019; Couch et al. 2019; Diaz et al. 2019)

Different investigations report that common mechanism for interacting between organisms are signs and sending signals. As per Witzany (2010), specific combinatorial rules (syntax), content-specific rules (semantic) and contextual rules (pragmatic) are crucial for successful communication processes. Signals may come from either in the form of biotic (the *Mitwelt*) or abiotic (the *Umwelt*) components which are received by certain receptors and communication may be generated both within cells (intracellular) or between cell (extracellular) (Witzany 2006, 2010). Bacteria are imperative hosts for multi-viral colonization within its cell or surrounding environment by determining the order of nucleic acid sequences and mechanisms. In bacteria, quorum sensing (sign-mediated chemical interactions 'semiochemicals') is well reported that control their population density (Greenberg 2003). Similar signalling behaviour is apparent in bacteria during changes in abiotic components in environmental situations (Daniels et al. 2004). It is apparent that, communication in bacteriophages can also have the similarity in signs and context.

Wastewater, being a complex environment having multiple physico-chemical factors, determining targeted phage activity is still a challenge. Similarly, wastewater treatment plants (WWTPs) are the hotspots for microbial communications and gene exchange. However, due to high cost (both economic and environmental) of disinfecting wastewater, phage biocontrol and wastewater treatment becoming an upcoming research area to overcome the environmental factors that may affect phage-host interactions (Jassim et al. 2016). Exploiting bacterial metabolic machinery, a particular bacteriophage can only replicate within specific bacteria. Thus, host specificity of bacteriophage is an interesting character for the researchers to counter the bacterial load. Although it was a century-old discovery, application of bacteriophages is becoming more popular in the post-antibiotic era. Researchers are working with various environmental applications by influencing lysis of targeted pathogenic bacteria in wastewater treatment (Dou et al. 2018; Zhao et al. 2019; Sun et al. 2018, 2019).

This review emphasizes on the up to date literature survey on the new developments in phage communication, host infection strategies and their environmental applications (mostly with wastewater applications) in controlling pathogenic bacterial infection in natural environments.

22.2 Bacteriophage- the Most Abundant Component in Biosphere

Bacteriophages or phages ("eaters of bacteria") are viruses that infect bacteria. Bacteriophages are the most plentiful entities in biosphere, present in all the environments. Population wise, it is estimated that, more than 10^{31} bacteriophages are present on the planet with a total weight of 10^{9} tons (Dou et al. 2018; Ye et al. 2019).

They are having typical harmless properties for all organisms (including humans) except their targeted bacterial hosts, where, each type of bacterium can be infected by the specific bacteriophage(s) (Harada et al. 2018). They are simplest entity having genome (with at least one nucleic acid- either DNA or RNA molecule) enclosed by capsid (a protein coat) and some may contain extra features such as spikes and tails. At early years of the twentieth century, Twort (in 1915) and d'Herelle (in 1917) independently discovered bacteriophages (Hadley 1928; Sharma et al. 2016). The discovery of bacteriophages led to a definitive research to control and/or destroy the growth of couple of pathogenic bacteria which were otherwise problematic to control at that point of time (Sharma et al. 2016). Advent of antibiotics made a sharp decline on phage-based research, where, fast advancement of antibiotic-based therapies became popular throughout the world form 1950s onwards (Czajkowski et al. 2019). However, enormous number and vast variety of bacteriophages present in Nature can be a huge potential to explore their genes and related applications in controlling bacterial population. Thus, bacteriophages are becoming important tools for molecular studies on genes, genomes, infections, drug delivery and bioinformatics (Kropinski 2018). Different studies indicate the potential application of bacteriophages as a component for surface sterilization, environmental pathogen inactivation, food preservation, controlling biofilm formation, corrosion prevention, phage-based vaccination, phage display, phage-based biosensors development etc. (Keen 2015; Harada et al. 2018; Ye et al. 2019). Recent development of antibiotic resistant bacterial strains are encouraging researchers to utilise bacteriophage as a potential environmental tool to control selective bacterial pathogens of soil &/ or water.

22.3 General Features of Bacteriophages

Ernest Hanbury Hankin, a British bacteriologist, first reported (1896) antibacterial activity by then unidentified 'antibacterial agent in the waters of the river Ganga and the river Yamuna in India (Hankin 1896: http://icmr.nic.in/buapril02.pdf). This agent was reported to limit the spreading of cholera epidemic (Hankin 1896; Sulakvelidze et al. 2001). In 1915, FW Twort, working in the laboratory, found zones of clearance in plates of micrococcus bacteria and he also reported that the agent multiplied itself within bacteria by the process of killing the host bacteria (Twort 1915, 1922, 1949). In 1917, Félix d'Herelle reported phage effect on bacteria in case of severe haemorrhagic dysentery, similar to his earlier (1910) observations on an epizootic locust infection. Working on bacteria-free filtrates from fecal samples on Shigella, d'Herelle observed appearance of small, clear zones (plaques) (d'Herelle 1917a, b, 1930). d'Herelle also proposed the name of this phenomenon as 'bacteriophage' (the word came into existence on 18 October 1916) (d'Herelle 1949; Sulakvelidze et al. 2001; Sharma et al. 2016). As per the International Committee on Taxonomy of Viruses (ICTV), bacteriophages can be classified into Myoviridae, Podoviridae, Siphoviridae, and Filamentous bacteriophages.

Length wise bacteriophages varies widely (ranging from 20 to200 nm) with, T4 bacteriophages are among the largest bacteriophages (approximately 200 nm in length and 80–100 nm wide) (Sharma et al. 2016). A typical phage particle consists of ahead (having nucleic acid), tail (attached with a collar), and a base plate (having tail fibers and spikes) (Tolstoy et al. 2018). Genome of phage may consist either of double-stranded DNA (dsDNA), single-stranded DNA (ssDNA), double-stranded RNA (dsRNA), or single-stranded RNA (ssRNA) (Aiewsakun et al. 2018). dsDNA as genetic material is most common (for more than 96% of bacteriophages); however, variation in shape (eg. cubic, filamentous, and pleomorphic) is likely (Ackermann 2011). Based on genome integration with bacterial genome, bacteriophages can be either lytic or temperate bacteriophages. Lytic bacteriophages are responsible in lysis or destruction of the bacterial cell involving their six stages of reproductive cycle (viz., attachment, penetration, transcription, biosynthesis, maturation, and lysis) (Ofir and Sorek 2018; Ye et al. 2019). On the contrary temperate phages generally integrate phage genome into the host bacterial DNA and pass on their genomic information to the progeny of bacteria through host's reproduction, however, conversion to lytic cycle may take place under the circumstances when some troubling factors like, ionizing radiation, high temperature, antibiotics, ultraviolet, or heavy metals are present (Erez et al. 2017; Kim and Bae 2018).

22.4 Wastewater Bacteriophages

Phage-mediated amelioration of wastewater is a potential area of research for developing eco-friendly treatment systems. Wastewater contains different microbes and bacteriophages. Amongst, virulent somatic coliphages are abundantly present that generally follow lytic cycle for their reproduction. These somatic bacteriophages belong to the families Siphoviridae, Myoviridae, Podoviridae, and Microviridae that primarily infect members of Enterobacteriaceae family especially Escherichia coli (Sharma et al. 2016). However, the activities of bacteriophages vary with different factors like host bacterial density, pH and temperature of the medium, presence of varied ions, heavy metals and organic matters. It is estimated that, at least 10^4host bacteria per millilitre is required for effective phage replication (Goyal et al. 1987). Many faecal bacteria favours sewage wastewater. Reports suggest that phage titre against those bacteria for somatic coliphages was around to be 10^6–10^8per millilitre (Jofre et al. 2016).

22.5 Studies Related to Bacteriophage Applications in Wastewater Decontamination

Several researchers are working to mitigate bacterial population in wastewater. Some recent reports on environmental applications of phage-based treatments are reviewed here (Fig. 22.1).

Change of phenotype and/or activity of bacteria due to phage infection happens to various bacterial phyla. Either transduction or incorporation of phage-encoded proteins are responsible for such changes in the bacteria (Hargreaves et al. 2014). Escherichia coli, which is regarded as the indicator species for fecal contamination in wastewater research can naturally be infected with two groups of bacteriophages i.e., somatic and F-specific coliphages (Jofre et al. 2016). Again, bacteriophages are also considered as indicators; two groups of E. coli infecting bacteriophages (coliphages) can be treated indicator as viral and fecal contamination (Jofre et al. 2016).

Several class of antibiotics like sulfonamides, beta-lactams, trimethoprim macrolides, quinolones and tetracyclines are common in Wastewater treatment plants (WWTPs) which creates situations like multidrug efflux genes. Therefore, WWTPs are supposed to be feasible hotspots for propagation of antibiotic resistance in the environment. WWTPs provide suitable situations for proliferation of antibiotic resistant bacteria (ARB) and antibiotic resistance genes (ARGs) through horizontal gene transfer among different microorganisms (Pazda et al. 2019).

Zhang and his colleagues, working on Escherichia coli bacteriophages of municipal wastewater treatment plants, found that, many bacteriophages with wide-ranging hosts can harbour antibiotic resistant genes (like β-Lactam resistance genes e.g. blaCMY, blaSHV, blaTEM, blaOXA and blaCTX-M), which can also be transmitted via phage transduction (Gunathilaka et al. 2017). They even found that couple of bacteriophages were able to transduce blaTEM into E. coli (strain ATCC 13706), possibly would contributing to antibiotic resistance gene distribution in the environment (Gunathilaka et al. 2017).

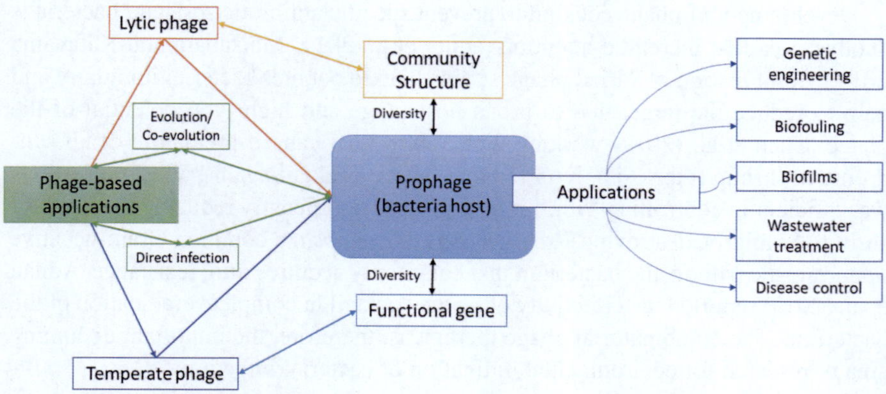

Fig. 22.1 Diagrammatic representation of phage based environmental applications

Kotay et al. (2011) working on filamentous bacterium *Haliscomenobacter hydrossis* on activated sludge process of wastewater treatment plant, demonstrated that the biomass bulking can be controlled through applying lytic bacteriophages. According to their report, the bacteriophage belongs to the Myoviridae family is stable at considerable higher temperature (42 °C) and pH (pH between 5–8), having titre with the bacterium was 5.2 ± 0.3 x 10^5 PFU/mL, burst size was 105 ± 7 PFU/infected cell, increased biomass settling and reduction in sludge volume index (SVI) (Kotay et al. 2011).

Escherichia coli infecting phage FRNAPHs (F-specific RNA bacteriophages; family Leviviridae) have two genera with each having two subtypes (Grabow 2001). Genus *Levivirus* classified into genotypes I and II (GI and GII) and Allolevivirus into genotypes III and IV (GIII and GIV); which are water contamination indicator having similar fates like other viruses in wastewater treatment plants (WWTPs) (Haramoto et al. 2015; Lee et al. 2019a). Raw municipal wastewater contains more GII and GIII than GI and GIV (Lee et al. 2019a, b). The strains of FRNAPH GI and GII genotypes showed more resistant to chlorine and ultraviolet disinfection treatment than other genotypes clearly indicating the presence of disinfectant-resistant viral strains in the environment (Boudaud et al. 2012; Lee et al. 2019b).

Activated sludge wastewater treatment processes require filamentous bacteria. However, their uncontrolled growth may cause foaming and bulking which cause disturbances in the process. Liu et al. (2015), working with filamentous bacteria Gordonia sp. and its phages (GordTnk2, Gmala1, and GordDuk1), reported significant reduction in Gordonia host levels in comparison with reactors without having phage treatment. Similarly, Khairnar et al. (2016) reported eco-friendly way to control foaming in WWTPS through the application of bacteriophage.

Bacteriophages may also be used as a potential component for sustainable operation and monitoring of membrane-based wastewater treatment systems. According to Wu et al. (2017), bacteriophages can have a great potential to evaluate membrane performance indicators which may directly be linked with membrane fouling reduction and membrane condition monitoring. However, there is a number of practical challenges to this prospective area.

Development of phage cocktail to prevent specific antibiotic-resistant bacteria is another area that increased attention (Culot et al. 2019; Matsubara and Katayama 2019). Combination of varied phage strains help to counter bacterial immunity and help to reduce the target due to broad host range and high lytic potential of the phage. Chen et al. (2019) working with Vibrio, that caused premature death syndrome in shrimp (vibriosis), reported that, phage cocktail against Vibrio infections was efficient in controlling Vibrio infections and significantly reducing the death of shrimp. Similarly, destroying *Pseudomonas aeruginosa*, a common Gram-negative opportunistic pathogenic bacterium that commonly acquires drug resistance, Adnan et al. (2019) reported the effectivity of phage cocktail in complete eradication of the bacterium. The combinatorial phage therapy, is, therefore, the important upcoming area of research for environmental mitigation of bacterial infection.

22.6 Phage Decisions on Lytic/ Lysogenic Cycles

In nature, phages evolve to manipulate themselves to evade bacterial immunity. Bacteria adopt several strategies to overcome the attacking phages, which include inhibition of adsorption, restriction-modification (R/E) systems, prokaryotic Argonaute, CRISPR-Cas systems, and abortive infection (Abi) (Safari et al. 2019; Liu Q et al. 2019). The lytic or the lysogenic cycles are two major pathways for infecting host bacteria by temperate phages. In the lysogenic cycle, in contrary to the lytic (lysis) cycle, phage integrates its genome to the host. However, lysogenized bacterium develops immunity against the same phage and therefore can evade further infection (Erez et al. 2017).

Bacteriophage Lambda that infects *Escherichia coli* was studied for its lytic–lysogenic cycles. It was found that, infected bacterial nutritional conditions and the quantity of co-infecting phage particles influence phage to take 'probabilistic' decision to go for either lytic (killing the host) or lysogenic cycle (keeping the host viable) (Oppenheim et al. 2005). Sorek and his colleagues first reported that SPbeta group of phages infecting *Bacillus* species, communicate with its fellow members by using a small molecule (formed by six amino acids) (Erez et al. 2017). This communication peptide was termed by Sorek and his as team as 'arbitrium' (Latin meaning 'decision'). This peptide is released into the medium in succeeding infections when progeny phages lyse the host.

22.7 Phage Communication and Arbitrium System

Arbitrium in Phi3T (SPbeta group) phage against *Bacillus* species is encoded by three phage genes: aimP (produces peptide), aimR (intracellular receptor for the peptide) and aimX (a negative regulator of lysogeny) (Erez et al. 2017). Sorek and his colleagues found that, each arbitrium of Phi3T is a six amino acid long peptide having Ser-Ala-Ile-Arg-Gly-Ala (SAIRGA) amino acids. They further reported that, while addition of the SAIRGA in the culture considerably raised the lysogeny by about 48% (±7.9%) within 60 min after infection (compared to 18% ±3.3%, grown without addition), but, one amino acid less truncated version of peptide (i.e. AIRGA or SAIRG) did not show likewise result of concentration-dependent consequences on the dynamics of phage infection (Erez et al. 2017). Concentration of arbitrium in the medium help new phages to take decisions on subsequent infection strategies, like, whether to lysogenize with the available host bacteria (Erez et al. 2017; Garneau et al. 2017; Dolgin 2019; Gallego Del Sol et al. 2019).

The process of lysogeny is phage specific, as they produce particular peptide that helps to communicate among own clan. It was also evident that, Phi3T secreted SAIRGA arbitrium peptide did not affect the SPbeta phage that secrete Gly-Met-Pro-Arg-Gly-Ala (GMPRGA) arbitrium peptide for promoting lysogeny of their own clan. The specificity of such lysogenic activities also lies with AimR receptor that helps in specific binding of arbitrium secreted from specific phage (Erez et al. 2017). AimR receptor (constituting a tetratricopeptide repeat (TPR) domain) present in the phage is similar to the quorum sensing receptors (the RRNPP family of intracellular peptide receptors) of the Gram-positive bacteria (Rocha-Estrada et al. 2010; Dunny and Berntsson 2016; Perez-Pascual et al. 2016).

Although the phage-derived arbitrium peptides SAIRGA and GMPRGA seem to be similar in activity, however, structural and functional analyses suggest that, SAIRGA modulates conformational changes in phAimR (corresponding receptor of the peptide in phi3T), while GMPRGA stabilizes spAimR (corresponding receptor of the peptide in SPbeta phage) in the dimeric state during the lysis-lysogeny regulation (Dou et al. 2018).

Comparable to the Gram-positive bacterial communication schemes, where, after attaching to the receptor, communication peptide directly influences the transcriptional activities, it was reported that, having the helix-turn-helix (HTH) motif in the N terminus the AimR receptor of the arbitrium system binds to phage genome directly at the downstream of the *aimP* gene (Erez et al. 2017). *aimR* and *aimP* genes are expressed immediately after first infection of a phage with bacteria and activates AimX which promotes lytic cycle of phages. However, accumulation of arbitrium in the external medium regulates the activity of AimR receptor, which deactivate expression of AimX, leading to phage preference of lysogeny (Peters et al. 2016; Erez et al. 2017; Gallego Del Sol et al. 2019). Thus, the option for lytic or lysogenic cycle by a phage is being communicated by the predecessors to its descendant members through small arbitrium peptides.

Wang et al. (2018), while working on SPbeta phage, reported crystal structures for AimR, which dimerizes through the C-terminal capping helix in presence or absence of the arbitrium peptide and forms a superhelical fold, accommodating the peptide within its tetratricopeptide repeats (similar to bacterial quorum-sensing RRNPP family members); however, AimR targets the upstream sequence of the aimX when the arbitrium is absent.

Guan et al. (2019) reported the molecular mechanisms of DNA recognition by AimR that activates aimX gene expression by precisely targeting the downstream DNA in a Bacillus bacteriophage. They reported that, binding of arbitrium peptide makes a closed-to-open conformational change of AimR, eliminating DNA targeting (Guan et al. 2019; Gallego Del Sol et al. 2019).

Stokar-Avihail et al. (2019) recently reported similar peptide-based communication in phages of soil and pathogenic microbes, namely, *Bacillus cereus*, *Bacillus thuringiensis* and *Bacillus anthracis*. They reported the diversity of arbitrium like peptides which often carry pathogenic toxins along with their mobile elements. However, regulation of lysogenic state is affected by an antisense RNA controlling peptide-based decision (Stokar-Avihail et al. 2019).

22.8 Trade-off Between Superinfection and Immunity in Phage-Bacterial Context

Even though lysogeny process of temperate phages could evolutionarily be helpful, however, possibility of second infection by other related homotypic (closely related), mesotypic (moderately related) or heterotypic (unrelated) phages to the resident prophage is enormous in nature (Dedrick et al. 2017; Doron et al. 2018; Mavrich and Hatfull 2019). Phage develops gene circuitry based immune systems which control lysogeny, evading another prophage defence and counteracting against superinfecting phages (Mavrich and Hatfull 2019). These genetic circuits are complex, occasionally involving of multiple genetic loci. One of the mostly elucidated phages, λ coliphage, is having two DNA binding transcriptional regulatory genes, *cI* and *cro* present on a single genetic locus. For lysogenised condition, CI-operator bonding blocks the transcription initiation of cro (for lytic growth) (Campbell 1994; Doron et al. 2018).

Superinfection of temperate phages is an interesting phenomenon, where, a related phage tries to infect bacterium already infected with lysogenised resident prophage. This kind of infection immunity interplay may happen in different contexts. Complete infection or complete immunity related binary phenotype may be the occurrence for superinfecting phages, which may either be genetically identical or closely identical (homotypic) ensuing homoimmunity (symmetric immunity phenotypes) or unrelated (heterotypic) leading to heteroimmunity (Mavrich and Hatfull 2019). However, genetically linked but discrete mesotypic phage variants (nonidentical genetic elements displaying nonorthogonal characteristics) can also be a part of superinfection, resulting mesoimmunity (incomplete immunity phenotypes) (Mavrich and Hatfull 2019; Gentile et al. 2019). Coliphages λ and HK97 are homoimmune phages, where prophage confers immunity to host to counter superinfection by other homotypes (Juhala et al. 2000). In this process, it may so happen that, superinfecting phage gaining mutations and escape homoimmunity (Ravin 2015). Campbell (1994) reported that, coliphage λ could superinfect a λ-lysogenised condition, when its operator acquired minimum 03 point-mutations. While, in heteroimmune coliphages (coliphages λ and 434), *cro* expression in the opposing phage could not stopped and CI repressors display specificity for dissimilar operator sequences (Youderian et al. 1983; Campbell 1994; Mavrich and Hatfull 2019).

Engelberg-Kulka and Kumar (2015) demonstrated that, there is an intricate mechanism related to λcI repressor in switching from lysogenic to lytic cycle in λ phage. Under DNA damage, λcI repressor becomes cleaved and deactivated through SOS response within the cell triggering lytic cycle. However, mazEF pathway of *Escherichia coli* inhibits the SOS response (by producing toxin); but, lysogenic prophage triggers λrexB, which plays a crucial role in safeguarding lysogenised phase by degrading the antitoxin MazE (Engelberg-Kulka and Kumar 2015).

Evolution of homo or hetero conditions in phages is yet to be fully elucidated. Interactions between prophage and phage that generates immunity or defence mechanisms involves considerable interactions for developing immune specificity.

A number of regulatory components needs to be changed to switch immune specificities from a homoimmune to heteroimmune configuration. Although, substitutions of amino acid (point mutation or small number substitution) can alter repressor DNA-binding site, however, changes in immune specificity involves couple of mutations at different levels (operators, secondary immunity loci, and phage-encoded protein translation) (Campbell 1994).

In natural conditions, this kind of phage immunity evolution is very complex where, change in the susceptibility with homotypic or heterotypic phages can be addressed. Temperate phage immune activities, therefore, may be mesotypic (in-between immune characteristics) as well, having moderate relation but markedly different genetic elements (Montgomery et al. 2019).

22.9 Evading CRISPR Based Immunity by Phage

Ishino et al. (1987) first reported anomalous palindromic repeat structures in the genome of one of the K12 strains of *Escherichia coli* (AN234), which later termed as clustered regularly interspaced short palindromic repeats (CRISPR). CRISPR, present in approximately 40% of bacteria, is composed of a short direct repeat CRISPR array, segregated by short mutable DNA sequences; these spaces are flanked by diverse cas (CRISPR-associated) genes (Makarova et al. 2011; Liu Q et al. 2019). CRISPR-Cas mechanism in bacteria is an adaptive immune system to counter the phage infection. It is 'an ancient evolutionary arms race' (Borges et al. 2018) between bacteria and bacteriophages resulting in diversity of the existing system. CRISPR diversity is expanded by acquiring tiny fragments of phage genomes and aligned it into the CRISPR array (Koonin et al. 2017). Bacteria subsequently process the small fragments as small CRISPR RNAs which guide the destruction of the phage genome (Levy et al. 2015; Borges et al. 2018). CRISPR-Cas immune system is key to survival of the bacteria in phage infested environment through prevention of plasmids acquisition, prophage integration and phage lytic replication (van Houte et al. 2016; Borges et al. 2018). To counter the bacterial CRISPR based immunity, few phages produce inhibitory anti-CRISPR (Acr) proteins (Borges et al. 2018). Several researchers have identified inhibitory types of Acr proteins (as for example, inhibitors for type I-D, I-E, and I-F CRISPR-Cas3 systems or type II-A and II-C CRISPRCas9 systems), which significantly help phages in replication in the presence of CRISPR-Cas immunity (van Houte et al. 2016; Pawluk et al. 2014, 2018; Borges et al. 2018). Bondy-Denomy et al., (2013), working with Pseudomonas aeruginosa type I-F system, first reported anti-CRISPRs AcrIF1–5. Afterwards, more I-F anti-CRISPRs (AcrIF6–10) were reported from many mobile genetic elements (Pawluk et al. 2016a, b).

In bacteria, I-F Cascade (also called, I-F Csy surveillance complex, comprised of 4 proteins that assemble with CRISPR RNA (crRNA)) helps in finding out foreign DNA, targeting complimentary to the crRNA and activating foreign DNA degrading enzymes (helicase/nuclease) (Chowdhury et al. 2017; Guo et al. 2017; Rollins et al. 2017). Anti-CRISPR (Acr) proteins interact directly with the I-F Csy surveillance complex and prevent nuclease-mediated degradation. It has further been discovered that, I-F Acr proteins (eg. AcrIF1, AcrIF2, AcrIF3, and AcrIF10) are mechanistically dissimilar inhibitors which can strongly bind to the targets (Chowdhury et al. 2017).

It is not yet known about the deployment and utilization of Acr proteins by phage for infection, however, the protein production doesn't have direct link to phage replication when challenged by CRISPR-Cas immunity. Borges et al. (2018) revealed that that phage population numbers (related to adequate Acr dose to a single bacterium leading to its immunosuppression) plays an important role in infection after facing CRISPR-Cas immunity. In an independent study, Landsberger et al. (2018), working on *Pseudomonas* bacteria, reported that to evade CRISPR-Cas resistance, Acr-producing phages often team up with each other. In the sequential event, the authors reported that, the first infecting phage blocks the host CRISPR-Cas immune system to help a second Acr-phage to replicate efficaciously (Landsberger et al. 2018). However, this phage collaborations lead to 'epidemiological tipping points' related to Acr-phage density (from death of phage to a phage epidemic) which controls the population dynamics of bacteria-phage density.

There are phages that do not produce Acr proteins as they lack acr genes. While working with Acr-negative phages, Chevallereau et al. (2019) pointed out that in a composite environment cooperative behaviour ('cooperative offerings') between Acr negative and Acr positive phages help them to win critical situations. During pairwise competitions, Acr negative phages get benefit from Acr positive phages through the presence of Acr proteins. It has been demonstrated by Westra and his team that, potency of Acr proteins ('strong' or 'weak') determines the exploitable benefit for both types of phages; stronger Acr proteins, more advantage to the phages to counter bacterial immunity. However, 'weak' Acr proteins mostly assist Acr-positive phages by providing a greater fitness (Chevallereau et al. 2019; Dolgin 2019).

Recently Fineran and his colleagues, studying on a Serratia jumbo phage, reported that the phage forms a nucleus-like structure to escape type I CRISPR-Cas (type I-E, I-F and III-A) DNA targeting (Malone et al. 2020). But the jumbo phage is vulnerable to type III-A CRISPR-Cas system (involving Cas7, Cas10 and an accessory nuclease) by targeting phage RNA-based immunity, which seems to be a common phenomenon as reported by the authors according to global viral sequence datasets (Malone et al. 2020).

The initial attack by phage helps the way for other phages to overcome the microbial immunity (by generating immunosuppressed microbial cells), in a way, which can be designated as altruistic behaviour demonstrating inter-viral cooperation (Díaz-Muñoz et al. 2017; Landsberger et al. 2018; Chevallereau et al. 2019; Dolgin 2019).

22.10 Phage Behaviour in Environment

Languages for interacting among viral particles are universal but highly diverse. Even some viruses need another one (helper virus) to get either activated or suppressed in completing infection cycle which is being conveyed through the activity of proteins and peptides. Coinfection is a common phenomenon to the host bacteria, where, multiple viral genomes can be co-transmitted in the same transferrable component (Díaz-Muñoz et al. 2017). Social evolution theory for viruses is a rather new concept, though in 1940s, Max Delbrück and Alfred Hershey separately demonstrated two viral particles could concurrently enter the same cell and exchange genes, for which they received Nobel Prize for 'their discoveries concerning the replication mechanism and the genetic structure of viruses' (Díaz-Muñoz et al. 2017; Dolgin 2019; https://www.nobelprize.org/prizes/medicine/1969/press-release/). Viral cooperation might be for genetically identical individual (like self-reproducing) or for relatives that benefit in fitness of the individual by sharing genes (kin selection) (Díaz-Muñoz et al. 2017). Due to higher rate of mutation, RNA viruses quickly spread into quasi-species of associated genotypes, that facilitate them in adaptation, and better fitness of the population (Xue et al. 2016). Vital social evolution in virus directs circumvention of innate immunity and virulence through antiviral interferences (Domingo-Calap et al. 2019).

Using the plaque assay (a common method for measuring virus concentrations where each plaque represents a single infectious unit), Aguilera et al. (2017) found that more than one virus can contribute to plaque formation. Viral stocks may comprise of both aggregates and single particles, where aggregated virions are proficient in inducing coinfection and chimeric plaque formation. Aguilera et al. (2017) further reported that, this increase in frequency of coinfection was directly correlated with mutation load. They found that, heavily mutagenized viruses were involved in significant numbers of chimeric plaque formation, indicating the involvement of more than one virus in plaque formations (Aguilera et al. 2017). Altan-Bonnet and his team, working on rotavirus and norovirus, reported that some viruses are even transported as vesicle-cloaked viral clusters which are more virulent in causing infection. The vesicles are of plasma membrane or exosomal origin, may further help in infecting new host (Santiana et al. 2018). Altuvia et al. (2018) reported small RNA (sRNAs) from bacteria and prophage are having direct influence on regulating infection, bacterial metabolism and lysogenic- lytic states. Sponge RNAs are prophage-encoded sRNAs that affect posttranscription phases in bacteria (Altuvia et al. 2018).

In a recent review, Taylor et al. (2019) elaborated how a phage contribute its bacterial host to stay fit. Temperate phage encodes genes to produce fitness factors (called morons) that help in expanding the environmental niche and influence the bacterial fitness in various ways, like, bacterial diseases, motility, adhesion, antibiotic resistance, and quorum sensing (Taylor et al. 2019).

22.11 Future Prospects

The potential of application of phage-based therapy is enormous. As phage can target and control specific bacteria in a complex environment, it could be a helpful tool in many areas of environmental application, especially wastewater treatment. Liu Q et al. (2019) recently reviewed the potential genetic engineering applications of anti-CRISPR proteins (Acrs) in inactivating CRISPR-Cas systems, which is important in future implementation of phage-based therapy. Application of phage in various aspects requires more insights on phage behaviour, its semiotics at trans-. inter-, and intra-organismic (inter- and intracellular) levels. However, tapping yet to discover huge source of phages and their genomes are mostly important to make best use of the advantages of the phages. Phage diversity study will not only help to discover new phages from different niches, but will give information on composition, structure, and function of the microbial community. Activity and transmission of functional genes may further help in eco-friendly ways of pollution remediation, wastewater treatments and nutrient cycling. Dissimilar blends of phage types require to be tested, to understand relationship with host, its infection potential and immunity. Studying on polyvalent or cocktail of phages for specific application is another area, that requires to be explored to assure phage-based mitigation (Keen 2015; Ye et al. 2019). Being very minute in structure and its relative criticality in research, diverse phage-based research is still in its early phase. However, fast developments of related technology and instrumentations, including different biophysical and metagenomic approaches, may serve the purpose to explore the most abundant but tiniest entity and better understand their social language that existed in the planet billions of years before human evolved.

Acknowledgments Authors wish to acknowledge to all phage researchers of the world who are trying to contribute towards this immense important tiniest entity. Authors also wish to acknowledge Director, DRL Tezpur for support.

References

Ackermann HW (2011) The first phage electron micrographs. Bacteriophage 1(4):225–227

Adnan M, Ali Shah MR, Jamal M, Jalil F, Andleeb S, Nawaz MA, Pervez S, Hussain T, Shah I, Imran M, Kamil A (2019) Isolation and characterization of bacteriophage to control multidrug-resistant Pseudomonas aeruginosa planktonic cells and biofilm. Biologicals 63:89–96. pii: S1045-1056(19)30109-5. https://doi.org/10.1016/j.biologicals.2019.10.003

Aguilera ER, Erickson AK, Jesudhasan PR, Robinson CM, Pfeiffer JK (2017) Plaques formed by mutagenized viral populations have elevated Coinfection frequencies. MBio 8(2). pii: e02020-16. https://doi.org/10.1128/mBio.02020-16

Aiewsakun P, Adriaenssens EM, Lavigne R, Kropinski AM, Simmonds P (2018) Evaluation of the genomic diversity of viruses infecting bacteria, archaea and eukaryotes using a common bioinformatic platform: steps towards a unified taxonomy. J Gen Virol 99:1331–1343

Altuvia S, Storz G, Papenfort K (2018) Cross-regulation between Bacteria and phages at a posttranscriptional level. Microbiol Spectr 6(4). https://doi.org/10.1128/microbiolspec.RWR-0027-2018

Borges AL, Zhang JY, Rollins MF, Osuna BA, Wiedenheft B, Bondy-Denomy J (2018) Bacteriophage cooperation suppresses CRISPR-Cas3 and Cas9 immunity. Cell 174(4):917–925.e10. https://doi.org/10.1016/j.cell.2018.06.013

Boudaud N, Machinal C, David F, Fréval-Le Bourdonnec A, Jossent J, Bakanga F, Arnal C, Jaffrezic MP, Oberti S, Gantzer C (2012) Removal of MS2, Qβ and GA bacteriophages during drinking water treatment at pilot scale. Water Res 46:2651–2664

Campbell A (1994) Comparative molecular biology of lambdoid phages. Annu Rev Microbiol 48:193–222. https://doi.org/10.1146/annurev.mi.48.100194.001205

Chen L, Fan J, Yan T, Liu Q, Yuan S, Zhang H, Yang J, Deng D, Huang S, Ma Y (2019) Isolation and characterization of specific phages to prepare a cocktail preventing Vibrio sp. Va-F3 infections in Shrimp (Litopenaeus vannamei). Front Microbiol 10:2337. https://doi.org/10.3389/fmicb.2019.02337

Chevallereau A, Meaden S, Fradet O, Landsberger M, Maestri A, Biswas A, Gandon S, van Houte S, Westra ER (2019) Exploitation of the cooperative behaviours of anti-CRISPR phages. bioRxiv. https://doi.org/10.1101/574418

Chowdhury S, Carter J, Rollins MF, Golden SM, Jackson RN, Hoffmann C, Nosaka L, Bondy-Denomy J, Maxwell KL, Davidson AR, Fischer ER, Lander GC, Wiedenheft B (2017) Structure reveals mechanisms of viral suppressors that intercept a CRISPR RNA-guided surveillance complex. Cell 169:47–57

Couch M, Agga GE, Kasumba J, Parekh RR, Loughrin JH, Conte ED (2019) Abundances of tetracycline resistance genes and tetracycline antibiotics during anaerobic digestion of swine waste. J Environ Qual 48:171–178

Culot A, Grosset N, Gautier M (2019) Overcoming the challenges of phage therapy for industrial aquaculture: a review. Aquaculture 513:734423. https://doi.org/10.1016/j.aquaculture.2019.734423ff.ffhal-02291468f

Czajkowski R, Jackson RW, Lindow SE (2019) Editorial: environmental bacteriophages: from biological control applications to directed bacterial evolution. Front Microbiol 10:1830. https://doi.org/10.3389/fmicb.2019.01830

D'Herelle F (1917a) Sur un microbe invisible antagoniste des bacillesdysente ́riques. C R Acad Sci (Paris) 165:373–375

d'Herelle F (1917b) On an invisible microbe antagonistic to dysentery bacilli. C R Acad Sci 165:373–375

d'Herelle F (1930) The bacteriophage and its clinical applications. Charles C Thomas, Springfield

d'Herelle F (1949) The bacteriophage. Sci News 14:44–59

Daniels R, Vanderleyden J, Michiels J (2004) Quorum sensing and swarming migration in bacteria. FEMS Microbiol Rev 28:261–289

Dedrick RM, Jacobs-Sera D, Bustamante CA, Garlena RA, Mavrich TN, Pope WH, Reyes JC, Russell DA, Adair T, Alvey R, Bonilla JA, Bricker JS, Brown BR, Byrnes D, Cresawn SG, Davis WB, Dickson LA, Edgington NP, Findley AM, Golebiewska U, Grose JH, Hayes CF, Hughes LE, Hutchison KW, Isern S, Johnson AA, Kenna MA, Klyczek KK, Mageeney CM, Michael SF, Molloy SD, Montgomery MT, Neitzel J, Page ST, Pizzorno MC, Poxleitner MK, Rinehart CA, Robinson CJ, Rubin MR, Teyim JN, Vazquez E, Ware VC, Washington J, Hatfull GF (2017) Prophage-mediated defence against viral attack and viral counter-defence. Nat Microbiol 2:16251. https://doi.org/10.1038/nmicrobiol.2016.251

Diaz H, Graham K, Moreland R, Liu M, Ramsey J (2019) Complete genome sequence of Escherichia coli phage Pisces. Microbiol Resour Announc 8(39). pii: e01054-19. https://doi.org/10.1128/MRA.01054-19

Díaz-Muñoz SL, Sanjuán R, West S (2017) Sociovirology: conflict, cooperation, and communication among viruses. Cell Host Microbe 22(4):437–441. https://doi.org/10.1016/j.chom.2017.09.012

Dolgin E (2019) The secret social lives of viruses. Nature 570:290–292

Domingo-Calap P, Segredo-Otero E, Durán-Moreno M, Sanjuán R (2019) Social evolution of innate immunity evasion in a virus. Nat Microbiol 4(6):1006–1013. https://doi.org/10.1038/s41564-019-0379-8

Doron S, Melamed S, Ofir G, Leavitt A, Lopatina A, Keren M, Amitai G, Sorek R (2018) Systematic discovery of antiphage defense systems in the microbial pangenome. Science 359(6379). pii: eaar4120. https://doi.org/10.1126/science.aar4120

Dou C, Xiong J, Gu Y, Yin K, Wang J, Hu Y, Zhou D, Fu X, Qi S, Zhu X, Yao S, Xu H, Nie C, Liang Z, Yang S, Wei Y, Cheng W (2018) Structural and functional insights into the regulation of the lysis-lysogeny decision in viral communities. Nat Microbiol 3(11):1285–1294. https://doi.org/10.1038/s41564-018-0259-7

Dunny GM, Berntsson RP-A (2016) Enterococcal sex pheromones: evolutionary pathways to complex, two-signal systems. J Bacteriol 198:1556–1562

Engelberg-Kulka H, Kumar S (2015) Yet another way that phage λ manipulates its Escherichia coli host: λrexB is involved in the lysogenic-lytic switch. Mol Microbiol 96(4):689–693. https://doi.org/10.1111/mmi.12969

Erez Z, Steinberger-Levy I, Shamir M, Doron S, Stokar-Avihail A, Peleg Y, Melamed S, Leavitt A, Savidor A, Albeck S, Amitai G, Sorek R (2017) Communication between viruses guides lysis-lysogeny decisions. Nature 541(7638):488–493. https://doi.org/10.1038/nature21049

Gallego Del Sol F, Penadés JR, Marina A (2019) Deciphering the molecular mechanism underpinning phage Arbitrium communication systems. Mol Cell 74(1):59–72. https://doi.org/10.1016/j.molcel.2019.01.025

Garneau JR, Depardieu F, Fortier L-C, Bikard D, Monot M (2017) PhageTerm: a tool for fast and accurate determination of phage termini and packaging mechanism using next-generation sequencing data. Sci Rep 7:8292. https://doi.org/10.1038/s41598-017-07910-5

Garner E, Inyang M, Garvey E, Parks J, Glover C, Grimaldi A, Dickenson E, Sutherland J, Salveson A, Edwards MA, Pruden A (2019) Impact of blending for direct potable reuse on premise plumbing microbial ecology and regrowth of opportunistic pathogens and antibiotic resistant bacteria. Water Res 151:75-86

Gentile GM, Wetzel KS, Dedrick RM, Montgomery MT, Garlena RA, Jacobs-Sera D, Hatfull GF (2019) More evidence of collusion: a new prophage-mediated viral defense system encoded by mycobacteriophage Sbash. mBio 10:e00196-19. https://doi.org/10.1128/mBio.00196-19

Goyal SM, Gerba CP, Bitton G (1987) Phage ecology. Wiley, New York, p. 321

Grabow WOK (2001) Bacteriophages: update on application as models for viruses in water. Water SA 27:251–268

Greenberg EP (2003) Bacterial communication: tiny teamwork. Nature 424:134

Guan Z, Pei K, Wang J, Cui Y, Zhu X, Su X, Zhou Y, Zhang D, Tang C, Yin P, Liu Z, Zou T (2019) Structural insights into DNA recognition by AimR of the arbitrium communication system in the SPbeta phage. Cell Discov 28(5):29. https://doi.org/10.1038/s41421-019-0101-2

Gunathilaka GU, Tahlan V, Mafiz AI, Polur M, Zhang Y (2017) Phages in urban wastewater have the potential to disseminate antibiotic resistance. Int J Antimicrob Agents 50(5):678–683. https://doi.org/10.1016/j.ijantimicag.2017.08.013

Guo TW, Bartesaghi A, Yang H, Falconieri V, Rao P, Merk A, Eng ET, Raczkowski AM, Fox T, Earl LA, Patel DJ, Subramaniam S (2017) Cryo-EM structures reveal mechanism and inhibition of DNA targeting by a CRISPR-Cas surveillance complex. Cell 171:414–426

Hadley P (1928) The Twort-D'Herelle phenomenon: a critical review and presentation of a new conception (homogamic theory) of bacteriophage action. J Infect Dis 42:263-434

Hankin EHL (1896) Action bactericide des Eaux de la Jumna et duGange sur le vibrion du cholera. Ann Inst Pasteur 10:511

Harada LK, Silva EC, Campos WF, Del Fiol FS, Vila M, Dąbrowska K, Krylov VN, Balcão VM (2018) Biotechnological applications of bacteriophages: state of the art. Microbiol Res 212–213:38–58

Haramoto E, Fujino S, Otagiri M (2015) Distinct behaviors of infectious F-specific RNA coliphage genogroups at a wastewater treatment plant. Sci Total Environ 520:32–38

Hargreaves KR, Kropinski AM, Clokie MRJ (2014) Bacteriophage behavioral ecology: how phages alter their bacterial host's habits. Bacteriophage 4:e29866. https://doi.org/10.4161/bact.29866

Ishino Y, Shinagawa H, Makino K, Amemura M, Nakata A (1987) Nucleotide sequence of the iap gene, responsible for alkaline phosphatase isozyme conversion in Escherichia coli, and identification of the gene product. J Bacteriol 169:5429–5433

Jassim SAA, Limoges RG, El-Cheikh H (2016) Bacteriophage biocontrol in wastewater treatment. World J Microbiol Biotechnol 32(4). https://doi.org/10.1007/s11274-016-2028-1

Jofre J, Lucena F, Blanch AR, Muniesa M (2016) Coliphages as model organisms in the characterization and management of water resources. Water 8:199. https://doi.org/10.3390/w8050199

Juhala RJ, Ford ME, Duda RL, Youlton A, Hatfull GF, Hendrix RW (2000) Genomic sequences of bacteriophages HK97 and HK022: pervasive genetic mosaicism in the lambdoid bacteriophages. J Mol Biol 299:27–51. https://doi.org/10.1006/jmbi.2000.3729

Keen EC (2015) A century of phage research: bacteriophages and the shaping of modern biology. BioEssays 37(1):6–9. https://doi.org/10.1002/bies.201400152

Khairnar K, Chandekar R, Nair A, Pal P, Paunikar WN (2016) Novel application of bacteriophage for controlling foaming in wastewater treatment plant- an eco-friendly approach. Bioengineered 7(1):46–49. https://doi.org/10.1080/21655979.2015.1134066

Kim MS, Bae JW (2018) Lysogeny is prevalent and widely distributed in the murine gut microbiota. ISME J 12:1127–1141

Koonin EV, Makarova KS, Wolf YI (2017) Evolutionary genomics of defense systems in archaea and bacteria. Annu Rev Microbiol 71:233–261

Kotay SM, Datta T, Choi J, Goel R (2011) Biocontrol of biomass bulking caused by Haliscomenobacter hydrossis using a newly isolated lytic bacteriophage. Water Res 45:694–704

Kropinski AM (2018) Bacteriophage research – what we have learnt and what still needs to be addressed. Res Microbiol 169:481–487. https://doi.org/10.1016/j.resmic.2018.05.002

Landsberger M, Gandon S, Meaden S, Rollie C, Chevallereau A, Chabas H, Buckling A, Westra ER, van Houte S (2018) Anti-CRISPR phages cooperate to overcome CRISPR-Cas immunity. Cell 174(4):908–916.e12. https://doi.org/10.1016/j.cell.2018.05.058

Lee S, Suwa M, Shigemura H (2019a) Metagenomic analysis of infectious F-specific RNA bacteriophage strains in wastewater treatment and disinfection processes. Pathogens 8(4). pii: E217. https://doi.org/10.3390/pathogens8040217

Lee S, Suwa M, Shigemura H (2019b) Occurrence and reduction of F-specific RNA bacteriophage genotypes as indicators of human norovirus at a wastewater treatment plant. J Water Health 17:50–62

Levy A, Goren MG, Yosef I, Auster O, Manor M, Amitai G, Edgar R, Qimron U, Sorek R (2015) CRISPR adaptation biases explain preference for acquisition of foreign DNA. Nature 520:505–510

Liu M, Gill JJ, Young R, Summer EJ (2015) Bacteriophages of wastewater foaming-associated filamentous Gordonia reduce host levels in raw activated sludge. Sci Rep 5:13754. https://doi.org/10.1038/srep13754

Liu Q, Zhang H, Huang X (2019) Anti-CRISPR proteins targeting the CRISPR-Cas system enrich the toolkit for genetic engineering. FEBS J. https://doi.org/10.1111/febs.15139

Liu K, Sun MM, Ye M, Chao HZ, Zhao YC, Xia B, Jiao WT, Feng YF, Zheng XX, Liu MQ, Jiao JG, Hu F (2019) Coexistence and association between heavy metals, tetracycline and corresponding resistance genes in vermicomposts originating from different substrates. Environ Pollut 44:28–37

Makarova KS, Haft DH, Barrangou R, Brouns SJ, Charpentier E, Horvath P, Moineau S, Mojica FJ, Wolf YI, Yakunin AF (2011) Evolution and classification of the CRISPR-Cas systems. Nat Rev Microbiol 9:467–477

Malone LM, Warring SL, Jackson SA, Warnecke C, Gardner PP, Gumy LF, Fineran PC (2020) A jumbo phage that forms a nucleus-like structure evades CRISPR-Cas DNA targeting but is vulnerable to type III RNA-based immunity. Nat Microbiol 5(1):48–55. https://doi.org/10.1038/s41564-019-0612-5

Matsubara K, Katayama H (2019) Development of a portable detection method for enteric viruses from ambient air and its application to a wastewater treatment plant. Pathogens 8(3). pii: E131. https://doi.org/10.3390/pathogens8030131

Mavrich TN, Hatfull GF (2019) Evolution of superinfection immunity in cluster a mycobacteriophages. mBio 10:e00971-19. https://doi.org/10.1128/mBio.00971-19

Montgomery MT, Guerrero Bustamante CA, Dedrick RM, Jacobs-Sera D, Hatfull GF (2019) Yet more evidence of collusion: a new viral defense system encoded by Gordonia phage CarolAnn. mBio 10:e02417-18. https://doi.org/10.1128/mBio.02417-18.. 1983

Ofir G, Sorek R (2018) Contemporary phage biology: from classic models to new insights. Cell 172:1260–1270

Oppenheim AB, Kobiler O, Stavans J, Court DL, Adhya S (2005) Switches in bacteriophage lambda development. Annu Rev Genet 39:409–429

Pawluk A, Bondy-Denomy J, Cheung VHW, Maxwell KL, Davidson AR (2014) A new group of phage anti-CRISPR genes inhibits the type I-E CRISPR-Cas system of Pseudomonas aeruginosa. MBio 5:e00896

Pawluk A, Amrani N, Zhang Y, Garcia B, Hidalgo-Reyes Y, Lee J, Edraki A, Shah M, Sontheimer EJ, Maxwell KL et al (2016a) Naturally occurring off-switches for CRISPR-Cas9. Cell 167:1829–1838

Pawluk A, Staals RHJ, Taylor C, Watson BNJ, Saha S, Fineran PC, Maxwell KL, Davidson AR (2016b) Inactivation of CRISPR-Cas systems by anti-CRISPR proteins in diverse bacterial species. Nat Microbiol 1:16085

Pawluk A, Davidson AR, Maxwell KL (2018) Anti-CRISPR: discovery, mechanism and function. Nat Rev Microbiol 16:12–17

Pazda M, Kumirska J, Stepnowski P, Mulkiewicz E (2019) Antibiotic resistance genes identified in wastewater treatment plant systems – a review. Sci Total Environ 697:134023. https://doi.org/10.1016/j.scitotenv.2019.134023

Perez-Pascual D, Monnet V, Gardan R (2016) Bacterial cell–cell communication in the host via RRNPP peptide-binding regulators. Front Microbiol 7:706

Peters JM, Colavin A, Shi H, Czarny TL, Larson MH, Wong S, Hawkins JS, Lu CHS, Koo BM, Marta E, Shiver AL, Whitehead EH, Weissman JS, Brown ED, Qi LS, Huang KC, Gross CA (2016) A comprehensive, CRISPR-based functional analysis of essential genes in bacteria. Cell 165:1493–1506

Prasad RK, Chatterjee S, Mazumder PB, Gupta SK, Sharma S, Vairale MG, Datta S, Dwivedi SK, Gupta DK (2019) Bioethanol production from waste lignocelluloses: a review on microbial degradation potential. Chemosphere 231:588–606. https://doi.org/10.1016/j.chemosphere.2019.05.142

Ravin NV (2015) Replication and maintenance of linear phage-plasmid N15. Microbiol Spectr 3(1):PLAS-0032-2014. https://doi.org/10.1128/microbiolspec.PLAS-0032-2014

Rocha-Estrada J, Aceves-Diez AE, Guarneros G, de la Torre M (2010) The RNPP family of quorum-sensing proteins in gram-positive bacteria. Appl Microbiol Biotechnol 87:913–923

Rollins MF, Chowdhury S, Carter J, Golden SM, Wilkinson RA, Bondy-Denomy J, Lander GC, Wiedenheft B (2017) Cas1 and the Csy complex are opposing regulators of Cas2/3 nuclease activity. Proc Natl Acad Sci U S A 114:E5113–E5121

Safari F, Sharifi M, Farajnia S, Akbari B, Karimi Baba Ahmadi M, Negahdaripour M, Ghasemi Y (2019) The interaction of phages and bacteria: the co-evolutionary arms race. Crit Rev Biotechnol:1–19. https://doi.org/10.1080/07388551.2019.1674774

Santiana M, Ghosh S, Ho BA, Rajasekaran V, Du WL, Mutsafi Y, De Jésus-Diaz DA, Sosnovtsev SV, Levenson EA, Parra GI, Takvorian PM, Cali A, Bleck C, Vlasova AN, Saif LJ, Patton JT, Lopalco P, Corcelli A, Green KY, Altan-Bonnet N (2018) Vesicle-cloaked virus clusters are optimal units for inter-organismal viral transmission. Cell Host Microbe 24(2):208–220.e8. https://doi.org/10.1016/j.chom.2018.07.006

Sharma S, Chatterjee S, Datta S, Prasad R, Dubey D, Prasad RK, Vairale MG (2016) Bacteriophages and its applications: an overview. Folia Microbiologica (Springer) 62:17–55. https://doi.org/10.1007/s12223-016-0471-x

Stokar-Avihail A, Tal N, Erez Z, Lopatina A, Sorek R (2019) Widespread utilization of peptide communication in phages infecting soil and pathogenic Bacteria. Cell Host Microbe 25(5):746–755.e5. https://doi.org/10.1016/j.chom.2019.03.017

Sulakvelidze A, Alavidze Z, Morris JG Jr (2001) Bacteriophage therapy. Antimicrob Agents Chemother 45:649–659

Sun M, Ye M, Jiao W, Feng Y, Yu P, Liu M, Jiao J, He X, Liu K, Zhao Y, Wu J, Jiang X, Hu F (2018) Changes in tetracycline partitioning and bacteria/phage comediated ARGs in microplastic contaminated greenhouse soil facilitated by sophorolipid. J Hazard Mater 345:131–139

Sun M, Ye M, Zhang Z, Zhang S, Zhao Y, Deng S, Kong L, Ying R, Xia B, Jiao W, Cheng J, Feng Y, Liu M, Hu F (2019) Biochar combined with polyvalent phage therapy to mitigate antibiotic resistance pathogenic bacteria vertical transfer risk in an undisturbed soil column system. J Hazard Mater 365:1–8

Taylor VL, Fitzpatrick AD, Islam Z, Maxwell KL (2019) The diverse impacts of phage morons on bacterial fitness and virulence. Adv Virus Res 103:1–31. https://doi.org/10.1016/bs.aivir.2018.08.001

Tolstoy I, Kropinski AM, Brister JR (2018) Bacteriophage taxonomy: An evolving discipline. Methods Mol Biol 1693:57–71

Twort FW (1915) An investigation on the nature of ultra-microscopic viruses. Lancet 2:1241–1243

Twort FW (1922) The bacteriophage: the breaking down of bacteria by associated filter passing lysins. Br Med J 2:293–296

Twort FW (1949) The discovery of the bacteriophage. Sci News 14:33–34

van Houte S, Ekroth AKE, Broniewski JM, Chabas H, Ashby B, Bondy-Denomy J, Gandon S, Boots M, Paterson S, Buckling A, Westra ER (2016) The diversity-generating benefits of a prokaryotic adaptive immune system. Nature 532:385–388

Wang Q, Guan Z, Pei K, Wang J, Liu Z, Yin P, Peng D, Zou T (2018) Structural basis of the arbitrium peptide-AimR communication system in the phage lysis-lysogeny decision. Nat Microbiol 3(11):1266–1273. https://doi.org/10.1038/s41564-018-0239-y

Witzany G (2006) From umwelt to Mitwelt: natural laws versus rule-governed sign-mediated interactions (rsi's). Semiotica 2006:425–438

Witzany G (2010) Uniform categorization of biocommunication in bacteria, fungi and plants. World J Biol Chem 1(5):160–180. https://doi.org/10.4331/wjbc.v1.i5.160

Wu B, Wang R, Fane AG (2017) The roles of bacteriophages in membrane-based water and wastewater treatment processes: a review. Water Res 110:120–132. https://doi.org/10.1016/j.watres.2016.12.004

Xue KS, Hooper KA, Ollodart AR, Dingens AS, Bloom JD (2016) Cooperation between distinct viral variants promotes growth of H3N2 influenza in cell culture. eLife 5:e13974. https://doi.org/10.7554/eLife.13974

Ye M, Sun M, Huang D, Zhang Z, Zhang H, Zhang S, Hu F, Jiang X, Jiao W (2019) A review of bacteriophage therapy for pathogenic bacteria inactivation in the soil environment. Environ Int 129:488–496

Youderian P, Vershon A, Bouvier S, Sauer RT, Susskind MM (1983) Changing the DNA-binding specificity of a repressor. Cell 35:777–783. https://doi.org/10.1016/0092-8674(83)90110-1

Zhao Y, Ye M, Zhang X, Sun M, Zhang Z, Chao H, Huang D, Wan J, Zhang S, Jiang X, Sun D, Yuan Y, Hu F (2019) Comparing polyvalent bacteriophage and bacteriophage cocktails for controlling antibiotic-resistant bacteria in soil-plant system. Sci Total Environ 657:918–925

The manufacturer's authorised representative in the EU is Springer Nature Customer Service Centre GmbH, Europaplatz 3, 69115 Heidelberg, Germany. If you have any concerns regarding our products, please contact ProductSafety@springernature.com

Printed and bound by CPI Group (UK) Ltd, Croydon, CR0 4YY

30/09/2026

02224059-0005